WATER PLANTS OF THE WORLD

WATER PLANTS OF THE WORLD

A manual for the identification of the genera of freshwater macrophytes

CHRISTOPHER D. K. COOK
BERNARDO J. GUT
E. MARTYN RIX
JAKOB SCHNELLER
MARTA SEITZ

Dr. W. Junk b.v., Publishers, The Hague 1974

ISBN 90 6193 024 3

Set in 10/11pt IBM Press Roman at The Pitman Press, Bath, Avon, England

CONTENTS

PREFACE

Freshwater macrophytes play a very important role in aquatic ecosystems. They provide, either directly or indirectly, food, shelter and a variety of habitats for a large number of organisms, including wildfowl and economically important fish. It must be mentioned that rice, an aquatic plant, is the most important single crop species in the world. Many other aquatic plants are also of direct use to man as food, raw materials for industrial processes, building materials and manure in agriculture. Aquatic plants absorb dissolved minerals and enrich water with oxygen produced during photosynthesis. These properties are of benefit to man as they assist in the maintenance of clean water and the help in the recovery of polluted water. However, in disturbed or newly constructed bodies of water rampant growth of aquatic plants may interfere with man's use of freshwater. They may obstruct water-flow, navigation or water intakes: they may interfere with fish production and crops in aquatic environments or on irrigated land; they may also create conditions favourable for pests, diseases and vectors affecting humans, animals and crop plants; they may also upset recreation pursuits.

Before one can communicate any information about a particular plant it must be identified and named. The name of a plant is the key to all we know about it. Unfortunately, the identification of aquatic plants is particularly difficult. Firstly, aquatic plants are easily modified by environmental conditions and they are frequently found without reproductive structures. This often makes it difficult to determine which group or groups they belong to because the present classification of plants is largely based on reproductive structures and so are the manuals for their identification. Secondly, aquatic plants are often very mobile and are constantly leaving or invading new areas with the consequence that local Floras are often out of date. Thirdly, it would appear that botanists on the whole do not like getting their feet wet because the state of knowledge on aquatics, in comparison to other plant groups, is very incomplete. With these difficulties in mind it was recommended at a meeting of experts on the ecology and control of aquatic vegetation convened by UNESCO through the Co-ordinating Council for the IHD (International Hydrological Decade) with the IBP (International Biological Programme) at Paris in December 1968 that a manual for the identification of aquatic plants on a world-wide scale should be written. I agreed to undertake this task.

This work has been designed chiefly to be a source for the identification of all genera of aquatic macrophytes. It is intended that the non-botanist should be able to use this manual, therefore all genera have been illustrated, the botanical terminology has been kept to a minimum and where ever possible vegetative features and easily seen floral characters have been used in keys and descriptions.

Notes on distribution, number of species, ecology and uses have been given and selected references for the identification of species within the genera have been cited where they are of value to the user.

The fonds national suisse de la recherche scientifique generously supplied a grant* for three years for a plant illustrator and a full-time post-doctoral research fellow. From 1970 to 1971 the research fellowship was filled by Bernardo J. Gut and from 1971 to 1973 by E. Martyn Rix. It is difficult to single out the particular parts undertaken by each co-worker as the project has been essentially team work and each has modified the final draft. Jakob Schneller, an assistant of the Institute for Systematic Botany of the University of Zürich wrote and illustrated the part on aquatic mosses and contributed to the final preparation stages.

Most of the illustrations are by Marta Seitz, initialed "MS". Jakob Schneller illustrated the mosses, initialed "JS" and I have contributed a few illustrations, initialed "CC". I would very much like to thank the co-workers for their efforts and humour in times of crisis. Thanks are also due to the Co-ordinating Council for the IHD (UNESCO) and the Council of the "Georges and Antoine Claraz-Schenkung" for financial assistance for fieldwork. The following institutions kindly and very promptly lent us herbarium material: Conservatoire et Jardin Botanique Genève, Royal Botanic Gardens Kew, Botanische Staatssammlung München, Smithsonian Institution Washington and Eidgenössische Technische Hochschule Zürich. I would like to thank the following for help, either with their specialist knowledge or for sending plant material: R. C. Bakhuizen van den Brink (Leiden), F. D. Bennett (Trinidad), H. W. E. van Bruggen (Heemskerk), W. D. Clayton (Kew), W. G. Dore (Ottawa), J. B. Hall (Ghana), B. Hellquist (Boston), S. Hooper (Kew), D. J. Keil (Ohio), N. G. Marchant (Perth, Australia), H. Merxmüller (München), R. van der Meijden (Leiden), D. S. Mitchell (Salisbury, Rhodesia), H. Moldenke (New Jersey), P. Raven (St. Louis), V. P. Rao (Bangalore), P. Schneider (Zuzgen, Switzerland), R. M. Schuster (Massachusetts), C. G. G. J. van Steenis (Leiden) and P. Taylor (Kew). I would particularly like to thank L. Constance (Berkley) for advice and help with the Apiaceae, C. E. Hubbard (Kew) for the grasses and E. V. Watson (Reading) for the mosses. Special thanks are due to Anni Küpfer who has done the administrative work and the greater part of the typing; last minute typing was kindly undertaken by Luise Engler. I would also like to thank Julian Rzoska (IBP, London) who has constantly given encouragement and kept a fatherly eye on the project. Finally I would like to thank my wife for putting up with a physically or mentally absent husband.

CHRISTOPHER D. K. COOK

Institut für systematische Botanik
der Universität Zürich

*Number 3.98.69.

INTRODUCTION

Our aim has been to provide a manual for the identification of freshwater macrophytes of the world. The term "freshwater macrophyte" is to be interpreted as all Charophyta (Stoneworts), Bryophyta (Mosses and Liverworts), Pteridophyta (Ferns and Fern Allies) and Spermatophyta (Seed-bearing Plants) whose photosynthetically active parts are permanently or, at least, for several months each year submerged in freshwater or floating on the water surface. Marine and exclusively brackish water plants have been excluded.

We hope that all macrophytes encountered growing in permanent or seasonally semipermanent water have been included in this book. However, following the above definition means that the majority of plants described are not obligately submerged aquatics. Many plants (for example *Marsilea, Boisduvalia, Eryngium, Thorella, Dopatrium, Glossostigma, Microcarpaea*, etc.) reach sexual maturity only when the habitat has dried out but they are, nevertheless, normally submerged during their juvenile phases of growth. There are many plants, particularly in the tropics, that grow on rocks or trees by streams, rivers and waterfalls. After rain they may be submerged in swiftly flowing water, sometimes for only a few hours at a time. These plants are called rheophytes and, although specialised for this particular kind of habitat, they have been excluded from this book because their vegetative growth and reproduction takes place in the terrestrial milieu. In the colder regions of the world and on high mountains many normally terrestrial plants may be found growing submerged in water; on the whole these plants have been excluded.

Two large general identification keys are presented. The first (page 6), is a more or less traditional key to families based on reproductive structures. The second (page 16), is a key to genera and other higher taxonomic categories based, as far as possible, on easily seen vegetative characters. The non-specialist is recommended to use the second key. With the exception of monotypic families, further keys to each genus are provided after the family description. From the point of view of keys the Bryophyta are treated as a single family. The genera within each family and the families within each higher taxonomic category (Charophyta, Bryophyta, Pteridophyta and Spermatophyta) are arranged alphabetically.

Botanical nomenclature is unfortunately still somewhat unstable so the commonest synonyms are cited in square brackets after the accepted name. Frequently the name in square brackets is not a synonym in a strictly nomenclatural sense. It is often the case that the cited synonym is a perfectly valid genus but the aquatic species have been removed from it. This kind of synonym is followed by "pro parte" which indicates that this name is only a synonym from the point of view of the aquatic species. For example,

Hydropectis [*Pectis* pro parte] indicates that the aquatic species has been taken out of the genus *Pectis* and placed in *Hydropectis,* the other species of *Pectis* are terrestrial plants. The folly of citing local names was realised at an early stage in the preparation of this book. For example, the name "pyle" is used by some people to refer to either *Salvinia* or *Pistia* or *Eichhornia* and by other people to all or some of them collectively. We have therefore given scientific names only and beg the users of this book to adopt them.

The family and generic descriptions are intended to be diagnostic and refer, as a rule, only to the aquatic members. As far as possible, the generic descriptions have been based on living or pressed plant specimens and care has been taken to describe the juvenile and sterile submerged parts. We have tried to keep the botanical terminology down to a minimum. A glossary is provided on page 551.

The illustrations have been based, where possible, on living or dried plant material. Like the descriptions the illustrations are intended to be diagnostic. For each genus we have chosen one or occasionally two representative species that show the important features for the identification of the genus as a whole. Time, printing space and expense have prevented us making full comprehensive drawings of each species. The length of the scale line accompanying each drawing is given in brackets in the legend.

It is often thought that the simplest way to identify a plant is to examine a set of pictures until one is found that resembles it. This method becomes impractical when many plants are involved. Moreover, many aquatic plants are superficially alike (natural selection is a strong mistress) and even if the pictures show the difference the user must be directed by keys and descriptions as to what points to observe.

To keep the size of this book within reasonable limits we have had to make a compromise with the references. On page 4 a selected bibliography of the larger works of general value to the taxonomy of aquatic plants is given. The references following each family or generic description have been carefully chosen to serve as an introduction to the taxonomic literature of the group in question. Occasionally the reference chosen is of limited value in itself but may include a full bibliography. The references cited in this book should therefore be considered as no more than a guide to the literature on aquatic plants. However, as a particularly valuable source book we would like to single out Sculthorpe, C. D. The biology of aquatic vascular plants. London (1967).

The observations following the generic descriptions give information on the number of species, geographical distribution, the kinds of habitats where the plants grow and notes on economic importance. If any species are known to be weeds or pests this is mentioned. However, when undesirable qualities of a species are described it does not mean that this species must always be considered a pest. A weed is best defined as a plant that grows where it is not wanted. In some man-made or artificially maintained waters a particular plant species may be regarded as a pest but in other, often natural, waters the same species may be highly desirable. For example, *Typha* (Reedmace) a widespread and common reedswamp plant is very important as a shelter and food plant for a wide variety of animals, it also collects silt and is very valuable in stabilizing the banks of rivers and lakes and thus prevents flooding. However, *Typha* in a man-made irrigation ditch is far from desirable as it reduces water-flow and its silt collecting properties may eventually cause flooding. Another example is *Eichhornia crassipes* (Water Hyacinth); it is without doubt a disaster on the River Nile in the Sudan but in some regions of S.E. Asia it is considered desirable and is even cultivated as pig and fish food.

While we hope that this manual will prove useful, we are fully aware that it has many deficiencies and will doubtless be found to contain errors. We would be most grateful if users of the book who detect errors or omissions would inform us.

GENERAL REFERENCES

The following is a selected bibliography of the more important local Floras and general works on aquatic plants that are of importance to the taxonomist.

Arber, A. Water plants: A study of aquatic angiosperms. Univ. Press Cambridge (1920); reprint as Hist. Nat. Classica 23, Weinheim (1963)

Aston, H. I. Aquatic plants of Australia. Melbourne University Press (1973)

Beal, E. O. and Monson, P. H. Marsh and aquatic angiosperms of Iowa. State Univ. Iowa Studies Nat. Hist. 19 (5): 1–195 (1954)

Benl, G. A key to the genera of aquarium plants based on vegetative characters. *Baileya* 18 (4): 121–132 (1971)

Biswas, K. and Calder, C. C. Handbook of common water and marsh plants of India and Burma. Health Bull. 24, Delhi (1955)

Brünner, G. Aquarienpflanzen, Stuttgart (1964) Transl. as Aquarium Plants, London (1966)

Correll, D. S. and Correll, H. B. Aquatic and wetland plants of southwestern United States. Washington (1973)

Fassett, N. C. A manual of aquatic plants. Madison, Wisconsin, ed. 2 (1967)

Gessner, F. Hydrobotanik. Berlin. 1: (1955), 2: (1959)

Glück, H. Biologische und morphologische Untersuchungen über Wasser- und Sumpfgewächse. Jena 3: (1911), 4: (1924)

Glück, H. in Pascher, A. Die Süsswasser-Flora Mitteleuropas, 15: Jena (1936)

Hejný, S. Oekologische Charakteristik der Wasser- und Sumpfpflanzen in den slowakischen Tiefebenen. Bratislava (1960)

Hoehne, F. C. Plantas aquaticas. São Paulo (1955)

Hotchkiss, N. Common marsh, underwater and floating-leaved plants of the United States and Canada. New York (1972)

Kruijne, A. A. Vegetative herkenning der voornaamste water – en oeverplanten. Zwolle, Netherlands (1963)

Mason, H. L. A flora of the marshes of California. Berkeley and Los Angeles (1957)

Mitchell, D. S. and Thomas, P. A. Ecology of Water Weeds in the Neotropics. UNESCO, Technical Papers in Hydrology, No. 12: 1–50 (1972)

Mitchell, D. S. ed. 1974 Aquatic vegetation and its use and control. A contribution to the Intern. Hydrol. Decade. UNESCO-Paris: p.1–135 ISBN 92-3-101082-4. Printed in France.

Muenscher, W. C. Aquatic plants of the United States. Ithaca, New York (1944)

Otto, N. E. and Bartley, T. R. Aquatic pests on irrigation systems, identification guide. Washington (1965)

Prescott, G. W. The aquatic plants. Dubuque, Iowa (1969)

Sculthorpe, C. D. The biology of aquatic vascular plants. London (1967)

Smith, G. G. and Marchant, N. G. A census of aquatic plants of Western Australia. Western Austr. Nat. 8 (1): 5–17 (1961)

Stewart, A. N., La Rea, J. D. and Gilkey, H. M. Aquatic plants of the Pacific Northwest Oregon, ed. 2 (1963)
Stodola, J. Encyclopedia of water plants. New Jersey (1967)
Subramanyam, K. Aquatic angiosperms: A systematic account of common Indian aquatic angiosperms. Bot. Monog. 3, C.S.I.R., New Delhi (1962)
Wendt, A. Die Aquarienpflanzen in Wort und Bild. Stuttgart (1952–1958)
Wild, H. Harmful aquatic plants in Africa and Madagascar. Kirkia 2: 1–66 Repr. as CCTA/CSA Publ. 73 (1961). Transl. to French by Germain, R. (1964)
Wit, H. C. D. De. Aquariumplanten, I and II. Belmontia, ser. 3, 2: 125 pp. and 193 pp. (1958). Repr. as Het Handboek voor de Aquarium lief hebber, 5 and 6: Baam (1957, 1958). Transl. as Aquarium plants, by Schuurman, J. A. and Higgins, V. London (1964)
Wit, H. C. D. De. Aquarienpflanzen. Stuttgart (1971)

KEY TO FAMILIES AND OTHER MAJOR GROUPS BASED ON REPRODUCTIVE ORGANS

A key to families and genera based on vegetative structures is to be found on page 16.

The numbers following the plant names refer to the page numbers in the text.

1A Egg cell surrounded by 5, elongate spirally wound cells; internodes with 1, elongate, axial cell

Characeae p. 40

1B Egg cell surrounded by a multicellular jacket or tissue; internodes multicellular

2A Sporophyte parasitic on gametophyte (mosses)

Bryophyta p. 47

2B Sporophyte not parasitic on gametophyte

3A Sporophytes reproducing by spores which give rise to small gametophytes (ferns)

4A Sporangia embedded on the adaxial leaf surface near the base

Isoetaceae p. 84

4B Sporangia not embedded in the leaf, borne either adaxially on peltate scales, abaxially on leaf surface, or in hardened nut like structures

5A Sporangia borne adaxially on peltate scales; stems grooved and jointed

Equisetaceae p. 85

5B Sporangia not borne adaxially on peltate scales; stems not grooved, not jointed

6A Sporangia borne on the abaxial surface of the leaf; leaves pinnately divided

Parkeriaceae p. 90

6B Sporangia borne in hardened, nut-like structures; leaves simple or with 2 or 4 leaflets

7A Plants free-floating

8A Leaves opposite; abaxial leaf surface hairy

Salviniaceae p. 93

8B Leaves alternate; abaxial leaf surface glabrous

Azollaceae p. 87

7B Plants not free-floating; with elongate creeping stems and erect or floating leaves

Marsileaceae p. 88

3B Sporophytes reproducing by seeds (flowering plants)

9A Plants not clearly differentiated into stem with leaves
10A Plants not attached to a solid substrate
Lemnaceae p. 286

10B Plants firmly attached to a solid and hard substrate by a band-, disc- or irregularly-shaped thallus (usually in swiftly flowing water)
11A Flowers bisexual; sepals 2, or more (widespread, mostly tropical)
Podostemaceae p. 445

11B Flowers unisexual, sepals absent (Madagascar and S. Africa)
Hydrostachyaceae p. 270

9B Plants clearly differentiated into stem with leaves
12A Petals present (see p. 12)
13A Carpels embedded in the flat top of the fleshy receptacle (herbs with large, erect, peltate leaves)
Nelumbonaceae p. 332

13B Carpels not embedded in the flat top of the fleshy receptacle or receptacle not fleshy
14A Sepals sepaloid and petals petaloid (perianth clearly differentiated into petals and sepals, see p. 11)
15A Petals free (see p. 9)
16A Ovary superior (see p. 9)
17A Ovaries and styles 2 or more, free, or united only at base of ovary
18A Leaves in opposite pairs, simple, entire linear to ovate; plant fleshy
Crassulaceae p. 184

18B Leaves alternate or divided or lobed
19A Sepals 5 or occasionally more
Ranunculaceae p. 502

19B Sepals 3
20A Plant with milky latex; seeds scattered on inner walls of ovary
Limnocharitaceae p. 297

20B Plant without milky latex; seeds solitary or a few on an axillary placenta
Alismataceae p. 99

17B Ovaries and styles united, or carpels solitary

21A Petals 3, or inner perianth 3, petal-like

22A Flowers in a more or less globose head, subtended by an involucre of bracts; outer perianth segments scarious

23A Flowers unisexual; inner perianth segments inconspicuous

Eriocaulaceae p. 233

23B Flowers bisexual; inner perianth segments conspicuous and showy

Xyridaceae p. 543

22B Flowers not in a globose head, not subtended by an involucre of bracts; outer perianth segments not scarious

24A Pedicels subtended by a lanceolate bract

25A Leaves linear, notched at apices, not enclosing the stem at bases

Mayacaceae p. 321

25B Leaves lanceolate to ovate, entire at apices, enclosing the stem at bases

Commelinaceae p. 180

24B Pedicels not subtended by a lanceolate bract

26A Flowers in spikes; fruit a nut

Polygonaceae p. 481

26B Flowers solitary or clustered in leaf axils; fruit a capsule

Elatinaceae p. 231

21B Petals more than 3

27A Stamens more than 2 times as many as petals

28A Leaves with long petioles; stigmas radiate

Nymphaeaceae p. 334

28B Leaves sessile; stigmas capitate

Clusiaceae p. 179

27B Stamens 2 times as many as petals or fewer

29A Petals unequal in shape and size

30A Leaves compound (1- or 2-pinnate); fruit a pod

Fabaceae p. 242

30B Leaves simple; fruit a capsule

Balsaminaceae p. 161

29B Petals more or less equal in shape and size

31A Flowers unisexual

Euphorbiaceae p. 239

31B Flowers bisexual

32A Leaves with traps and terminal bristles; plants free-floating

Droseraceae p. 229

32B Leaves without traps and without terminal bristles; plants bottom rooted

33A Anthers opening by terminal pores; stamens distinctly jointed

Melastomaceae p. 323

33B Anthers opening by longitudinal slits; stamens straight

34A Fruit a 3-angled nut

Polygonaceae p. 481

34B Fruit a capsule

35A Petals and stamens inserted on the sepal tube

Lythraceae p. 310

35B Petals and stamens inserted on the receptacle; sepals not tubular

36A Flowers sessile or shortly stalked, solitary or in clusters in leaf axils

Elatinaceae p. 231

36B Flowers distinctly stalked, in terminal racemes

Brassicaceae p. 163

16B Ovary inferior or semi-inferior

37A Petals 3

Hydrocharitaceae p. 254

37B Petals more than 3

38A Petals numerous, usually more than 8

Nymphaeaceae p. 334

38B Petals usually 4 or 6

39A Fruit a capsule

Onagraceae p. 343

39B Fruit indehiscent, nut-like or a schizocarp

40A Leaves often with swollen, floating petioles; fruit with 2 or 4 thorn-like processes

Trapaceae p. 537

40B Leaves without swollen, floating petioles; fruit without thorn-like processes

41A Flowers in an umbel or head

Apiaceae p. 117

41B Flowers axillary or in racemes or spikes

Haloragaceae p. 246

15B Petals united

42A Ovary superior

43A Flowers in a globose head subtended by an involucre of bracts; perianth 3-merous

Eriocaulaceae p. 233

43B Flowers not in a globose head not subtended by an involucre of bracts; perianth 4- or 5-merous

44A Flowers actinomorphic (radially symmetrical)

45A Stamens opposite the petals; placentation free-central

46A Sepals 2, usually free
Portulacaceae p. 493
46B Sepals 5, united into a tube below
Primulaceae p. 497
45B Stamens opposite the sepals, or more than petal lobes; placentation not free-central
47A Ovary 3-locular
Polemoniaceae p. 480
47B Ovary 2- or rarely 4-locular
48A Placentation parietal; petals usually with hairs or lamellae on the surface
Menyanthaceae p. 326
48B Placentation axile; petals usually glabrous and smooth
49A Fruit of 4, united nutlets; style gynobasic (minute herbs)
Tetrachondraceae p. 535
49B Fruit a capsule; style terminal
50A Ovules 1 or 2 in each loculus; petal tube funnel-shaped
Convolvulaceae p. 182
50B Ovules many; petal tube lobed
51A Styles 2, united only at base
Hydrophyllaceae p. 269
51B Style 1, sometimes 2-lobed at apex
Scrophulariaceae p. 511
44B Flowers zygomorphic (bilaterally symmetrical)
52A Plant bearing bladder-like traps
Lentibulariaceae p. 292
52B Plant not bearing bladder-like traps
53A Fruit of 4 nutlets; style gynobasic
Lamiaceae p. 280
53B Fruit a capsule; style terminal
54A Seeds on papilliform or hook-like projections
Acanthaceae p. 95
54B Seeds borne directly on the placenta
Scrophulariaceae p. 511
42B Ovary inferior or semi-inferior
55A Petals 3; some stamens sterile and petaloid
56A Ovary 1-locular; 2 anther loculi fertile
Marantaceae p. 317
56B Ovary 3-locular; 1 anther loculus fertile
Cannaceae p. 172

55B Petals 4 or more; stamens not petaloid

57A Inflorescence a head subtended by an involucre of bracts
Asteraceae p. 155

57B Inflorescence a spike or panicle or flowers solitary

58A Leaves opposite or apparently whorled

59A Flowers in branched cymes; lower leaves whorled
Rubiaceae p. 505

59B Flowers solitary in leaf axils; lower leaves in opposite pairs
Trapellaceae p. 539

58B Leaves alternate or all basal

60A Inflorescence densely spicate, terminal
Sphenocleaceae p. 533

60B Inflorescence lax or flowers axillary

61A Flowers actinomorphic; anthers free
Primulaceae p. 497

61B Flowers zygomorphic; anthers united
Lobeliaceae p. 302

14B Sepals and petals petaloid (perianth not clearly differentiated into sepals and petals)

62A Perianth segments free; ovary always superior

63A Fruit a solitary, 3-sided nutlet
Polygonaceae p. 481

63B Fruit of several carpels or a dehiscent capsule

64A Flowers in simple or forked spikes
Aponogetonaceae p. 135

64B Flowers solitary, or in variously pedicellate inflorescences

65A Flowers zygomorphic; stamen 1
Philydraceae p. 346

65B Flowers actinomorphic; stamens more than 1

66A Inflorescence umbel-like; leaves linear, arising in 2 rows from a rhizome
Butomaceae p. 166

66B Inflorescence not umbel-like; leaves not linear, not in 2 rows

67A Carpels united; stigma radiate
Nymphaeaceae p. 334

67B Carpels free; stigma not radiate

68A Leaves sheathing at base, never peltate, never divided into capillary segments
Ranunculaceae p. 502

68B Leaves not sheathing at base, peltate or divided into capillary segments
Cabombaceae p. 168

62B Perianth segments united; ovary inferior or superior

69A Ovary superior

70A Inflorescence subtended by 2 spathes; style 1; leaves parallel nerved

Pontederiaceae p. 482

70B Inflorescence not subtended by spathes; styles 2 or 3; leaves with midrib and lateral, pinnate nerves

Polygonaceae p. 481

69B Ovary inferior

71A Fruit a 2-seeded schizocarp

Apiaceae p. 117

71B Fruit a capsule

72A Inner and outer perianth whorls dissimilar in shape

Iridaceae p. 273

72B Inner and outer perianth whorls similar

73A Perianth segments united for at least half their length, forming a tube below; style long

Amaryllidaceae p. 115

73B Perianth segments united only at base; style short

Hypoxidaceae p. 271

12B Petals absent

74A Perianth sepaloid, scarious or membranous (see p. 14)

75A Perianth sepaloid (not scarious or membranous, see p. 13)

76A Flowers crowded on to a fleshy spike (spadix) subtended by a fleshy bract (spathe)

Araceae p. 137

76B Flowers not crowded on to a spadix subtended by a fleshy spathe

77A Ovary superior

78A Fruit a many seeded capsule or of many seeded carpels

79A Sepals united into a tube

Lythraceae p. 310

79B Sepals free

80A Stamens 15, dimorphic

Nymphaeaceae p. 334

80B Stamens 6, monomorphic

Brassicaceae p. 163

78B Fruit a 1-seeded capsule or indehiscent

81A Flowers solitary, sessile in the axils of whorled, forked capillary leaves

Ceratophyllaceae p. 177

81B Flowers in spikes or stalked; leaves entire

82A Carpets 3, united; fruit 1, lenticular or 3-angled nut; perianth enclosing fruit

Polygonaceae p. 481

82B Carpels usually more than 3, or solitary; fruit of 1 or many nutlets; perianth not enclosing fruit

83A Sepals 3

Alismataceae p. 99

83B Sepals 2, 4 or 6

84A Leaves mostly cauline; fruit of 4, free nutlets; stamens 4

Potamogetonaceae p. 494

84B Leaves basal, linear; fruit of 4, united 1-seeded carpels, or 6, more or less free, 1-seeded carpels at maturity

Juncaginaceae p. 278

77B Ovary inferior

85A Inflorescence umbellate, or contracted into a head

Apiaceae p. 117

85B Inflorescence spike-like or flowers in leaf-axils

86A Flowers emerging from a bifid spathe-like bract or a pair of bracts

Hydrocharitaceae p. 254

86B Inflorescence without spathe-like bract or bracts

87A Fruit a many-seeded capsule

Onagraceae p. 343

87B Fruit nut-like or a schizocarp

Haloragaceae p. 246

75B Perianth scarious or membranous

88A Perianth a 2-lipped membranous sheath; whole plant including flowers submerged; leaf margin toothed

Najadaceae p. 331

88B Perianth of 3 to 6 segments; flowers emersed; leaf margin not toothed

89A Flowers in a compact, globose head, subtended by an involucre of bracts

Eriocaulaceae p. 233

89B Flowers in elongate spikes or panicles

90A Perianth segments 6; stamens 3 or 6

91A Flowers unisexual; leaves ovate to lanceolate with distinct cross veins

Hanguanaceae p. 251

91B Flowers bisexual; leaves lanceolate to linear without cross veins

Juncaceae p. 275

90B Perianth segments 3 to 5; stamens 4 or 5

92A Leaves linear, basal and spirally arranged

Plantaginaceae p. 348

92B Leaves lanceolate or oblanceolate, cauline and in opposite pairs

Amaranthaceae p. 112

74B Perianth reduced to hairs, minute scales, or glumes, or absent

93A Perianth present, of hairs, minute scales or glumes (leaves usually linear or ligulate, or grass-like)

94A Styles heteromorphic, of lower flowers (hidden in leaf bases) at least 2 cm long

Lilaeaceae p. 295

94B Styles not heteromorphic, less than 1 cm long

95A Flowers totally submerged, sessile or short-stalked, in leaf axils; perianth hairs absent; nutlets 2 or more in each flower

Zannichelliaceae p. 546

95B Flowers usually emersed, usually stalked, or not in leaf axils, if submerged then with perianth of hairs (ie *Websteria*, Cyperaceae), nutlets usually solitary

96A Flowers unisexual, densely arranged in unisexual, superposed spikes or unisexual globose heads borne on the same axis

97A Flowers arranged in superposed, cylindrical spikes; ovary on a hairy stalk

Typhaceae p. 540

97B Flowers arranged in globose heads; ovary sessile or subsessile

Sparganiaceae p. 531

96B Flowers bisexual or if unisexual then mixed in heads or borne on separate axes

98A Flowers in terminal heads subtended by distichous bracts; styles simple, usually 3 or more, if 1 then stamen also 1 (minute, tufted annuals, mainly in Australasia and 1 species in S.E. Asia)

Centrolepidaceae p. 174

98B Inflorescence various; styles 2 or 1, with 2 or 3 plumose stigmas; stamens usually 3 or more

99A Flowers usually subtended by 1 glume; stems usually solid between nodes; leaves in 3 or rarely 2 rows

Cyperaceae p. 185

99B Flowers usually subtended by 2 glumes; stems usually hollow between nodes; leaves in 2 rows

Poaceae p. 349

93B Perianth absent (leaves very rarely linear, ligulate or grass-like)

100A Flowers borne on a fleshy spike (spadix) subtended by a fleshy bract (spathe)

Araceae p. 137

100B Flowers not borne on a spike or spike not fleshy and subtending bract membranous or absent

101A Plant free-floating; leaves orbicular, somewhat swollen in the centre (Amazon Basin)

Euphorbiaceae p. 239

101B Plant submerged or rooted in the substrate

102A Leaves whorled

Hippuridaceae p. 253

102B Leaves in opposite pairs or alternate

103A Inflorescence a spike or spike-like; each flower subtended by a bract

104A Leaves linear (small scapose annuals with female flowers in leaf axils and bisexual and male flowers in spikes)

Lilaeaceae p. 295

104B Leaves with distinct petiole and blade; axillary female flowers absent

105A Leaves cauline, blades usually cordate

Saururaceae p. 509

105B Leaves basal, blades linear to oblong-elliptic

Aponogetonaceae p. 135

103B Inflorescence not a spike or spike-like; flowers axillary or in few-flowered clusters or umbels

106A Fruit splitting into 4 nutlets at maturity

Callitrichaceae p. 171

106B Fruit of 1 nutlet

107A Flowers sessile or subsessile in leaf axils

Zannichelliaceae p. 546

107B Flowers on stalked umbels, with stalks elongating in fruit

Ruppiaceae p. 507

KEY BIASED IN FAVOUR OF EASILY SEEN VEGETATIVE CHARACTERS.

1A Plants thalloid (not differentiated into stem with leaves or main axis with lateral appendages)

2A At least basal portion of thallus firmly attached to and flattened against a hard substrate (usually in swiftly flowing water)

3A Flowers bisexual; sepals 2 or more, usually small, occasionally united below; stamens 1 or more (widespread in the tropics and subtropics)

Podostemaceae p. 445

3B Flowers unisexual; sepals absent; stamen 1 (confined to Madagascar and S. Africa)

Hydrostachys p. 270

2B Thalli not attached to or flattened against a hard substrate, free-floating or attached to soft substrate (usually in still or slowly flowing water)

4A Thallus distinctly asymmetrical (thickened "axis" with wing on one side and over the apex), usually partly embedded in a soft substrate

Riella p. 55

4B Thalli symmetrical or slightly asymmetrical, free-floating or free-swimming

5A Thalli repeatedly and regularly forked

6A Thalli broad, not elongate, tinged with purple, fringed with tongue-shaped, toothed, scales

Ricciocarpus p. 53

6B Thalli narrow, elongate, not tinged with purple, not fringed with scales

Riccia p. 53

5B Thalli not repeatedly and regularly forked, elongate to orbicular, flat or globose, with or without roots

Lemnaceae p. 286

1B Plants not thalloid (differentiated into stem with leaves or main axis with lateral processes)

7A Internodes of main axis unicellular or with unicellular medulla surrounded by sheath of cortical cells; oogonium ("fruit") with spirally ridged wall (obligate aquatics, usually with unpleasant smell)

Characeae p. 40

7B Internodes of main axis distinctly multicellular; fruit without spiral ridges

8A Sporophyte or diploid generation parasitic on the gametophyte or haploid generation (usually small plants with leaves spiral or in 2, 3 or 5 rows, the leaves usually 1 cell layer thick, attached to the substrate by rhizoids)

Bryophyta p. 47

8B Sporophyte free living; gametophyte reduced or parasitic on sporophyte (ferns and flowering plants)

9A Leaves reduced to non-photosynthetic scales (the stem being photosynthetic)

10A Scales in regular whorls, united below

Equisetum p. 85

10B Scales opposite or alternate, not united below

11A Fruit of 4, 1-seeded mericarps ("nutlets"); scales usually opposite

Myriophyllum p. 248

11B Fruit of 1, 1-seeded nutlet or a many-seeded capsule; scales solitary or alternate

12A Perianth of 6, scarious scales; fruit a capsule

Juncus p. 275

12B Perianth absent or of hairs, bristles or 1 to 3 scales; fruit a nutlet

Cyperaceae p. 185

9B Leaves photosynthetic, not reduced to scales

13A Leaves jointed between sheath and blade (the base of the leaf is a cylindrical sheath, free or united at edges, enveloping the stem; above the sheath the blade is elongate and free; between the blade and sheath is a joint or petiole usually bearing a membranous ligule, a swelling or row of hairs)

14A Leaves pinnately nerved, with straight, parallel nerves diverging from the midrib (large emergent herbs, tropical)

15A Petiole (or joint) distinctly swollen; leaves in 2 rows

Marantaceae p. 317

15B Petiole (or joint) not swollen; leaves spirally arranged

Canna p. 172

14B Leaves not pinnately nerved, or if pinnate then nerves not straight and parallel

16A Leaf inserted at base of sheath; nerves irregularly branched

Polygonum p. 481

16B Leaf inserted at or towards apex of sheath; nerves indistinct or parallel

17A Leaves in 3 rows

Cyperaceae p. 185

17B Leaves in 2 rows or spiral

18A Styles of lower flowers at least 2 cm long, simple with capitate stigma (scapose annuals; female flowers sessile in leaf axils; bisexual and male flowers in a simple, stalked spike)

Lilaea p. 295

18B Styles less than 1 cm long or styles branched with plumose stigmas

19A Perianth present (petaloid, petaloid and sepaloid, or sepaloid and appearing as attached to stamens)

20A Perianth entirely petaloid (inflorescence subtended by a distinct spathe)

Pontederiaceae p. 482

20B Perianth with petals and sepals, or entirely sepaloid

21A Petals present; anthers distinctly stalked; leaves with lanceolate or ovate blade

Murdannia p. 180

21B Petals absent; anthers sessile; leaves linear or capillary

22A Leaves basal; stamens 6 or in some flowers less

Juncaginaceae p. 278

22B Leaves cauline; stamens 4

Potamogeton p. 496

19B Perianth absent, or reduced to hairs, bristles or scales

23A Flowers in a few-rayed umbel

Ruppia p. 507

23B Flowers in spikes or heads

24A Flowers not in the axils of dry, scarious bracts (inflorescence a simple, bract-less spike or raceme)

Juncaginaceae p. 278

24B Flowers in the axils of dry, scarious bracts (grass- or sedge- like plants)

25A Flowers each enclosed by 2 bracts (lemma outside, palae inside), arranged in spikelets each usually subtended by 1 or 2 empty bracts (glumes) (stems usually hollow, not triangular in transverse section, with swollen nodes; leaves mostly cauline and in 2 rows)

Poaceae p. 349

25B Flowers usually each enclosed by 1 bract (glume), variously arranged in spikelets (stems usually solid, usually triangular in transverse section, without swollen nodes; leaves mostly basal and in 3 rows)

Cyperaceae p. 185

13B Leaves not jointed between sheath and blade

26A Plants Salvinia – or Azolla-like (free-floating; stems creeping; floating leaves sessile or subsessile, in 2 rows, leaf blades parallel to stem)

27A Floating leaves opposite, upper surface with multicellular hairs

Salvinia p. 93

27B Floating leaves alternate, upper surface without multicellular hairs

28A Leaves orbicular with 2, blister-like swellings in the centres (Tropical S. America)

Phyllanthus p. 239

28B Leaves 2-lobed; the lower lobe in contact with the water (Cosmopolitan)

Azolla p. 87

26B Plants not Salvinia- or Azolla-like

29A Leaves peltate

30A Petiole bearing spines; leaf blade more than 10 cm diameter

31A Leaf blade without spines; plant with milky latex

Nelumbo p. 332

31B Leaf blade with spines on undersurface; plants without milky latex

32A Edge of leaf turned upwards; spines absent on upper leaf surface (S. America)

Victoria p. 341

32B Edge of leaf not turned upwards; spines on both leaf surfaces (Asia)

Euryale p. 336

30B Petiole without spines; leaf blade less than 10 cm diameter

33A Plant bearing bladder-like traps; flowers zygomorphic; petals united below

Lentibulariaceae p. 292

33B Plant not bearing bladder-like traps; flowers actinomorphic; petals absent or free

34A Finely divided capillary leaves present; blades of peltate leaves elongate, occasionally with short sinus

Cabomba p. 170

34B All leaves simple and peltate; blades more or less oblong to orbicular

35A Submerged parts of plant coated in mucilage; carpels superior, free

Brasenia p. 168

35B Submerged parts of plant not coated in mucilage; carpels inferior, united

Hydrocotyle p. 127

29B Leaves not peltate

36A Leaves, or whorls or bunches of leaves regularly dispersed along the stem (leaves cauline, see p. 32)

37A Leaves 3 or more at each node (see p. 22)

38A Leaves terminating in an orbicular lobe hinged along the midrib and 4 to 6 bristles

Aldrovanda p. 229

38B Leaves not terminating in an orbicular, hinged lobe and terminal bristles less than 4 or absent

39A Leaves compound or divided into secondary linear or capillary segments

40A Leaves pinnately divided

41A Leaves 2-pinnate or 1-pinnate with at least lower segments forked; petals tubular

Limnophila p. 524

41B Leaves 1-pinnate, all segments simple; petals free, caducous

Myriophyllum p. 248

40B Leaves repeatedly forked, or segments repeatedly forked

42A Ultimate segments with small, marginal, thorn-like projections and terminal bristles; all divisions 2-fid (simply forked)

Ceratophyllum p. 177

42B All segments smooth; lower divisions 5- or 3-fid

Cabomba p. 170

39B Leaves simple

43A Leaves lanceolate to ovate or orbicular

44A Stems semi-woody, whip-like, spongy at base, emergent above; flowers pink, in clusters in leaf axils

Decodon p. 314

44B Stems not woody, flaccid, not spongy at base; flowers not pink and not in clusters in leaf axils

45A Petals yellow; fruiting head of numerous, free nutlets; leaves heteromorphic (usually capillary below and with ovate blades above; E. Europe)

Ranunculus p. 502

45B Petals not yellow or absent; fruit a capsule; leaves not distinctly heteromorphic

46A Leaves not translucent, light green, with entire margins; plants usually amphibious

Hemianthus p. 520

46B Leaves translucent, dark green, with minute or large marginal teeth; plants entirely submerged

47A Petals about 3 times as long and as broad as sepals

Egeria p. 257

47B Petals rarely exceeding sepals or absent

48A Male spathes globose, spiny; stamens 3; leaves usually more than 4 in a whorl

Hydrilla p. 259

48B Male spathes ovoid or cylindrical, smooth; stamens usually 6 or 9, rarely 3; leaves usually less than 4 in a whorl

Elodea p. 258

43B Leaves linear to capillary

49A Leaves not equally spaced around the node (leaves in bunches at each node)

50A Leaves flattened, translucent, with toothed margin

Najas p. 331

50B Leaves capillary, not translucent, with entire margin

51A Leaf sheaths free above or leaves with 2 partly free stipules

52A Anther 1-locular, without distinct connective (W. Old World)

Althenia p. 546

52B Anther 2-locular, or appearing 6-locular, with distinct connective (Australia, New Zealand)

Lepilaenia p. 548

51B Leaf sheathes not free above or stipules absent

53A Flowers subtended by a 2-lobed sheat, paired in leaf axils (numerous hair-like, unsheathed leaves arising in axil of 1, larger leaf, with sheating base, Brazil)

Hydrothrix p. 487

53B Flowers not subtended by a 2-lobed sheat, in an ovoid head, with numerous spirally arranged glumes

Eleogiton p. 206

49B Leaves equally spaced around the node (there is a remarkable amount of convergent evolution among aquatics with this particular growth-form, identification without flowers is not practical)

54A Flowers when young enclosed within a spathe

55A Petals about 3-times as long and as broad as sepals

Egeria p. 257

55B Petals rarely exceeding sepals or absent

56A Male spathes globose, spiny; stamens 3; leaves usually more than 4 in a whorl

Hydrilla p. 259

56B Male spathes ovoid or cylindrical, smooth; stamens usually 6 or 9, rarely 3; leaves usually less than 4 in a whorl

Elodea p. 258

54B Flowers not enclosed within a spathe

57A Petals present, united into a tube below

58A Petal tube split adaxially; leaves toothed at least towards apex (NW. America)

Howellia p. 304

58B Petal tube not split adaxially; leaves not toothed

59A Inflorescence repeatedly forked; ovary inferior (Trop. S. America)

Limnosipanea p. 505

59B Inflorescence not repeatedly forked; ovary superior

60A Inflorescence very dense, spike-like, terminal; fruit of 4 nutlets

Dysophylla p. 280

60B Inflorescence not dense, not spike-like, terminal or axillary, or flowers solitary; fruit a capsule

61A Petal tube yellow; stamens 4, 2 fertile and 2 sterile (Madagascar)

Hydrotriche p. 520

61B Petal tube blue, purple or white; stamens 2 to 5, all fertile

62A Anther lobes free, divergent

Limnophila p. 524

62B Anther lobes united, parallel

Bacopa p. 514

57B Petals absent or present and free

63A Fruit a capsule containing several seeds

64A Sepals tubular, often appendaged between lobes; petals and stamens inserted on the inner surface of the sepal tube **Rotala p. 316**

64B Sepals free, not appendaged between lobes; petals and stamens free from sepals

Elatine p. 231

63B Fruit of 1, or 4 or more, 1-seeded nutlets

65A Petals yellow; fruiting head globular, with more than 4 nutlets

Ranunculus p. 502

65B Petals not yellow or absent; fruiting head flat or with 1 nutlet

66A Stamens 4 or 8; sepals and/or petals distinct, free

Myriophyllum p. 248

66B Stamens 1; sepals and petals not distinct, reduced to a rim around the nutlet

Hippuris p. 253

37B Leaves 1 or 2 at each node

67A Stems creeping; rooting at almost every internode (see p. 24)

68A Leaves differentiated into elongate petioles with flattened blades (see p. 24)

69A Leaves compound

70A Leaves 2-pinnate, with numerous leaflets; stem usually somewhat woody and spongy

Neptunia p. 244

70B Leaves with 2, 3 or 4 leaflets; stems not woody and not spongy

71A Leaflets 3, with midrib and netlike nerves; petiole base sheathing the stem

Menyanthes p. 329

71B Leaflets 2 or 4, without midrib and with repeatedly forked nerves; petiole base not sheathing the stem

72A Leaflets 4 (Widespread)

Marsilea p. 88

72B Leaflets 2 (S. Brazil and N. Argentina)

Regnellidium p. 90

69B Leaves simple

73A Petiole with transverse septae; petals 3, large and yellow

Hydrocleis p. 297

73B Petiole without transverse septae; petals not 3 or if 3 then not large and yellow

74A Flowers embedded in a fleshy spike (spathe); fruits bright red, berries

Calla p. 142

74B Flowers not embedded in a fleshy spike; fruits not bright red, not berries

75A Flowers inconspicuous; fruit of 2, 1-seeded, inferior carpels

Hydrocotyle p. 127

75B Flowers conspicuous; fruit of 1 or several, 1-seeded, superior carpels or fruit many-seeded

76A Fruit of several, many-seeded free carpels (follicles); outer perianth segments often somewhat sepaloid, the inner petaloid

Caltha p. 502

76B Fruit of 1 or several, 1-seeded nutlets or a many-seeded capsule; perianth entirely petaloid or differentiated into sepals and petals

77A Perianth entirely petaloid

78A Stamens 3; 1 anther much larger than the other 2

Heteranthera p. 486

78B Stamens 6; anthers almost equal in size

79A Fruit a 1-seeded nutlet, with spiny ridges

Reussia p. 490

79B Fruit a many-seeded, 3-locular capsule with smooth walls

Eichhornia p. 483

77B Perianth differentiated into sepaloid sepals and petaloid petals

80A Petals free to base; stamens 6 or more

81A Sepals 4; petals numerous; fruit a many-seeded capsule (flowers large with yellow petals)

Nymphaea p. 337

81B Sepals 3 or 5; petals 3 or 5; fruit of several 1-seeded nutlets

82A Sepals 3; petals 3; leaf margins entire

Luronium p. 109

82B Sepals 5; petals 5; leaf margins lobed or notched

Ranunculus p. 502

80B Petals united at least at base; stamens 5

83A Petals tubular, trumpet-shaped, up to 5 cm or more long, usually pink or pale lilac with a purple throat, rarely white; leaves usually sagittate

Ipomoea p. 182

83B Petals distinctly lobed above, rarely more than 2 cm long, usually white or yellow; leaves linear to orbicular, not sagittate

84A Flowers solitary or in clusters below the leaf blade

Nymphoides p. 329

84B Flowers in an open panicle

Villarsia p. 330

68B Leaves not clearly differentiated into an elongate petiole with blade

85A Leaves spirally wound when young

Pilularia p. 90

85B Leaves not spirally wound when young

86A Leaves with transverse septae

Lilaeopsis p. 127

86B Leaves without transverse septae

87A All leaves in opposite pairs; petals united below; fruit a capsule

Scrophulariaceae p. 511

87B At least leaves in flowering portion of stem alternate; petals free; fruit of several, free nutlets

Ranunculus p. 502

67B Stems erect (emergent or submersed) or floating, not rooting regularly at each internode

88A Leaves in opposite pairs (see p. 27)

89A Leaves compound or divided into linear segments

90A Leaves pinnately divided

91A Fruit a 2-valved capsule; stamens 4; emergent leaves aromatic (tropical regions)

Hygrophila p. 95

91B Fruit of 4 nutlets; stamens 2; emergent leaves not aromatic (temperate regions)

Lycopus p. 280

90B Leaves palmately divided

92A Leaf segments simple to base of leaf (S. Africa)

Cotula p. 155

92B Leaf segments repeatedly branched

93A Leaf segments lying in one plane; flowers solitary with 3 sepals and 3 petals

Cabomba p. 170

93B Leaf segments lying in more than one plane; flowers in composite heads with numerous ray florets

Megalodonta p. 157

89B Leaves simple

94A Leaves capillary, with sheathing stipular bases (obligately submerged aquatics)

95A Flowers sessile or subsessile in leaf axils, unisexual

Zannichelliaceae p. 546

95B Flowers stalked, borne in an umbel-like inflorescence, bisexual

Ruppia p. 507

94B Leaves not capillary, without sheathing stipular bases (submerged or partly emergent herbs)

96A Leaves with undulate margin, longitudinally folded at least when young, dark green and translucent

Groenlandia p. 494

96B Leaves without undulate margin, not longitudinally folded, not dark green and translucent

97A Flowers in heads subtended by an involucre of bracts

98A Leaf margin serrate; green parts of plant clothed with whitish, appressed hairs (widespread)

Eclipta p. 157

98B Leaf margin entire; green parts not clothed with whitish, appresssed hairs (Mexico)

Hydropectis p. 160

97B Flowers not in heads or when in heads then not subtended by an involucre of bracts

99A Flowers naked (sepals, petals or perianth absent)

Callitriche p. 171

99B Flowers not naked (sepals, petals or perianth present)

100A Fruit a cylindrical nutlet bearing up to 5 processes at apex (processes either 5 spirally rolled at tips, or 3 long and spirally rolled and 2 thorn-like, or 2 or 3 wing-like)

Trapella p. 539

100B Fruit not a cylindrical nutlet or when so then without apical processes

101A Ovary completely inferior
102A Seeds numerous in each loculus
103A Sepals persisting in fruit; petals not 2-lobed or absent
Ludwigia p. 343
103B Sepals not persisting in fruit; petals 2-lobed
Boisduvalia p. 343
102B Seeds 1 in each loculus
Haloragaceae p. 246
101B Ovary superior or semi-inferior
104A Carpels several, free or splitting into 4 nutlets at maturity
105A Petals free or absent
106A Leaves not differentiated into petiole and blade, with entire margin
Crassula p. 184
106B Leaves with petiole and blade; blade with lobed margin
Ranunculus p. 502
105B Petals united at base; leaves not succulent
107A Leaves not exceeding 1 cm in length; flowers solitary; sepals 4, united below
Tetrachondra p. 535
107B Leaves more than 1 cm long; flowers in axillary cymes or terminal heads; sepals 5, united below
Lamiaceae p. 280
104B Carpel 1, or more, united, not splitting into 4 nutlets at maturity
108A Perianth of 5, dry, membranous segments
109A Stems conspicuously grooved; outer perianth segment with recurved, subulate tip: anthers 4-locular
Centrostachys p. 114
109B Stems not conspicuously grooved; outer perianth segment without recurved, subulate tip: anthers 2-locular
Alternanthera p. 112
108B Perianth sepaloid and/or petaloid, 5 or not, not dry and membranous
110A Sepals 2, free or joined below (small, fleshy, creeping herbs)
Montia p. 493
110B Sepals more than 2, or if 2-lobed then distinctly tubular below
111A Petals free to base
112A Anthers opening by terminal pores; connective prolonged into 2 lobes

113A Petals obovate, with blunt apex; anthers dimorphic; inflorescence small and few-flowered

Acisanthera p. 323

113B Petals ovate, with acute to acuminate apex; anthers isomorphic; inflorescence freely branched and many flowered

Nepsera p. 323

112B Anthers opening by longitudinal slits; connective not prolonged into 2-lobes

114A Sepals united into a perigynous tube

Lythraceae p. 310

114B Sepals free or united at base

115A Stamens numerous (more than 10), grouped together in fascicles; petals conspicuous, yellow or orange

Hypericum p. 179

115B Stamens 10 or less, not grouped in fascicles; petals inconspicuous, whitish or pinkish

116A Sepals 5, acute; flowers in dense clusters

Bergia p. 231

116B Sepals 2, 3 or 4, obtuse; flowers solitary or few together

Elatine p. 231

111B Petals tubular or united at base

117A Fruit opening by 5 valves; stamens 5, inserted opposite the petal lobes

Lysimachia p. 501

117B Fruit opening by 2 valves; stamens usually 4 or less, rarely 5, inserted between the petal lobes

118A Seeds inserted on hardened, hook-like out-growth from the central placenta (retinacula)

Acanthaceae p. 95

118B Seeds inserted directly on the central placenta, hardened, hook-like outgrowth (retinacula) absent

Scrophulariaceae p. 511

88B Leaves alternate or spiral (1 leaf at each internode)

119A Leaves compound or simple and divided into lobes or segments

120A Leaves laciniate, irregularly divided into thin, translucent, flattened, band-like segments

121A Stems hollow; flowers in umbels (N.W. Europe)

Oenanthe p. 129

121B Stems solid flowers in composite heads or solitary

122A Flowers in composite heads (Mexico)

Erigeron p. 157

122B Flowers solitary (China)

Ranunculus p. 502

120B Leaves not laciniate, not divided into thin, translucent, flattened, band-like segments

123A Leaves palmately divided

124A Segments simple to base of leaf

Cynosciadium p. 126

124B Segments repeatedly forked

Ranunculus p. 502

123B Leaves pinnately divided

125A Leaves 2- or more -pinnate

126A Stems somewhat woody with soft pithy cortex; leaves 2-pinnate, with regular, ovate leaflets (Tropics)

Neptunia p. 244

126B Stems not woody, without pithy cortex; leaflets usually irregular (Temperate Regions)

127A Inflorescence racemose; fruit containing several seeds

Rorippa p. 163

127B Inflorescence umbellate; fruits of 2, inferior, 1-seeded carpels

Apiaceae p. 117

125B Leaves 1-pinnate

128A Inflorescence umbellate; fruits of 2, inferior, 1-seeded carpels (stems hollow, often constricted at nodes)

Apiaceae p. 117

128B Inflorescence racemose or flowers grouped in leaf axils or flowers solitary

129A Leaflets or pinnae linear

130A Pinnae tipped with spines; petals united at base

Navarretia p. 480

130B Pinnae not tipped with spines; petals free or absent

131A Fruit splitting into 1-seeded nutlets; stamens 4 or 8

Myriophyllum p. 248

131B Fruit nut-like, 3-angled, 3-seeded; stamens 3

Proserpinaca p. 250

129B Leaflets or pinnae lanceolate to ovate or suborbicular

132A Leaflets or pinnae varying in shape and size in any single leaf; stem not woody, without pithy cortex

133A Capsules flattened in transverse section; valves of capsule coiling from base at dehiscence

Cardamine p. 163

133B Capsules round in transverse section; valves of capsule not coiling

Rorippa p. 163

132B Leaflets or pinnae very constant in shape in any single leaf; stem somewhat woody, usually with pithy cortex

134A Fruit indehiscent, breaking into regularly shaped portions

Aeschynomene p. 242

134B Fruit marginally dehiscent, not breaking into portions

Sesbania p. 244

119B Leaves simple and entire

135A Leaves pinnately nerved, with straight, parallel nerves (large emergent herbs)

136A Petiole indistinct, not swollen; leaves spiral

Canna p. 172

136B Petiole distinctly swollen; leaves in 2 rows

137A Flowers in short, dense spikes

Phrynium p. 317

137B Flowers in lax, spreading panicles

138A Ovary 3-locular; petaloid staminodes 2

Donax p. 317

138B Ovary 1-locular; petaloid staminodes 1

Thalia p. 320

135B Leaves not pinnately nerved or if pinnate then nerves not straight and parallel

139A Flowers in dense, more or less globose heads; heads subtended by an involucre of bracts

Eriocaulaceae p. 233

139B Flowers not in dense, more or less globose heads; heads not subtended by an involucre of bracts

140A Leaves notched at apex (leaves linear, clothing the stem, habit like a spiral-leaved Hippuris)

Mayaca p. 321

140B Leaves not notched at apex

141A Nodes completely surrounded above the insertion of the leaf by a membranous or scarious sheath (ochrea)

Polygonum p. 481

141B Nodes not surrounded by an ochrea

142A Inflorescence a simple, fleshy spike (spadix) of inconspicuous flowers subtended by a large, thick bract (spathe)

Araceae p. 137

142B Inflorescence without spadix, spathe absent or if present then small and membranous

143A Perianth segments scarious (dry and brownish), not clearly differentiated into sepals and/or petals

144A Perianth of 4, clawed scales, attached to each stamen

Potamogeton p. 496

144B Perianth of 6, more or less lanceolate scales, not attached to stamens

Juncus p. 275

143B Perianth absent or segments not scarious, sepaloid and/or petaloid

145A Perianth absent

146A Leaves linear to capillary (obligately submerged aquatics)

147A Flowers and fruits sessile or subsessile in leaf axils

Zannichelliaceae p. 546

147B Flowers and fruits distinctly stalked, in umbels

Ruppia p. 507

146B Leaves with petiole and blade; blade ovate to reniform; inflorescence and some foliage leaves emergent

Saururaceae p. 509

145B Perianth present (sepaloid and/or petaloid)

148A Ovary inferior or semi-inferior or plants dioecious

149A Fruit indehiscent, nut-like or splitting into 1-seeded nutlets

150A Fruit splitting into 1-seeded nutlets

Myriophyllum p. 248

150B Fruit indehiscent, nut-like, not splitting into 1-seeded nutlets

151A All flowers bisexual, sessile or shortly stalked

Haloragis p. 246

151B Some flowers unisexual, males distinctly stalked

Laurembergia p. 248

149B Fruit dehiscent, many-seeded

152A Flowers arranged in a bifid spathaceous bract or between 2 opposite bracts (obligately submerged aquatics)

Hydrocharitaceae p. 254

152B Flowers not arranged in a bifid spathaceous bract or between 2 opposite bracts (inflorescence and some foliage leaves emergent)

153A Stamens twice as many as sepal lobes; flowers usually 4-merous

Onagraceae p. 343

153B Stamens as many as sepal lobes; flowers usually 5-merous

154A Anthers united in a ring around the style

Lobeliaceae p. 302

154B Anthers free

155A Inflorescence a densely packed spike; capsule opening by an apical lid

Sphenoclea p. 533

155B Inflorescence not a spike; capsule opening by 5 valves

Samolus p. 501

148B Ovary superior; plants not dioecious

156A Ovary of several free carpels

157A Sepals 5 or rarely more; petals 5 or more, yellow

Ranunculus p. 502

157B Sepals 3; petals 3, white with a yellow spot in the centre

Luronium p. 109

156B Ovary of 1 or more united carpels

158A Flowers with spur; leaf bases bearing paired glands; fruit purplish red

Hydrocera p. 161

158B Flowers without spur; leaf bases without glands; fruit not purplish red

159A Perianth segments 1-seriate, entirely petaloid

Pontederiaceae p. 482

159B Perianth segments 2-seriate, sepaloid and petaloid (in female flowers petals occasionally absent)

160A Flowers 3-merous; staminodes at apex 3-lobed or sagittate

Murdannia p. 180

160B Flowers not 3-merous; staminodes absent or simple at apex

161A Sepals 4 and petals 4

Cardamine p. 163

161B Sepals 5; petals 5 or sometimes absent in unisexual flowers

162A Leaves sagittate; petals trumpet-like, c. 5 cm long

Ipomoea p. 182

162B Leaves linear to ovate, not sagittate; petals not trumpet-like, less than 5 cm long

163A Leaves tipped with spines; blades more or less terete

Navarretia p. 480

163B Leaves not tipped with spines; blades flattened

164A Flowers unisexual; leaves with caducous stipules

Caperonia p. 239

164B Flowers bisexual; leaves without stipules

165A Fruit opening irregularly or splitting by valves (usually robust herbs with swollen, spongy stems)

Hydrolea p. 269

165B Fruit opening by an apical lid (usually small and delicate with stems not swollen and not spongy)

Anagallis p. 497

36B Leaves not regularly dispersed along the total length of the stem, most leaves in basal or terminal rosettes or clusters

166A Leaves compound or at least some leaves lobed to base or midrib

167A Leaves 3-foliate; petals hairy inside

Menyanthes p. 329

167B Leaves not 3-foliate; petals not hairy inside

168A Leaves palmately lobed; fruit of several, free nutlets

Ranunculus p. 502

168B Leaves pinnately lobed; fruit not of several, free nutlets

169A Young leaves spirally rolled; mature leaves bearing sporangia on the abaxial surface

Ceratopteris p. 90

169B Young leaves not spirally rolled; mature leaves not bearing sporangia

170A Leaves regularly ovate in outline, usually 1-pinnate; pinnae linear, mostly simple some occasionally forked; inflorescence verticillate

Hottonia p. 501

170B Leaves more or less triangular or irregular in outline; most pinnae pinnately divided or repeatedly forked or pinnae not linear

Apiaceae p. 117

166B Leaves simple

171A At least some foliage leaves clearly differentiated into an elongate petiole and flattened blade (see p. 35)

172A Leaf blades forming a circular, whorled rosette floating on the surface

173A Petioles inflated; fruit not cylindrical, with 2, 3 or 4 hard horns

Trapa p. 537

173B Petioles not inflated; fruit cylindrical

174A Fruit 1-seeded, with 3 or more, long, spirally rolled processes; petals 5, tubular below (E. Asia)

Trapella p. 539

174B Fruit many-seeded, without processes; petals 4, free (S. America)

Ludwigia p. 343

172B Leaf blades not forming a circular, whorled rosette floating on the surface

175A Inflorescence a simple fleshy spike (spadix) of inconspicuous flowers subtended by a large thick bract (spathe) or if spathe absent then spike yellow

Araceae p. 137

175B Inflorescence without spadix and spathe

176A Carpels 2 or more, free or united at extreme base

177A Inflorescence a simple or branched spike; leaf nervature almost regularly rectangular

Aponogetonaceae p. 135

177B Inflorescence not a spike; leaf nervature not regularly rectangular

178A Flowers distinctly 3-merous; leaf margins entire

179A Ovaries containing numerous seeds; plants with milky latex; ovules scattered over inner surface of carpels

Limnocharitaceae p. 297

179B Ovaries 1-seeded, or if more-seeded then fruit star-shaped; plants without milky latex; ovules solitary or axile

Alismataceae p. 99

178B Flowers not 3-merous (petals and sepals 5 or numerous); leaf margins lobed or notched

Ranunculaceae p. 502

176B Carpels 1 or more united

180A Stigmas united into a disc with radiating rays; stamens more than 6

Nymphaeaceae p. 334

180B Stigmas free, not united into a disc with radiating rays; stamens 6 or less

181A Fruit of 2, 1-seeded mericarps; inflorescence umbellate

Apiaceae p. 117

181B Fruit of 1, 1-seeded carpel or a many seeded capsule

182A Ovary superior

183A Perianth clearly differentiated into sepals and petals; flowers 5-merous

Menyanthaceae p. 326

183B Perianth not clearly differentiated into sepals and petals; flowers 3-merous

184A Perianth petaloid, showy; leaves not distinctly ribbed

Pontederiaceae p. 482

184B Perianth sepaloid or bract-like, not showy; leaves stiff and distinctly ribbed

Hanguana p. 251

182B Ovary inferior

185A Petioles without stipules; stolons absent

Ottelia p. 264

185B Petioles with 1 or 2 transparent stipules at base; stolons present

186A Petals absent or when present up to 1½ times as long as sepals

Limnobium p. 262

186B Petals present, more than 1½ times as long as sepals

Hydrocharis p. 260

171B Leaves without a distinct petiole; blade flattened or not

187A Leaves distinctly flattened throughout most of their length (parallel sided in transverse section, not terete, semicircular, triangular or rectangular, see p. 37)

188A Leaves widest towards apex (spathulate, obovate or obtriangular)

189A Leaves densely pubescent; plants usually free-floating rosettes
Pistia p. 154

189B Leaves glabrous; plants attached to bottom
190A Flowers without petals or sepals; fruit splitting into 4, 1-seeded nutlets
Callitriche p. 171

190B Flowers with petals and sepals; fruit a capsule
191A Rhizome fleshy; inflorescence terminal, 1-flowered
Liparophyllum p. 326

191B Rhizome absent or not fleshy; inflorescence axillary or if terminal then more than 1-flowered

192A Plants with forked and spirally twisted traps descending into substrate; petals with abaxial spur
Genlisea p. 292

192B Plants without traps, petals without spur

193A Leaves opposite; stamens alternating with petal lobes
Scrophulariaceae p. 511

193B Leaves alternate; stamens opposite petal lobes
Samolus p. 501

188B Leaves widest at or towards base, or leaves parallel-sided

194A Leaves with distinct spinose-serrate margins
195A Entire plant floating to surface to flower; petals white, showy
Stratiotes p. 266

195B Entire plant not floating to surface to flower; stout, erect stem, covered with black fibrous remains of leaves; petals scarious (S. Africa)
Prionium p. 277

194B Leaves with entire or faintly toothed margins

196A Flowers in more or less spherical heads

197A Heads terminal, solitary or in umbels, bisexual, subtended by an involucre of bracts

198A Perianth dry, papery and inconspicuous, greyish or blackish; flowers unisexual
Eriocaulaceae p. 233

198B Perianth pataloid, showy, yellow, blue or rarely white; flowers bisexual
Xyridaceae p. 543

197B Heads in simple or branched racemes, unisexual (male above, female below), without an involucre of bracts.
Sparganium p. 531

196B Flowers not in spherical heads

199A Flowers densely packed in cylindrical spikes

200A Spike appearing to arise laterally from a foiliage leaf; flowers bisexual
Acorus p. 140

200B Spike terminal; flowers unisexual (male above, female below)
Typha p. 540

199B Flowers not densely packed in cylindrical spikes

201A Leaves monofacial, equitant; style petaloid
Iris p. 273

201B Leaves not monofacial, not equitant; style not petaloid

202A Petals united, tubular below

203A Petal tube 2-lipped, stamens 4; rootstock not a bulb
Craterostigma p. 516

203B Petal tube equally 6-lobed; stamens 6; rootstock a bulb
Crinum p. 115

202B Petals free or absent

204A Perianth scarious, bract-like
Juncus p. 275

204B Perianth with sepaloid sepals and petaloid petals (petals occasionally reduced or absent)

205A Ovary inferior; carpels united; fruit a many seeded capsule

206A Greater part of leaf parallel-sided, without spines (occasionally minutely serrate); male flowers minute, breaking off and floating on the surface
Vallisneria p. 267

206B Leaves not parallel-sided or if so then with spines; male flowers not minute, not breaking off and floating on the surface

207A Leaves tapering to a fine point; spathes not inflated; not spiny, not winged but occasionally ribbed

Blyxa p.256

207B Leaves lanceolate or if linear then spiny; spathes inflated and distinctly winged or spiny

Ottelia p. 264

205B Ovary superior; carpels free; fruit of many 1-seeded nutlets

Alismataceae p. 99

187B Leaves not flattened throughout most of their length (not parallel-sided in transverse section)

208A Leaves capillary (hair-like and flaccid)

209A Flowers in tightly packed, globose heads; heads subtended by an involucre of bracts

210A Perianth dry, papery, inconspicuous, greyish or blackish; flowers unisexual

Eriocaulaceae p. 233

210B Perianth petaloid, showy, yellow, blue or rarely white; flowers bisexual

Xyridaceae p. 543

209B Flowers not in tightly packed, globose heads; heads not subtended by an involucre of bracts

211A Perianth present; perianth segments 6

212A Perianth segments scarious; flowers in clusters

Juncus p. 275

212B Perianth segments sepaloid; flowers in an elongate raceme or spike

Juncaginaceae p. 278

211B Perianth replaced by bracts, scales or bristles

213A Fruit dehiscent (minute herbs, S. Hemisphere)

Centrolepidaceae p. 174

213B Fruit indehiscent (small or large herbs, Cosmopolitan)

Cyperaceae p. 185

208B Leaves not capillary

214A Leaves subulate, rarely more than 10 cm long, usually rigid and distinctly curved

215A Plants not connected by persistent rhizomes or stolons, not turf-forming

216A Leaves with 4, longitudinal, septate air canals; leaf bases swollen, often bearing spores

Isoetes p. 84

216B Leaves without 4, longitudinal, septate air canals; leaf bases not swollen, never bearing spores

217A Leaf apices somewhat obtuse; sepals 5, united below
Lobelia p. 308

217B Leaf apices acute; sepals 4, free
Subularia p. 165

215B Plants connected by persistent rhizomes or stolons, usually forming a closed turf

218A Fruit of numerous, 1-seeded nutlets; sepals free; petals free

219A Sepals 3; petals 3, without nectaries; flowers occasionally unisexual
Alismataceae p. 99

219B Sepals 5; petals 5, with nectaries at base; flowers bisexual
Ranunculus p. 502

218B Fruit a many seeded capsule or a solitary, 1- or 2-seeded nut-like capsule

220A Flowers unisexual; petals scarious; fruit nut-like
Littorella p. 348

220B Flowers bisexual; petals petaloid; fruit dehiscent
Limosella p. 524

214B Leaves not subulate, frequently more than 10 cm long; flaccid or rigid and erect

221A Leaves with 4, longitudinal, septate air canals; leaf bases swollen, often bearing spores
Isoetes p. 84

221B Leaves without 4, longitudinal, septate air canals; leaf bases not bearing spores

222A Perianth dry and scarious, or reduced to bristles or scales, or absent

223A Flowers in spherical heads (heads male above, female below)
Sparganium p. 531

223B Flowers not in spherical heads

224A Styles heteromorphic, those of lower (female) flowers at least 2 cm long (scapose annuals)
Lilaea p. 295

224B Styles not heteromorphic, not exceeding 1 cm in length

225A Leaves in 2 rows, equitant; inflorescence of 2, superposed spikes (male above, female below)
Typha p. 540

225B Leaves in 3 rows or spiral, not equitant; inflorescence not of 2, superposed spikes

226A Perianth of 6 scarious, segments
Juncus p. 275

226B Perianth absent or of scales or bristles
Cyperaceae p. 185

222B Perianth sepaloid and/or petaloid, not scarious, not reduced to bristles or scales

227A Flowers zygomorphic; stamen 1 (S. and E. Asia, N. Australia)
Philydrum p. 346

227B Flowers actinomorphic; stamens 5 or more

228A Perianth segments entirely petaloid

229A Ovary inferior; carpels 3, united; perianth not persisting (warmer regions)

230A Inflorescence umbellate; rootstock a bulb; perianth united below
Crinum p. 115

230B Inflorescence racemose or flowers solitary; rootstock a rhizome; perianth free to top of ovary
Hypoxis p. 271

229B Ovary superior; carpels 6 or more, almost free; perianth persisting (temperate N. hemisphere)
Butomus p. 166

228B At least outer perianth segments sepaloid

231A Fruit of numerous, 1-seeded nutlets in a spherical head
Sagittaria p. 109

231B Fruit of 2, 1-seeded mericarps or 1 or more many-seeded carpels

232A Flowers solitary; petals united below
Liparophyllum p. 326

232B Flowers many in a inflorescence, petals absent or free

233A Inflorescence a hemispherical head subtended by spinous bracts
Eryngium p. 126

233B Inflorescence not a hemispherical head, not subtended by spinous bracts

234A Inflorescence umbellate; fruits winged
Oxypolis p. 131

234B Inflorescence a raceme or spike; fruits not winged
Juncaginaceae p. 278

CHAROPHYTA

Aquatic cryptogams; cells becoming multinucleate at maturity. The thallus consists of a central axis ("stem") with whorls of laterals arising at more or less regular intervals from nodes. The axis consists of a chain of alternating long and short cells, single long cells forming internodes and very short, discoid cells forming nodes. The node cells divide periclinally (around the outer margin) producing a ring of peripheral nodal cells from which lateral branches (branchlets) arise. In addition, the branchlets in some genera are subtended by 1-celled processes known as stipulodes; such 1-celled processes may arise at branchlet nodes and are then known as bract-cells or they arise from gametangial laterals and are then known as bractlets. Highly modified bract-cell- or stipulode-like processes may develop from the nodes and remain appressed to the outer face of the internodial cells forming a cortex one cell in thickness. The single axial internodal cells are usually 1 to 4 cm long but sometimes reach 15 cm long and 0.3 cm wide. Size of thallus variable (0.5) 15 to 30 (200) cm. The entire shoot is anchored by nongreen, single-celled rhizoids. The gametangia are enclosed within a jacket of sterile cells. In the female, the jacket (oogonium) consists of 5 elongate cells which are tightly wound around the egg: either one or two small cells terminate each cell and they form a coronula. The male comprises 8 (rarely 4) nearly triangular plates which fit together into a globose antheridium; the gametes are produced, one in each cell, in uniseriate filaments within the antheridium.

CHARACEAE

6 genera, *c.* 80 species; cosmopolitan. Taxonomically complex.

Wood, R. D. and Imahori, K. A revision of the Characeae. J. Cramer, Weinheim, pt. 1 Monograph 1–904, pt. 2 Icones pl. 1–395 (1965)

The Characeae, although classified as algae, constitute a group of freshwater macrophytes which forms a significant part of the submerged vegetation in streams, ponds and lakes throughout the world. Lime accumulates in some species and the plants may become brittle and coarse to the touch. Extensive growths may solidify directly into a marl layer or into a curious tufa rock. The Characeae are of considerable economic importance (see Zaneveld, J. S., Blumea 3 (2): 35 (1940) and Wood, R. D., Bot. Rev. 18 (5): 319 (1952)) and have been used, for example, in water purification, as food for fish and farm stock, in the manufacture of polishes and as an agent in settling silt; they have also, in some areas, become a problem by clogging channels and reservoirs.

The identification of members of the Characeae is difficult and requires the use of a microscope; it is recommended to send material preserved in alcohol to an expert.

1A Coronula of 5 cells in 1 tier
  2A Bract-cells normally 2 at a node (occasionally 1 or 3), anterior only, similar in size and shape to branchlet segment; branchlet segments 2 to 5; without cortex, dioecious; bulblets developed, 1- or more celled; coronula cells spherical
  Nitellopsis
  2B Bract-cells normally 4 or more at a node, including both anteriors and posteriors; branchlet segments 3 or more; with or without cortex; bulblets uncommon, 1-celled, only known in corticated species; coronula cells elongate
    3A Gametangia each from a separate peripheral nodal cell, generally with central oogonium flanked by 2 lateral antheridia
    Lychnothamnus
    3B Gametangia from a single peripheral nodal cell
      4A With or without cortex; stipulodes in 1 or 2 tiers, 1 or 2 (rarely 3 or 4) per branchlet (if 1, then alternate or rarely opposite); monoecious or dioecious; conjoined gametangia with oogonium above antheridium (in dioecious species antheridium replaced by bractlet which subtends oogonium)
      Chara
      4B Without cortex; stipulodes in 1 tier (rarely irregular, sparse or absent), 1 per branchlet, opposite, strongly decumbent; monoecious; conjoined gametangia with oogonium usually below antheridium
      Lamprothamnium

1B Coronula of 10 cells in 2 tiers
  5A Branchlets divided at nodes into 2 or more similar rays, rays whorled; oospores laterally compressed; generally without lime incrustation or, if present, then in bands or at the tips of branchlets
  Nitella
  5B Branchlets monopodial, not furcate; rays bilateral, with laterals differentiated from ab- or adaxial rays; oospores terete, not compressed
  Tolypella

Chara L. em. Braun, Hooker's Journ. Bot. 1: 195, 292 (1849)
Fig. 1.
Plants dioecious or monoecious, axes and branchlets with or without cortex. Stipulodes developed, sometimes rudimentary. Branchlets of 1 to 18 internodal cells separated by nodes, distal internodes, 1- or more-celled. Bract-cells

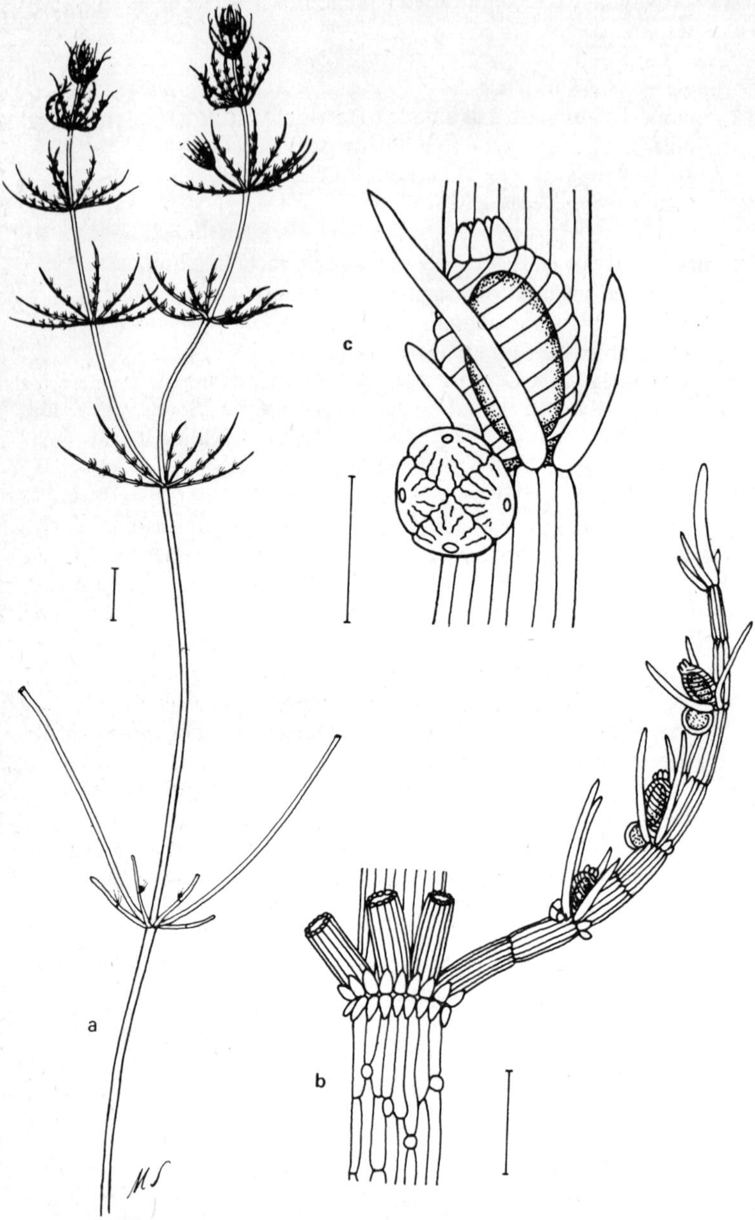

Fig. 1. *Chara vulgaris* L.: a, habit (1 cm); b, axial node with cortex and stipulodes in 2 tiers and 1 branchlet with bract cells and conjoined gametangia (1 mm); c, branchlet node with gametangia, male below and female above (0.5 mm).

5 to 7 at branchlet nodes, occasionally rudimentary. Bractlets usually 2, arising from gametangial stalk. Gametangia, when conjoined, arising from single peripheral nodal cell. Antheridium below oogonium. Coronula of 5 cells in 1 tier.

19 species: *C. vulgaris* L., *C. globularis* Thuill. and *C. braunii* Gm. are cosmopolitan, the majority of species are found in the African–S. Asian–Australian arc. The species of *Chara* occur in both acid and alkaline water, but the vast majority of habitats are alkaline (pH 7.2 to 9.5) (see Wood and Imahori, 1965, op. cit. I: 54 for references).

Lamprothamnium J. Groves, Journ. Bot. (London) 54: 336 (1916)
Fig. 2A.
Plants monoecious, entirely without cortex. Stipulodes in 1 tier, as numerous as the branchlets, inserted one below each branchlet. Bract-cells well developed and in whorls. Bractlets usually 2 but occasionally absent. Gametangia, when conjoined, arising from single peripheral nodal cell. Antheridium above oogonium (except in *L. succinctum* (A. Br.) R. D. W.). Coronula of 5 cells in 1 tier.

3 species; *L. papulosum* (Wallr.) J. Gr. widespread along coastal Europe and N. Africa, and Australia and New Zealand, with isolated reports from S. Africa and China, *L. succinctum* (A. Br.) R. D. W. through Africa and S. Asia to New Caledonia and *L. hansenii* (Sond.) Cor. known from two localities on the Baltic coast of Germany. *Lamprothamnium* is confined to brackish water.

Lychnothamnus (Rupr.) Leonh. em A. Braun in Braun and Nordstedt, Abh. K. Akad. Wiss. Berlin 1882: 100 (1882)
Fig. 2B.
Plants monoecious, with or without cortex, cortex imperfect. Stipulodes in 1 tier but twice as numerous as branchlets. Bract-cells well developed and in whorls. Bractlets 2. Gametangia, when conjoined, arising from separate peripheral nodal cells. Antheridia lateral, flanking oogonium. Coronula of 5 cells in 1 tier.

1 species, *L. barbatus* (Meyen) Leonh.; primarily known from N. Germany but records exist from Poland, France, Italy and India.

Nitella Ag. em Leonh., Lotos 13: 69 (1863)
Fig. 3A.
Plants monoecious or dioecious, entirely ecorticate, generally without lime

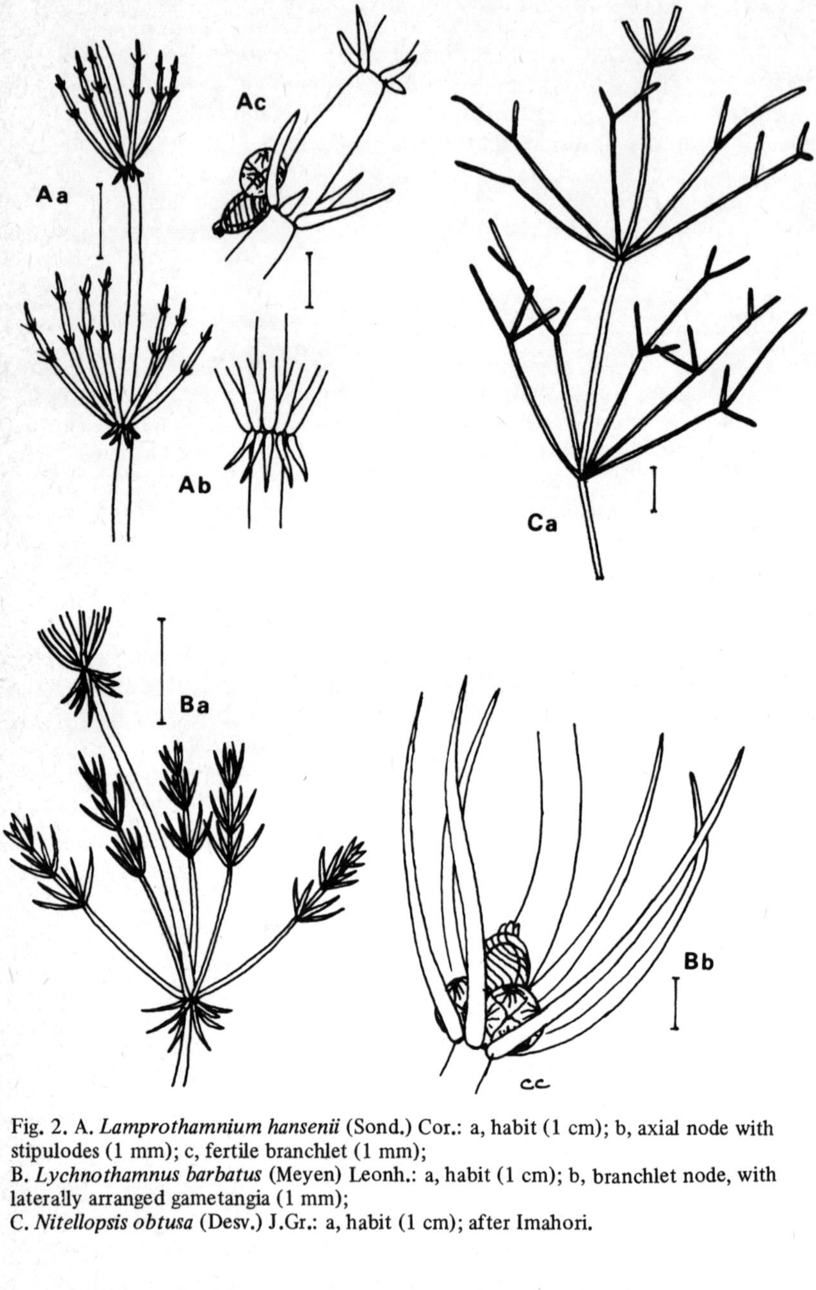

Fig. 2. A. *Lamprothamnium hansenii* (Sond.) Cor.: a, habit (1 cm); b, axial node with stipulodes (1 mm); c, fertile branchlet (1 mm);
B. *Lychnothamnus barbatus* (Meyen) Leonh.: a, habit (1 cm); b, branchlet node, with laterally arranged gametangia (1 mm);
C. *Nitellopsis obtusa* (Desv.) J.Gr.: a, habit (1 cm); after Imahori.

incrustation or, if present, then in bands or at tips of branchlets. Branchlets not or slightly dimorphic, divided at the nodes into 2 or more similar rays. Gametangia, when conjoined, arising from single peripheral nodal cell or at the branches. Antheridium is generally terminal at end of branchlet ray; oogonia are lateral at branches. Oospores compressed in transverse section. Coronula of 10 cells in 2 tiers.

Approximately 53 species; *N. gracilis* (Sm.) Ag. and *N. hyalina* (DC.) Ag. are cosmopolitan and *N. flexilis* (L.) Ag. and *N. furcata* (Bruz.) Ag. nearly so, the other species are of more limited range but temperate and tropical regions of all continents are represented. The nitellas occur primarily in mildly acid or neutral water (pH 4.6 to 7.0) and are rather rare in alkaline conditions; coastal species are restricted to fresh or nearly fresh water.

Nitellopsis Hy, Bull. Soc. Bot. France 36: 397 (1889)
Fig. 2C.
Plants dioecious, entirely without cortex. Stipulodes absent or rudimentary. Branchlets of 2 to 5 internodes, occasionally bearing stellate bulblets. Bract-cells 1 to 3 at each branchlet node. Bractlets absent. Gametangia arising individually, each from separate peripheral nodal cell. Coronula of 5 cells in 1 tier.

3 species; *N. obtusa* (Desv.) J. Gr. widespread throughout Europe and Asia from Scandinavia to Japan, *N. sarcularis* Zaney from Indonesia and *N. bulbilifera* C. Dont. from Argentina.

Tolypella (A. Braun) A. Braun, Abh. K. Akad. Wiss. Berlin 1856: 338 (1857)
Fig. 3B.
Plants monoecious, entirely without cortex, often lime encrusted. Stipulodes absent. Branchlets dimorphic, sterile large and fertile small and generally compacted into coarse heads somewhat resembling bird nests, both simple or monopodial; fertile branchlet axis of 5 to 8 cells with 1 or 2 nodes bearing 2 to 4 rays, 2 of which are lateral, 1 abaxial and 1 adaxial; sterile branchlet, when divided with 3 similar rays, 1 central and 2 lateral. Gametangia, when conjoined, arising from single peripheral nodal cell. Antheridium usually central, adaxial, flanked by 2 to 4 lateral oogonia. Oospores terete in transverse section. Coronula of 10 cells in 2 tiers.

2 species; *T. nidifica* (O. Müll.) A. Br. and *T. intricata* (Roth.) Leonh. distribution similar, widespread in temperature regions of N. Hemisphere extending sporadically into the S. Hermisphere.

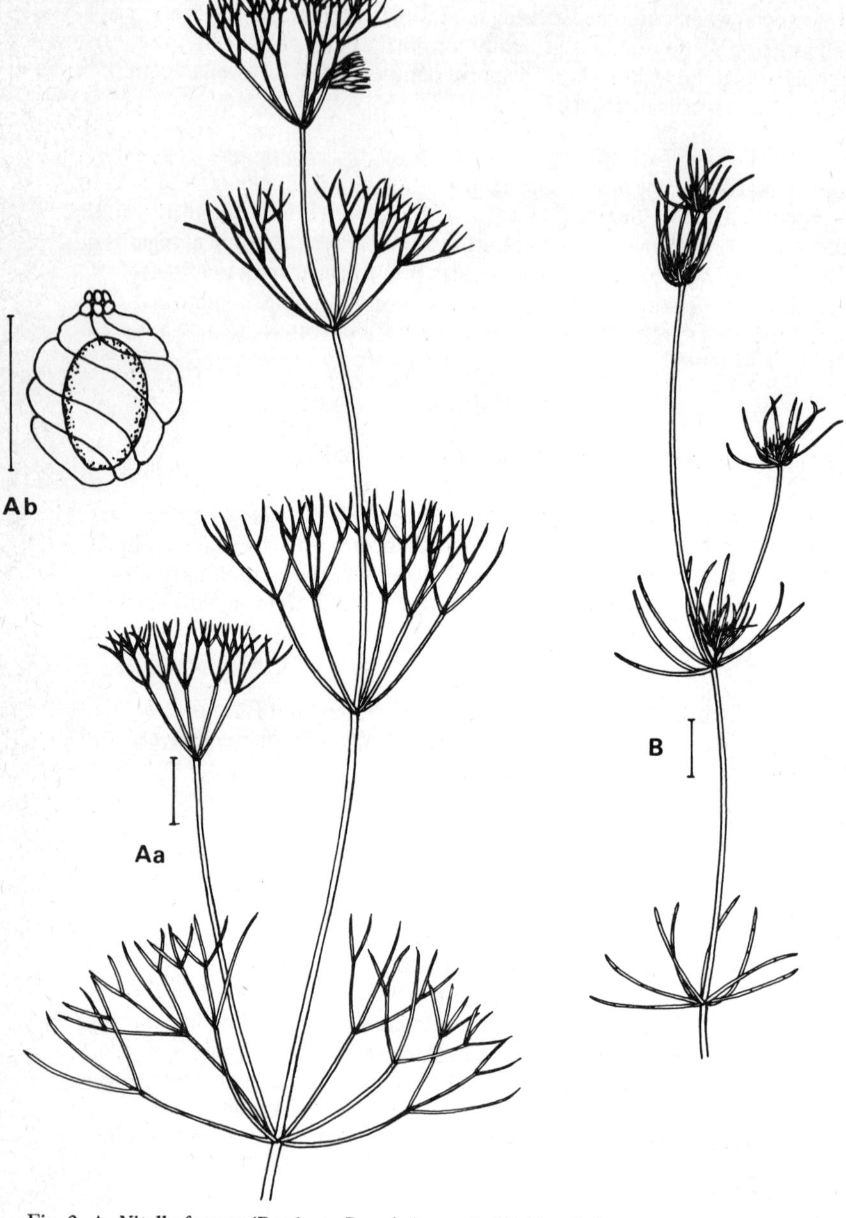

Fig. 3. A. *Nitella furcata* (Roxb. ex Bruz.) Ag.: a, habit (1 cm); b, mature oogonium (0.5 mm);
B. *Tolypella nidifica* (O. Muell.) A.Br.: a, habit (1 cm); after Imahori.

BRYOPHYTA (Mosses and Liverworts)

Cryptogams with dominant gametophyte (haploid phase). Gametophyte rootless, thalloid or differentiated into stem with leaves. Developing gametes surrounded by a sterile jacket one cell layer thick. Male gametes uninuclear and biflagellate. Sporophyte (diploid phase) parasitic on gametophyte, consisting of a capsule containing spores and a stalk that may be long, short or occasionally absent.

Brotherus, V. F., Paul, H. and Ruhland, W. in Engler, Nat. Pflanzenfam. ed. 2, 10: (1924), 11: (1925)

Wijk, R. van der, Margadant, W. D. and Florschütz, P. A. Index Muscorum Reg. Veg. 17: (1959), 26: (1962), 33: (1964), 48: (1967), 65: (1969)

The delimitation of families and genera in the Bryophyta presents many problems. In this account the treatment of Wijk (1959–1969) has been used as a guide.

There are about 23 500 species in the Bryophyta, most of them are small and many inhabit wet or damp places and are capable of withstanding temporary flooding. In this account an attempt has been made to describe only the bryophytes that normally grow either permanently submerged in water or those regularly submerged for several months each year, or those that are free floating on the surface of water.

1A Plants thallose, not differentiated into stem and leaves
- 2A Plants submerged, attached to substrate; thallus asymmetrical (an "axis" winged on one side and over the apex)
  Riella **p. 55**
- 2B Plants free-floating; thallus symmetrical repeatedly forked
  - 3A Thalli broad, 5 to 10 mm long, *c*. 5 mm wide, tinged purple, fringed with tongue-shaped, toothed, scales
    Ricciocarpus **p. 53**
  - 3B Thalli elongate, 10 to 50 mm long, 0.5 to 1.0 mm wide, not tinged purple, not fringed with tongue-shaped, toothed scales
    Riccia **p. 53**

1B Plant with stem and leaves; not free floating
- 4A Side branches produced in groups at regular intervals along the main axis; leaf cells distinctly dimorphic, small and green alternating with large and hyaline
  Sphagnum **p. 79**
- 4B Side branches not produced in groups or absent; leaf cells not distinctly dimorphic

5A Leaves in 2 rows (some genera e.g. *Vesicularia* have flattened stems and, superficially, leaves appear to be in 2 rows but are spiral); each leaf 2-lobed
6A Both leaf lobes broadly rounded and flattened
Scapania p. 55
6B Each leaf lobe consists of a boat-shaped, clasping lobe beyond which there extends a wing of tissue
Fissidens p. 69
5B Leaves not in 2 rows; each leaf with simple blade
7A Leaves in 3 or 5 rows; stems 3- or 5-angled
8A Leaves in 3 rows
9A Leaves directed to one side of stem
Dichelyma p. 73
9B Leaves not directed to one side of stem
10A Leaves obovate to oblanceolate, 1.0 to 2.25 mm long (Tropical S. America)
Hydropogonella p. 77
10B Leaves ovate, lanceolate or linear, 2 to 8 (−10) mm long (Temperate N. Hemisphere, N. Andes)
11A Leaves without nerve, flat, concave or folded, with or without keel
Fontinalis p. 71
11B Leaves with nerve, folded, with keel
Brachelyma p. 71
8B Leaves in 5 rows
12A Ends of branches club-shaped; leaves 2 to 3 mm long, nerve short usually unequally forked; auricles absent (Tropical S. America)
Hydropogon p. 75
12B Ends of branches not club-shaped; leaves 1 to 2 mm long, nerve absent; auricles distinct (Cape of Good Hope)
Wardia p. 82
7B Leaves not in rows, spirally arranged
13A Branches and often main axis distinctly dorso-ventrally flattened
14A Rhizoids grouped in bunches at regular intervals along stem; leaves distinctly dimorphic (lateral ones large and spreading, upper and lower ones smaller and more or less appressed to stem); leaf margin usually finely toothed
15A Cells of leaf-blade rhomboidal, 2 to 8 times as long as wide, not papillose
Vesicularia p. 77
15B Cells of leaf-blade linear, 10 to 15 times as long as wide, papillose
Glossadelphus p. 79
14B Rhizoids not grouped in bunches at regular intervals along stem; leaves not distinctly dimorphic; leaf margin not toothed **Potamium p. 79**

13B Branches and main axis not dorso-ventrally flattened

16A Leaves with distinct border due to special cells or thickening

17A Cells at margin and base of leaf without chlorophyll
Dendrocryphaea p. 69

17B Cells at margin and base of leaf with chlorophyll

18A Cells of leaf-blade about as long as wide (hexagonal or square), capsules terminal; creeping part of stem often with numerous reddish-brown rhizoids

19A Leaves distinctly keeled; nerve on abaxial surface only
Hydrogrimmia/ Schistidium p. 73

19B Leaves not keeled; nerve not on abaxial surface only

20A Operculum (lid) persistently attached to columella; nerve not quite reaching leaf apex
Scouleria p. 75

20B Operculum not persistently attached to columella; nerve usually reaching or exceeded leaf apex
Cinclidotus p.66

18B Cells of leaf-blade 2 to 6 times as long as wide; capsules lateral; rhizoids few or absent

21A Leaf margin not toothed (widespread)
Sciaromium p. 61

21B Leaf margin (particularly near apex) toothed

22A Plants robust; nerve distinctly contracted at leaf apex or not reaching apex (China and Japan)
Sciaromiopsis p. 63

22B Plants delicate; nerve reaching leaf apex, not contracted (N. America)
Platylomella p. 63

16B Leaves without distinct border due to special cells or thickening

23A Nerves simple or forked, less than half as long as leaf or nerve absent

24A Stolons present; walls of leaf-blade cells papillose
Braunia p. 75

24B Stolons absent; walls of leaf-blade cells not papillose

25A Leaves strongly concave giving shoots a stout swollen appearance; side branches usually curved
Scorpidium p. 63

25B Leaves flat or concave but not giving shoots a stout swollen appearance; side branches not curved

26A Leaf apex long and pointed, curved and directed away from stem apex
27A Leaves directed to one side of stem, 2 to 5 mm long
Drepanocladus p. 59
27B Leaves not directed to one side of stem, 1 to 3 mm long
Campylium p. 56
26B Leaf apex blunt and rounded, not long and pointed, not directed away from stem apex
28A Auricles elongate and pointed; leaves 2 to 4 mm long
Calliergon p. 56
28B Auricles small, not elongate and pointed, leaves 1 to 2 mm long
Hygrohypnum p. 59
23B Nerve distinct, simple, reaching leaf apex or at least more than half as long as leaf
29A Leaves directed to one side of stem
30A Paraphyllia (minute simple or branched, green structures borne among the leaves on the stem) present
Cratoneuron p. 56
30B Paraphyllia absent
31A Lower part of stems densely clothed in rhizoids
Philonotis p. 66
31B Stems not clothed in rhizoids
32A Nerve exceeding leaf apex; cells of leaf blade in 2 layers
Hydrodicranum p. 69
32B Nerve not exceeding leaf apex; cells of leaf blade in 1 layer
33A Leaf apex acute or blunt, not drawn out to a fine point
Hygrohypnum p. 59
33B Leaf apex acute, drawn out to a fine point
34A Leaves strongly falcate; leaf apices directed away from stem apex
Drepanocladus p. 59
34B Leaves not falcate; leaf apices directed towards stem apex or spreading but not directed away from stem apex
35A Leaf apices directed towards stem apex; nerves reaching or almost reaching leaf apex; cells at centre of leaf blade 2 to 4 (–6) times as long as wide
Hygroamblystegium p. 59

35B Leaf apices spreading away from axis; nerves rarely more than ¾ the length of the leaf, cells at centre of leaf-blade 4 to 8 times as long as wide
Leptodictyum p. 61

29B Leaves not directed to one side of stem
36A Stem densely clothed in rhizoids (appearing felty)
37A All cells of leaf blade linear, thick-walled; leaves shiny, with longitudinal folds
Breutelia p. 63
37B Some cells of leaf blade oblong or square; leaves not shiny, without longitudinal folds
Philonotis p. 66
36B Stem not densely clothed in rhizoids (not appearing felty)
38A Nerve widened above and filling apical portion of leaf; cells of leaf blade in 2 layers
Hygrodicranum p. 69
38B Nerve not widened above, not filling apical portion of leaf; cells of leaf blade in 1 layer
39A Capsule short-stalked, immersed in surrounding leaves; cells in centre of leaf blade less than 3 times as long as wide; stem usually repeatedly forked and thus not showing well developed main axis with lateral branches
40A Leaf keeled; nerve on abaxial leaf surface only
Hydrogrimmia/ Schistidium p. 73
40B Leaf not keeled; nerve not on abaxial surface only but in centre of leaf
41A Auricles distinct
Dendrocryphaea p. 69
41B Auricles indistinct or absent
Cyptodon p. 66
39B Capsule long-stalked, not immersed in surrounding leaves; cells in centre of leaf blade at least 3 times as long as wide; stem usually not repeatedly forked but with well developed main axis with lateral branches
42A Paraphyllia (minute simple or branched, green structures borne among the leaves on the stem) present
Cratoneuron p. 56
42B Paraphyllia absent
43A Leaf apex long and pointed
44A Cells of leaf-blade usually more than 6 times as long as wide
45A Auricles distinct
Drepanocladus p. 59
45B Auricles indistinct or absent
Leptodictyum p. 61
44B Cells of leaf-blade less than 6 times as long as wide

46A Leaf base cordate; leaves spreading
Campylium p. 56

46B Leaf base not cordate; leaves directed towards stem apex
Hygroamblystegium p. 59

43B Leaf apex not long and pointed

47B Margin of leaf distinctly toothed from base to apex
Platyhypnidium p. 61

47B Margin of leaf entire or minutely toothed at apex only

48A Auricles large and elongate; cells of auricles large and loosely packed; leaves 2 to 4 mm long
Calliergon p. 56

48B Auricles small and not elongate; cells of auricles small and tightly packed; leaves 1 to 2 mm long
Hygrohypnum p. 59

HEPATICAE

Many genera grow on rocks near water, particularly in mountainous districts and may be submerged after rain or snow melt. Sexual reproduction apparently takes place on land and these genera are essentially terrestrial but tolerate flooding and have been excluded from this account. Such genera are: *Chiloscyphus* Corda emend. Dum., *Gymnocolea* (Dum.) Dum., *Marchantia* (L.) Raddi, *Marsupella* Dum., *Nardia* Gray, *Riccardia* Gray, and *Solenostoma* Mitten. For an account of these genera in C. Europe see Paul, H., Mönkemeyer, W. and Schiffner, V. in Pascher, A. Die Süsswasser-Flora Mitteleuropas, Heft 14, Jena (1931).

RICCIACEAE

Riccia L., Sp. Pl. 1138 (1753)
Fig. 4A.
Thallus uniformly green, flat, *c.* 1.5 to 6 times as wide as thick, elongate 10 to 50 mm long, 0.5 to 1.0 mm wide, repeatedly forked; lower surface without scales; rhizoids few or absent.

Rivola, M. Příspěvky k poznání játrovek rodu Riccia I. Riccia rhenana Lorbeer v. jižnich Čechách. Preslia 39: 72–82 (1967)

c. 4 somewhat similar species found floating on or near the surface of water. These species are frequently grouped together under the name *R. fluitans* L. *R. fluitans* sensu lato is almost cosmopolitan and is characteristically found in eutrophic water.

Ricciocarpus Corda in Opiz., Beitr. 651 (1829)
Fig. 4B.
Thallus green but tinged with purple, flat, broad, 4 to 10 mm long, 2 to 5 mm wide, usually no more than 2 or 3 times forked, forming small rosettes; lower surface bearing toothed, ribbon-shaped, brown to violet scales, projecting beyond margin of thallus.

Rieth, A. Bemerkungen über Ricciocarpus natans (L.) Corda. Die Kulturpflanze 7: 207–217 (1959)

1 species, *R. natans* (L.) Corda: in the warmer parts of the world. Free-floating with the habit of *Lemna*, characteristically found in eutrophic water. In tropical regions it is occasionally very abundant and in Kenya has been reported to interfere with fishing.

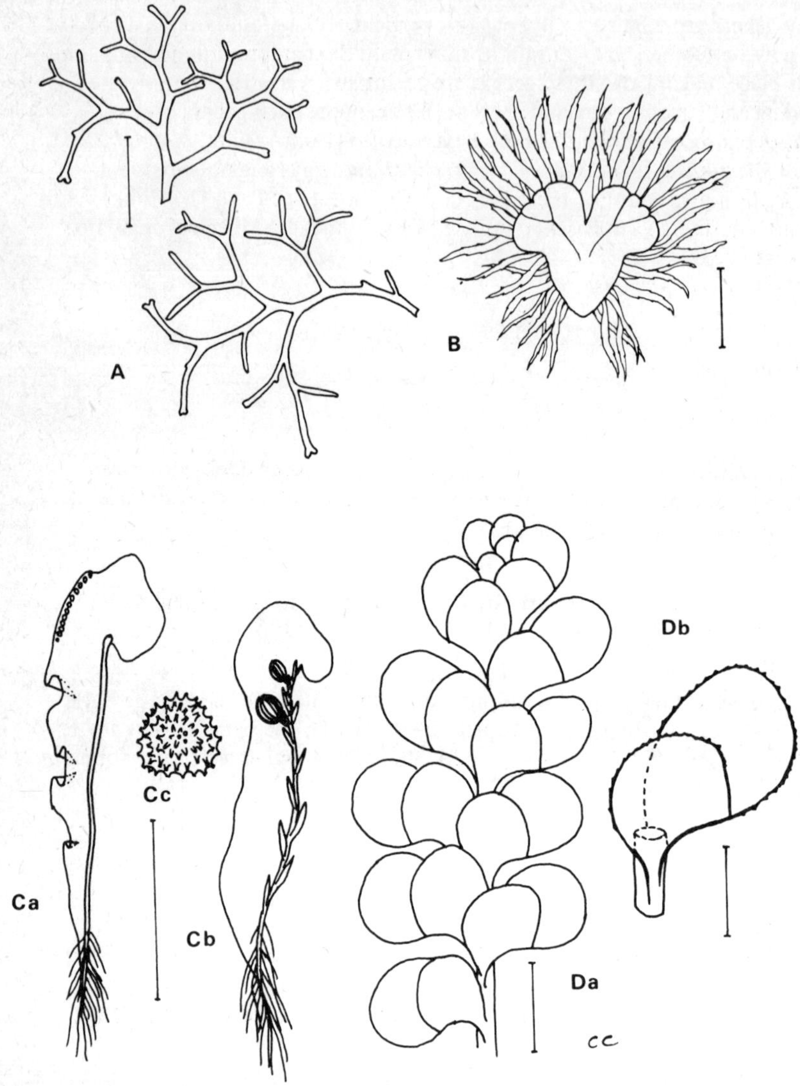

Fig. 4. A. *Riccia fluitans* L.: habit (5 mm);
B. *Ricciocarpus natans* (L.) Corda: habit (5 mm);
C. *Riella*: a, *R. americana* Howe and Underwood, male plant (5 mm); b, *R. cossoniana* Trabut, female plant (5 mm); c, spore of b.
D. *Scapania undulata* (L.) Dum.: a, stem apex (1 mm); b, leaf (0.5 mm).

RIELLACEAE

1 genus

Riella Montagne, Ann. Sci. Nat. sér. 3, 18: 12 (1852)
Fig. 4C.
Dioecious or monoecious annuals or rarely perennials. Stem thickened, usually erect, up to 20 cm long (usually much less), bearing a variously formed wing on 1 side and over the top, with occasionally scale-like or leaf-like lobes on the axis. Antheridia borne in cavities sunk in the margin of the wing. Archegonia and sporogonia borne in large involucres on the axis; spores distinctly spinous or papillose, not remaining in tetrads.

Studhalter, R. A. Germination of spores and development of juvenile thallus of Riella americana. Bot. Gaz. 92: 172–191 (1931)

c. 12 species: Mediterranean, N. America, S. Africa and ? India. Mostly found gregariously in vernal pools, the spores germinate underwater, sexual reproduction apparently takes place underwater and as the water recedes the plants usually die and pass through the dry season as spores. Some species are found in brackish pools.

SCAPANIACEAE

Scapania Dum., Rec. d'Observ. 14 (1835)
Fig. 4D.
Stems erect, creeping or floating, scarcely branched, up to 12 cm long (usually much less). Leaves 1 cell thick, in 2 rows; each leaf 2-lobed, the smaller lobe folded over the larger; both lobes broadly rounded, with entire or toothed margins. Capsule long-stalked, ovoid, opening by 4, straight valves.

Gline, J. M. An observation on the vegetative reproduction of Scapania undulata. Bryologist 73 (3): 624–625 (1970)

c. 50 species: mostly in the temperate North Hemisphere. Although probably not an obligate aquatic, *S. undulata* (L.) Dum. is commonly found in swiftly flowing water in mountainous districts in the Boreal Zone.

MUSCI

AMBLYSTEGIACEAE

c. 16 genera; 12 genera contain aquatics. The form of the amphibious and aquatic members of this family is very much influenced by environmental factors which make the identification of species and genera difficult. The Amblystegiaceae is also much in need of a world-wide revision.

Calliergon (Sull.) Kindb., Canad. Rec. Sci. 6 (2): 72 (1894)
Fig. 5A.
Stems prostrate or floating, regularly or irregularly branched or sparingly branched when submerged; rhizoids common on terrestrial stems, rare on submerged. Leaves directed towards stem apex, concave, ovate to lanceolate, not markedly attenuate at apex, 2 to 4 mm long; nerve distinct, occasionally reaching leaf apex; margin entire; apex apiculate; auricles distinct and elongate; cells of auricles large and loosely packed; cells of leaf-blade variable. Capsule long-stalked, cylindrical, distinctly curved; peristome in 2 rows.

Karczmarz, K. A. A monograph of the genus Calliergon (Sull.) Kindb. Monog. Bot. (Warszawa) 31: 1–209 (1971)

c. 15 species: scattered in climatically temperate and sub-arctic zones of the world. Found in a variety of aquatic and semi-aquatic habitats.

Campylium (Sull.) Mitten, Journ. Linn. Soc. London, Bot. 12: 631 (1869)
Fig. 5B.
Stems creeping, prostrate or partly floating, usually tufted, regularly or irregularly branched; rhizoids sparse but occasionally in bunches on creeping stems. Leaves cordate at base, attenuate towards apex, spreading; nerve variable, almost absent, short and forked or distinct but not reaching leaf apex; margin usually entire, occasionally minutely toothed; apex long and pointed, usually directed away from stem apex; auricles distinct; cells of leaf-blade variable, square to elongate. Capsule long-stalked, cylindrical, distinctly curved; peristome in 2 rows.

c. 36 species; almost cosmopolitan, absent in the humid tropics. Most species are terrestrial but some, such as *C. elodes* (Lindb.) Kindb. are marsh plants that are frequently found submerged.

Cratoneuron (Sull.) Spruce, Cat. Musc. Amaz. And. 21 (1867), [*Cratoneuropsis* (Broth.) Fleischer]
Fig. 6A.
Stems prostrate, floating or usually erect, often pinnately branched, frequently thickly clothed in brownish rhizoids. Leaves ovate to lanceolate, usually cordate at base, attenuate at apex, spreading or directed to one side of stem; nerves distinct, reaching or almost reaching leaf apex; margin distinctly toothed at base; apex long and pointed; auricles distinct and free of chlorophyll; cells of leaf-blade usually more than 6 times as long as wide; paraphyllia (minute simple

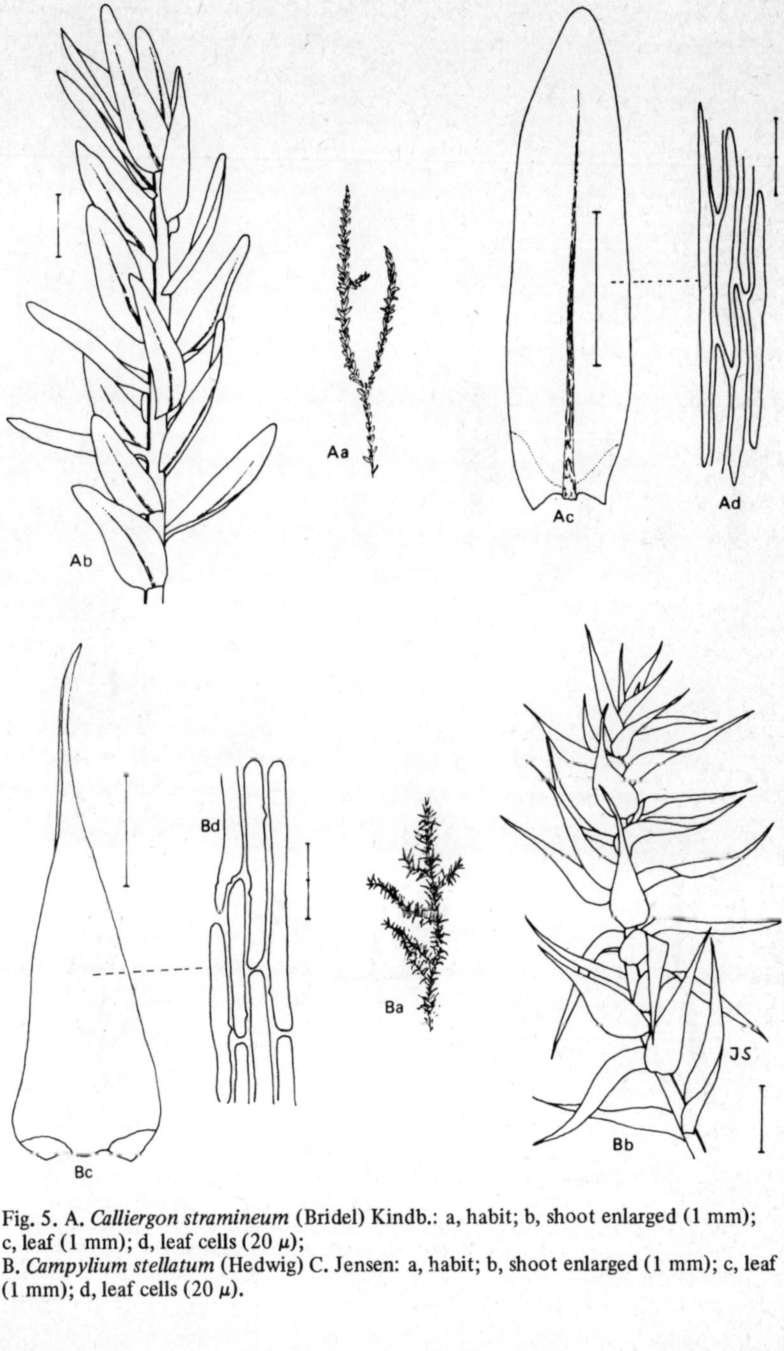

Fig. 5. A. *Calliergon stramineum* (Bridel) Kindb.: a, habit; b, shoot enlarged (1 mm); c, leaf (1 mm); d, leaf cells (20 μ);
B. *Campylium stellatum* (Hedwig) C. Jensen: a, habit; b, shoot enlarged (1 mm); c, leaf (1 mm); d, leaf cells (20 μ).

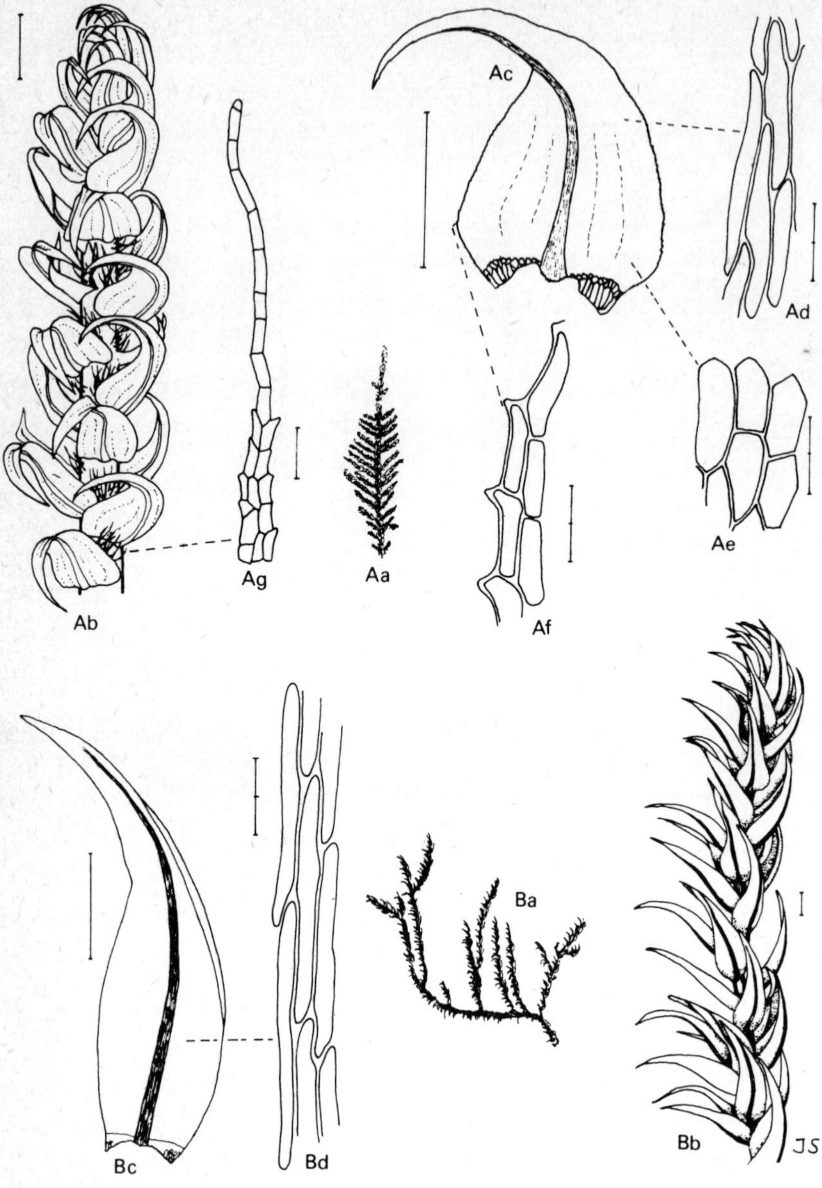

Fig. 6. A. *Cratoneuron commutatum* (Hedwig) Roth: a, habit; b, shoot enlarged (1 mm); c, leaf (1 mm); d, cells in centre of leaf (20 μ); e, cells of basal part of leaf (20 μ); f, leaf margin (20 μ); g, paraphyllium (50 μ);
B. *Drepanocladus exannulatus* (B.S.G.) Warnst.: a, habit; b, shoot enlarged (1 mm); c, leaf (1 mm); d, leaf cells (20 μ).

or branched, green structures borne among the leaves on the stem) present. Capsule long-stalked, cylindrical, distinctly curved; peristome in 2 rows.

c. 21 species: almost cosmopolitan, lacking in Malaysia, Australia and Antarctica. Found in or near water in a variety of habitats.

Drepanocladus (C. Müller) Roth, Hedwigia 38, Beibl. 6 (1899)
Fig. 6B.
Stems creeping, prostrate, erect or floating, regularly or irregularly branched; tips of branches usually curved; rhizoids at base only. Leaves triangular to lanceolate, distinctly falcate, usually directed to one side of stem; nerves indistinct and reaching middle of leaf or distinct and reaching or almost reaching leaf apex, very rarely exceeding leaf apex; margin entire or minutely toothed; leaf apex long and pointed, usually directed away from step apex; auricles usually present; cells of leaf-blade more than 6 times as long as wide. Capsule long-stalked cylindrical, distinctly curved; peristome in 2 rows.

Danilov, A. N. and Ladygenskaia, C. I. Variability of the moss Drepanocladus aduncus. Sov. Bot. 6: 3 (1934), in Russian.

c. 36 species: cosmopolitan. Found in or near water in a variety of habitats.

Hygroamblystegium Loeske, Moosfl. des Harz 298 (1903), [*Amblystegium* B.S.G. pro parte]
Fig. 7A.
Stems creeping or floating; rhizoids in bunches on creeping stems or confined to base on floating stems. Leaves ovate to lanceolate, attenuate towards apex; nerve or nerves distinct, reaching or almost reaching leaf apex, margin entire or minutely toothed; apex long and pointed; auricles present, distinct or indistinct; cells of leaf-blade 2 to 4 or at most 6 times as long as wide. Capsule long-stalked, cylindrical, distinctly curved; peristome in 2 rows.

c. 23 species: almost cosmopolitan, absent in the tropical regions. Found in or near water in a variety of habitats.

Hygrohypnum Lindb., Act. Soc. Sci. Fenn. 10: 277 (1872)
Fig. 7B.
Stems prostrate or floating, freely but irregularly branched, often denuded of leaves at base but with leaf bases persisting; rhizoids confined to base. Leaves spreading or occasionally directed to one side of stem, oval to lanceolate, somewhat attenuate at apex, 1 to 2 mm long; nerve variable, short and forked or long and simple but not reaching leaf apex; margin entire or minutely toothed; apex blunt and rounded; auricles distinct but small; cells of auricles rectangular and tightly packed; cells of leaf-blade more than 5 times as long as wide. Capsule long-stalked, cylindrical, distinctly curved; peristome in 2 rows.

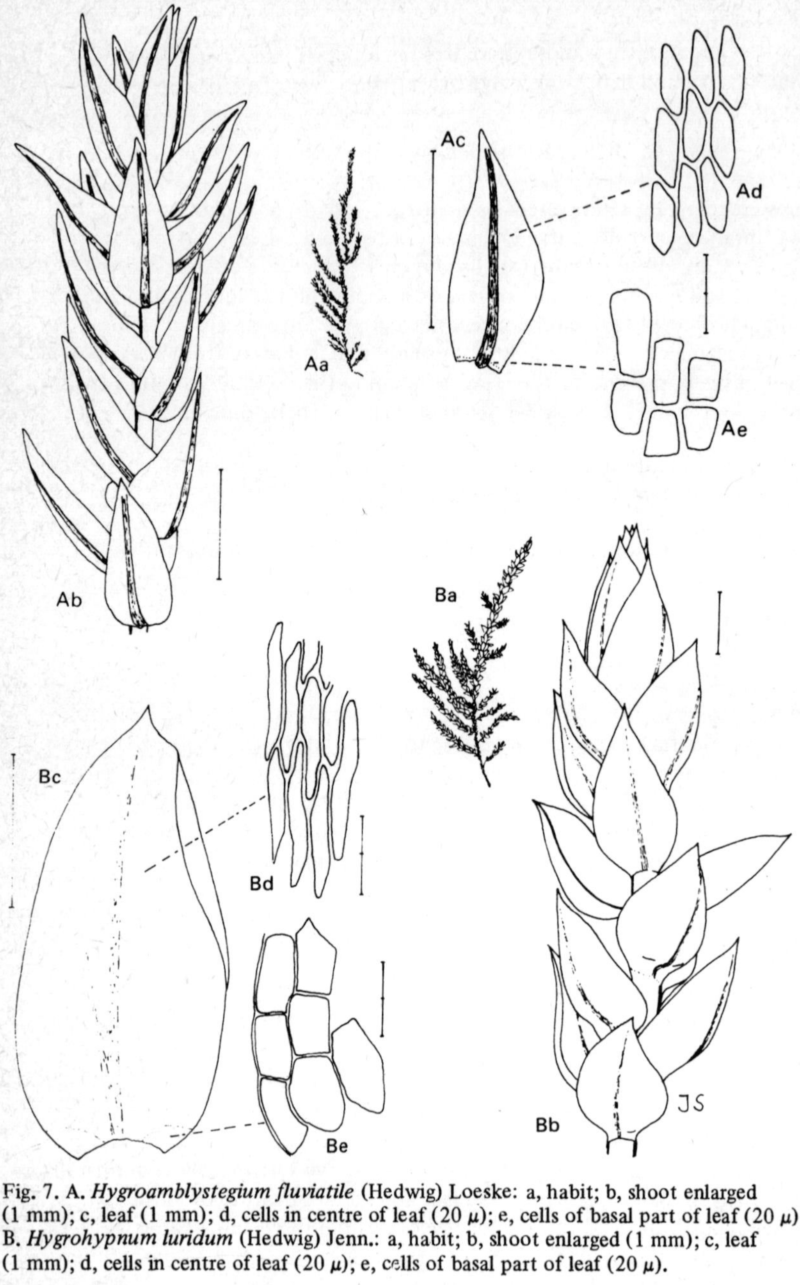

Fig. 7. A. *Hygroamblystegium fluviatile* (Hedwig) Loeske: a, habit; b, shoot enlarged (1 mm); c, leaf (1 mm); d, cells in centre of leaf (20 μ); e, cells of basal part of leaf (20 μ); B. *Hygrohypnum luridum* (Hedwig) Jenn.: a, habit; b, shoot enlarged (1 mm); c, leaf (1 mm); d, cells in centre of leaf (20 μ); e, cells of basal part of leaf (20 μ).

c. 36 species: almost cosmopolitan, absent in S. E. Asia and Australasia. Found in a variety of aquatic or semi-aquatic habitats.

Leptodictyum (Schimper) Warnst., Krypt. Fl. Brandenburg 2: 840, 847 (1906). [*Amblystegium* B. S. G. pro parte]
Fig. 8A.
Stems very variable, creeping, prostrate and tufted or sparsely branched and floating; rhizoids confined to base. Leaves spreading (in or out of water), ovate, attenuate towards apex, nerve distinct, ending before leaf apex, rarely more than ¾ as long as leaf; margin entire; apex long and pointed, spreading away from main axis but not pointing away from stem apex; auricles absent or indistinct; cells of leaf-blade 4 to 8 times as long as wide. Capsule long-stalked, cylindrical, distinctly curved; peristome in 2 rows.

c. 14 species: almost cosmopolitan, absent in Australasia and the Pacific Islands. Found attached to stones, wood, etc. in a variety of aquatic habitats, most frequently in flowing water.

Platyhypnidium Fleischer, Musci Fl. Buitenzorg 4: 1536 (1923), [*Eurhynchium* B. S. G. pro parte, *Oxyrrhynchium* (B. S. G.) Warnst. pro parte, *Rhynchostegium* B. S. G.]
Fig. 8B.
Stems creeping, prostrate or floating, up to 15 cm or more long, sparingly branched, often denuded of leaves at base but with leaf bases persisting; rhizoids confined to base of stem. Leaves spreading but directed towards stem apex, ovate, slightly attenuate at apex, 1.5 to 2.5 mm long; nerve distinct but tapering above, about ¾ as long as leaf; margin distinctly toothed from base to apex; apex acute or blunt, not drawn out to a point; auricles indistinct or absent; cells of leaf-blade usually more than 5 times as long as wide. Capsule long-stalked, cylindrical, distinctly curved; lid with long beak; peristome in 2 rows.

c. 21 species: almost cosmopolitan. Found on rocks in and by flowing water.

Sciaromium (Mitten) Mitten, Journ. Linn. Soc. London, Bot. 12: 571 (1869)
Fig. 9A.
Stems creeping, prostrate or floating, robust, much branched. Leaves ovate to lanceolate, crowded, concave; nerve thick and strong reaching leaf apex; margin clearly thickened, entire; apex blunt; auricles absent; cells of leaf-blade up to 6 times as long as wide. Capsule long-stalked, cylindrical, distinctly curved, peristome in 2 rows.

c. 27 species: N. and S. America, S. Africa, India, Australia, New Zealand and Fiji Islands. Usually found in flowing water attached to rocks, wood etc.

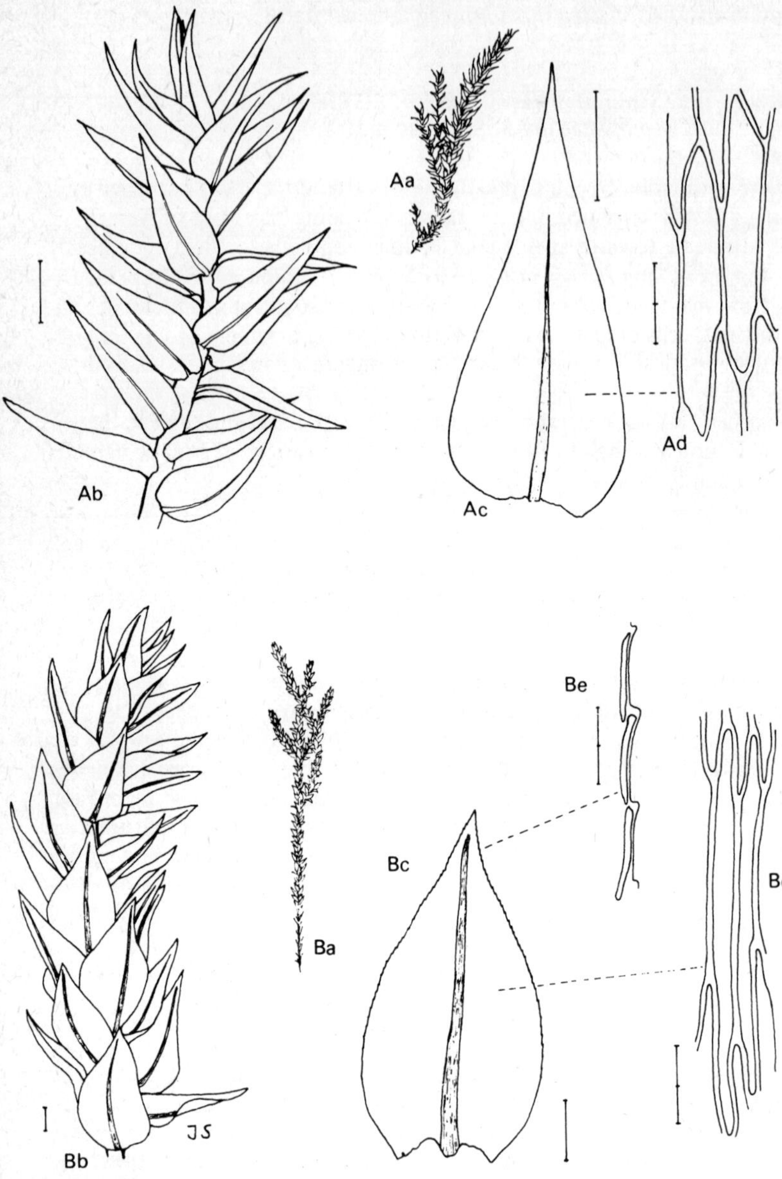

Fig. 8. A. *Leptodictyum riparium* (Hedwig) Warnst.: a, habit; b, shoot enlarged (1 mm); c, leaf (1 mm); d, leaf cells (20 μ);
B. *Platyhypnidium riparioides* (Hedwig) Dixon: a, habit; b, shoot enlarged (1 mm); c, leaf (1 mm); d, cells in centre of leaf (20 μ); e, leaf margin (20 μ).

Sciaromiopsis Broth., Sitzungsber. Akad. Wiss. Wien Math. Nat. Kl. Abt.1, 133: 580 (1924)
Fig. 9B.
Like *Sciaromium* but nerve distinctly contracted at leaf apex or not reaching leaf apex and leaf margin near apex minutely toothed.

3 species: China and Japan.

Platylomella Andrews, Bryologist 53: 58 (1950)
Fig. 9C.
Like *Sciaromium* but delicate and complete leaf margin distinctly toothed.

1 species: *P. lescurii* (Sull.) Andrews, N. America.

Scorpidium (Schimper) Limpr., Laubm., Deutschl. 3: 570 (1899)
Fig. 10A.
Stems prostrate or floating, robust, up to 30 cm long; main axis sparsely branched, with numerous regular or irregular, somewhat curved, side branches. Leaves ovate to lanceolate, 2 to 4 mm long, crowded, strongly concave giving shoots and side branches a stout swollen appearance, at least apical leaves directed to one side of stem; nerves short and forked or absent; margin entire; apex blunt to pointed; auricles small, distinct; cells of auricles oval to rectangular; cells of leaf-blade 10 to 15 times longer than wide. Capsule long-stalked, cylindrical, distinctly curved; peristome in 2 rows.

3 species: Temperate N. Hemisphere and mountains of Bolivia. Amphibious plants usually found in oceanic mountainous regions in bogs and streams.

BARTRAMIACEAE

c. 9 genera: *Breutelia* and *Philonotis* contain aquatics.

Breutelia Schimper, Coroll, 85 (1856)
Fig. 10B.
Stems creeping or erect, usually tightly packed and cushion forming, densely clothed in yellowish-brown rhizoids which give the stems a felty appearance. Leaves shiny, lanceolate to linear, drawn out and pointed towards apex, with longitudinal folds; nerve thin, reaching or exceeding leaf apex; apex long and pointed; auricles indistinct or absent; cells of leaf-blade linear only, thick-walled, often papillose. Capsule long-stalked; peristome in 2 rows.

c. 125 species: almost cosmopolitan lacking in N. America and N. E. Asia. Mostly found in wet areas, some species frequently found submerged in water.

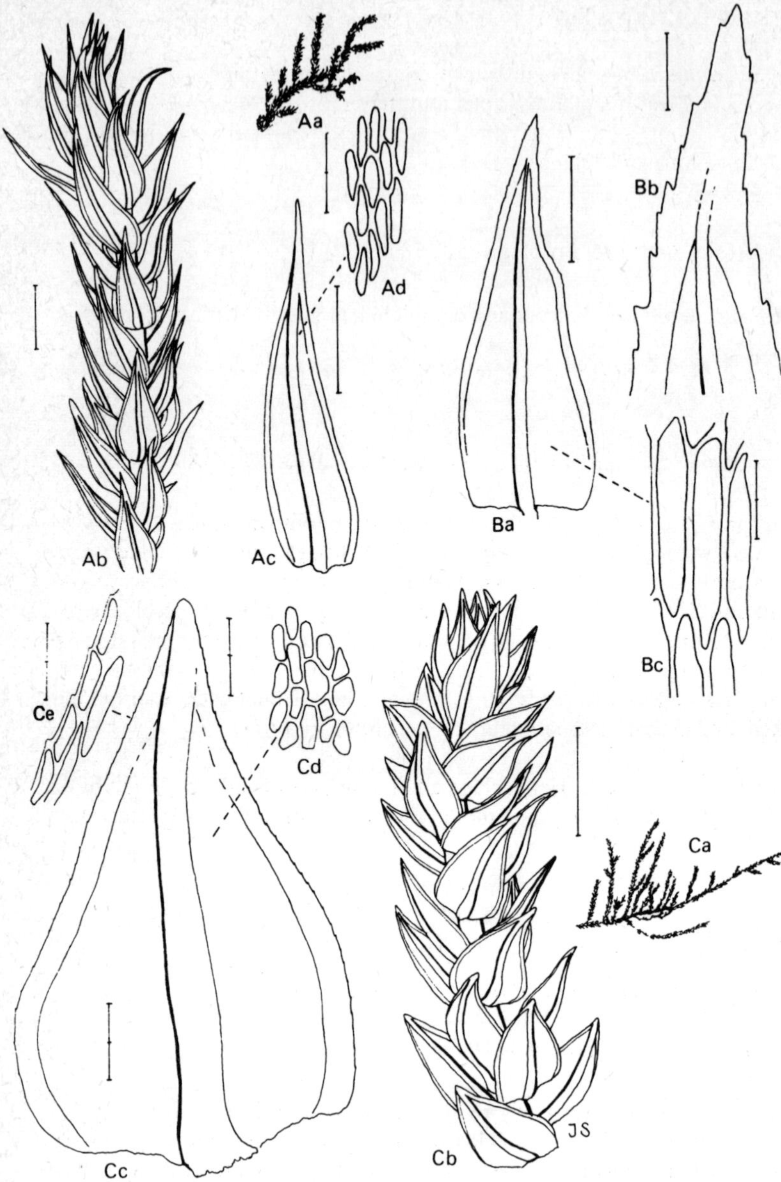

Fig. 9. A. *Sciaromium confluens* (Mueller) Par.: a, habit; b, shoot enlarged (1 mm); c, leaf (1 mm); d, leaf cells (20 μ);
B. *Sciaromiopsis sinensis* (Broth.) Broth.: a, leaf (1 mm); b, apex of leaf (50 μ); c, leaf cells (20 μ);
C. *Platylomella lescurii* (Sull.) Andrews: a, habit; b, shoot enlarged (1 mm); c, leaf (0.1 mm); d, leaf cells (20 μ); e, leaf margin (20 μ).

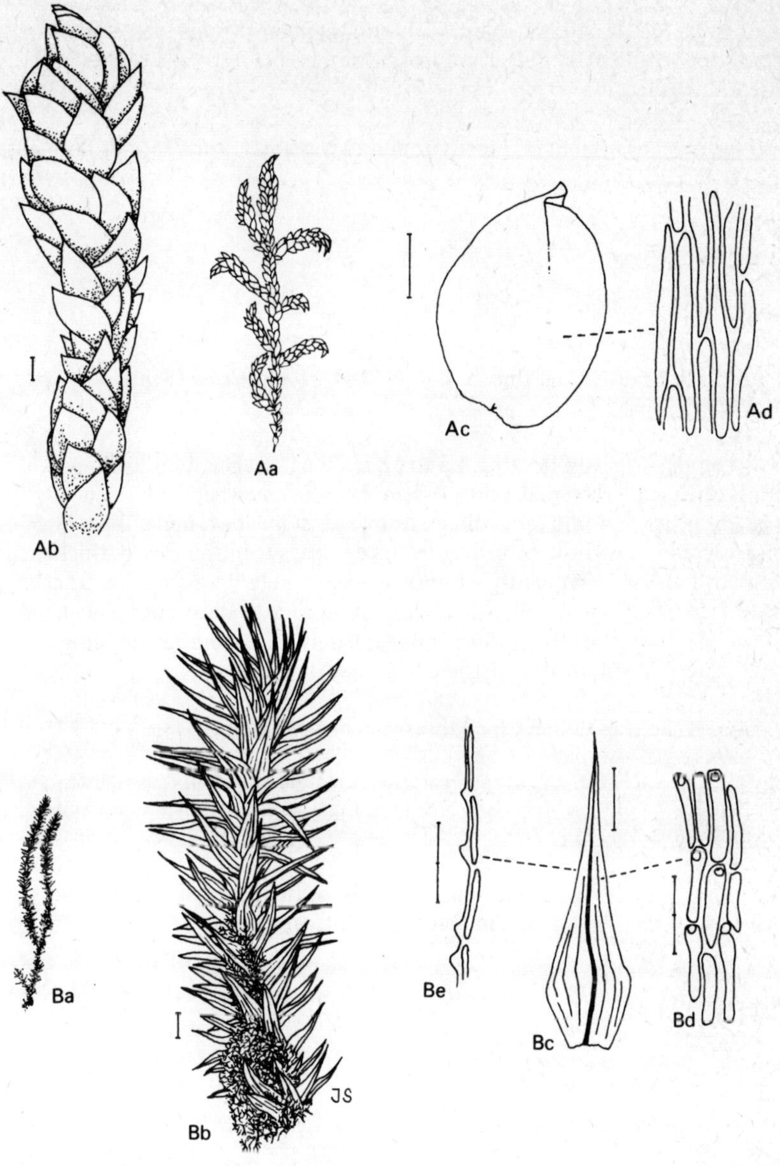

Fig. 10. A. *Scorpidium scorpioides* (Hedwig) Limpr.: a, habit; b, shoot enlarged (1 mm); c, leaf (1 mm); d, leaf cells (20 μ);
B. *Breutelia chrysocoma* (Hedwig) Lindb.: a, habit; b, shoot enlarged (1 mm); c, leaf (1 mm); d, leaf cells with papillae (20 μ); e, leaf margin (20 μ).

Philonotis Bridel, Bryologia Universa 2: 15 (1827)
Fig. 11A.
Like *Breutelia* but leaves not shiny, without longitudinal folds, occasionally directed to one side of stem; cells of leaf-blade variable but some, oblong to square, thin walled.

c. 170 species: cosmopolitan. Mostly found in wet areas, some species frequently submerged in water.

CINCLIDOTACEAE

1 genus.

Cinclidotus P. Beauv., Mag. Enc. 5: 319 (1804), [*Dialytrichia* (Schimper) Limpr.]
Fig. 11B.
Stems creeping or partly floating, up to 18 cm long, irregularly branched; rhizoids confined to creeping part of stem. Leaves somewhat fleshy, lanceolate to linear, up to 4 mm long, directed towards stem apex; nerve thick and strong reaching or slightly exceeding leaf apex; margin entire, clearly thickened; apex blunt but nerve frequently extends as a short point; auricles absent; cells of leaf-blade nearly square in outline with rather thick walls. Capsule terminal on short lateral branch, erect, short-stalked, barely exceeding surrounding leaves, ovoid to cylindrical; peristome in 1 row.

Hörmann, H. Beitrag zu den Cinclidotus-Arten. Nova Hedwigia 9: 233–235 (1965)
Philippi, G. Zur Kenntnis des Wassermooses Cinclidotus danubicus Schiffn et Baumgartn. und seiner Verbreitung in Europa. Beitr. Naturk. Forsch. Südw. Dtl. 26 (2): 77–81 (1967)

c. 10 species: Asia (tropical and temperate), N. Africa, Europe and N. America. Attached to stones, wood, etc. in and at edge of streams.

CRYPHAEACEAE

c. 8 genera: *Cyptodon* and *Dendrocryphaea* are aquatic.

Cyptodon (Broth.) Paris and Schimper ex Fleischer, Hedwigia 55: 284 (1914)
Fig. 12A.
Stems floating, branched; main axis sparsely leafy, often denuded of leaves below; branches lateral, regular, densely covered with leaves. Leaves ovate, spreading under water, appressed to stem out of water; nerve reaching but not exceeding leaf apex; margin toothed or entire; apex blunt; auricles indistinct or absent; cells of leaf-blade square or rectangular, often wider than long. Capsule immersed in leaves; peristome in 2 rows.

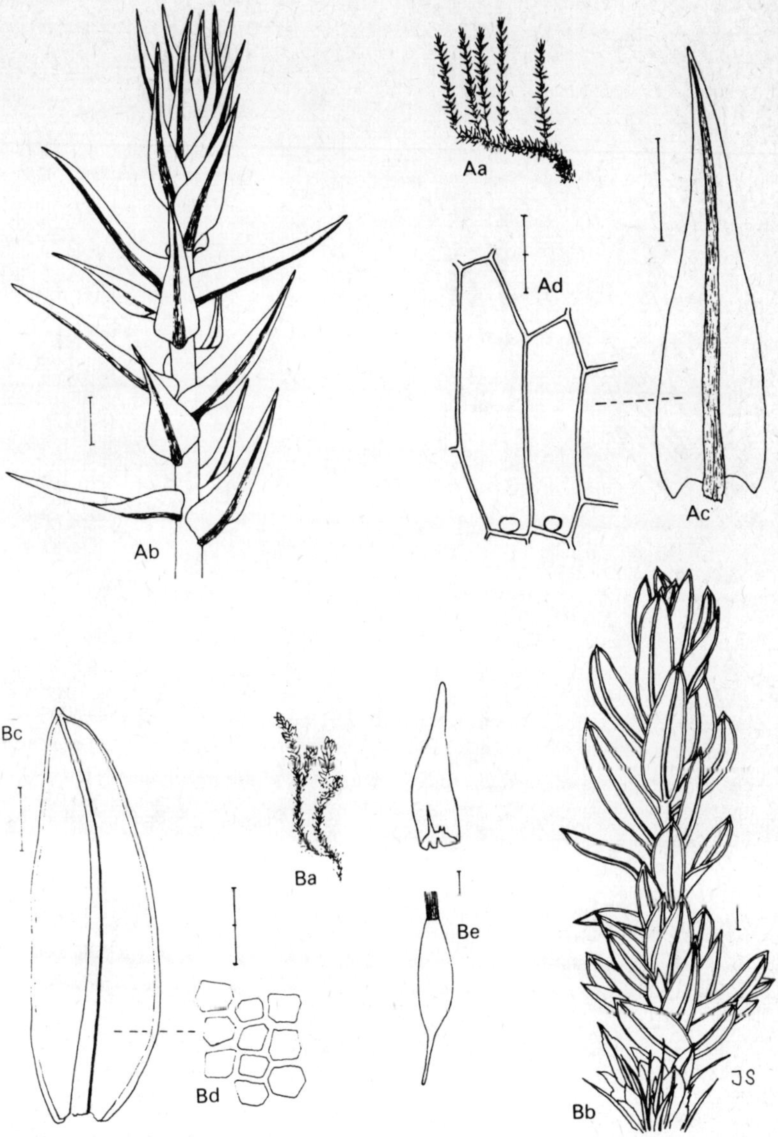

Fig. 11. A. *Philonotis fontana* (Hedwig) Bridel: a, habit; b, shoot enlarged (1 mm); c, leaf (1 mm); d, leaf cells (20 μ);
B. *Cinclidotus fontinaloides* (Hedwig) P. Beauv.: a, habit; b, shoot enlarged (1 mm); c, leaf (1 mm); d, leaf cells (20 μ); e, capsule and calyptra (1 mm).

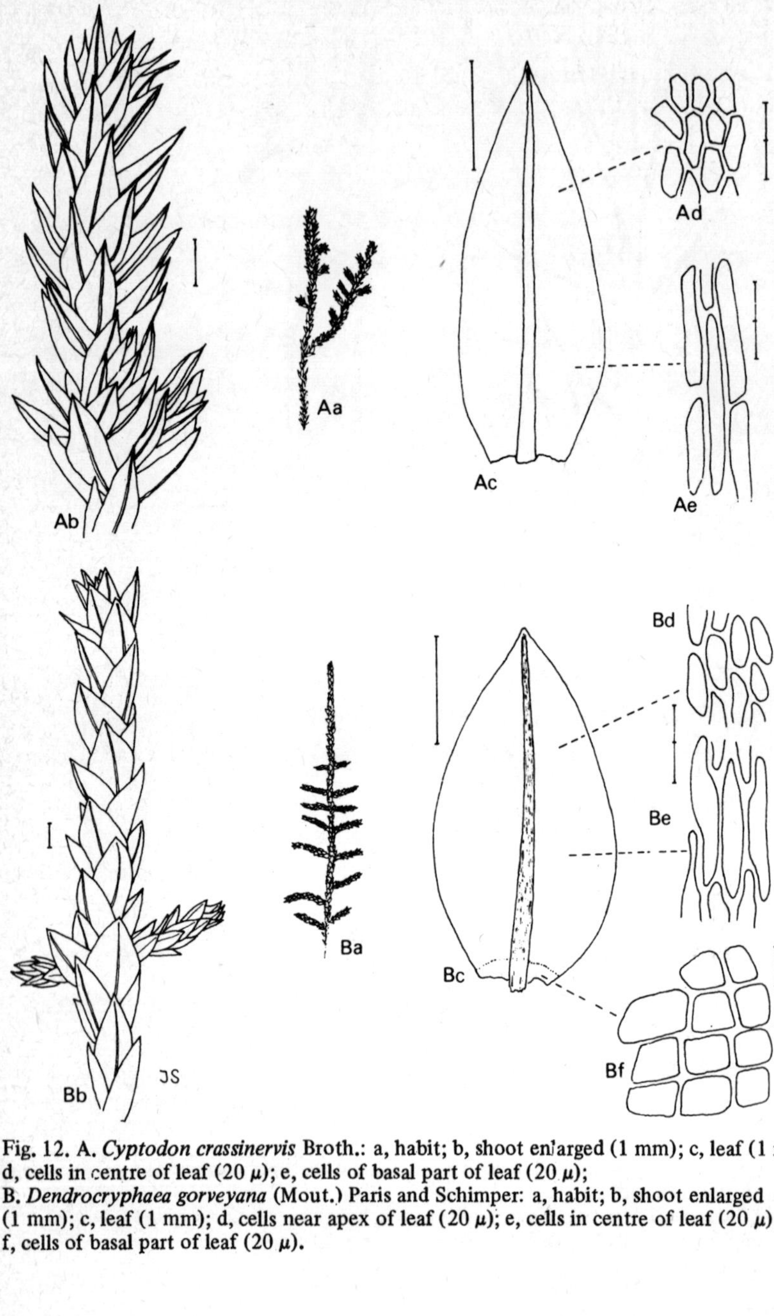

Fig. 12. A. *Cyptodon crassinervis* Broth.: a, habit; b, shoot enlarged (1 mm); c, leaf (1 mm); d, cells in centre of leaf (20 μ); e, cells of basal part of leaf (20 μ);
B. *Dendrocryphaea gorveyana* (Mout.) Paris and Schimper: a, habit; b, shoot enlarged (1 mm); c, leaf (1 mm); d, cells near apex of leaf (20 μ); e, cells in centre of leaf (20 μ); f, cells of basal part of leaf (20 μ).

c. 6 species: Temperate S. Hemisphere (S. America, Australia and New Zealand). Found attached to stones, stems, roots, etc. in flowing water.

Dendrocryphaea Paris and Schimper ex. Broth. in Engler, Nat. Pflanzenfam. 1 (3): 743, 556 (1905)

Fig. 12B.

Like *Cyptodon* except auricles distinct and cells at edge and base of leaf lacking chlorophyll.

c. 4 species: Temperate S. America. Found attached to stones, stems, roots, etc. in flowing water.

DICRANACEAE

c. 48 genera: *Hygrodicranum* is aquatic.

Hygrodicranum Cardot, Rev. Bryol. 38: 51 (1911)

Fig. 13A.

Stems creeping or floating, somewhat delicate, up to 15 cm long. Leaves sparsely distributed along stem, directed to one side, broad at base tapering gradually to a finely drawn out tip; margin entire; nerve thick and strong, widened above and filling leaf apex; cells each side of nerve elongate and thickened; cells of leaf-blade rectangular and in 2 layers. Reproductive organs unknown.

3 species: S. America, *H. falklandicum* Cardot from the Falkland Isles, *H. bolivianum* Herz from between 4000 and 4600 m in Bolivia and *H. herrerai* Williams in Peru, all species are aquatic.

FISSIDENTACEAE

4 genera: *Fissidens* contains aquatics.

Fissidens Hedwig, Sp. Musc. 152 (1801) [*Octodiceras* Bridel]

Fig. 13B.

Stems erect, prostrate, creeping or floating; branching irregular; rhizoids usually confined to base. Leaves in 2 rows; each leaf consists essentially of a boat-shaped clasping portion, beyond which there extends a wing of tissue, the nerve being prolonged out into this wing. Capsule terminal or lateral, long-stalked; peristome in 1 row.

Iwatsuki, Z. Critical or otherwise interesting Fissidens species in Japan. Journ. Hattori Bot. Lab. 30: 91–104 (1967)

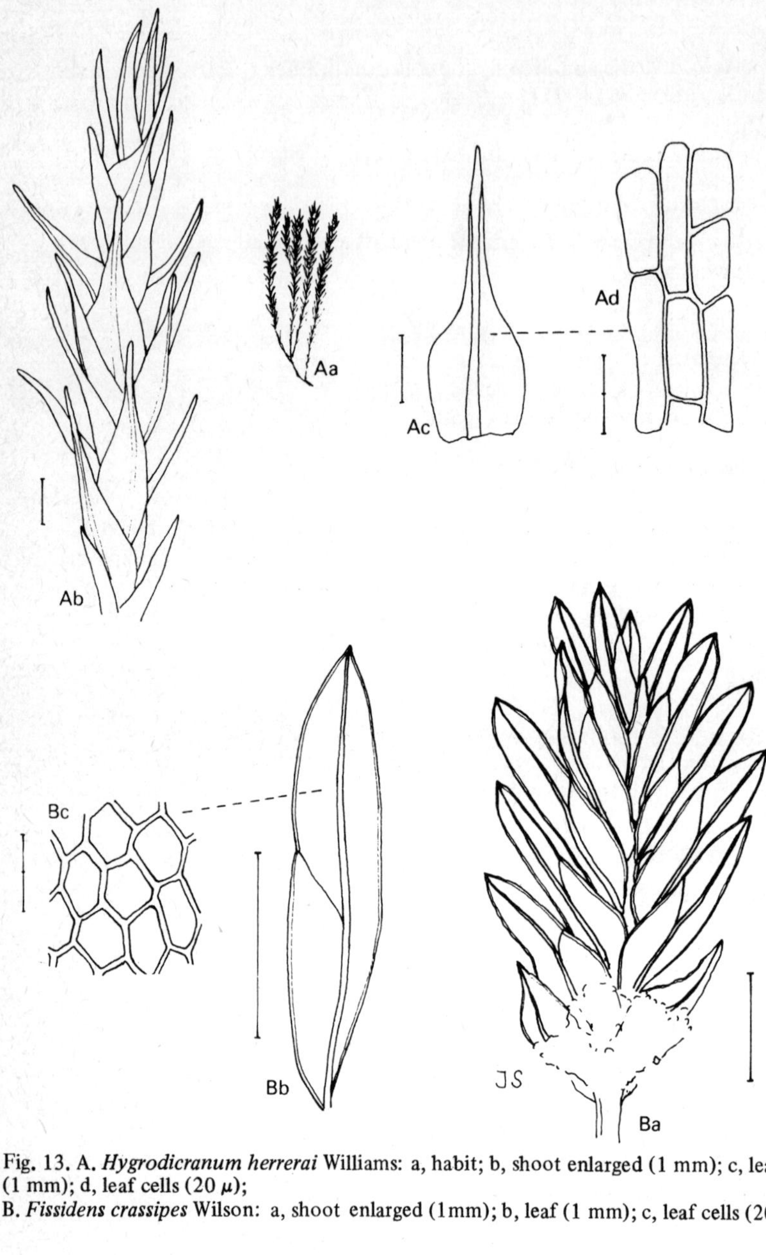

Fig. 13. A. *Hygrodicranum herrerai* Williams: a, habit; b, shoot enlarged (1 mm); c, leaf (1 mm); d, leaf cells (20 μ);
B. *Fissidens crassipes* Wilson: a, shoot enlarged (1mm); b, leaf (1 mm); c, leaf cells (20 μ).

Whittier, H. O. and Miller, H. A. Mosses of the Society Islands. Fissidens, Bryologist 70: 76–93 (1967)

c. 900 species: cosmopolitan (large number of local endemic species described). Found in many different habitats, several species are submerged aquatics.

FONTINALACEAE

3 genera: all aquatics.

Fontinalis (Dill.) Hedwig, Sp. Musc. 298 (1801)
Fig. 14A.
Stems floating, flaccid or rigid, up to 90 cm or more long, 0.2 to 0.5 mm diameter, regularly or irregularly branched. Leaves in 3 rows, not directed to one side of stem, flat, curved or folded, lanceolate, ovate or rhomboidal, 2.0 to 8.0 (–10.0) mm long, 0.35 to 6.5 (–8.5) mm wide; nerves absent; margin entire; apex attenuate, auricles absent; old leaves frequently split or tear lengthwise through the centre of the blade, these leaf halves have often lead to incorrect determinations. Capsule sessile or short-stalked, often concealed among the leaves or emergent; peristome in 1 or 2 rows or absent.

Welch, W. H. A monograph of the Fontinalaceae. Martinus Niijhoff, The Hague (1960)

c. 29 species: mostly from the temperate and sub-arctic regions of the N. Hemisphere. *F. antipyretica* L. is the most common and widely distributed species, it is found throughout the N. Temperate Zone, Abyssinia and S. Africa. *F. bogotensis* Hamp. is confined to Colombia. *Fontinalis* grows attached to a solid substrate (boulders, logs, tree-roots, etc.) partly emergent, or usually completely submerged in a wide range of aquatic environments. Some species can tolerate long periods out of water. A useful field character for the identification of most species when removed from the water is that the appressed leaves give the stem the appearance of being 3-angled.

Brachelyma Schimper, Syn. Musc. 557 (1876)
Fig. 14B.
Like *Fontinalis* except stems rarely exceed 20 cm long; leaves always folded longitudinally along the centre, subulate to elliptic-lanceolate, 2.0 to 4.0 mm long, 0.5 to 1.5 mm wide; longitudinal nerves present; capsule always completely immersed in leaves.

2 species: N. America, confined to the southeastern part of the USA. Ecology as for *Fontinalis*.

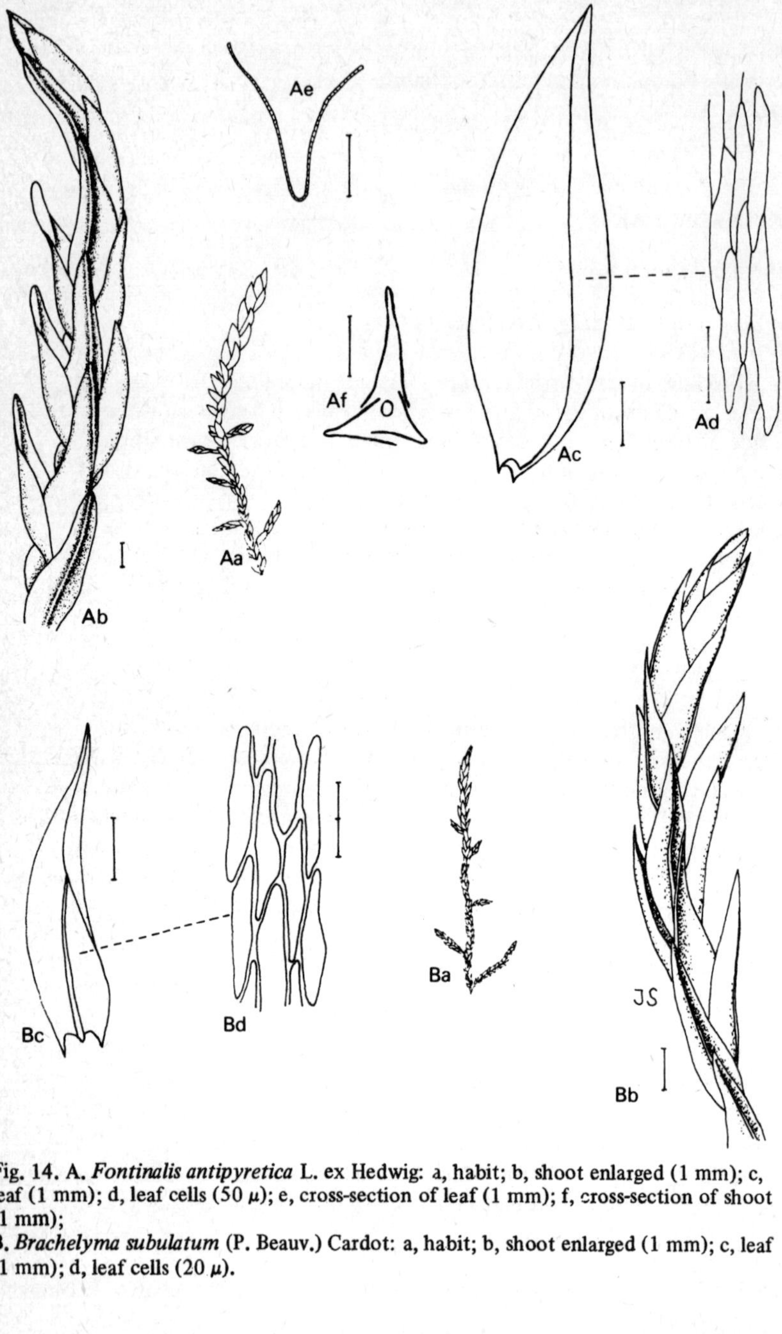

Fig. 14. A. *Fontinalis antipyretica* L. ex Hedwig: a, habit; b, shoot enlarged (1 mm); c, leaf (1 mm); d, leaf cells (50 μ); e, cross-section of leaf (1 mm); f, cross-section of shoot (1 mm);
B. *Brachelyma subulatum* (P. Beauv.) Cardot: a, habit; b, shoot enlarged (1 mm); c, leaf (1 mm); d, leaf cells (20 μ).

Dichelyma Myrin, Act. Reg. Acad. Sci. Holm 1832: 274 (1833)
Fig. 15A.
Like *Fontinalis* except stems rarely exceeding 20 cm; leaves directed to one side of stem, always folded up the centre, 2.75 to 7.0 mm long, 0.4 to 1.5 mm wide, subulate to oblong-lanceolate, usually curved, longitudinal nerves present; capsule surpassing surrounding leaves.

5 species: temperate regions of the N. Hemisphere. Ecology as for *Fontinalis*.

GRIMMIACEAE

c. 8 genera: *Hydrogrimmia, Schistidium* and *Scouleria* contain aquatics.

Hydrogrimmia (Hagen) Loeske, Stud. Morph. Syst. Laubm. 108 (1910) [*Grimmia* Ehrh. ex Hedw. pro parte]
Fig. 15B see *Schistidium*
Stems creeping or partly floating, freely branched, forming small cushions. Leaves distinctly concave, ovate, 2 to 3 mm long, somewhat attenuate at apex, *c.* 1 mm wide; nerve on abaxial surface, reaching or almost reaching apex; margin entire, frequently incurved; apex blunt or acute, rounded; auricles absent; cells of leaf-blade square to rectangular. Capsule short-stalked but exceeding surrounding leaves, ovoid; peristome in 1 row.

Loeske, L. Monographie der europäischen Grimmiaceen. Bibl. Bot. 101: 44–45 (1930)

1 species: *H. mollis* (B. S. G.) Loeske [*Grimmia mollis* B. S. G.], mountains (1800–3200 m) of the N. Hemisphere. Found in a variety of aquatic and subaquatic habitats.

Schistidium Bridel, Mant. Musc. 20 (1819). [*Grimmia* Ehrh. ex Hedw. pro parte]
Fig. 15B.
Like *Hydrogrimmia* but capsule barely exceeding surrounding leaves. Suitable differences between *Hydrogrimmia* and *Schistidium* based on vegetative characters have not been found.

Loeske, L. Monographie der europäischen Grimmiaceen. Bibl. Bot. 101: 49–77 (1930)

c. 19 species of which two are aquatic: *S. apocarpum* (Hedw.) B. S. G. [*Grimmia apocarpa* Hedw.] and *S. alpicola* (Hedw.) Limpr. [*G. alpicola* Hedw.] Found in temperate and sub-tropical regions of the world on rocks in and near streams, most frequent in mountain streams.

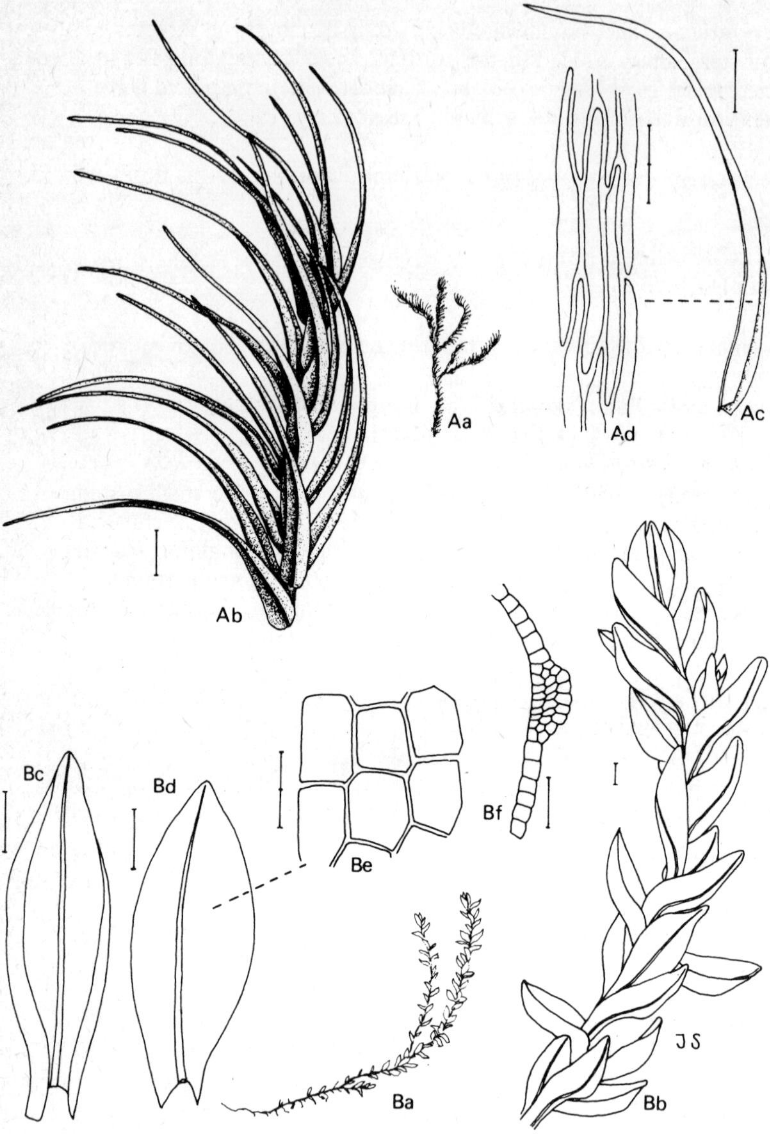

Fig. 15. A. *Dichelyma capillaceum* (With.) Myrin: a, habit; b, shoot enlarged (1 mm); c, leaf (1 mm); d, leaf cells (20 μ);
B. *Schistidium alpicola* (Hedwig) Limpr.: a, habit; b, shoot enlarged (1 mm); c, leaf seen from above (1 mm); d, leaf seen from below (1 mm); e, leaf cells (20 μ); f, cross-section of leaf with midrib (50 μ).

Scouleria Hooker in Drumm., Musci. Bor. Amer. 63 (1828)
Fig. 16A.
Stem creeping or partly floating, irregularly branched; rhizoids confined to creeping part of stem. Leaves somewhat fleshy, lanceolate to linear, up to 4 mm long, usually curved with apices directed away from stem apex; nerve thick and strong almost reaching leaf apex; margin clearly thickened, toothed towards apex; apex blunt, rounded, cells of leaf-blade square or shortly rectangular. Capsule short-stalked, barely exceeding surrounding leaves, ovoid; operculum persistently attached to columella; peristome in 1 row or absent.

c. 5 species: N. E. Asia, N. W. America and Argentina. Found on rocks and stones in streams.

HEDWIGIACEAE

c. 7 genera: *Braunia* contains aquatics.

Braunia B. S. G., Bryol. Eur. 3: 159 (1846)
Fig. 16B.
Stems creeping or partly floating, regularly or irregularly branched, stolons frequent. Leaves concave, clasping the stem, oval, somewhat attenuate at apex; nerve absent but delicate longitudinal folds evident; margin entire, often rolled inwards; apex usually short, occasionally hair-like; cells of leaf-blade square to rectangular with papillae over the lumen. Capsule stalked, cylindrical to urn-shaped, somewhat curved; peristome absent.

c. 22 species: warmer parts of the world. Most species are terrestrial. *B. exserta* C. Müller and *B. cirrhifolia* (Wilson) Jaeger from S. America are known to be aquatic and are found in flowing water.

HYDROPOGONACEAE

2 genera: 2 species: tropical S. America.

Hydropogon (Hooker) Bridel, Bryologia Universa. 1: 770 (1826)
Fig. 17A.
Stems flaccid, thread-like, 5 to 20 (–30) cm long, robust in general appearance because of numerous short, close, pinnately arranged branches; ends of branches and main stem characteristically club-shaped; upper branches up to 1.5 cm long, lower ones up to 2.5 cm long. Leaves in 5 rows, loosely overlapping, up to 0.25 mm apart, oblong to ovate, 2.0 to 3.0 mm long; nerves usually unequally forked; margin entire; apex somewhat attenuate; auricles absent. Capsule sessile or nearly so, concealed among the leaves; peristome in 1 row.

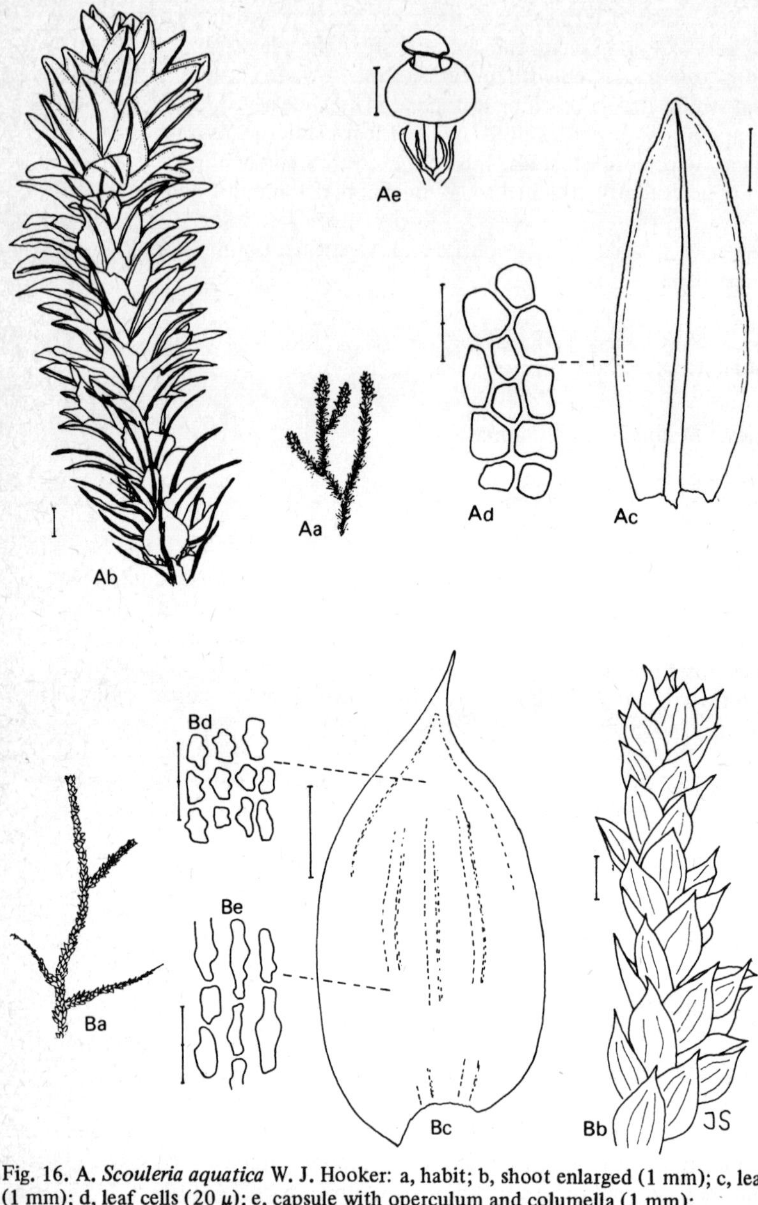

Fig. 16. A. *Scouleria aquatica* W. J. Hooker: a, habit; b, shoot enlarged (1 mm); c, leaf (1 mm); d, leaf cells (20 μ); e, capsule with operculum and columella (1 mm); B. *Braunia cirrhifolia* (Wilson) Jaeger: a, habit; b, shoot enlarged (1 mm); c, leaf (1 mm); d, cells near apex of leaf (20 μ); e, cells of basal part of leaf (20 μ).

Welch, W. H. The systematic position of the genera *Wardia, Hydropogon,* and *Hydropogonella,* Bryologist 46 (2): 25–46 (1943)

1 species: *H. fontinaloides* (Hooker) Bridel: known from Brazil, French Guiana and Venezuela. It is found in water attached to rocks, roots and branches of trees.

Hydropogonella Cardot, Rev. Bryol. 22: 18 (1895)
Fig. 17B.
Like *Hydropogon* except stems slender and delicate, up to 15 cm long; branches not club-shaped at ends, up to 1 cm long; leaves in 3 rows, obovate to oblanceolate, 1.0 to 2.25 mm long, without nerves; capsule not completely concealed among leaves; peristome absent.

1 species: *H. gymnostoma* (Bruch and Schimper) Cardot: known from Bolivia, Brazil, Guiana and Venezuela. It is found in water attached to rocks, roots and branches of trees.

HYPNACEAE

c. 30 genera, *Vesicularia* contains aquatics.

Vesicularia (C. Müller) C. Müller, Bot. Jahrb. 23: 330 (1896)
Fig. 18A.
Stems creeping or floating, oval in transverse section; branches pinnate or irregular; rhizoids sparse or produced in bunches at regular intervals along the stem. Leafy stems dorso-ventrally flattened. Leaves ovate to lanceolate, dimorphic; lateral leaves spreading or weakly directed to one side, upper and lower leaves smaller, more or less clasping the stem; margin entire or slightly toothed near apex; apex blunt or attenuate; nerve short, forked, simple or absent; cells in the centre of leaf-blade rhomboidal, 2 to 8 times as long as wide, papillae absent; auricles absent. Capsule lateral, long-stalked, oval to urn-shaped, rarely cylindrical, horizontal or penduluous; peristome in 2 rows.

Hörmann, H. Beitrag zur Anatomie von Vesicularia graminicolor (Angström) Brotherus und zu V. graminicolor fo. fluitans. Nova Hedwigia 9: 237–243 (1965)
Sauer, H. F. Das Javamoos. Die Aquarien- und Terrarien-Zeitschr. 4: 110–111 (1957)

c. 135 species: warmer regions of the world. A few species from S. E. Asia and Africa are known to be submerged aquatics. *V. dubyana* (C. Müller) Broth. is frequently cultivated as a decorative aquarium plant.

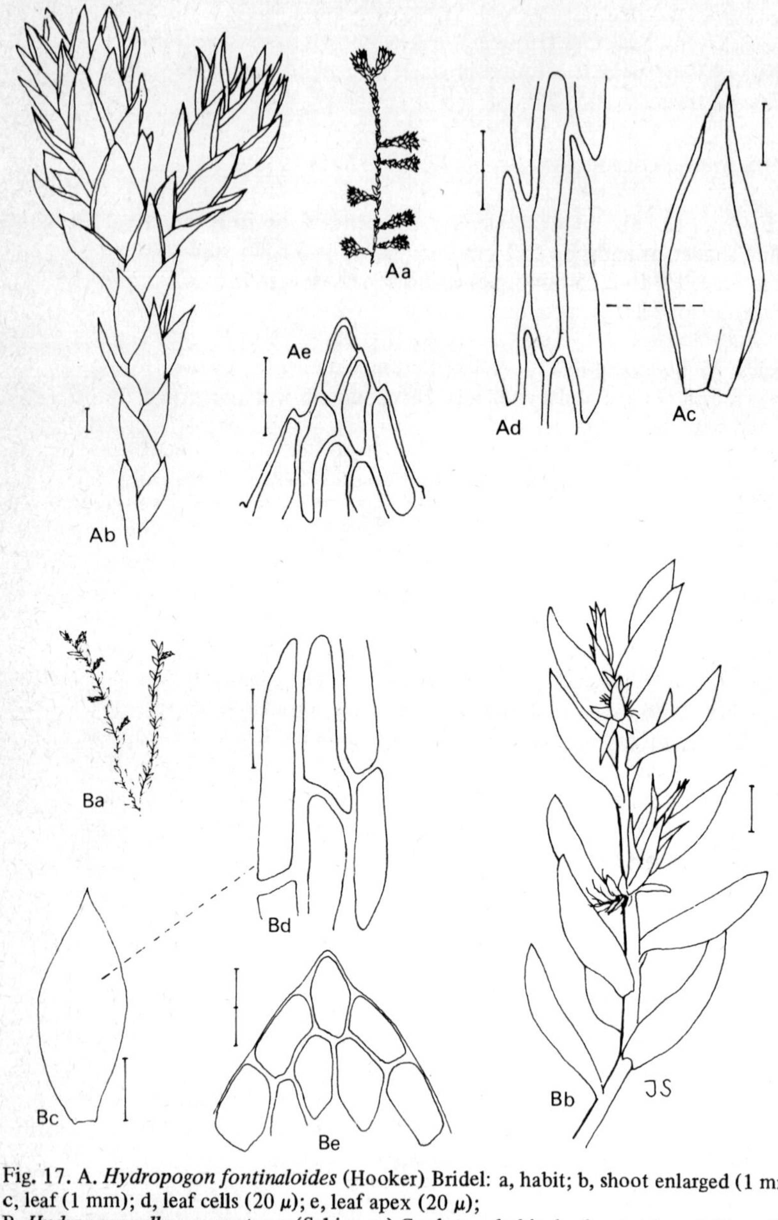

Fig. 17. A. *Hydropogon fontinaloides* (Hooker) Bridel: a, habit; b, shoot enlarged (1 mm); c, leaf (1 mm); d, leaf cells (20 μ); e, leaf apex (20 μ);
B. *Hydropogonella gymnostoma* (Schimper) Cardot: a, habit; b, shoot enlarged (1 mm); c, leaf (1 mm); d, leaf cells (20 μ); e, leaf apex (20 μ).

SEMATOPHYLLACEAE

c. 36 genera: *Glossadelphus* and *Potamium* contain aquatics.

Glossadelphus Fleischer, Fl. Buitenzorg 4: 1351 (1923)
Fig. 18B.
Stems creeping or floating; rhizoids grouped in bunches at regular intervals along stem. Leafy stems dorso-ventrally flattened. Leaves ovate to elongate-ovate, often asymmetrical, dimorphic; lateral leaves concave and wider than the upper and lower ones; nerve short, forked, simple or absent; margin finely toothed; apex blunt, rarely attenuate; cells in centre of leaf-blade elongate, 10 to 15 times longer than wide, papillae present particularly on cells near leaf apex and margin. Capsule lateral, long-stalked, pendulous, asymmetrically ovoid; peristome in 2 rows.

Benl. G. Glossadelphus zollingeri (C. Müller) Fleischer, ein zweites "Javamoos". Die Aquarien- und Terrarien-Zeitschr. 22 (12): 369–372 (1969)

c. 66 species: tropical and subtropical regions of the world. A few species are aquatic. *G. zollingeri* (C. Müller) Fleischer is cultivated as a decorative aquarium plant.

Potamium Mitten, Journ. Linn. Soc. London Bot.: 472 (1869)
Fig. 19A.
Stems usually floating, long, regularly pinnately branched; rhizoids at base of stem. Leafy stems somewhat flattened. Leaves ovate to lanceolate, concave, not dimorphic; nerves absent; margin entire; apex blunt; cells in centre of leaf-blade rhomboidal, 2 to 8 times longer than wide, papillae absent. Capsule lateral, long-stalked, erect to pendulous, ovoid to obovoid; peristome in 2 rows.

c. 8 species: tropical America and Isle St. Marie (off Madagascar). Most species are robust aquatics.

SPHAGNACEAE

1 genus.

Sphagnum L., Sp. Pl. 1106 (1753)
Fig. 19B.
Stems erect or floating, bearing simple, leafy branches arranged in bunches at regular intervals along main axis; rhizoids absent on mature stems. Leaves in 3 rows, ovate, covering main axis and branches (leaves on main axis somewhat smaller than those on branches); cells of leaf-blade dimorphic, small and green alternating with large and hyaline. Capsule terminal, globose, without chlorophyll; peristome absent; greater part of capsule stalk gametophytic in origin.

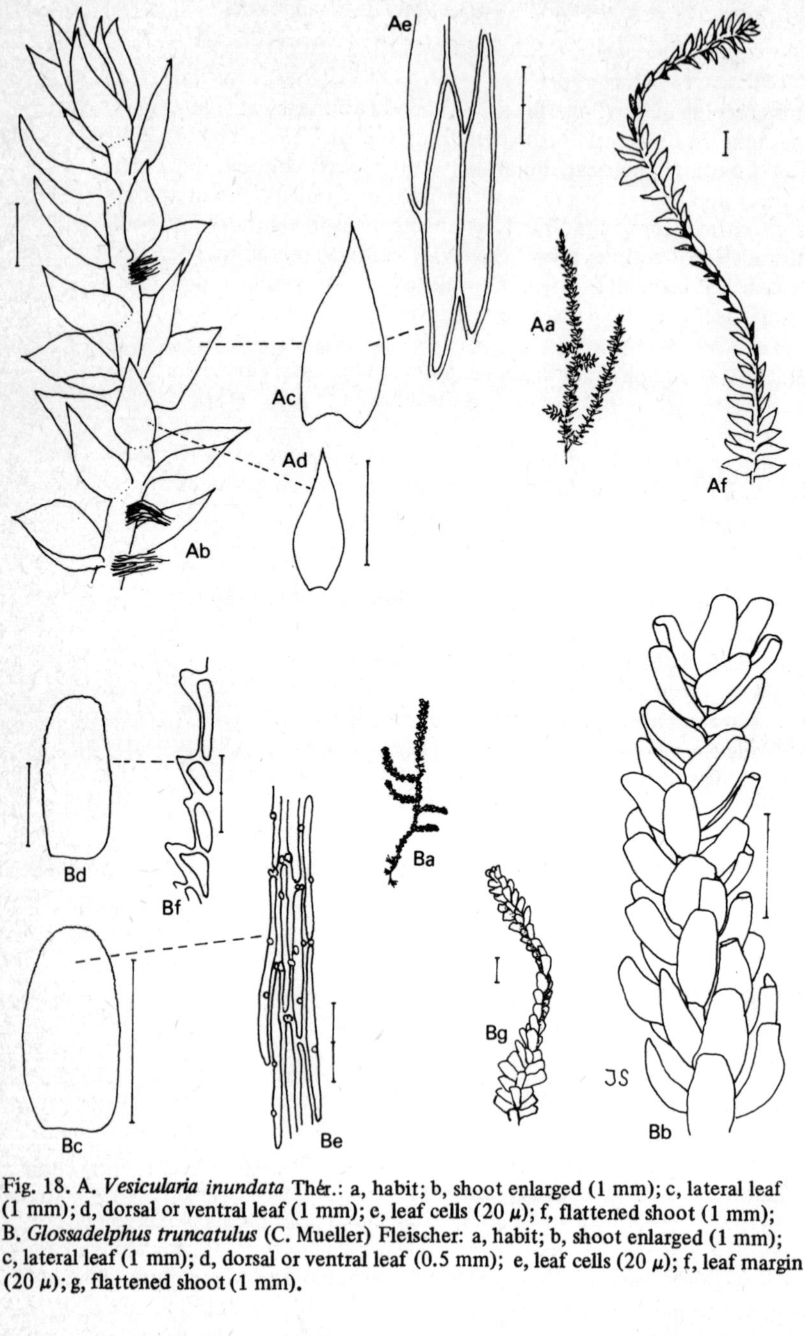

Fig. 18. A. *Vesicularia inundata* Thér.: a, habit; b, shoot enlarged (1 mm); c, lateral leaf (1 mm); d, dorsal or ventral leaf (1 mm); e, leaf cells (20 μ); f, flattened shoot (1 mm); B. *Glossadelphus truncatulus* (C. Mueller) Fleischer: a, habit; b, shoot enlarged (1 mm); c, lateral leaf (1 mm); d, dorsal or ventral leaf (0.5 mm); e, leaf cells (20 μ); f, leaf margin (20 μ); g, flattened shoot (1 mm).

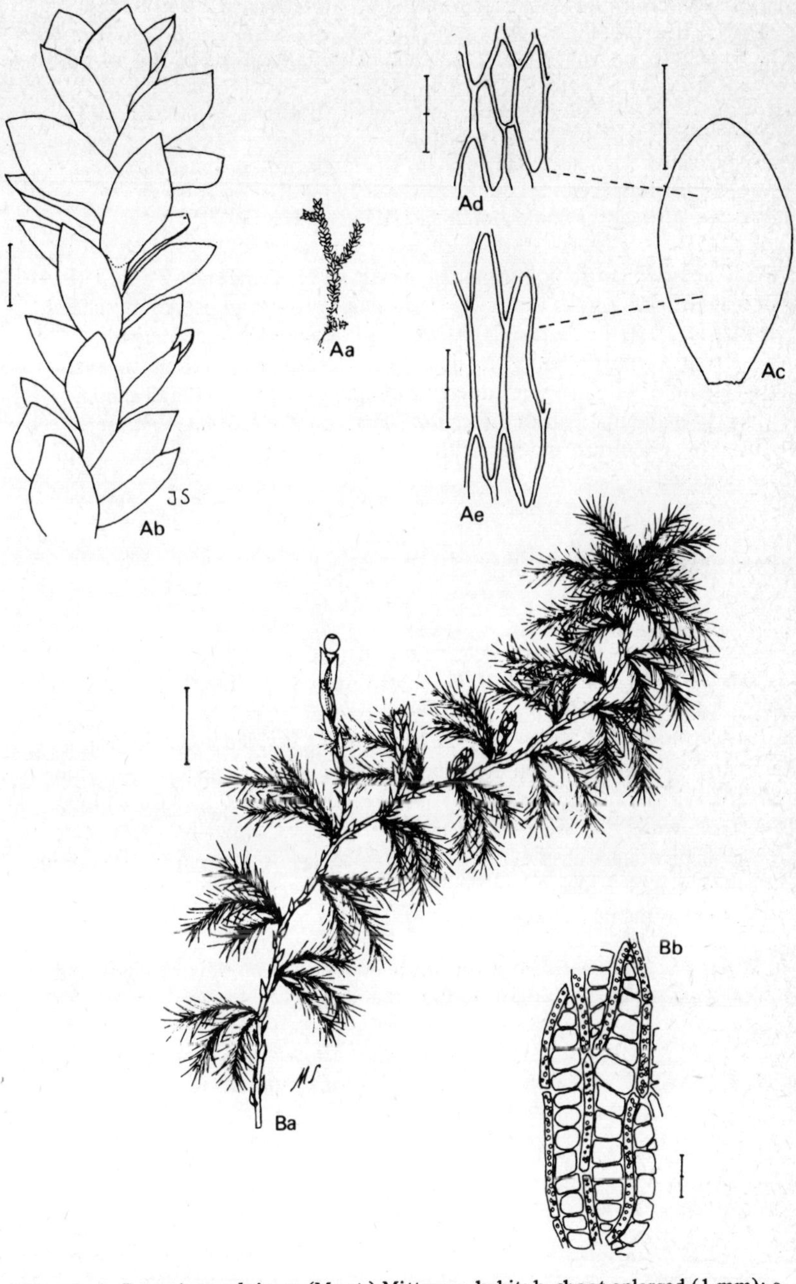

Fig. 19. A. *Potamium vulpinum* (Mout.) Mitten: a, habit; b, shoot enlarged (1 mm); c, leaf (1 mm); d, cells of apical part of leaf (20 μ); e, cells of basal part of leaf (20 μ); B. *Sphagnum cuspidatum* Ehrh.: a, habit (1 cm); b, empty hyaline cells and green cells (20 μ).

Hansen, B. Studies in the flora of Thailand. 37. Sphagnaceae. Dansk Bot. Arkiv 23: 295–300 (1966)
Isoviita, P. Studies on Sphagnum L. I. Nomenclatural revision of the European taxa. Ann. Bot. Fenn. 3: 199–264 (1966)
Maass, W. S. G. Studies on the taxonomy and distribution of Sphagnum. IV. Nova Hedwigia 14: 187–214 (1968)
Savicz-Ljubitzkaja, L. I. and Smirnova, Z. N. Handbook of Sphagnaceae of the USSR. Edit. Nauka Leningrad 1–112 (1968) – in Russian.
Warnstorf, C. in Engler, Pflanzenreich 51 (III): 1–546 (1911)

200 to 350 species: cosmopolitan, most species in N. Temperate Zone. They grow in mats or cushions in wet and often calcium free regions. A few species (*S. cuspidatum* Ehrh. ex Hoffm., *S. recurvum* P. Beauv.) are submerged aquatics. *Sphagnum* is probably the most important plant group in the autogenic succession of oligotrophic waters in temperate regions of high rainfall. Peat is largely made up of dead *Sphagnum.* Peat and fresh *Sphagnum* are widely used in agriculture and horticulture, in some regions dried peat is still an important fuel.

WARDIACEAE

1 genus.

Wardia Harvey and Hooker, in W. J. Hooker, Comp. Bot. Mag. 2: 183 (1837)
Fig. 20.
Stems rigid, robust, up to 9 cm long, at base simple, divided above into short rigid clusters of branches, often denuded of leaves below branches. Leaves in 5 rows, closely overlapping, up to 0.25 mm apart, ovate, 1.0 to 2.0 mm long, 0.5 to 1.0 mm wide; nerves absent; margin entire; apex somewhat attenuate; auricules conspicuous. Capsule long-stalked, borne at tip of stem or main branch, ovoid to orbicular, 1.5 to 2.0 mm long; operculum persistently attached to columella; peristome in 1 row, short, rudimentary.

Welch, W. H. The systematic position of the genera *Wardia, Hydropogon* and *Hydropogonella.* Bryologist 46 (2): 25–46 (1943)

1 species: *W. hygrometrica* Harvey and Hooker: known only from Cape of Good Hope, S. Africa. It is found submerged in small mountain rivulets.

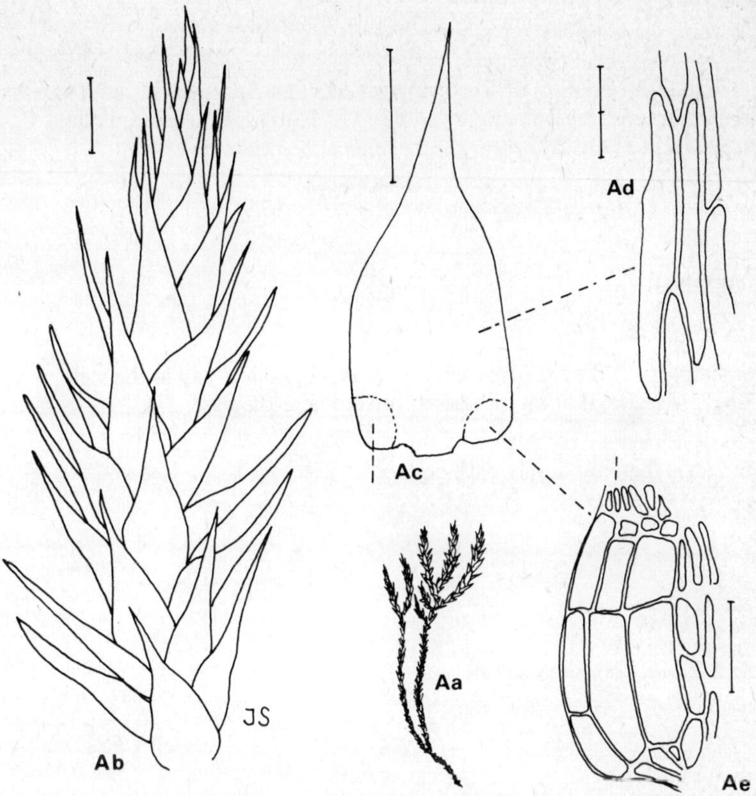

Fig. 20. *Wardia hygrometrica* Harvey and Hooker: a, habit; b, shoot enlarged (1 mm); c, leaf (1 mm); d, leaf cells (20 μ); e, auricles (20 μ).

PTERIDOPHYTA

Plants with an alternation of free-living generations. Sporophytes with vascular tissue, herbaceous, reproducing by spores which give rise to small thalloid gametophytes (prothalli) bearing archegonia and antheridia.

LYCOPSIDA

Stems simple or forked. Leaves spirally arranged, no leaf-gap in the stele. Sporangia in leaf axil or on its adaxial surface near the base.

Isoetaceae is the only family with aquatics.

ISOETACEAE

2 genera: *Isoetes* contains aquatics.

Isoetes L., Sp. Pl. 1100 (1753)
Fig. 21A.
Stems short, tuber-like, 2- or 3-lobed; roots forked. Leaves in a rosette, linear to subulate, each with 4 transversely septate, longitudinal air-channels and a central vascular strand; leaf-base spoon-shaped, with membranous margin and delicate ligule on adaxial surface at region where leaf narrows. Sporangia near leaf base, sessile, solitary, naked or covered with thin tissue (velum). Spores dimorphic, in separate sporangia; megaspores tetrahedral; microspores bilateral. Gametophytes remain within spore wall.

Fuchs, H. P. Nomenclatur, Taxonomie und Systematik der Gattung Isoetes Linnaeus in geschichtlicher Betrachtung. Beih. Nova Hedwigia 3: 1–103 (1962)
Pant, D. D. and Srivastava, G. K. The genus Isoetes in India. Proc. Nat. Inst. Sci. India, ser. B. 28: 242–280 (1962)
Pfeiffer, N. E. Monograph of the Isoetaceae. Ann. Mo. Bot. Gard. 9: 79–232 (1922)
Wanntorp, H.-E. The genus Isoetes in S. W. Africa. Svensk Bot. Tidskr. 64 (2): 141–157 (1970)

c. 75 species: cosmopolitan. The majority of the species are aquatic or amphibious and are found in a wide variety of habitats. The stems are readily eaten by water birds and rodents.

SPHENOPSIDA

Stems grooved and jointed. Leaves in whorls, united into a sheath at base. Sporangia borne on adaxial surface of peltate sporangiophores; sporangiophores in cones. Spores overlaid by 2 or 4 spiral bands (elaters).

1 family

EQUISETACEAE

1 genus

Equisetum L., Sp. Pl. 1061 (1753)
Fig. 21B.
Perennials, rhizome creeping, giving rise to erect stems. Erect stems green (in aquatic species), grooved, simple or with irregular whorls of branches. Leaves small, as many as grooves, united into a sheath at base. Cone at stem apex, consisting of numerous whorled, peltate sporangiophores. Sporangia borne several together around adaxial surface of sporangiophore. Spores all alike, overlaid by 2 or 4 spiral bands (elaters). Gametophytes unisexual, with cushion-like base and lobed apex.

Hauke, R. A taxonomic monograph of the genus Equisetum subgenus Hippochaete. Beih. Nova Hedw. 8: 1–123 (1963)
Page, C. N. An assessment of inter-specific relationships in Equisetum. New Phytol. 71 (2): 355–369 (1972)
Reed, C. F. Index to Equisetophyta. Reed Herbarium, Baltimore, Maryland, U.S.A. (1971)

c. 29 species: mainly N. Temperate Zone. A few species and hybrids are found in water. *E. fluviatile* L. is normally found in water at the edges of lakes, ponds and ditches; it is often dominant and can form extensive stands.

FILICOPSIDA

Leaves with leaf gap in stele. Sporangia usually grouped in sori, borne on undersurface of the leaves, on special leaf-segments or in hardened nut-like structures (sporocarps).

The Azollaceae, Marsileaceae, Parkeriaceae and Salviniaceae have species that are obligate aquatics. However, among the ferns there are several genera in the tropics that grow on rocks or trees by streams, rivers and waterfalls. These plants may, after rain, be submerged in swiftly flowing water, sometimes for only a few hours at a time. These plants are called rheophytes and although

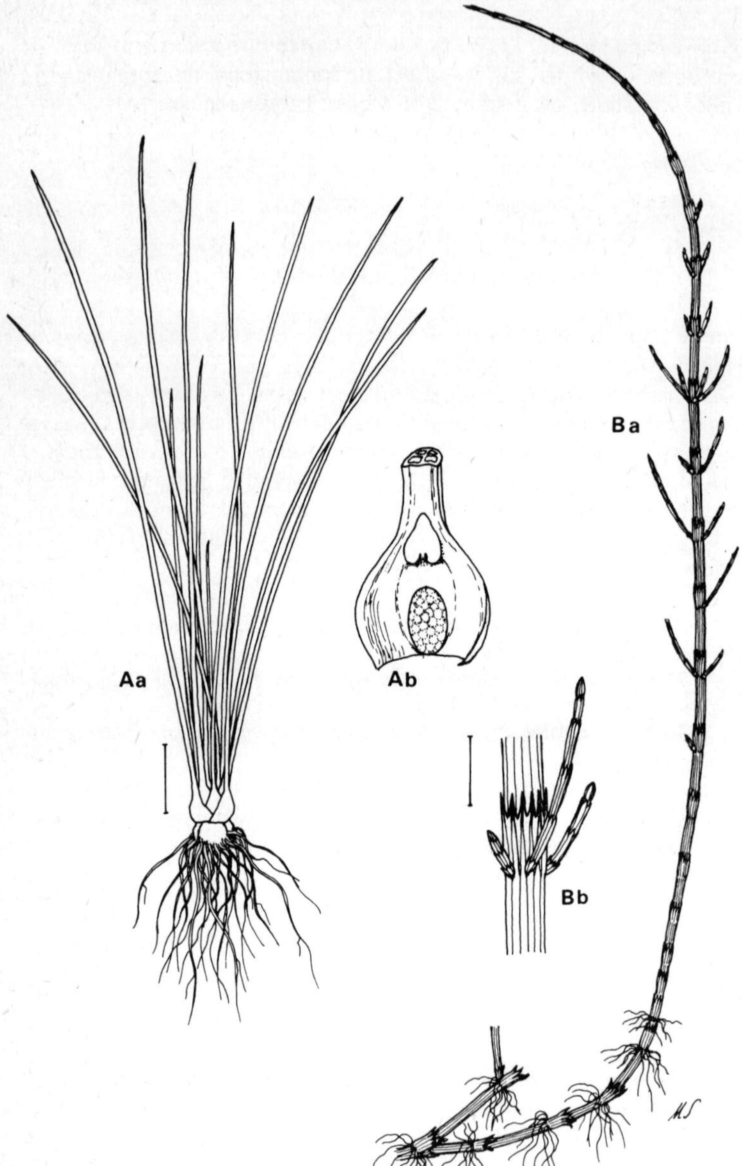

Fig. 21. A. *Isoetes setacea* Lam.: a, habit (2 cm); b, adaxial surface of leaf base. B. *Equisetum fluviatile* L.: a, habit; b, a node with lateral branches (5 mm).

they are specialised for this particular habitat they are being excluded from this account as their reproduction and perhaps vegetative growth takes place in the terrestrial milieu. They are regarded as terrestrial plants that tolerate flooding rather than aquatics that tolerate emergence. The following fern genera contain rheophytes: *Asplenium* L., *Blechnum* L. [*Lomaria* Willd.], *Bolbitis* Schott. [*Campium* Presl.], *Dipteris* Renoir, *Egenolfia* Schott., *Hymenophyllum* Sm. sensu lato, *Lindsaea* Dryand ex Sm., *Microsorium* Link, *Thelypteris* Schmid. sensu lato, and *Trichomanes* L. sensu lato. Some species, such as *Bolbitis heudelotii* (Bory ex Fée) Alston [see Benl, G., Heine, H. and Michel, F. Die Aquarien und Terrarien-Zeitschrift 23 (5): 146–150 (1970)] and *Microsorium pteropus* (Blume) Ching [see Benl, G., op. cit. 14: 210–212 (1961)], can, with care, be maintained in a submerged state in aquaria.

Some tropical ferns grow in places where they are totally or partially submerged at high tide. The genus *Acrostichum* L. sensu stricto (3 species) and *Tectaria semibipinnata* (Wall.) C. Chr. are almost confined to intertidal zones. Like other tidal and salt water plants they are excluded from this account.

AZOLLACEAE

Azolla Lamarck, Enc. Méth. 1: 343 (1783)
Fig. 22A.
Stems floating, more or less pinnately branched, densely covered with small, overlapping, alternate leaves; at some nodes branches and simple roots hang down into the water. Leaves divided into 2 lobes; upper (dorsal) lobe thick, green, borne above water; lower (ventral) lobe thin, colourless, in contact with water on lower surface only, with cavities usually inhabited by a blue-green alga (*Anabaena azollae*). Sporocarps born in pairs or rarely in quartets, in axil of first leaf of a branch on older parts of plant; megasporangial sporocarp small, producing one megasporangium and one megaspore; microsporangial sporocarp larger, producing 7 to 100 microsporangia, each containing 32 or 64 microspores. Megaspores germinate at water surface and produce minute prothalli.

Hall, J. W. Studies on fossil Azolla. Amer. Journ. Bot. 56 (10): 1173–1180 (1969)
Moore, A. W. Azolla: biology and agronomic significance. Bot. Rev. 35 (1): 17–34 (1969)

6 species: through introductions now almost cosmopolitan. In different situations *Azolla* is considered to be an undesirable weed or a very beneficial plant. The *Azolla*-blue-green alga combination fixes atmospheric nitrogen and is widely used for green manure and forage. The agronomic significance is excellently reviewed by Moore (1969) op. cit. 25 fossil species of *Azolla* have been described.

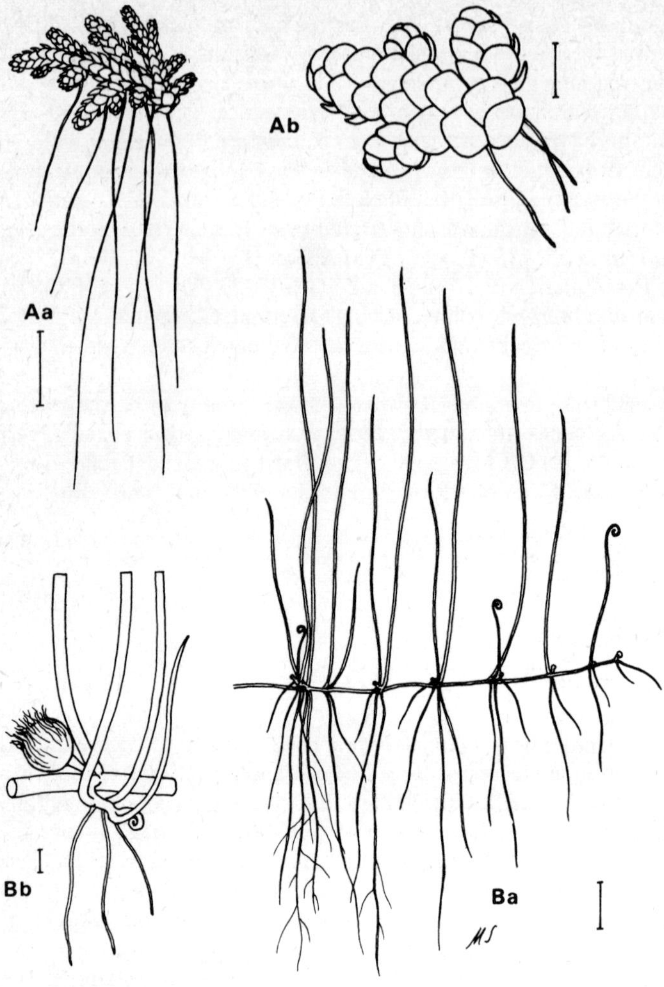

Fig. 22. A. *Azolla filiculoides* Lam.: a, habit (5 mm); b, leafy stem (1 mm); B. *Pilularia globulifera* L.: a, habit (1 cm); b, sporocarp (1 mm).

MARSILEACEAE

Marsilea L., Sp. Pl. 2: 1099 (1753)
Fig. 23.
Stems creeping or floating, rooting at nodes. Leaves rolled when young, compound; leaf stalk long, terminating in 4 leaflets; leaflets arranged symmetrically cross-wise, obdeltate or obovate, submerged, floating or emergent; veins repeatedly forked. Sporangia in closed sporocarps, borne on short stalks arising from leaf stalks. Spores of 2 kinds; solitary megaspores and numerous microspores (megasporangia and microsporangia found together in sporocarps).

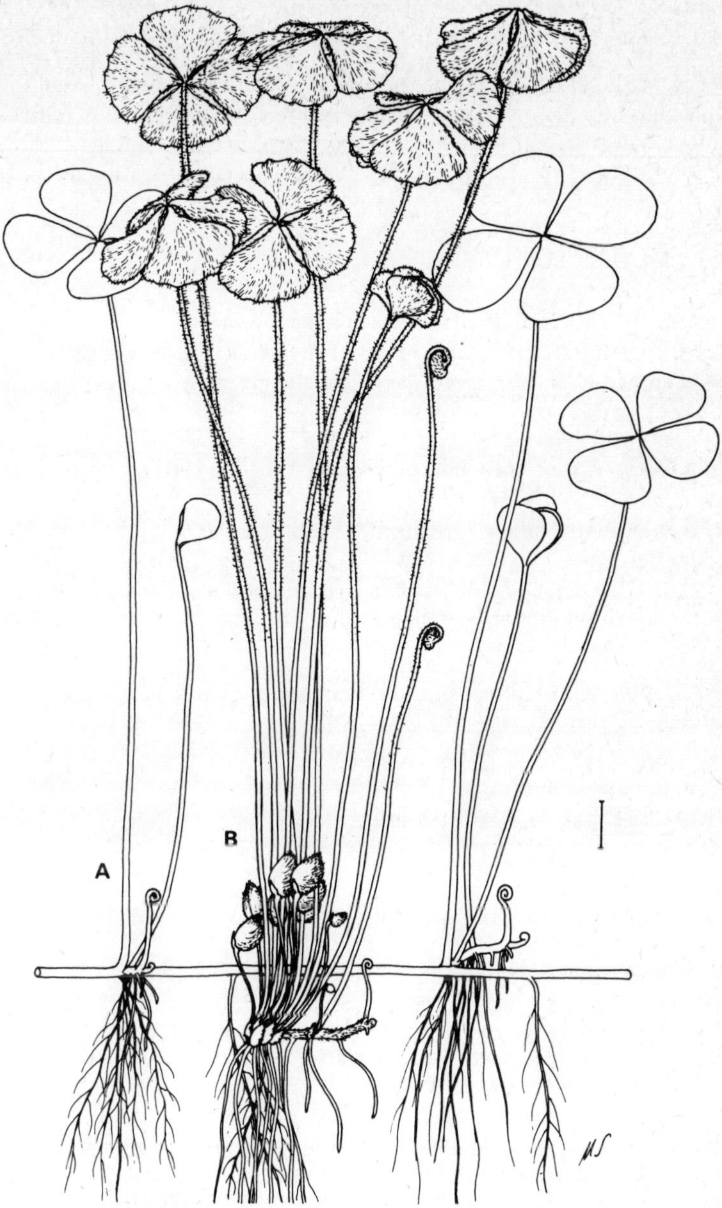

Fig. 23. *Marsilea*: A. *M. minuta* L.: habit of aquatic growth (1 cm); B. *M. hirsuta* R.Br.: terrestrial plant with sporocarps (1 cm).

Launert, E. A monographic survey of the genus Marsilea L. I. The species of Africa and Madagascar. Senckenbergiana Biol. 49 (3/4): 273–315 (1968)

c. 65 species: throughout the warmer regions of the world. Aquatic or semi-aquatic plants found in pools, pans, ditches, swamps and paddy fields. Most species require an emergent period before sporocarps will develop. Several species are regarded as troublesome weeds in rice fields and irrigation ditches.

Pilularia L., Sp. Pl. 2: 1100 (1753)
Fig. 22B.
Like *Marsilea* but with leaflets absent; leaf simple, linear.
6 species: scattered in temperature regions of the world. Small grass-like plants in swamps and at the edges of lakes and streams.

Regnellidium Lindmann, Arkiv Bot. Stockholm 3 (6): 2 (1904)
Fig. 24.
Like *Marsilea* but with leaflets 2, reniform to suborbicular.

Sota, E. R. de la, Distribución de las lacticíferos en el esporofito de Regnellidium diphyllum Lindm. Arquivos Jardim Bôt. Rio de Janeiro 18: 5–7 (1962)

1 species: *R. diphyllum* Lindm., a local endemic species in S. Brazil and N. Argentina.

PARKERIACEAE [ADIANTACEAE pro parte]

Only aquatic genus: *Ceratopteris.*

Ceratopteris Brongn., Bull. Soc. Sci. Philom. Paris 1821: 186 (1822)
Fig. 25.
Annuals or short-lived perennials; stem erect, very reduced, bearing few scales. Leaves 1- or more-pinnate, in a rosette; lower ones sterile and usually less divided than upper ones, often bearing buds which develop into new plants; fertile leaves finely divided, erect, bearing scattered sporangia on adaxial surface; margin of fertile leaves inflexed, partly covering sporangia.

Benedict, R. C. The genus Ceratopteris: a preliminary revision. Bull. Torrey Bot. Club 36: 463–476 (1909)
Donselaar, J. van. On the distribution and ecology of Ceratopteris in Surinam. Amer. Fern. Journ. 59 (1): 3–8 (1969)
Johnson, A. The genus *Ceratopteris* in Malaya. Gard. Bull. Singapore 18: 76–81 (1961)
Lloyd, R. M. Sexual and vegetative reproduction in Hawaiian Ceratopteris thalictroides. Amer. Fern. Journ. 63(1): 12–18 (1973)

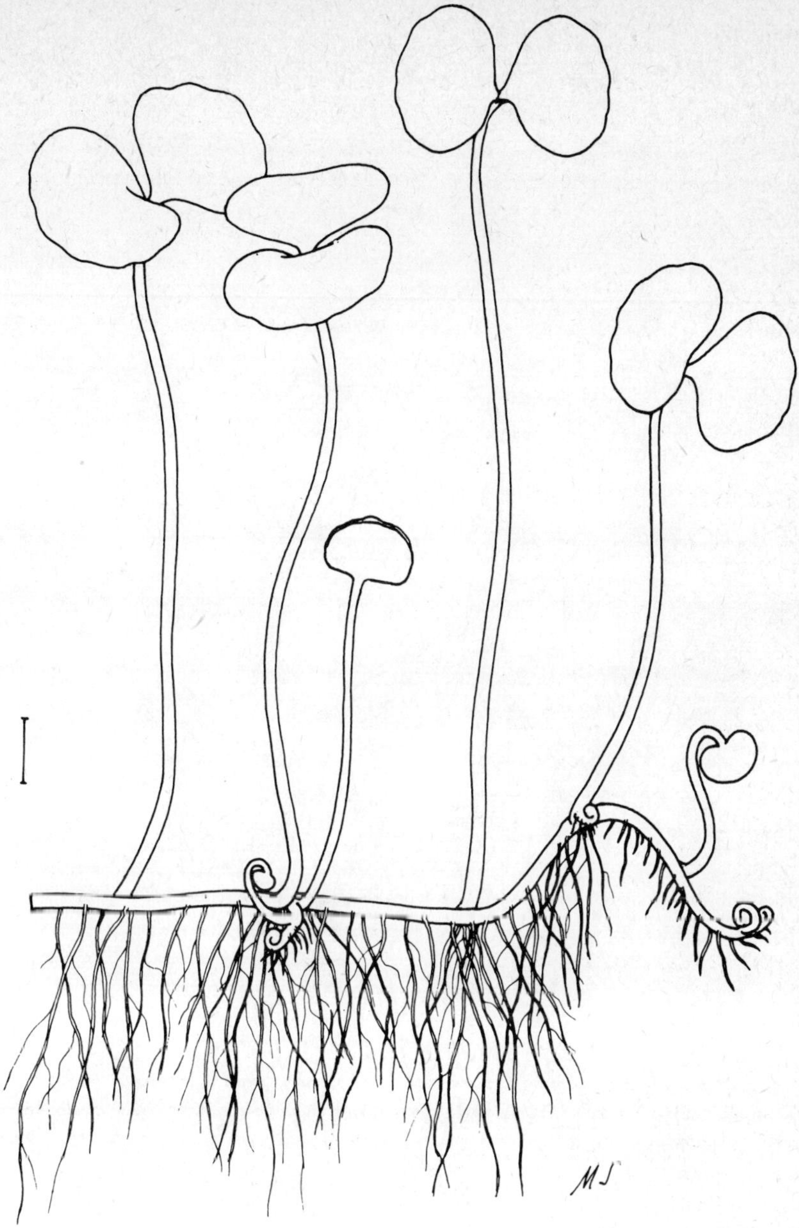

Fig. 24. *Regnellidium diphyllum* Lindm.: habit (1 cm).

c. 6 species: warmer regions of the world. Some species such as *C. cornuta* (P. Beauv.) Lepr. and *C. pteridoides* (Hooker) Hieron. are essentially free-floating, the other species are emergent aquatics or swamp plants. *C. pteridoides* is a potential danger in man made lakes and in 1966 colonised about 17 000 hectares of Brokopondo Lake in Surinam. *C. thalictroides* (L.) Brongn. is cultivated in Japan as a spring vegetable.

Fig. 25. *Ceratopteris cornuta* (P. Beauv.) Lepr.: habit (5 cm).

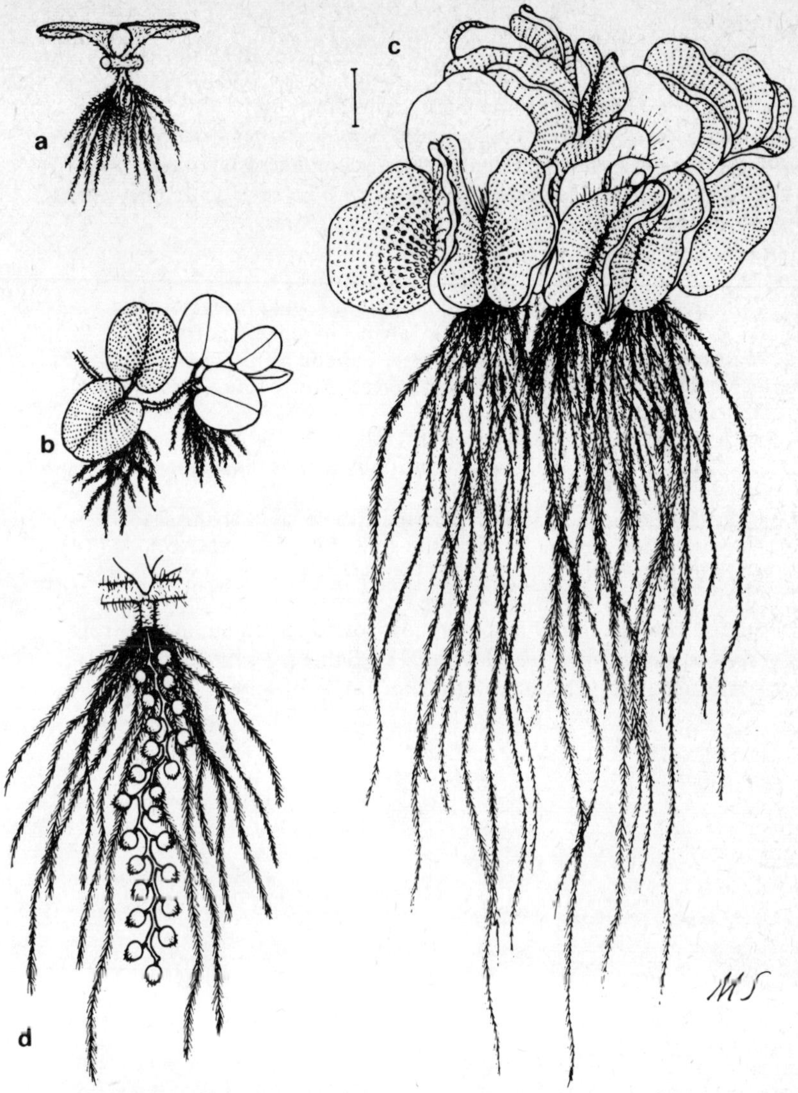

Fig. 26. *Salvinia molesta* Mitchell; a, internode showing whorl of three leaves; b, flat-phase growth; c, folded phase growth; d, chain of microsporangial sporocarps (1 cm).

SALVINIACEAE

Salvinia Séguier, Pl. Veron, 3: 52 (1754)
Fig. 26.

Stem floating, irregularly branched, without roots. Leaves in whorls of 3, 2 floating and 1 submerged; floating leaves photosynthetic, entire with peculiar unwettable hairs on adaxial surface and wettable hairs on abaxial surface; submerged leaves not photosynthetic, finely divided into linear segments that bear striking resemblance to roots. Sporocarps borne on modified segments of submerged leaves; megasporangial sporocarps developed first, producing up to 25 megasporangia each containing 1 megaspore; microsporangial sporocarps developed later, producing numerous microsporangia each containing usually 64 microspores. Prothalli develop inside floating spores.

Cook, C. D. K. and Gut, B. F. Salvinia in the State of Kerala, India. Pest Announc. News Service (PANS) 17 (4): 438–447 (1971)
Hossain, M. Spore-bearing structures in Salvinia nymphellula Desv. (Salviniaceae). Bot. Journ. Linn. Soc. London 64 (2): 141–148 (1971)
Jain, R. K. Pre-Tertiary records of Salviniaceae. Amer. Journ. Bot. 58 (6): 487–496 (1971)
Reed, C. F. Distribution of Salvinia and Azolla in South America and Africa in connection with studies for control by insects. Phytologia 12: 121–130 (1965)
Sota, E. R. De la. Contribucion al conocimiento de las "Salviniaceae" neotropicales. Darwiniana 12 (3): 465–498, 499–513, 514–520, 12 (4): 612–623 (1962); 13: 528–536 (1964)

c. 12 species: through introductions almost cosmopolitan but absent from colder regions. *Salvinia* is a free-floating fern. Some races have an enormous growth rate and are capable of covering large areas of water. *Salvinia* has become a very serious pest in Southeast Africa, Ceylon and South India.

SPERMATOPHYTA

ANGIOSPERMAE

Ovules enclosed in an ovary which is usually crowned by a style and stigma. Pollen grains (microspores) adhering to stigma and fertilization effected by means of a pollen tube.

ACANTHACEAE

c. 250 genera: *Justicia* and *Hygrophila* contain aquatics.

Herbs. Leaves opposite, simple. Inflorescence terminal or axillary, cymose. Flowers bisexual, zygomorphic. Sepals tubular 4- or 5-lobed. Petals tubular, 5-lobed or 2-lipped. Stamens 2 or 4. Ovary superior, 2-locular; ovules 2 or more in each loculus, produced on hook-like projections (retinacula); fruit a 2-valved capsule.

Long, R. W. The genera of Acanthaceae in the south-eastern United States. Journ. Arnold. Arb. 51 (3): 257–309 (1970)

1A Stamens 2 (sterile stamens absent); anther sacs separated, more or less superposed; nectariferous scale conspicuous, cup-shaped, 2-, 3- or 5-lobed **Justicia**

1B Stamens 4 (occasionally 2 fertile and 2 sterile); anther sacs at same level, parallel; nectariferous scale inconspicuous **Hygrophila**

Hygrophila R. Br., Prodr. Fl. Nov. Holl. 1: 479 (1810), [*Asteracantha* Nees, *Cardanthera* Buch-. -Ham ex. Bentham, *Nomaphila* Blume, *Synnema* Bentham]

Fig. 27.

Perennials or annuals. Stems erect or ascending. Leaves opposite, either monomorphic and entire, lanceolate to obovate, or heteromorphic with pinnately divided basal or submerged leaves and entire cauline or emergent leaves. Flowers borne in axillary cymes or rarely in terminal heads; bracts linear, ciliate; bracteoles lanceolate shorter than sepals. Sepals tubular, 5- or rarely 4-lobed, deeply lobed or united to above the middle. Petals white to purple, tubular, either almost equally 5-lobed or 2-lipped; adaxial lip, when present, erect, concave, 2-lobed or 2-toothed; abaxial lip 3-lobed. Stamens 4 (occasionally 2 sterile), not exserted, adnate to petal tube above the middle; filaments united in pairs by membrane at base; anthers 2-locular; anther sacs at same level and parallel; pollen 3- or rarely 4-colporate. Nectariferous disc inconspicuous. Ovary superior, 2-locular; fruit a capsule, opening by 2-valves; seeds axile, 6 to many in each loculus, produced on papilliform or hook-like projections (retinacula).

Heine, H. Some West African Acanthaceae. Kew Bull. 16 (2): 161–183 (1962)
Heine, H. Notes sur les Acanthacées africaines. Adansonia sér. 2, 11 (4): 641–659 (1971)

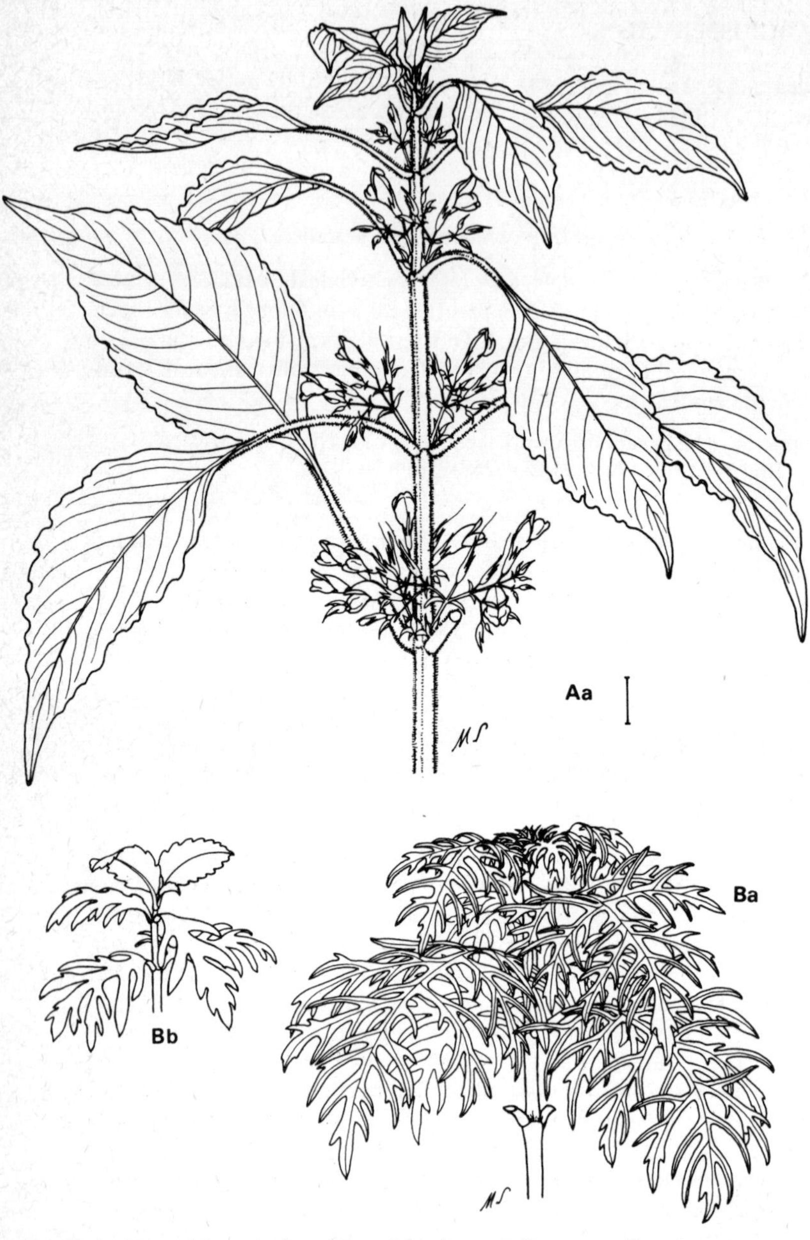

Fig. 27. A. *Hygrophila corymbosa* (Blume) Lindau: a, inflorescence (1 cm); B. *H. difformis* (L. fil.) Blume: a, stem apex of submerged plant; b, showing change of leaf shape on emergence.

80 to 90 species: cosmopolitan in warmer parts, most species in the Old World. Most species are marsh plants many of which are occasionally found in water and are frequently reported as weeds in rice fields and irrigation ditches. *H. auriculata* (Schumach.) Heine [*Asteracantha longifolia* (L.) Nees, *H. spinosa* T. Anderson] is occasionally troublesome as a semi-aquatic weed, it is particularly nasty to remove as it bears very sharp spines. Some species such as *H. difformis* (L. fil) Blume [*Synnema triflorum* (Roxb. ex Nees) O. Kuntze] and *H. africana* T. Anderson are more or less obligate aquatics and develop finely divided submerged leaves. Several species, such as *H. corymbosa* (Blume) Lindau [*Nomaphila stricta* (Vahl) Nees], *H. lacustris* (Schlecht.) Nees, *H. polysperma* (Roxb.) T. Anderson, *H. salicifolia* (Vahl) Nees and *H. difformis* are frequently cultivated as decorative aquarium plants.

Generic delimitation in the tribe Hygrophileae on a world-wide scale is much in need of revision. There is little doubt that *Asteracantha* and *Nomaphila* are congeneric with *Hygrophila. Synnema* and *Hygrophila* appear to be reasonably well separated in Asia (*Synnema* with almost free sepals, papilliform retinacula and many, small seeds and *Hygrophila* with united sepals, hook-like retinacula and few large seeds) but this combination of diagnostic characters breaks down in Africa.

Justicia L., Sp. Pl. 1: 15 (1753), [*Dianthera* L.]
Fig. 28.
Perennials. Stems erect or ascending. Leaves opposite, entire, linear to oblong. Flowers borne in axillary or terminal cymes, spikes or panicles; bracts and bracteoles various, from very small and inconspicuous to large and conspicuous. Sepals tubular, 5- or rarely 4-lobed, or divided almost to base. Petals white to reddish or purple, shortly tubular, 2-lipped; adaxial lip erect or concave, incurved, spreading, entire or 2-toothed, external in bud; abaxial lip 2-lobed. Stamens 2 (staminodes absent), usually slightly exserted but not longer than petal lobes; filaments adnate to throat of petals; anther 2-locular; anther sacs separated, more or less superposed, one or both sacs pointed or produced into a spur or tail; pollen 2-porate, 3-colporate or subporate. Nectariferous disc conspicuous, cup-shaped, 2-, 3- or 5-lobed. Ovary superior, 2-locular; fruit a capsule opening by 2 valves; seeds axile, 2 in each loculus, produced on hook-like projections (retinacula).

Datta, P. and Maiti, R. Study of floral anatomy of the tribe Justicieae with an aim of taxonomical interpretation. Acta Biol. Acad. Sci. Hung. 20: 311 317 (1969)
Penfound, W. T. The biology of Dianthera americana L. Amer. Midl. Nat. 24: 242–247 (1940)

c. 300 species: pantropical and subtropical. Many species are marsh plants and some, such as *J. americana* (L.) Vahl [*Dianthera americana* L.] are aquatic and the cause of some concern in southeastern N. America because dense stands constitute a favourable breeding place for malarial mosquitoes.

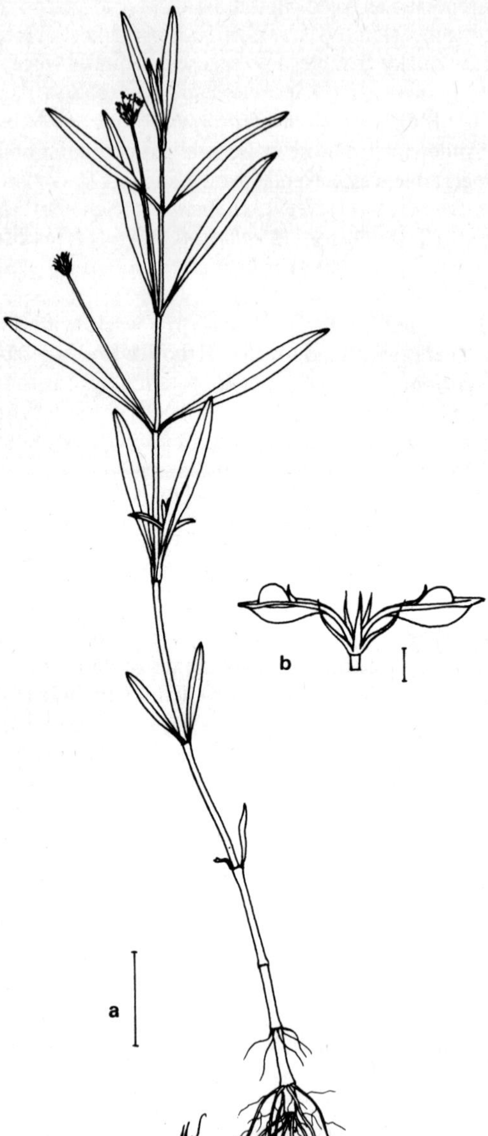

Fig. 28. *Justicia americana* (L.) Vahl: a, habit (3 cm); b, fruit (2 mm).

ALISMATACEAE

11 genera: *c.* 100 species: cosmopolitan: all aquatic.

Perennials or rarely annuals. Stems corm-like or stoloniferous. Leaves entire, linear to ovate, occasionally sagittate; petiole with expanded sheathing base; juvenile leaves usually linear and submerged. Inflorescence usually compound or rarely simple, of whorls of branches, rarely umbel-like or with solitary flowers; bracts 2 or 3 at base of each whorl. Flowers regular, bisexual or unisexual. Sepals 3. Petals 3, caducous or rarely absent. Stamens 3, 6, 9 or more; anthers 2-celled. Carpels superior, free or joined at base, 3 to numerous, in a whorl or spiral; fruit a head of indehiscent nutlets (in *Damasonium* dehiscent).

Buchenau, F. in Engler, A. Pflanzenreich 15 (IV.15): 1–66 (1903)

1A Carpels in 1 whorl
- 2A Carpels more or less united at base, long-beaked, when ripe spreading stellately on an elongated receptacle, usually containing more than 1 seed
  Damasonium
- 2B Carpels not united, not long-beaked, when ripe not spreading stellately, on flat receptacle, containing 1 seed
  - 3A Nutlets flattened, when ripe forming a regular circular or triangular whorl
    Alisma
  - 3B Nutlets swollen, not flattened, when ripe forming an irregular, loose, flattened head
    Caldesia

1B Carpels spirally arranged
- 4A Inflorescence spike-like; flowers inconspicuous, in sessile or subsessile whorls
  Wisneria
- 4B Inflorescence not spike-like; flowers conspicuous, distinctly stalked, whorled or not
  - 5A Nutlets laterally flattened with a flange following the outline of the seed cavity on each side; petals minute and scale-like or absent; plants dioecious
    Burnatia
  - 5B Nutlets laterally flattened or not but without flanges; petals conspicuous; plants not dioecious
    - 6A Flowers arising in the axils of leaves; leaves not in distinct rosettes but produced regularly along a creeping stem
      Luronium

6B Flowers borne on a distinctly stalked terminal inflorescence; leaves in rosettes with or without creeping stem
  7A Fruiting heads globose and smooth (head made up of very many flattened nutlets which fit neatly together giving a smooth outline)
  Sagittaria
  7B Fruiting heads either globose and spiny or not globose
    8A Upper flowers male; nutlets swollen with lateral air chambers
    Limnophyton
    8B All flowers bisexual; nutlets not swollen or when swollen without lateral air chambers
      9A Stamens more than 12 (New World)
      Echinodorus
      9B Stamens 12 or less
        10A Flowers solitary or inflorescence simple (1 or more whorls); receptacle elongated in fruit
          11A Nutlets laterally flattened, with 1, distinct, dorsal wing; inflorescence with 1 to 3 flowers (Tropical Old World)
          Ranalisma
          11B Nutlets more or less swollen, without 1, distinct, dorsal wing; inflorescence with usually more than 3 flowers
            12A Stamens 6; style terminal or nearly so (Europe and N. Africa)
            Baldellia
            12B Stamens usually more than 6; style lateral (New World)
            Echinodorus
        10B Flowers in a compound inflorescence; receptacle flat in fruit
          13A Nutlets with winged ridges (New World)
          Echinodorus
          13B Nutlets with unwinged ridges or warty (Old World)
          Caldesia

Alisma L., Sp. Pl. 342 (1753)
Fig. 29.
Perennials with stem reduced to small corm. Leaves erect; leaf-blades linear to ovate; apex acute; base decurrent to cordate. Inflorescence compound, pyramidal, consisting of whorls of usually more than 3 branches. Flowers bisexual. Petals 3, larger than sepals, white to purplish-pink, with yellowish spot near base, caducous. Stamens 6, in 3 pairs. Carpels numerous, in 1 circular or triangular whorl on flattened receptacle; style lateral; nutlets laterally compressed, ridged dorsally.

Fig. 29. *Alisma lanceolatum* With.: a, habit (5 cm); b, fruit (2 mm).

Björkqvist, I. Studies in Alisma I. Distribution, variation and germination. Opera Botanica 17: 1–128 (1967); II. Chromosome studies, crossing experiments and taxonomy. Opera Botanica 19: 1–138 (1968)

9 species: almost cosmopolitan but indigenous in the N. Hemisphere. Found in a wide variety of aquatic habitats. The fruits and leaves are eaten by a wide variety of animals and birds.

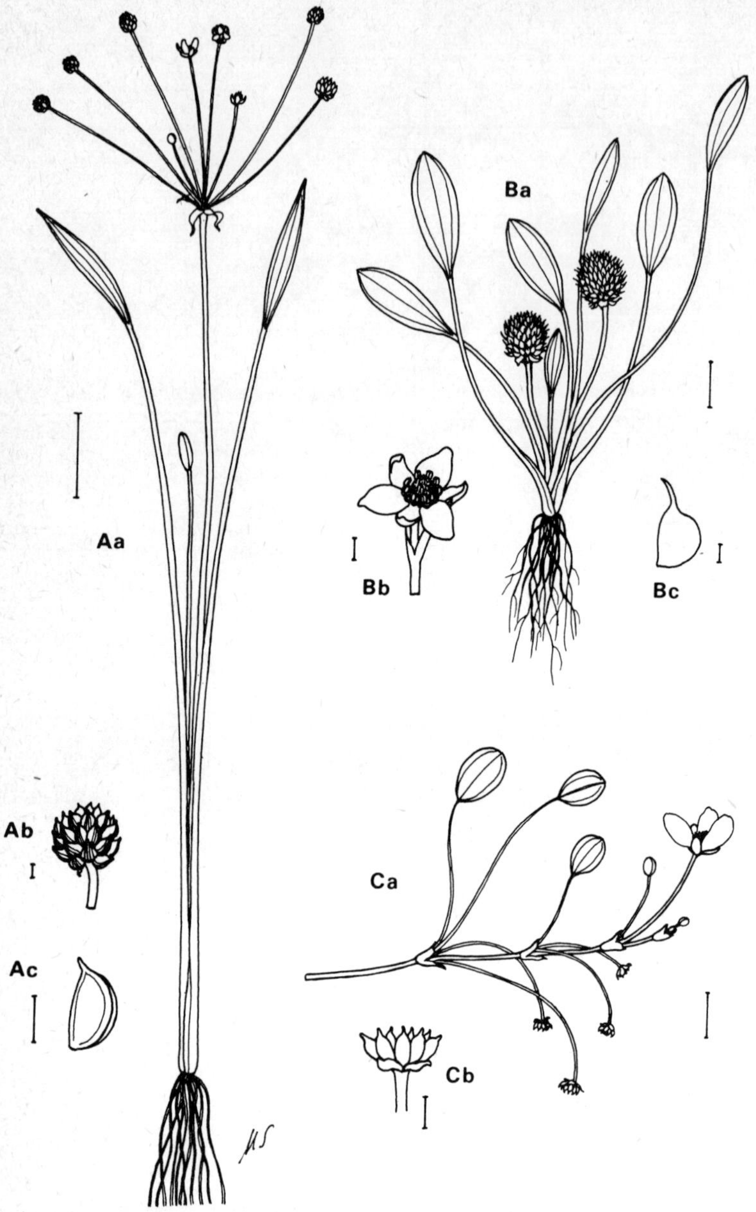

Fig. 30. A. *Baldellia ranunculoides* (L.) Parl.: a, habit (3 cm); b, fruit (1 mm); c, nutlet (1 mm);
B. *Ranalisma humile* (Kunth) Hutch.: a, habit (in fruit) (1 cm); b, flower (2 mm); c, nutlet (2 mm);
C. *Luronium natans* (L.) Rafin.: a, habit (1 cm); b, fruit (1mm).

Baldellia Parl., Nuovi Gen. Sp. 57 (1854)
Fig. 30A.
Stoloniferous perennial, very variable, erect or spreading, usually between 5 and 20 cm tall. Leaves with long petiole; leaf blade 2 to 4 cm long, linear to linear lanceolate, acute, narrowed gradually into petiole. Inflorescence usually umbel-like or rarely in 2 simple whorls or sometimes solitary. Petals 3, pale purplish, up to 7 mm long, longer than sepals. Stamens 6. Carpels numerous, free, forming a globose head; style terminal; nutlets *c.* 2 mm long, ovoid, curved, strongly 3-ribbed on the back with a double rib at ventral suture.

1 species: *B. ranunculoides* (L.) Parl.: Europe and N. Africa. Found in streams, ponds, lakes and ditches. A second species *B. alpestris* (Coss.) Lainz is sometimes recognised. *Baldellia* is very similar to *Echinodorus* and is occasionally included with in it.

Burnatia Micheli in DC., Mon. Phan. 3: 81 (1881), [*Rautanenia* Buch.]
Fig. 31A.
Dioecious perennial. Leaves submerged or erect; petioles up to 40 cm long; leaf blades linear-lanceolate to ovate, up to 16 cm long. Inflorescences of whorls of 3 branches or 3 flowers, up to 30 cm long; male inflorescences larger than female. Male flowers with 2 to 3 mm long sepals, 1 mm long petals. Stamens 9. Female flowers with sepals 1.5 mm long; minute scale-like petals or petals absent. Carpels obovoid, laterally flattened with a flange following the outline of the seed cavity on each side; style ventral, very short; nutlets (8–) 12 (–20), 1.5 to 2.5 mm long.

1 species: *B. enneandra* Micheli, tropical and southern Africa. Found in swamps, shallow lakes and slowly flowing water.

Caldesia Parl., Fl. Ital. 3: 599 (1858)
Fig. 31B.
Leaves submerged or floating, rarely erect; leaf blades broadly elliptic to broadly ovate; apex rounded to obtuse; base truncate to deeply cordate. Inflorescence pyramidal, consisting of whorls of branches or flowers; branches and flowers mostly in whorls of 3. Flowers bisexual. Stamens 6 (–11). Carpels 2 to 9 (–20) free, crowded on small, flat receptacle; style lateral; nutlets swollen, smooth, ridged or warty.

4 species: *C. parnassifolia* (Bassi) Parl. Europe, *C. reniformis* (D. Don) Makino, tropical and N. Africa, India, Malaysia, Australia, China and Japan, *C. oligococca* (F. Muell) Buch. India, Ceylon and N. Australia and *C. grandis* Samuelsson, N. India. Found in swamps, pools, lakes, rivers and canals.

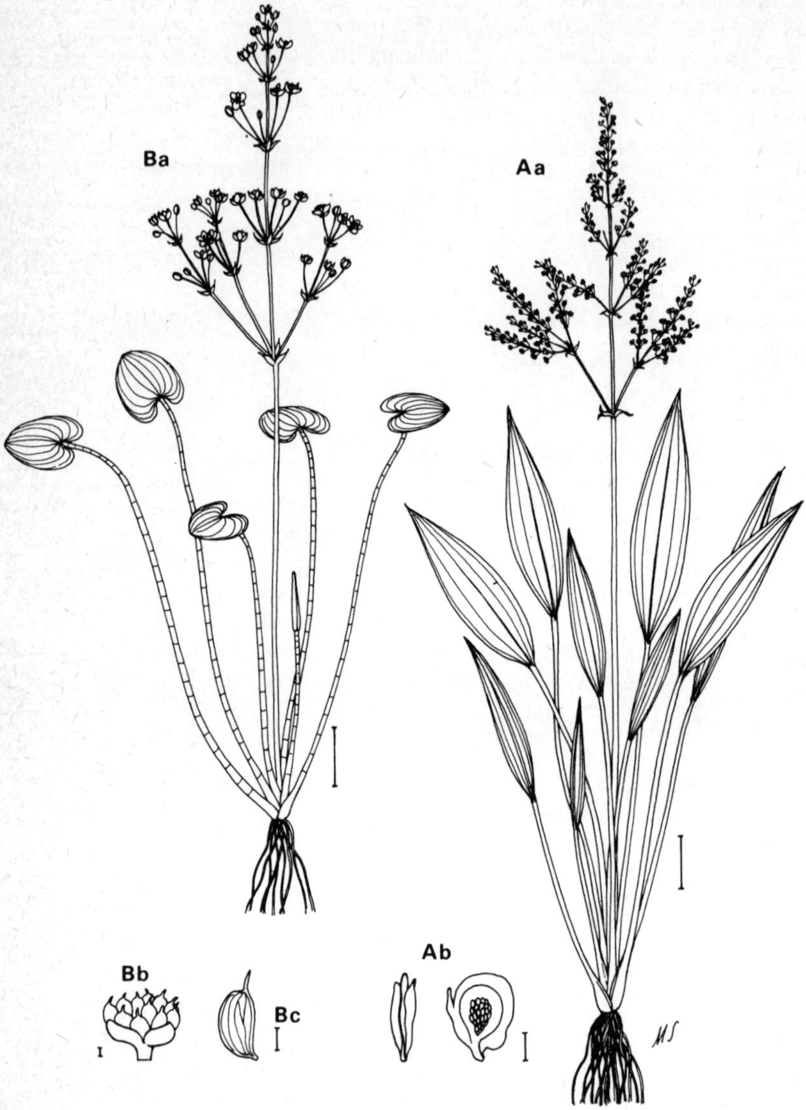

Fig. 31. A. *Burnatia enneandra* Micheli: a, habit (3 cm); b, nutlets (1 mm);
B. *Caldesia reniformis* (D. Don) Makino: a, habit (1 cm); b, fruit (1 mm); c, nutlet (1 mm).

Damasonium Miller, Gard. Dict. Abr. ed. 4 (1754), [*Machaerocarpus* Small]
Fig. 32.
Annuals or perennials with corm-like stem. Leaves submerged, floating or emergent; submerged leaves linear; floating and emergent leaves long stalked, with elliptic to ovate blade and cuneate to cordate base. Inflorescence of 1 to several simple whorls, the lower whorl sometimes compound. Flowers bisexual. Petals exceeding the sepals, white or pink with yellow spot in

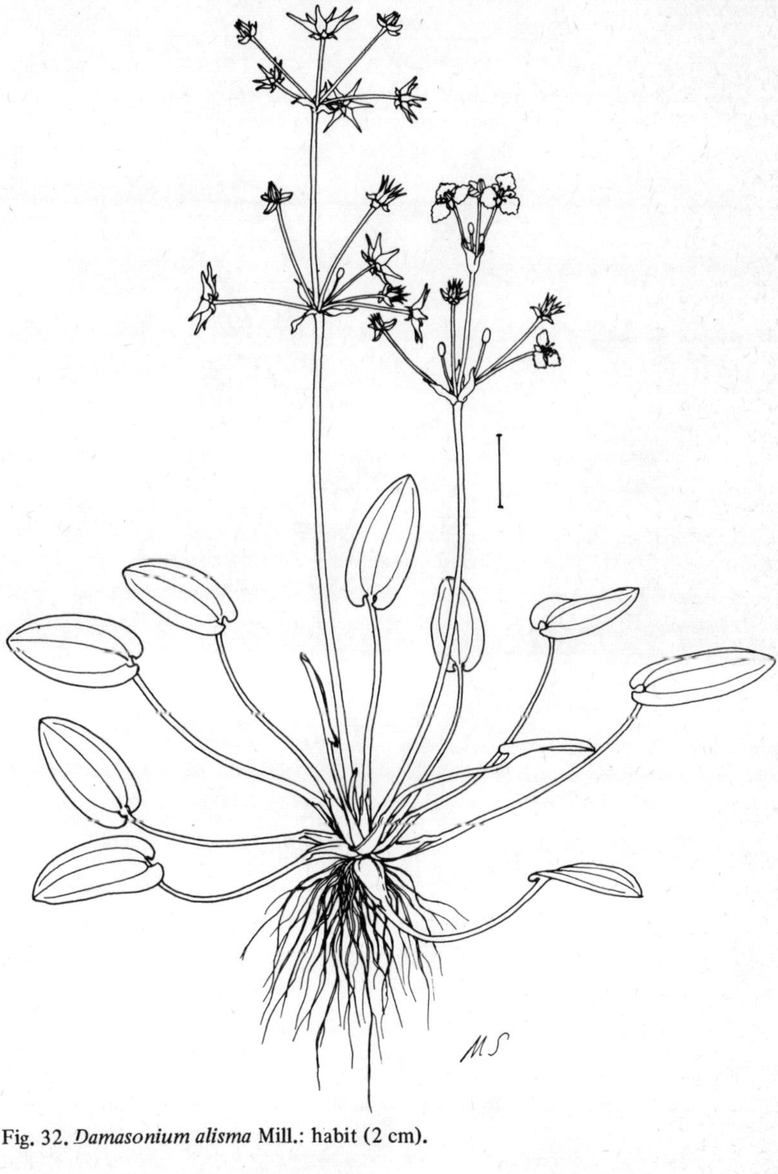

Fig. 32. *Damasonium alisma* Mill.: habit (2 cm).

centre. Stamens 6. Carpels 6 to 10, in 1 whorl, spreading stellately when ripe, long beaked, more or less united at base or adnate to the elongated receptacle; styles terminal; ripe carpels 1- to several-seeded, indehiscent or irregularly dehiscent at base.

c. 6 species: Europe, N. Africa, The Orient, Persia, S. Australia and W. N. America. Characteristically found in pools that dry out each year. The American species *D. californicum* Torr. is occasionally placed in the genus *Machaerocarpus.*

Echinodorus Richard ex Engelm. in A. Gray, Man. Bot. 460 (1848),[*Albidella* Pichon, *Helanthium* (Bentham and Hooker) Engelm. ex Britton]
Fig. 33.

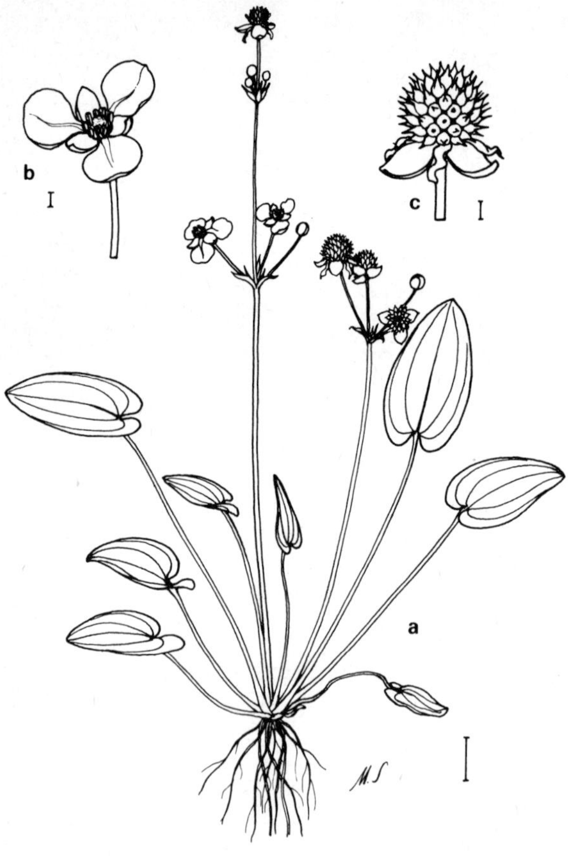

Fig. 33. *Echinodorus berteroi* (Spreng.) Fassett: a, habit (1 cm); b, flower (1 mm); c, fruit (1 mm).

Leaves submerged, floating or erect; juvenile leaves linear, without blade; mature leaves differentiated into blade and petiole; blade linear to broad-ovate; base attenuate to cordate. Inflorescence of several simple whorls or compound with branches bearing whorled flowers. Flowers bisexual. Stamens 9 to numerous (except for *E. nymphaeifolius* with 6). Carpels usually densely crowded on a conical receptacle (except for the *E. tenellus* group and *E. nymphaeifolius* which have flattened receptacles); style lateral to almost terminal, usually beaked in fruit; nutlets ribbed, often glandular.

Beal, E. O. Alismataceae of the Carolinas. Elisha Mitchell Sci. Soc. 76: 68–78 (1960)

Fassett, N. Echinodorus in the American tropics. Rhodora 57: 133–156, 174–188, 202–212 (1955)

Rataj, K. Taxonomic works on Echinodorus. Mitt. Bot. Staats. München 6: 613–619 (1967); Bull. Jard. Bot. Nat. Belg. 38: 401–408 (1968); Darwiniana 15: 183–188 (1969); Folia Geobot. Phytotax. Praha 4: 319–326, 331–336, 435–442 (1969); Preslia (Praha) 42: 264–266 (1970); Folia Geobot. Phytotax. Praha 5: 213–216 (1970); Preslia (Praha) 43: 10–16 (1971)

Schulze, J. Neue Echinodorus-Arten aus Südbrasilien. Die Aquar. u. Terr. Zeit. 21 (8): 244–248, 21 (9): 277–281, 21 (10): 309–312, 21 (11): 339–342 (1968)

c. 47 species: warmer regions of America. Emergent aquatics or marsh plants occurring in a wide variety of habitats. Many species are cultivated as decorative aquarium plants. In floral characteristics *E. nymphaeifolius* (Griseb.) Buch. resembles the Old World genus *Caldesia* but on vegetative characters resembles *Echinodorus.*

Limnophyton Miq., Fl. Ind. Bat. 3: 242 (1855)
Fig. 34A.
Leaves erect; leaf-blades glandular, with obtuse to rounded apex; base sagittate or cuneate. Inflorescence of 4 to 7 whorls of flowers, the lowest sometimes with 2 to 3 branches; upper whorls of male flowers, lower of bisexual and female flowers. Petals 3, white, larger than sepals. Stamens 6. Carpels 10 to 30, free, ovoid; style ventral; nutlets crowded on small receptacle, swollen with lateral air chambers, ridged.

3 species: *L. obtusifolium* (L.) Miq. Tropical Africa, Madagascar, India, Ceylon and the Malay Islands, *L. angolense* Buch. tropical Africa and *L. fluitans* Graebner tropical W. Africa. Found in swamps, pools, canals and irrigation ditches in still or slowly flowing water.

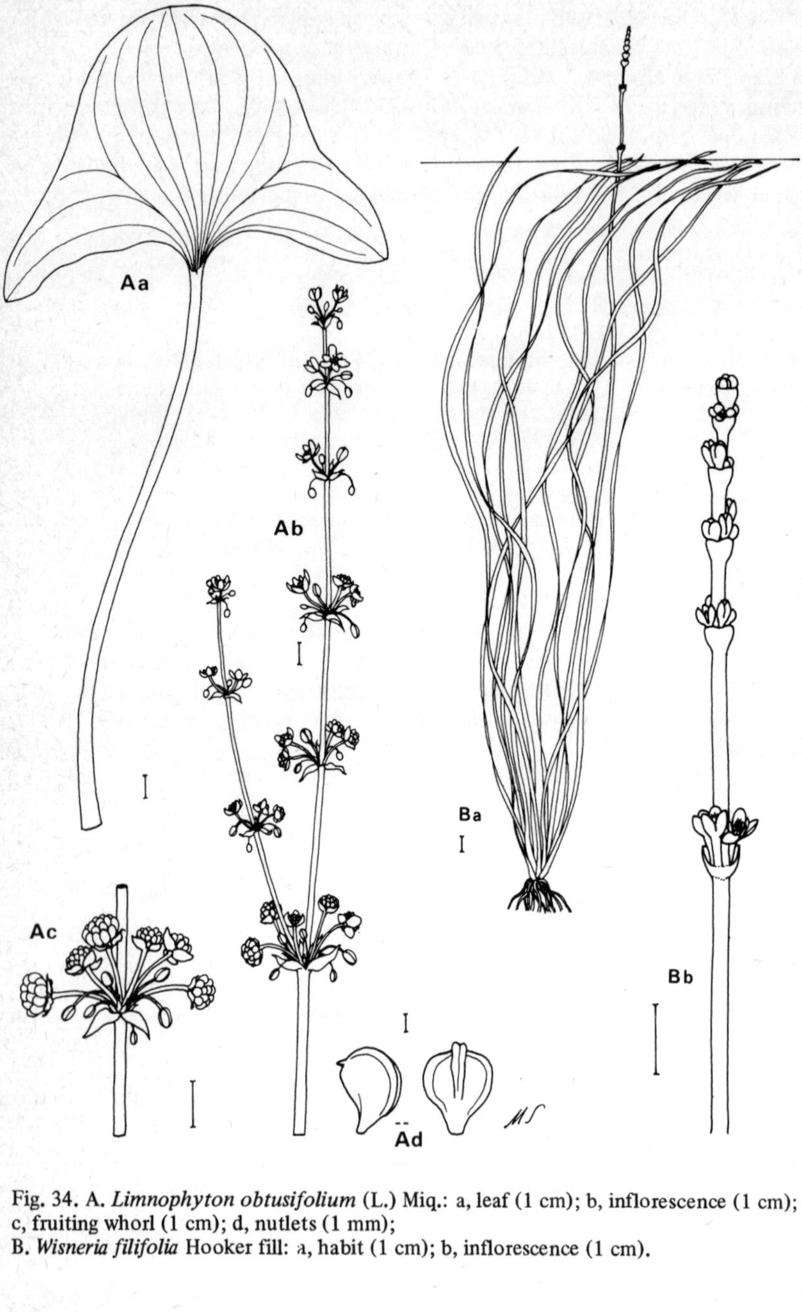

Fig. 34. A. *Limnophyton obtusifolium* (L.) Miq.: a, leaf (1 cm); b, inflorescence (1 cm); c, fruiting whorl (1 cm); d, nutlets (1 mm);
B. *Wisneria filifolia* Hooker fill: a, habit (1 cm); b, inflorescence (1 cm).

Luronium Rafin., Autikon Bot. 63 (1840), [*Elisma* Buch.]
Fig. 30C.
Perennial with stem creeping and rooting at nodes. Leaves submerged or floating; submerged leaves linear; floating leaves long stalked with blade 1 to 2.5 cm long, ovate to elliptic, rounded and obtuse at apex. Flowers 1 to 5 in axils of leaves, long stalked, bisexual. Petals 3, white with yellow spot in the centre, up to 10 mm long, longer than sepals. Stamens 6. Carpels 6 to 12, free, forming a hemispherical head; styles terminal; nutlets oblong-ovoid.

1 species, *L. natans* (L.) Rafin: C. and W. Europe. Found in a variety of aquatic habitats in usually acidic water. In Britain it is apparently increasing its range.

Ranalisma Stapf in Hooker, Ic. Pl. 27: t.2652 (1900)
Fig. 30B.
Stoloniferous herbs. Leaves erect or spreading; blade ovate to linear. Inflorescence umbel-like, 1- to 3-flowered with 2 membranous bracts. Flowers bisexual. Sepals persistent, reflexed at fruiting stage. Petals white, larger than sepals. Stamens (6–) 9 (–12). Carpels numerous, spirally arranged on a convex receptacle; style ventral; fruit a head of nutlets; nutlets strongly compressed laterally, glandular, winged, with terminal beak formed by remains of style.

2 species: *R. rostratum* Stapf Malay Peninsula and Indo China and *R. humile* (Kunth) Hutch. tropical Africa. Plants of seasonally inundated areas. *R. humile* is found as a weed in ricefields in W. Africa. *Ranalisma* is very similar to *Echinodorus* and is occasionally included within it.

Sagittaria L., Sp. Pl. 993 (1753), [*Lophotocarpus* Durand]
Fig. 35.
Perennials or occasionally annuals. Leaves erect, floating or submerged; blade of erect leaves linear to broadly ovate, often with sagittate base; blade of floating leaves linear to ovate, with cuneate to cordate base; submerged leaves rarely lanceolate, usually linear, flat or occasionally terete and spongy. Inflorescence erect or sometimes floating, with 1 to 12 whorls, each whorl usually 3-flowered, the upper usually male, the lower female or occasionally bisexual Petals 3, white, rarely pink or yellowish or white with purple spot near the base. Stamens 7 to numerous. Carpels numerous, free, spirally arranged on a convex receptacle; style central; nutlets laterally compressed, winged and beaked.

Adams, P. B. and Godfrey, R. K. Observations on the Sagittaria subulata complex. Rhodora 63: 247–266 (1961)

Bogin, C. Revision of the genus Sagittaria. Mem. New York Bot. Garden 9: 179–233 (1955)

Rataj, K. Revision of the genus Sagittaria. I, Old World species. Ann. Zool. Bot. Bratislava 76: 1–31 (1972); II, The species of the West Indies, Central and South America. op. cit. 78: 1–61 (1972)

Fig. 35. *Sagittaria sagittifolia* L.: a, habit (3 cm); b, inflorescence (1 cm).

Wooten, J. W. The monoecious and dioecious conditions in Sagittaria latifolia L. Evolution 25: 549–553 (1971)

20 or slightly more species: cosmopolitan with most species in the new world. Found in a wide variety of aquatic habitats. The tubers of *S. sagittifolia* L. and some other species are eaten, particularly by the Chinese who cultivate plants in many different parts of the world. The vegetative parts of *Sagittaria* are often fed to pigs. The fruits are eaten by water birds and fish.

Wisneria Micheli in DC., Mon. Phan. 3: 82 (1881)
Fig. 34B.
Leaves submerged or partly floating, linear to oblanceolate, up to 1 m long; blade and petiole often difficult to distinguish. Inflorescence emersed, spike-like, of unbranched whorls of sessile or sub-sessile flowers; upper whorls of male flowers, lower of female flowers; vegetative buds sometimes present in the lower whorls; bracts 3, united. Petals 3, much shorter than sepals or absent. Stamens 3. Carpels 3 to 6, free; styles terminal; nutlets ovate, swollen with lateral air-chambers.

3 species: *W. triandra* (Dalzell) Micheli, E. India, *W. filifolia* Hooker fil. and *W. schweinfurthii* Hooker fil. tropical Africa and Madagascar. In pools and shallow lakes.

AMARANTHACEAE

c. 60 genera: *c.* 850 species: cosmopolitan, most species in warmer regions. 2 genera (*Alternanthera* and *Centrostachys*) contain aquatics.

Annual or perennial herbs. Stems erect, creeping or floating (often swollen when floating), often rooting at internodes. Leaves opposite, entire, without stipules. Flowers small, in axils of thin bracts, each subtended by 2 membranous bracteoles. Perianth segments 3 to 5, free or nearly so, dry and membranous. Stamens (2–) 5, usually alternating with short pseudostaminodes; filaments united at base into a short tube. Ovary superior, 1-locular; ovule solitary, pendulous from a long funicle.

Cavaco, A. Les Amaranthaceae de l'Afrique au sud du tropique du Cancer et de Madagascar. Mém. Mus. Nat. Hist. Nat. N. S. sér. B, 13: 1–254 (1962)
Smith, L. B. and Downs, R. J. Resumo preliminar das Amarantaceas de Santa Catarina. Sellowia 12: 99–120 (1960)

1A Stems conspicuously grooved; outer perianth segment with recurved subulate tip; anthers 4-locular
Centrostachys
1B Stems not conspicuously grooved; outer perianth segment without recurved subulate tip; anthers 2-locular
Alternanthera

Alternanthera Forskål, Fl. Aegypt. 28 (1775), [*Achyranthes* L. pro parte, *Telanthera* R. Br.]
Fig. 36A.
Stems floating or ascending, when floating often swollen, not conspicuously grooved. Leaves lanceolate, ovate or obovate. Inflorescence axillary or rarely terminal, sessile or stalked heads or short spikes. Perianth segments 5, glabrous or hairy, equal or unequal, the outer one not terminating in a subulate tip. Stamens 2 to 5; anthers 2-locular.

Chodat, R. and Rehfous, L. La végétation du Paraguay. 14. Amarantacées. Bull. Soc. Bot. Genève, sér. 2, 18: 246–293 (1926)
Maheshwari, J. K. Alligator weed in Indian lakes. Nature, London. 206: 1270 (1965)
Penfound, W. T. The biology of Achyranthes philoxeroides (Mart.) Standley. Amer. Midl. Nat. 24: 248–252 (1940)

c. 170 species: in the warmer regions of the world, most species in tropical S. America; a few species are aquatic or subaquatic. *A. tetramera* R. E. Fries and *A. aquatica* (Parodi) Chodat are confined to S. America, *A. sessilis* (L.) DC. is pantropical and *A. philoxeroides* (Mart.) Griseb. (Alligator weed) is originally from S. America but has become naturalised in Asia, Australia and southern

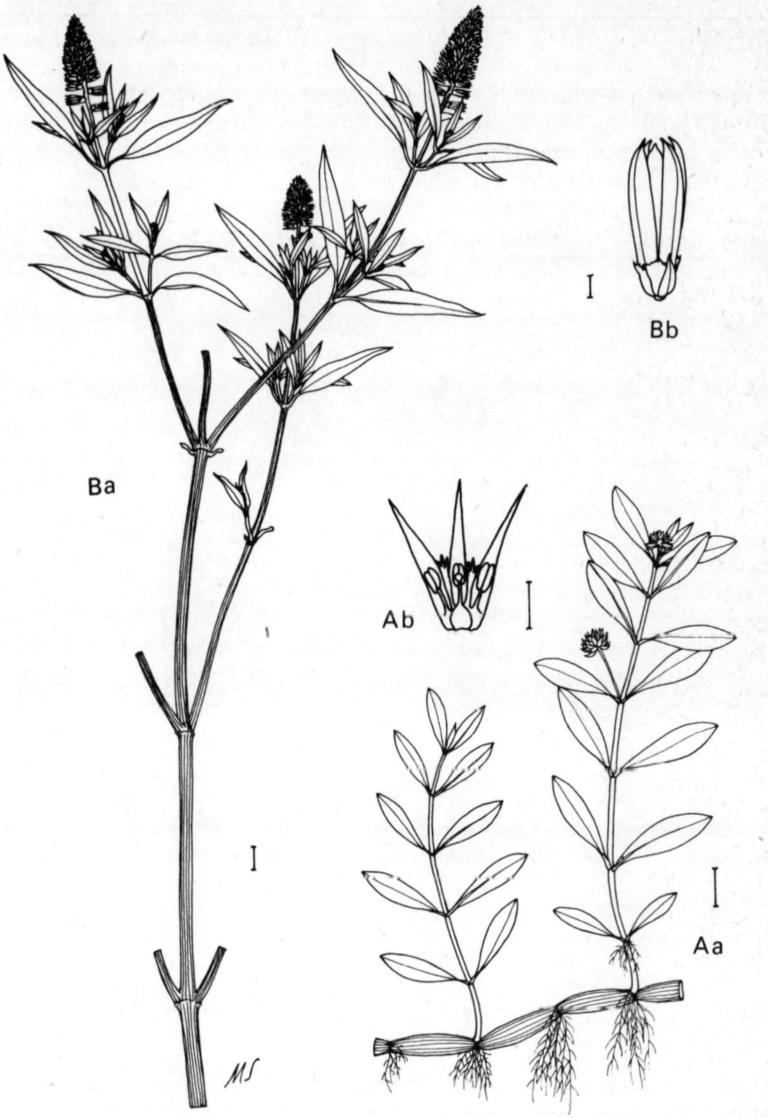

Fig. 36. A. *Alternanthera philoxeroides* (Mart.) Griseb.: a, habit (1 cm); b, flower (1 mm); B. *Centrostachys aquatica* (R.Br.) Wall.: a, habit (1 cm); b, section of flower (2.5 mm).

U.S.A. where it has frequently become a pest; it is difficult to control; see Little, E. C. S. Weed Res. 8: 79–105 (1968).

Centrostachys Wall. in Roxb., Fl. Ind. 2: 497 (1824), [*Achyranthes* L. pro parte]

Fig. 36B.

Stem floating or ascending, conspicuously grooved. Leaves lanceolate to ovate, cuneate at base, acute at apex, 7 to 18 cm long, 2 to 6 cm wide, pilose beneath with appressed white hairs when young. Inflorescence an erect spike, usually terminal, 7 to 45 cm long. Flowers at first crowded, later remote. Perianth segments 5, glabrous, unequal, 5 to 8 mm long, the outer longest one terminating in a firm, slightly recurved subulate tip. Stamens 5; anthers 4-locular.

1 species: *C. aquatica* (R. Br.) Wall. [*Achyranthes aquatica* R. Br.]. Found from tropical Africa to S. E. Asia in swampy or inundated localities.

AMARYLLIDACEAE

85 genera; *Crinum* has aquatic members.

Crinum L., Sp. Pl. 291 (1753)
Fig. 37.

Fig. 37. *Crinum defixum* Ker.: a, habit; b, inflorescence.

Perennial herb with tunicated, bulbous rootstock. Leaves basal, linear with parallel nerves, flat or terete. Flowers showy, pink to white, few to many in an umbel, subtended by a thick involucre which splits into two spathes. Perianth inserted above the ovary, tubular at the base, divided into 6 segments above, without a corona. Stamens 6, inserted at the mouth of the tube opposite the segments, free to the base; anthers versatile. Ovary sessile or subsessile, 3-locular; ovules few.

Baker, G. P. A synopsis of the known species of Crinum, Gard. Chron. 2, 15: 763, 786; 16: 39, 72, 180, 398, 495, 588, 760 (1881)

Nieuwenhuizen, A. von den. Einige Bemerkungen zum Biotop von Crinum natans Baker und Bolbitis heudelotii (Bory ex Feé) Alston in Liberia. Aquaterra 9 (7 + 8): 72–79 (1972) See also: Pryke, K. E. Journ. Roy. Hort. Soc. 91, 3: 130 (1966)

c. 110 species: pantropical. About 10 species are aquatic, notably *C. natans* Baker in tropical Africa and India, and *C. erubescens* Soland. in tropical S. America.

APIACEAE [UMBELLIFERAE]

c. 275 genera: *c.* 2850 species: cosmopolitan, chiefly N. Temperate. 15 genera contain aquatics.

Herbs. Stems usually hollow. Leaves often heterophyllous, alternate, usually compound, sometimes simple. Flowers small, actinomorphic, bisexual unisexual, or sterile, in simple or compound umbels, sometimes in heads, or occasionally proliferous; umbels usually subtended by bracts. Sepals 5, conspicuous, inconspicuous, or absent, adnate to ovary. Petals 5, usually with inflexed tip. Stamens 5, inserted on disc above ovary; anthers versatile. Ovary inferior, 2-locular; 1 ovule in each loculus; styles 2, united and often swollen at base; fruit dry, of 2, 1-seeded carpels, each carpel (mericarp) contiguous along the inner surface (commissure), with usually 5 longitudinal ribs on the exposed outer (dorsal) surface; dark oil tubes frequently found between ribs on dorsal surface and also on commissural side.

1A Fruits variously covered with scales or papillae; flowers sessile or subsessile, in hemispherical to cylindrical heads
Eryngium

1B Fruits without scales or papillae; flowers stalked, not in heads but usually in more or less flat-topped umbels

- 2A Umbels simple; all stems creeping, rooting at most nodes
  - 3A Leaves reduced to linear or spathulate, septate leaf stalks
    Lilaeopsis
  - 3B Leaves with entire or palmately lobed blades
    - 4A Mericarps 3-ribbed, without transverse nerves; leaves stipulate
      Hydrocotyle
    - 4B Mericarps 7- to 13-ribbed, with transverse nerves; leaves without stipules
      Centella
- 2B Umbels compound; at least some stems ascending or erect, rooting at lower nodes
  - 5A Leaves appearing whorled (1-pinnate, with leaflets palmately divided into horizontally spreading, capillary segments)
    Carum
  - 5B Leaves not appearing whorled (leaflets not palmately divided into horizontally spreading, capillary segments)
    - 6A Stem leaves palmately divided, with 3 or 5 leaflets; fruit beaked
      Cynosciadium
    - 6B Stem leaves not palmately divided; fruit not beaked
      - 7A Fruit strongly flattened dorsally (widest at commissure) with prominent lateral wings
        Oxypolis

7B Fruit not strongly flattened dorsally, without prominent lateral wings
- 8A Ribs of fruit wide and corky or obscured by corky pericarp in plants lacking bracts; fruit not constricted at commissure
  - 9A Robust biennials or perennials (widespread) **Oenanthe**
  - 9B Slender annuals (S. and E. N. America)
    - 10A Petals with inflexed tips; fruit not flattened dorsally **Ptilimnium**
    - 10B Petals flat, without inflexed tip; fruit slightly flattened dorsally **Limnosciadium**
- 8B Ribs of fruit not corky, fine and prominent or obscured by corky pericarp in plants with bracts; fruit constricted at commissure.
  - 11A Bracts large and conspicuous, occasionally leaf-like
    - 12A Basal rosette of leaves lacking; lower part of stem creeping and rooting at internodes
      - 13A Bracts divided or toothed; ribs of fruit obscured by corky pericarp; base of style conical **Berula**
      - 13B Bracts simple and entire; ribs of fruit prominent; base of style flat **Apium**
    - 12B Basal rosette of leaves present; stem erect and rooting at base (thin stolons may be present)
      - 14A Leaflets of stem leaves linear or spathulate, not serrate; lower leaves simple or 1-pinnate **Thorella**
      - 14B Leaflets of stem leaves lanceolate to ovate, serrate; lower leaves never simple, usually more than 1-pinnate **Sium**
  - 11B Bracts inconspicuous or absent
    - 15A Stem creeping, rooting at nodes; sepals minute and inconspicuous or absent **Apium**
    - 15B Stem erect, rooting at base; sepals conspicuous and persistent in fruit **Cicuta**

Apium L., Sp. Pl. 264 (1753), [*Helosciadium* Koch]
Fig. 38A.
Stems floating, creeping, or ascending; rooting at most nodes. Leaves 1-pinnate or trifoliate; leaflets suborbicular to capillary, entire, lobed or divided into linear or capillary segments. Umbels compound (some umbels occasionally

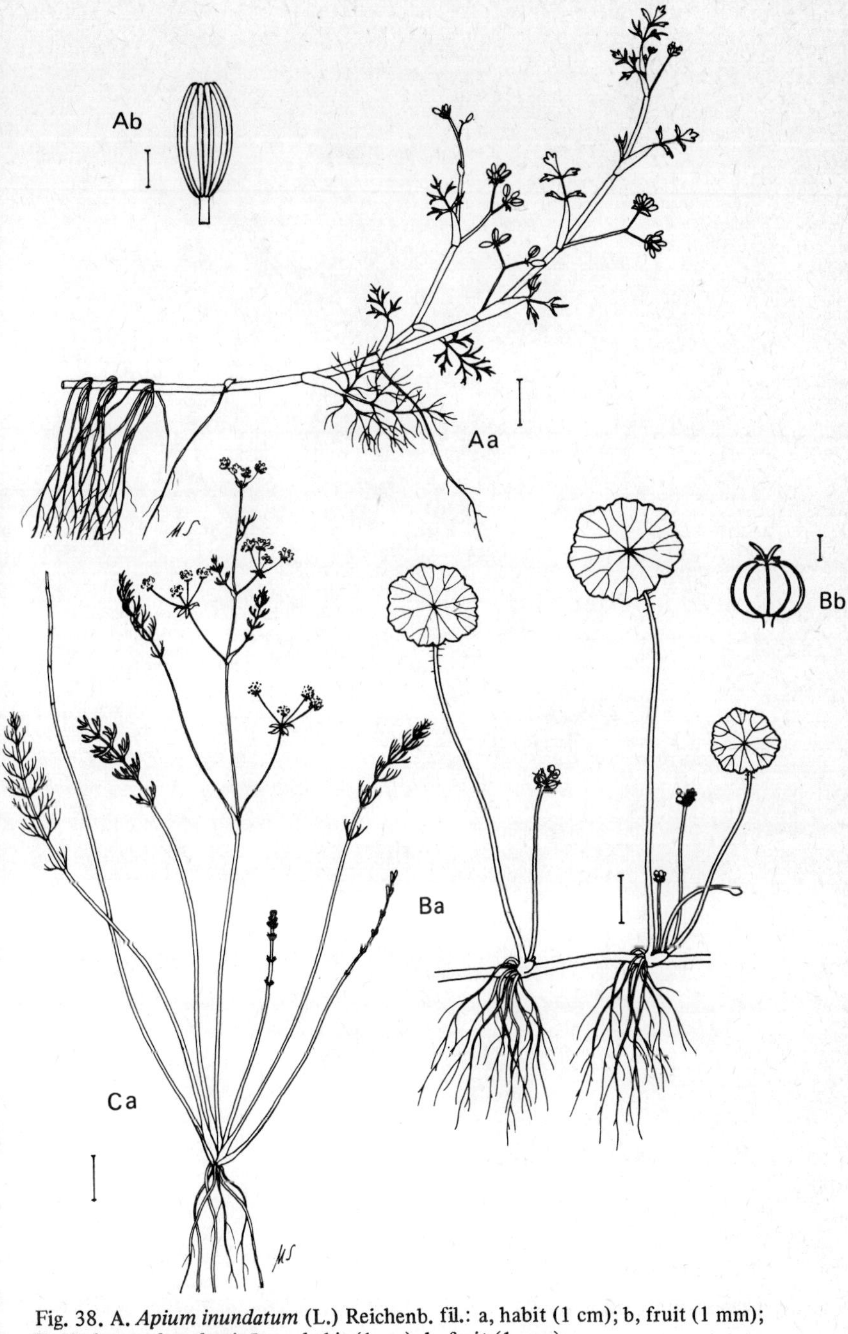

Fig. 38. A. *Apium inundatum* (L.) Reichenb. fil.: a, habit (1 cm); b, fruit (1 mm);
B. *Hydrocotyle vulgaris* L.: a, habit (1 cm); b, fruit (1 mm);
C. *Thorella verticillatinundata* (Thore) Briq.: a, habit (1 cm).

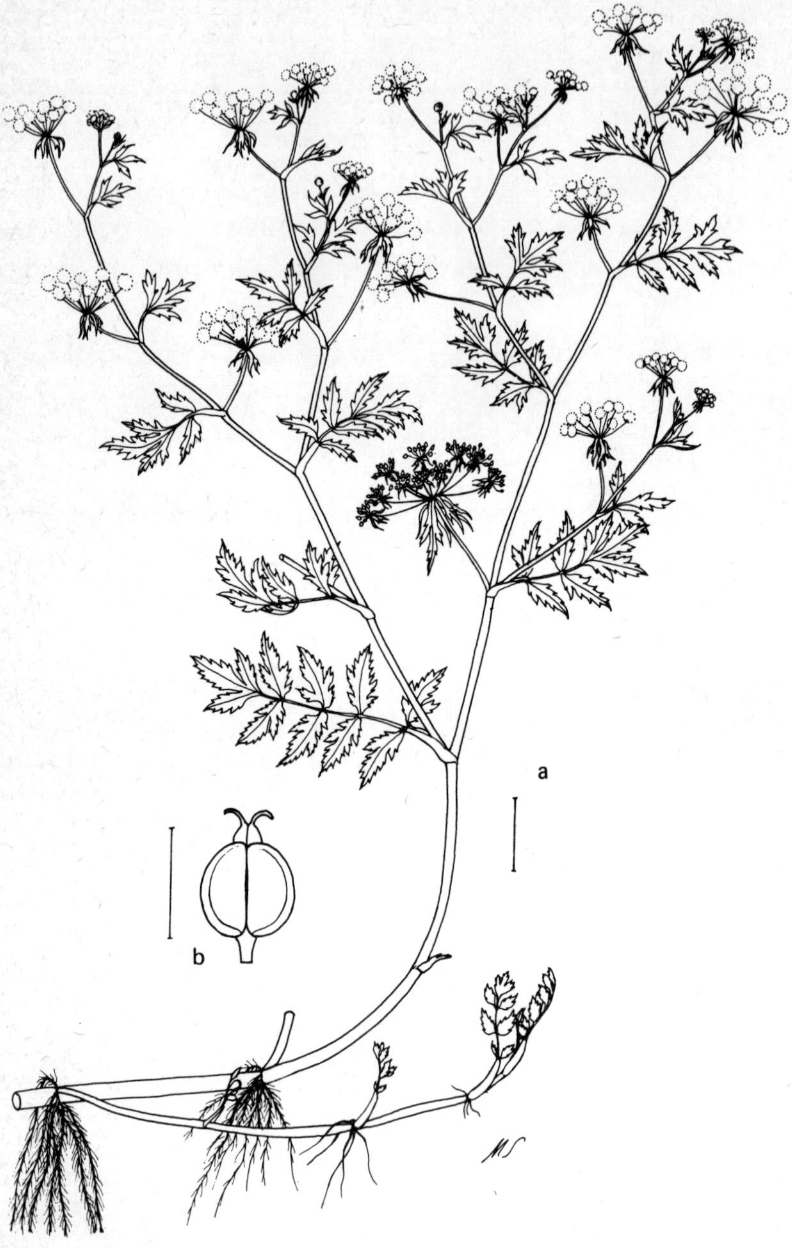

Fig. 39. *Berula erecta* (Hudson) Coville: a, habit (2 cm); b, fruit (1 mm).

simple), sessile or shortly stalked, opposite leaves; bracts absent or few, when present conspicuous and entire. Sepals minute or absent. Fruit oblong-ovoid to globose or ellipsoidal, flattened dorsally, somewhat constricted at the commissure; base of style flat; ribs prominent, subequal; oil tubes solitary between ribs.

Glück, H. Biologische und morphologische Untersuchungen über Wasser- und Sumpfgewächse. Jena 3: 445–461 (1911)
Wolff, H. in Engler, Pflanzenreich 90 (IV.288): 26–58 (1927)

c. 30 species: cosmopolitan. The aquatic or amphibious species (*c.* 4) are found in the section *Helosciadium* (Koch) Bab. These species are found in usually shallow water in pools and streams. They are confined to Europe and the Mediterranean regions with the exception of *A. nodiflorum* (L.) Lag. which is also found in Asia and has become naturalised in N. and S. America.

Berula Hoffm. in Besser, Enum. Pl. Volhyn. 44 (1822)
Fig. 39.
Stems floating, creeping or ascending, stolons frequent. Emergent leaves 1-pinnate, oblong-lanceolate to ovate, serrate (teeth frequently white in *B. thunbergii* (DC.) Wolff); submerged leaves similar to emergent ones except that the leaflets are frequently lobed. Umbels compound, stalked, opposite leaves; bracts numerous and often leaf-like, pinnatisect or ternate. Sepals inconspicuous, subulate. Fruit divided down the middle so that the mericarps are more or less separate at maturity, ovoid to globose, flattened dorsally; base of style conical; ribs not prominent, obscured by corky pericarp, fine; oil tubes deeply imbedded, numerous.

Glück, H. Biologische und morphologische Untersuchungen über Wasser- und Sumpfgewächse. Jena 3: 123–130 (1911)
Wolff, H. in Engler, Pflanzenreich 90 (IV. 288): 336–341 (1927)

2 species: *B. erecta* (Hudson) Colville, temperate and subtropical N. Hemisphere, *B. thunbergii* (DC.) Wolff, E. and S. Africa. Both species are usually found in shallow, clear, flowing water.

Carum L., Sp. Pl. 263 (1753)

20 to 30 species; Old World, N. Hemisphere, *C. carvi* L. widely introduced elsewhere. *C. verticillatum* (L.) Koch is the only species found as an aquatic.

C. verticillatum (L.) Koch
Fig. 40A.
Stems erect, up to 120 cm high, little branched, almost leafless. Basal leaves

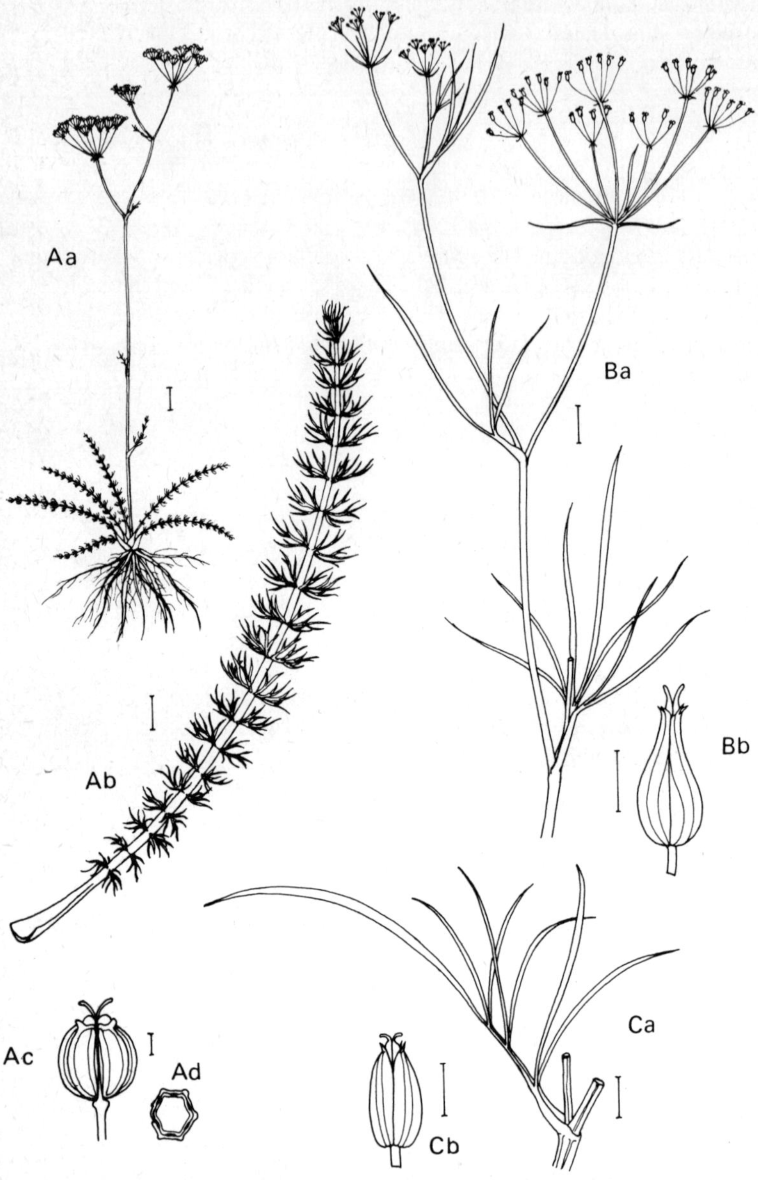

Fig. 40. A. *Carum verticillatum* (L.) Koch: a, habit (5 cm); b, leaf (1 cm); c, fruit; d, transverse section of fruit (1 mm);
B. *Cynosciadium digitatum* DC.: a, inflorescence (1 cm); b, fruit (1 mm);
C. *Limnosciadium pinnatum* (DC.) Mathias and Constance: a, stem leaf (1 cm); b, fruit (1 mm).

10 to 25 cm long, narrowly oblong in outline, 1-pinnate, with usually more than 20 pairs of deeply palmatisect leaflets; leaflet segments up to 10 mm long, capillary, appearing whorled; stem leaves few, small, reduced. Umbels compound, terminal; bracts linear-acuminate, reflexed. Sepals inconspicuous or absent. Fruit ellipsoidal, flattened dorsally; constricted at commissure; ribs, prominent, fine; oil tubes solitary between ribs.

Glück, H. Biologische und morphologische Untersuchungen über Wasser- und Sumpfgewächse. Jena 3: 328–332 (1911)
Wolff, H. in Engler, Pflanzenreich 90 (IV. 288): 143–167 (1927)

C. verticillatum, W. Europe from W. Spain northward to Scotland and the Netherlands. It is found in wet grassy places, marshes, ditches and pools submerged in winter and emergent or terrestrial in summer. It shows a preference for substrates poor in calcium.

Centella L., Pl. Afr. Rar. 28 (1760)
Fig. 41.
Stems creeping with long stolons, hairy when young. Leaves in rosettes, distinctly stalked, simple, reniform; margin crenate or crenate-dentate; stipules absent. Umbels simple, solitary or 2 to 5 together in the axils of *c.* 3 mm long bracts. Flowers usually 3, middle one sessile, lateral ones stalked. Petals red. Fruit ovoid to orbicular, flattened laterally; mericarps 7 to 13 ribbed, ribs connected by transverse veins; oil tubes absent.

c. 20 species of which only *C. asiatica* (L.) Urb. is frequently found in water. *C. asiatica* is found in the warmer regions of the world and grows in a variety of habitats including shallow water in streams, ditches, rice fields and pools.

Cicuta L., Sp. Pl. 255 (1753)
Fig. 42.
Stems erect, branching; base enlarged, ovoid to cylindrical, septate. Leaves 1- to 3-pinnate; leaflets linear to lanceolate, margin serrate, base asymmetrical Umbels compound, lateral and terminal; bracts inconspicuous or absent. Sepals conspicuous. Fruit ovoid to globose, slightly flattened dorsally; constricted at commissure; ribs prominent, corky, wide, subequal; oil tubes solitary between ribs.

Glück, H. Biologische und morphologische Untersuchungen über Wasser- und Sumpfgewächse. Jena 3: 103–106 (1911)
Mathias, M. E. and Constance, L. A synopsis of the American species of Cicuta. Madroño 6 (5): 145–151 (1942)
Wolff, H. in Engler, Pflanzenreich 90 (IV. 288): (75–86) (1927)

c. 10 species: N. Temperate Zone. Emergent herbs which grow in shallow, still or flowing water. *C. virosa* L. (Cowbane) is extremely poisonous and is a danger to grazing animals.

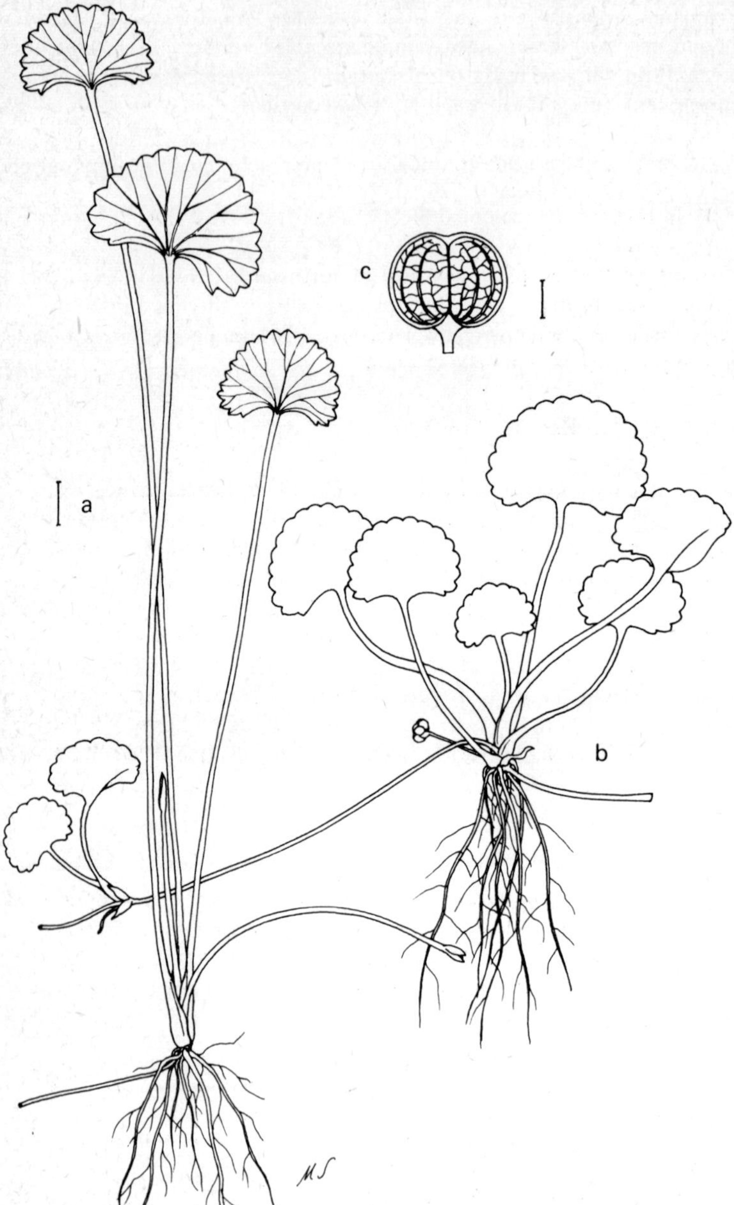

Fig. 41. *Centella asiatica* (L.) Urb.: a, aquatic habit (1 cm); b, terrestrial habit; c, fruit (1 mm).

Fig. 42. *Cicuta virosa* L.: a, inflorescence (2 cm); b, leaf (2 cm); c, fruit (1 mm).

Cynosciadium DC., Col. Mém. 5: 44 (1829)
Fig. 40B.

Annuals. Stems erect, branching up to 15 cm tall. Basal leaves entire, septate, linear-lanceolate; stem leaves palmate, segments linear. Umbels compound, terminal and lateral; bracts absent or few and linear. Sepals prominent. Fruit ovoid to urn-shaped, beaked at apex, rounded at base, 2 to 3 mm long, 1.5 to 2.5 mm wide; not constricted at commissure; base of style conical; ribs prominent, corky, laterals slightly larger than dorsals; oil tubes solitary between ribs.

1 species, *C. digitatum* DC.: Southern N. America, from E. Texas to S. Missouri. Amphibious herb.

Fig. 43. *Eryngium corniculatum* Lam.: a, habit of juvenile plant (1 cm); b, inflorescence (1 cm).

Eryngium L., Sp. Pl. 232 (1753)
Fig. 43.

Stems creeping, ascending, erect, or reduced to a tuber. Leaves very variable, entire, pinnately, or palmately lobed to divided, or simple, or reduced to linear septate leaf stalks; margins often spiny. Flowers usually sessile or subsessile in hemispherical to cylindrical heads; the heads solitary, or in cymes or racemes; bracts conspicuous, entire, or lobed. Sepals conspicuous, rigid. Fruit ovoid to globose, variously covered with scales or papillae; ribs inconspicuous; oil tubes inconspicuous.

Glück, H. Biologische und morphologische Untersuchungen über Wasser- und Sumpfgewächse. Jena 3: 338–403 (1911)
Mathias, M. E. and Constance, L. A synopsis of the North American species of Eryngium. Amer. Midl. Nat. 25: 361–387 (1941)
Wolff, H. in Engler, Pflanzenreich 61 (IV. 228): 106–305 (1913)

c. 230 species: almost cosmopolitan but lacking in tropical and S. Africa, most species in the Americas. Several species (tropical and temperate) are amphibious and are characteristically found in seasonally inundated regions.

Hydrocotyle L., Sp. Pl. 234 (1753)
Fig. 38B.
Stems floating or creeping; rooting at most nodes. Leaves distinctly stalked, simple, peltate or with basal sinus, orbicular to reniform, entire, toothed or lobed; stipules present. Umbels simple, sometimes proliferous, rarely an interrupted spike, axillary, usually inconspicuous. Fruit ovoid-ellipsoid to suborbicular, strongly flattened dorsally; mericarps with 3, subequal ribs; oil tubes absent or in main ribs only; fruit wall with a woody inner layer.

Mathias, M. E. The genus Hydrocotyle in northern South America. Brittonia 2: 201–237 (1936)
Péréz-Moreau, R. A. Revision de las "Hydrocotyle" Argentinas. Lilloa 2: 413–463 (1938)
Tseng, C. C. Anatomical studies of flower and fruit in Hydrocotyloideae (Umbelliferae). Univ. Calif. Publ. Bot. 42: 1–58 (1967)

c. 100 species: cosmopolitan. Many species are found in seasonally inundated regions. Some species, such as *H. ranunculoides* L. fil. (sombrerito de agua) form floating mats of vegetation and have been reported to be a serious nuisance particularly in C. America.

Lilaeopsis E. L. Greene, Pittonia 2: 192 (1821), [*Crantzia* Nutt.]
Fig. 44.
Stems creeping or floating; rooting at most nodes. Leaves erect, reduced to simple, septate leaf stalks, entire, subulate, linear or spathulate. Umbels simple, few-flowered, in axils of and shorter than leaves; bracts few and small, or absent. Sepals small. Fruit globose or ovoid, slightly flattened dorsally or terete; lateral ribs thick and corky; dorsal ribs smaller and thinner; oil tubes solitary between ribs.

Hill, A. W. The genus Lilaeopsis. Journ. Linn. Soc. London Bot. 47: 525–551 (1927)
Péréz-Moreau, R. A. Sinopsis de las Umbeliferas Argentinas del género "Lilaeopsis". Lilloa 1: 283–306 (1937)

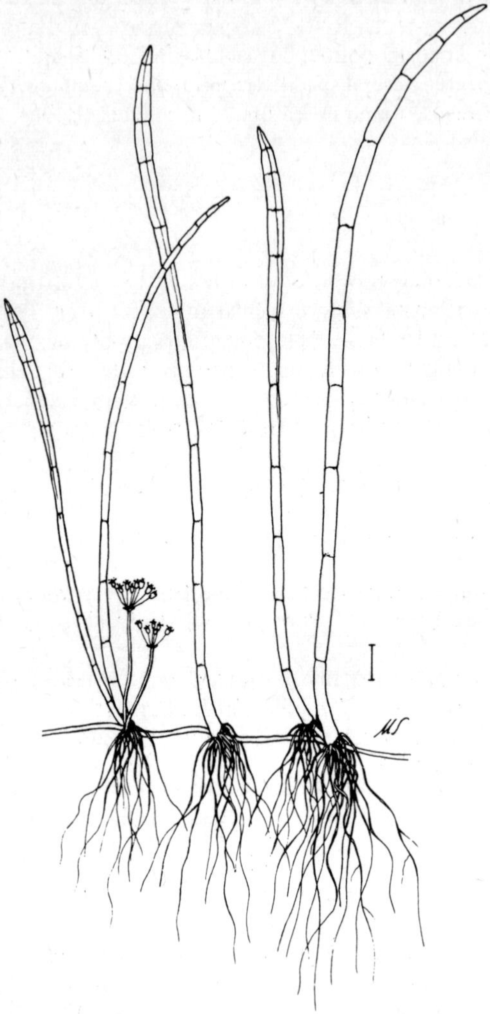

Fig. 44. *Lilaeopsis attenuata* (Hooker & Arnott) Fernald: habit (1 cm).

14 to 20 species: N. and S. America, Australia and New Zealand. *L. attenuata* (Hooker and Arnott) Fernald has become naturalised in Portugal. Most species are found on the shores of or submerged in rivers and lakes; some species are found in brackish water. Deeply submerged plants usually remain sterile (*L. lacustris* A. W. Hill is found in water up to 4 m deep in New Zealand).

Limnosciadium Math. and Const., Amer. Journ. Bot. 28: 162 (1941), [*Cynosciadium* DC. pro parte]

Fig. 40C.

Annuals. Stems procumbent to erect. Leaves pinnate, simple or reduced to septate, linear, leaf stalks; leaf segments linear. Umbels compound, terminal and lateral, occasionally sessile; bracts inconspicuous or conspicuous and entire. Sepals prominent. Petals oval to obovate, flat, without inflexed tip. Fruit oblong-oval to orbicular, rounded at base and apex, slightly flattened dorsally; not constricted at commissure; base of style conical; ribs prominent, dorsals thin, laterals thicker and corky; oil tubes solitary between ribs.

2 species: Southern N. America, from Texas to S. Missouri. Amphibious herbs.

Oenanthe L., Sp. Pl. 254 (1753)

Fig. 45.

Stems floating, ascending or erect; internodes often swollen; roots often swollen or bearing tubers. Leaves pinnatisect, 1- to 3-pinnate, or occasionally reduced to subulate leaf stalks, variable in shape; submerged leaves usually divided into capillary segments. Umbels compound, terminal or lateral; bracts inconspicuous, narrow, entire, or absent. Sepals conspicuous, acute, usually persistent in fruit. Fruit cylindrical, ovoid obconical or globose, terete, or slightly flattened dorsally; not constricted at commissure; base of style flat, conical or subglobose; ribs wide and corky or inconspicuous; oil tubes solitary between ribs.

Glück, H. Biologische und morphologische Untersuchungen über Wasser- und Sumpfgewächse. Jena 3: 412–444 (1911)

30 to 40 species: Temperate N. Hemisphere, S. Africa and mountains in Tropical Africa. *O. fluviatilis* (Bab.) Coleman from W. Europe and *O. aquatica* (L.) Poiret from Eurasia are species that spend the greater part of their generative history submerged. Several other species, such as *O. fistulosa* L., *O. lachenalii* C. C. Gmelin and *O. sarmentosa* Presl. although usually aquatic, may be found growing terrestrially in wet regions. The majority of the species are terrestrial.

Fig. 45. *Oenanthe fistulosa* L.: habit (2 cm).

Oxypolis Rafin., Neog. 2 (1825) [*Tiedmannia* DC.]
Fig. 46A.
Stems erect or ascending. Leaves 1-pinnate, or trifoliate, or occasionally reduced to septate leaf stalks; leaflets lanceolate to orbicular; margins toothed. Umbels compound, terminal or lateral; bracts few, narrow, entire, or absent. Sepals conspicuous or inconspicuous. Fruit ovoid to obovoid, strongly flattened dorsally; dorsal ribs very narrow; lateral ribs broadly winged, the wings nerved at inner margin, thus giving the appearance of 5 narrow dorsal ribs; oil tubes large, solitary between ribs.

c. 7 species: N. America. Most species are found in wet regions and may be found submerged during seasonal inundations. *O. occidentalis* Coulter and Rose (N. W. America) is occasionally found permanently submerged.

Ptilimnium Rafin., Amer. Month. Mag. 4: 192 (1819), [*Discopleura* DC.]
Fig. 46B.
Annuals. Stems erect, branching. Leaves 1-, 2- or occasionally 3-pinnate or reduced to linear, septate leaf stalks; leaf segments linear, capillary. Umbels compound, terminal and lateral; bracts conspicuous or inconspicuous, entire and linear or pinnate with capillary segments. Sepals small or conspicuous. Fruit ovoid to suborbicular, not flattened dorsally; not constricted at commissure; baes of style conical, ribs prominent, corky, laterals larger than dorsals; oil tubes solitary between ribs.

6 species: S. and E. N. America, from Texas to New England. Amphibious herbs in shallow water in swamps and rivers.

Sium L., Sp. Pl. 251 (1753)
Fig. 47.
Stems usually erect, occasionally creeping; rooting at lower nodes only. Submerged leaves when present, 1- to 5-pinnate; segments linear and usually entire; emergent leaves 1 pinnate or occasionally trifoliate; leaflets in pairs, lanceolate to ovate lanceolate, distinctly and sharply serrate. Umbels compound, lateral and terminal, long stalked; bracts conspicuous, numerous, persistent. Sepals inconspicuous or occasionally absent. Fruit ovoid to globose, slightly flattened dorsally, somewhat constricted at the commissure; base of style flat; ribs prominent, subequal, corky; oil tubes superficial, 1 to 3 between ribs.

Glück, H. Biologische und morphologische Untersuchungen über Wasser- und Sumpfgewächse. Jena 3: 403–412 (1911)
Wolff, H. in Engler, Pflanzenreich 90 (IV. 228): 341–358 (1927)

10 to 15 species: cosmopolitan except in S. America. Most species are found in wet areas and can tolerate flooding. Some species, such as *S. latifolium* L., have a submerged phase in their generative history and develop specialised underwater leaves. *S. sisarum* L. is cultivated for its edible roots.

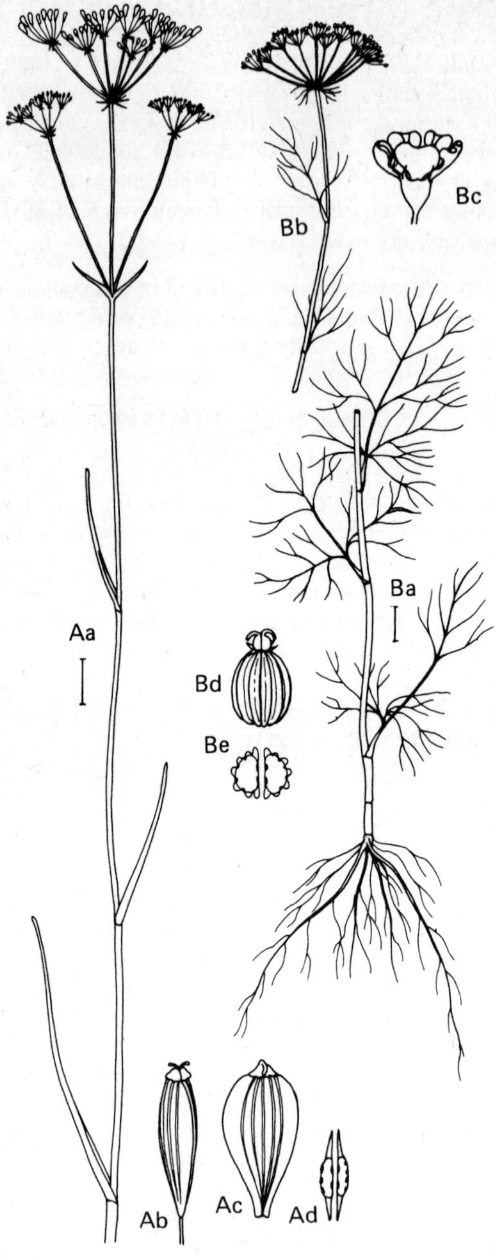

Fig. 46. A. *Oxypolis filiformis* Britton: a, habit (2 cm); b, fruit, side view; c, face view; d, transverse section;
B. *Ptilimnium nuttallii* (DC.) Britton: a, stem base (1 cm); b, stem apex; c, flower; d, fruit, face view; e, transverse section.

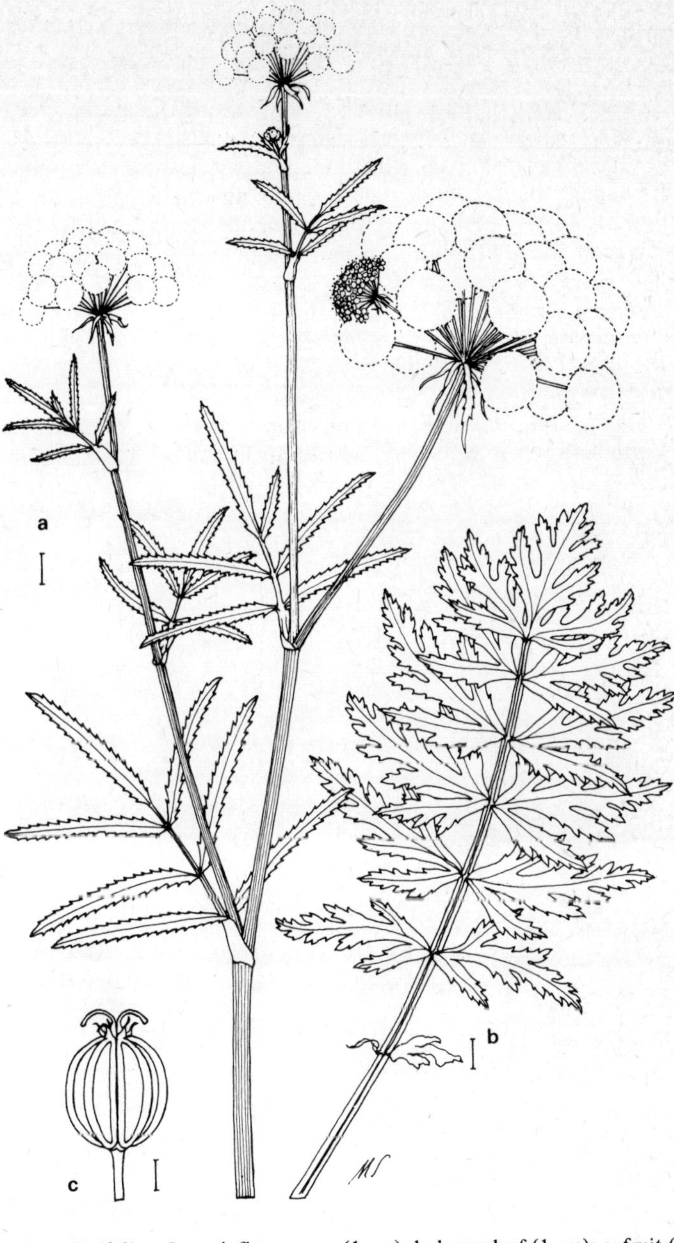

Fig. 47. *Sium latifolium* L.: a, inflorescence (1 cm); b, lower leaf (1 cm); c, fruit (1 mm).

Thorella Briq., Annu. Cons. Jard. Bot. Genève 17: 274 (1914), [*Ptychotis thorei* Godron and Gren.]

1 species; *T. verticillatinundata* (Thore) Briq.
Fig. 38C.
Stems erect or ascending, up to 20 cm high. Leaves mostly basal, the first usually reduced to septate, subulate leaf stalks; the others 1- or 2-pinnate; leaflets short, linear or spathulate, sometimes pinnatisect or ternate. Umbels compound, terminal; bracts conspicuous, entire or divided. Sepals inconspicuous. Fruit ovoid, *c.* 2 mm long, flattened dorsally; constricted at commisure; base of style flat; ribs prominent; oil tubes solitary between ribs.

Glück, H. Biologische und morphologische Untersuchungen über Wasser- und Sumpfgewächse. Jena 3: 333–337 (1911)
Wolff, H. in Engler, Pflanzenreich 90 (IV. 228): 136–137 (1927)

T. verticillatinundata, S. W. and W. C. France and W. Portugal. Found in sandy areas that are underwater in winter but which usually dry out in summer.

APONOGETONACEAE

1 genus.

Aponogeton L. fil.., Suppl. 32 (1781) nom. cons.
Fig. 48.

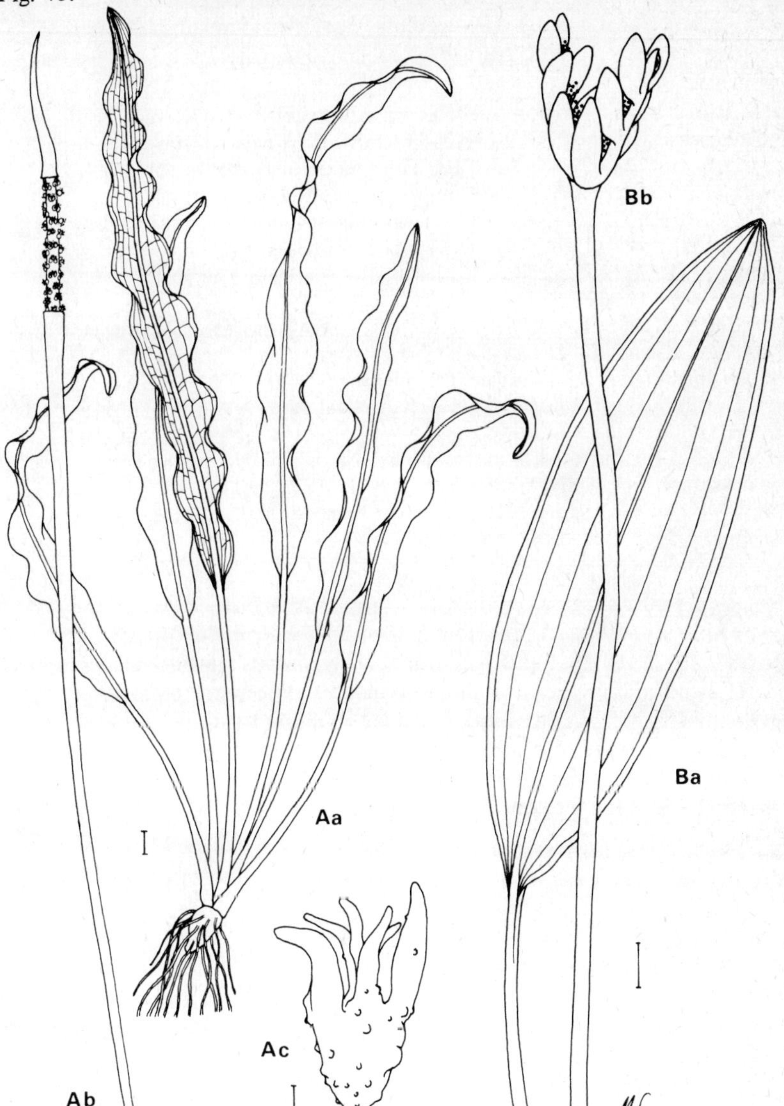

Fig. 48. A. *Aponogeton appendiculatus* Bruggen: a, habit (1 cm); b, inflorescence; c, ovary (1 mm);
B. *Aponogeton distachyon* L. fil.: a, leaf; b, inflorescence (1 cm).

Perennials. Stems tuberous or rhizomatous. Leaves linear or stalked; blades linear to oblong-elliptic; margin entire or crisped. Flowers bisexual or rarely unisexual borne in spikes; spikes stalked, simple or up to 10-forked, emerging above the water surface, in bud enveloped by a caducous or persistent spathe. Perianth segments (0 to 1) 2 (3 to 6), petal-like or bract like, mostly persistent, rarely caducous. Stamens 6 or more. Ovary superior; carpels (2 to) 3 (4 to 9), free, mostly sessile; each ovary narrowed into a style with an adaxial stigmatic ridge; ovules 1 to 8 per carpel; fruits dry, opening adaxially.

Bruggen, H. W. E. van. Revision of the genus Aponogeton. I. The species of Madagascar. Blumea 16: 243–263 (1968); II. A new species from India. op. cit. 16: 264–265 (1968); III. The species of Australia. op. cit. 17: 121–137 (1969); IV. The species of Asia and Malesia. op. cit. 18: 457–486 (1970); V. New data on Aponogeton tenuispicatus v. Bruggen. op. cit. 18: 487 (1970); VI. The species of Africa. Bull. Jard. Bot. Nat. Belg. 43(1–2): 193–233 (1973); see also articles in Het Aquarium (den Haag). (1967–1971)

Guillarmod, A. J. and Marais, W. A new species of Aponogeton (Aponogetonaceae). Kew Bull. 27 (3): 563–565 (1972)

Krause, K. and Engler, A. in Engler, Pflanzenreich 24 (IV. 13): 1–24 (1960)

Raynal, J. Un nouvel Aponogeton du Tchad. Adansonia sér. 2, 9 (4): 549–551 (1969)

Serguéeff, M. Contribution à la morphologie et à la biologie des Aponogetonacées. Thésis Inst. Bot. Univ. Genève sér. 7, 8: 1–132 (1907)

Toupin, G. Plantae Africanae. III. Aponogetonaceae (Nouvelle famille pour le Congo Belge). Bull. Jard. Bot. Etat. Brux. 23: 223–226 (1953)

c. 45 species: warmer regions of the Old World, most species in Africa and Madagascar; *A. distachyon* L. fil. from S. Africa has become naturalised in N. Australia. W. S. America and W. Europe. Many species are cultivated as decorative plants and further introductions may be expected. *Aponogeton* species are mostly local in distribution and are found in a variety of aquatic habitats; some species remain submerged throughout the vegetative phase, others produce floating leaves and others are amphibious. The tubers of several species are eaten by humans or their livestock.

ARACEAE

c. 115 genera; 16 with aquatics.

Perennial herbs with watery, bitter or milky juice, and usually an elongated or tuberous rhizome. Leaves mostly in basal or apical rosettes, with a membranous sheath at the base of the petiole. Flowers small or minute, crowded on a usually fleshy spike (spadix); usually enclosed in a large bract (spathe); bisexual or unisexual; when unisexual, the males in the upper part of the spadix, the females below. Small perianth often present in bisexual flowers, with 4 to 8 segments or united into a truncate cup. Stamens 2, 4 or 8, hypogynous, opposite the perianth segments. Ovary superior or immersed in the spadix; fruit usually a fleshy berry, occasionally a spongy berry, a nutlet, or a capsule.

Wilson, K. A. Genera of Arales in the S. E. United States. Journ. Arnold Arbor. 41: 47–72 (1960)
Engler, A. and Krause, K. Araceae in Engler, Pflanzenreich (IV. 23): (1905–20)

1A Plant free-floating; leaves cuneate and sessile
Pistia
1B Plant bottom-rooted or creeping; leaves not cuneate and sessile
- 2A Leaves monofacial, reed-like; spadix apparently borne on a leaf
  Acorus
- 2B Leaves not monofacial, differentiated into petiole and blade; spadix borne on a stem
  - 3A Spathe membranous, forming a sheath at the base of the scape, not overlapping the spadix
    Orontium
  - 3B Spathe not membranous, inserted at the base of the spadix and overlapping it
    - 4A Spathe broadly ovate, almost flat; flowers male above, bisexual below
      Calla
    - 4B Spathe usually lanceolate, rolled; flowers either all bisexual or all unisexual
      - 5A Male flowers separated from female by slender naked axis of spadix, and surrounded by an involute fold of the spathe
        - 6A Female flowers in a single whorl; fruit a capsule
          Cryptocoryne
        - 6B Female flowers in a spiral; fruit a berry
          Lagenandra

5B Male flowers separated from female by sterile flowers or not separated; spathe without an involute fold

7A Female flowers few, in a single whorl at the base of the male flowers **Aglaodorum**

7B Female flowers many, spirally arranged

8A Plants with an emergent stem with long (*c.* 30 cm) internodes or with a stem-like agglomeration of petioles, usually over 1 m high

9A Stem formed by an agglomeration of petioles **Typhonodorum**

9B Stem not an agglomeration of petioles, internodes *c.* 30 cm long **Montrichardia**

8B Stem usually less than 1 m high, if more then internodes short (less than 10 cm long)

10A At least some flowers bisexual

11A Petioles smooth; spathe hooded above **Dracontioides**

11B Petioles warty or spiny; spathe not hooded

12A Leaves sagittate-pinnatifid; spathe convolute above, open only at the base **Lasia**

12B Leaves sagittate; spathe open, or convolute only at the base **Cyrtosperma**

10B All flowers unisexual

13A Many sterile flowers crowded on narrow part of spadix between male and female flowers; leaves cordate, peltate **Colocasia**

13B Male and female flowers contiguous, or few sterile flowers present; leaves not peltate

14A Leaves annual from a short tuberous rhizome, always sagittate (temperate) **Peltandra**

14B Leaves evergreen, rarely sagittate (tropical)

15A Sterile flowers scattered on naked central part of spadix; female part of spadix equal to or longer than upper male part **Dieffenbachia**

15B Sterile flowers absent, or crowded between male and female flowers; female part of spadix much shorter than male **Anubias**

Fig. 49. *Acorus calamus* L.: habit (3 cm).

Acorus L., Sp. Pl. 324 (1753)
Fig. 49.
Rhizome thick, creeping. Leaves equitant, monofacial, glabrous. Spadix borne laterally on an axis resembling a foliage leaf; spathe absent (the prolongation of the fertile axis is sometimes interpreted as the spathe). Flowers bisexual, 3-merous, greenish. Perianth of 6 segments in 2 whorls of 3. Stamens 6; filaments linear. Ovary 2- or 3-locular; ovules numerous; fruit a gelatinous, few-seeded berry.

Wulff, H. D. Zur Zytologie, geographische Verbreitung und Morphologie des Kalmus. Arch. Pharm. 287: 529–541 (1954)

2 species: *A. calamus* L. occurs in America (possibly introduced into S. America), Europe and Asia, in marshes and the edges of lakes and streams. The American plants are diploid; the European (introduced from S. Asia in the 16th C.) are triploid; those from Siberia and Asia are tetraploid. *A. gramineus* Soland is restricted to C. and E. Asia.

Aglaodorum Schott, Gen. Aroid. 58 (1858)
Fig. 50A.
Rhizome creeping. Leaves erect, glabrous, fleshy, linear to oblong; petiole up to 60 cm long; blade up to 45 cm long. Inflorescence much shorter than leaves. Spathe convolute, pointed, cuspidate, pale green, *c.* 7.5 cm long, 2.5 cm wide. Spadix as long as the spathe; male part cylindric, stout, *c.* 6 cm long. Flowers unisexual; perianth absent; female flowers few, in 1 or 2 whorls at the base, with staminodes; ovary 1- or 2-locular; fruit ovoid, *c.* 4 cm long, green; fruit wall thick.

1 species, *A. griffithii* Schott: Malaysia, Sumatra and Borneo. Occurs in swamps, particularly among *Nypa*. In this genus the fruits are large and green, with a thick spongy wall, in contrast to the fleshy red fruits of the terrestrial but closely related genus *Aglaonema*; they appear to be adapted to water dispersal (Ridley, H. N. Fl. Malay Penins. 5: 100 (1925)).

Anubias Schott, Oesterr. Bot. Wochenbl. 7: 398 (1857)
Fig. 50B.
Rhizome creeping. Leaves erect, glabrous, submerged or emergent; petioles sheathing at base; blade usually lanceolate, up to 30 cm long, about equalling the petiole. Spathe convolute, tubular or expanded nearly to the base. Spadix longer or shorter than spathe, edges free. Flowers unisexual, male and female contiguous. Perianth absent. Stamens 5 or 6, united; staminodes 3 to 5 in sterile flowers. Ovary 2- or 3-locular; ovules numerous; style very short or absent; berries enclosed within the spathe, subglobose.

Hepper, F. Notes on Tropical African Monocots: 2. Kew Bull. 22: 456 (1968) *c.* 12 species: mainly west Africa. *A. afzellii* Schott from Sierra Leone is a submerged aquatic; the other species occur mainly in wet forest.

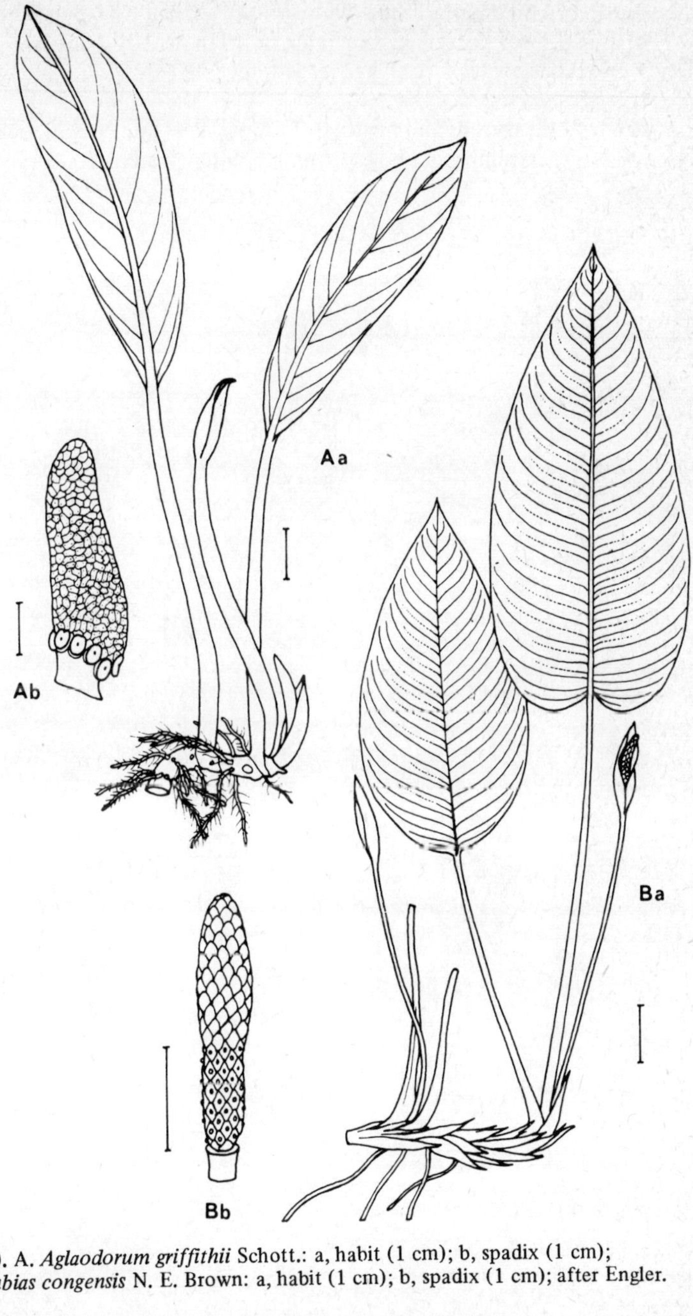

Fig. 50. A. *Aglaodorum griffithii* Schott.: a, habit (1 cm); b, spadix (1 cm); B. *Anubias congensis* N. E. Brown: a, habit (1 cm); b, spadix (1 cm); after Engler.

Calla L., Sp. Pl. 968 (1753)
Fig. 51A.

Rhizome long, creeping. Leaves long-petiolate, up to 30 cm high; petiole with long sheath; blade cordate. Spathe ovate, white on the adaxial side, persistent. Spadix *c.* 3 cm long. Lower flowers bisexual, upper often male. Filaments slender. Ovary 1-locular; ovules 5 to 9; berries red, few-seeded.

1 species, *C. palustris* L.: central and northern Europe, Asia, and N. America. It occurs in wet woods, swamps and bogs, often in shallow water.

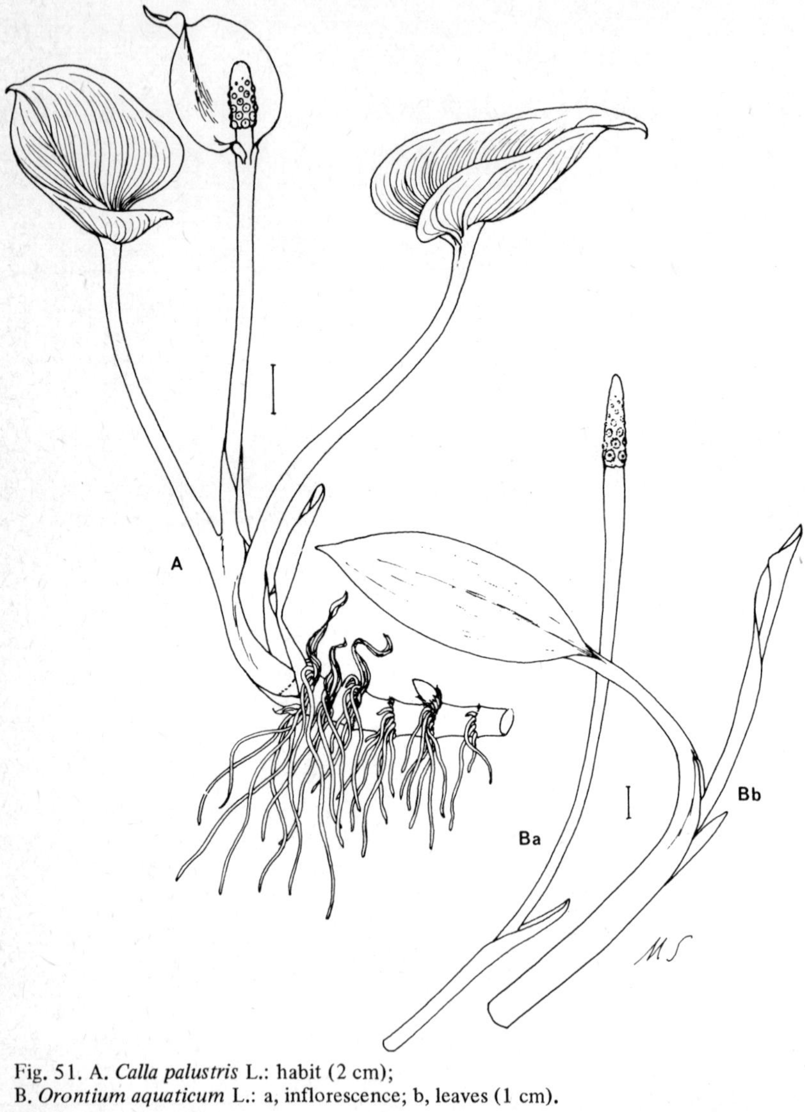

Fig. 51. A. *Calla palustris* L.: habit (2 cm);
B. *Orontium aquaticum* L.: a, inflorescence; b, leaves (1 cm).

Colocasia Schott. Melemata Bot. 18 (1832)
Fig. 52.
Rhizome tuberous, starch-filled. Stems thick, up to 1 m high, often spreading by stolons. Petioles up to 2 m long; blade almost peltate. Inflorescence shorter than the petioles. Spathe ovate-lanceolate, constricted below the middle. Spadix shorter than the spathe; male flowers above, female below, with sterile

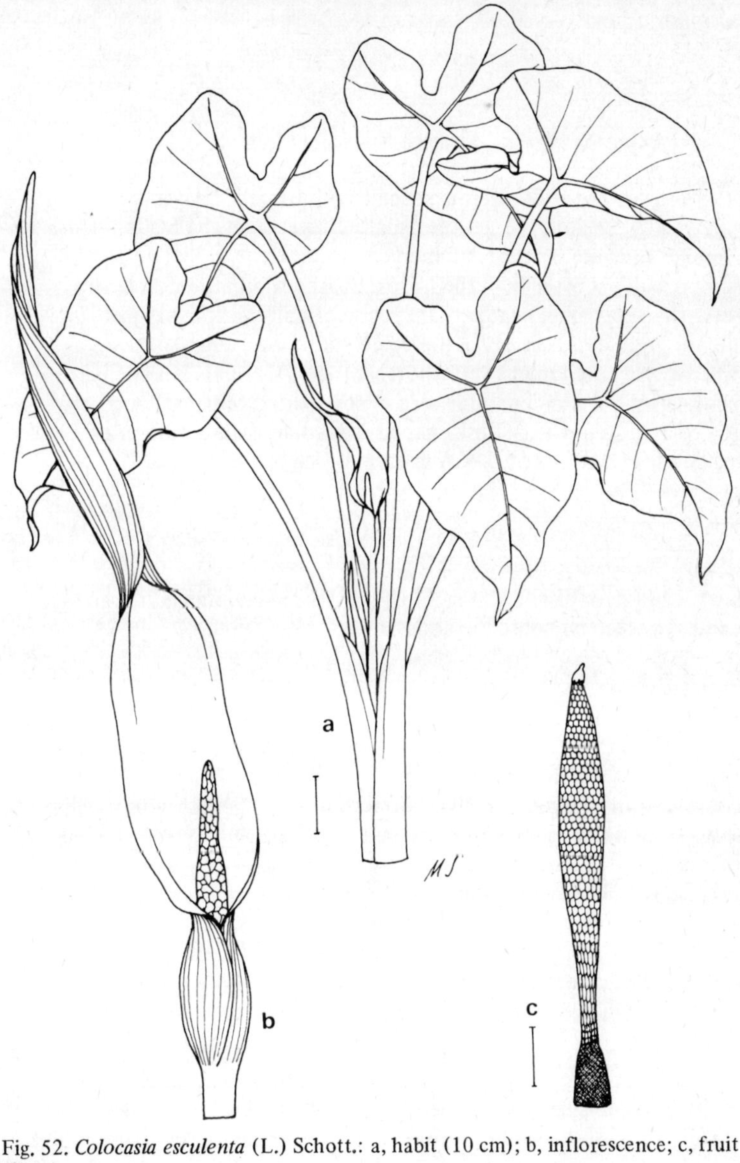

Fig. 52. *Colocasia esculenta* (L.) Schott.: a, habit (10 cm); b, inflorescence; c, fruit (2 cm).

flowers between and above male flowers. Perianth absent. Stamens united into a peltate mass. Ovary 1-locular; ovules numerous; style short; berry obconic or oblong.

c. 8 species: tropical Asia. *C. esculenta* (L.) Schott has a starch-filled rhizome which is often eaten. *C. esculenta* var. *aquatilis* Hasskarl. from Java is an aggressive weed in parts of S. E. N. America along streams, in marshes and by roadsides.

Cryptocoryne Fischer ex Wydler, Linnaea 5: 428 (1830)
Fig. 53A.
Rhizome creeping, often stoloniferous, usually submerged. Leaves up to 75 cm long, usually smaller, the petiole usually longer than the blade. Blades linear-lanceolate to ovate cordate, submerged or emergent, rolled when young. Inflorescence usually shorter than the leaves. Spathe tubular, limb usually much shorter than the tube, ovate or lanceolate, usually purplish. Spadix very short, enclosed in dilated base of spathe. Male part conic, wrapped in an involute fold of spathe. Stamens very short, anthers 2-locular. Female part below separated by the slender sterile axis of the spadix; female flowers few in a single whorl; united. Ovary flask-shaped, 1-locular; ovules numerous; fruit a capsule of several united, 2-valved carpels; seeds numerous.

Wit, H. C. D. de, Cryptocoryne in Aquarium Pflanzen. Ulmer Stuttgart (1971)

c. 60 species: Indo-malaya and China. Most species are submerged or emergent aquatics, and are often cultivated in aquaria.

Cyrtosperma Griffith, Notul. 3: 149 (1848)
Fig. 54.
Rhizome short, stout or tuberous. Plant up to 4 m high. Leaves with usually thorny or warty petioles up to 3 m high; blade sagittate, with strong lateral nerves running into the lobes, up to 1 m long. Inflorescence exceeding or equalling the leaves. Spathe persistent, not convolute, rarely twisted, ovate-lanceolate, whitish within, yellowish without. Spadix shorter than the spathe. Flowers bisexual. Perianth segments 4 to 8. Stamens 4 to 8; filaments short, flat. Ovary 1-locular; ovules 1 to many; stigma sessile; fruiting spadix *c.* 15 cm long; berries crowded, green or purple; seeds with 4 crests.

c. 18 species: tropical Africa, America and Asia. *C. lasioides* Griffith is a frequent aquatic in Malaysia and Borneo. *C. senegalense* (Schott) Engler occurs throughout West Africa in swamps. The other species are usually found in wet places.

Fig. 53. A. *Cryptocoryne wendtii* De Wit: a, habit (1 cm); b, spadix (1 mm); B. *Lagenandra thwaitesii* Engl. (after De Wit): a, habit (2 cm); b, spadix (1 mm).

Fig. 54. *Cyrtosperma afzelii* (Schott) Engl.: a, habit (10 cm); b, leaf (1 cm); c, flowers (1 mm).

Dieffenbachia Schott, Wiener Zeitschr. Kunst 3: 803 (1829)
Fig. 55A.
Subshrubby with stout stems, sometimes creeping at the base, up to 2 m high; internodes short. Leaves with long petioles sheathing to above their middle; blade oblong with a straight midrib and parallel lateral nerves. Inflorescence

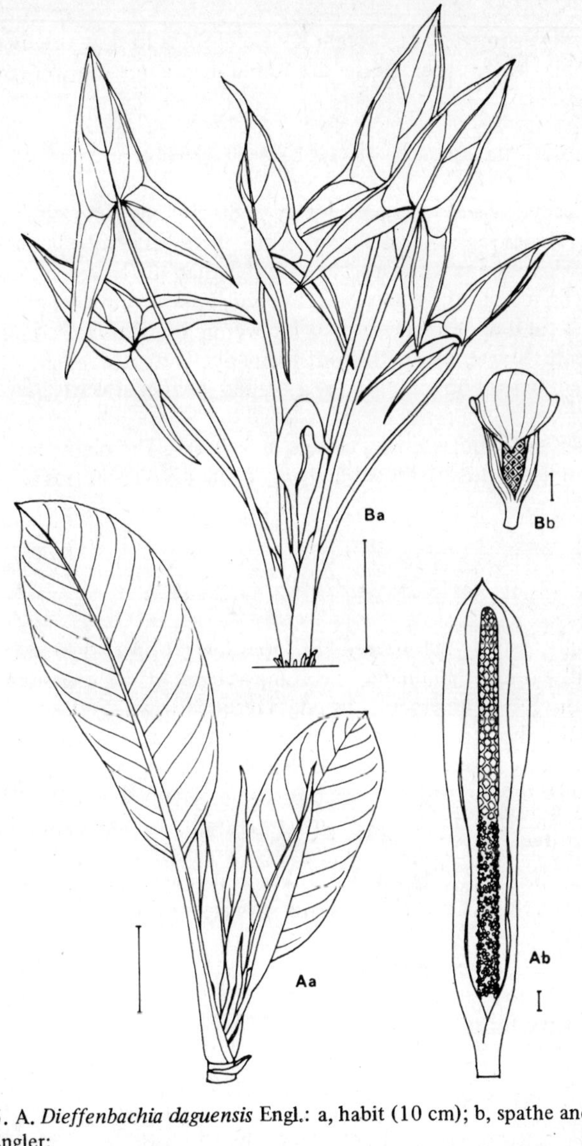

Fig. 55. A. *Dieffenbachia daguensis* Engl.: a, habit (10 cm); b, spathe and spadix (1 cm); after Engler;
B. *Dracontioides dehiscens* (Schott) Engl.: a, habit (10 cm); b, spadix and spathe (1 cm); after Engler.

shorter than the leaves. Spathe oblong, persistent, convolute below, open and sometimes recurved above. Spadix a little shorter than the spathe; the female part below usually enfolded by the spathe, the male part exposed above, the two separated by sterile flowers. Flowers unisexual. Stamens 4 to 5 united; female flowers with 4 to 5 staminodes; ovary 1- to 3-locular; ovules few; fruit a berry.

c. 30 species: Tropical America. *D. costata* Klotsch occurs in slow flowing streams in Colombia. Many other species are found in swamps and wet forests.

Dracontioides Engler, Pflanzenreich 48 (IV.23c): 36 (1911)
Fig. 55B.
Rhizome thick, fleshy. Leaves with petioles up to 80 cm long, smooth, sheathing at base; blades sagittate, the lobes acute, the median lobe larger than the laterals, up to 25 cm long. Inflorescence shorter than the leaves. Spathe open to the base, hooded in the upper half, brownish-purple outside, purplish inside. Spadix *c.* 4 cm long. Flowers bisexual, covering the spadix. Perianth segments 4, truncate. Stamens 4, with stout filaments. Ovary 2-locular; stigma small, short stalked; berry obovoid, usually 1-seeded; seed coat warty.

1 species, *D. dehiscens* (Schott) Engler occurs in S. Brazil. The plant flowers in the rainy season when it is partly submerged; in the dry season it is dormant.

Lagenandra Dalzell, Jour. Bot. Kew Gard. Misc. 4: 289 (1852)
Fig. 53B.
Herbs with creeping, usually submerged rhizome. Leaves up to 120 cm high; petiole usually longer than blade, sheathing at base; blades linear-lanceolate to ovate, submerged to emergent, the edges inrolled when young. Inflorescence usually shorter than the leaves. Spathe tubular with edges united at base, the limb usually about equal to the tube. Spadix very short, enclosed in the dilated base of the spathe; the male part conic, wrapped in an involuted fold of the spathe. Stamens very short, 1 or 2 per flower. Female part below, separated by the slender sterile axis of the spadix; female flowers many, free, spirally arranged at the base of the spadix. Ovary 1-locular, widest at the top; ovules 1 to 6; fruit an indehiscent berry; seeds many, very small.

c. 6 species: mainly in southern India. Occurring in marshes, ponds and streams in forests; very similar to *Cryptocoryne* in habit, differing mainly in details of the arrangement of the female flowers.

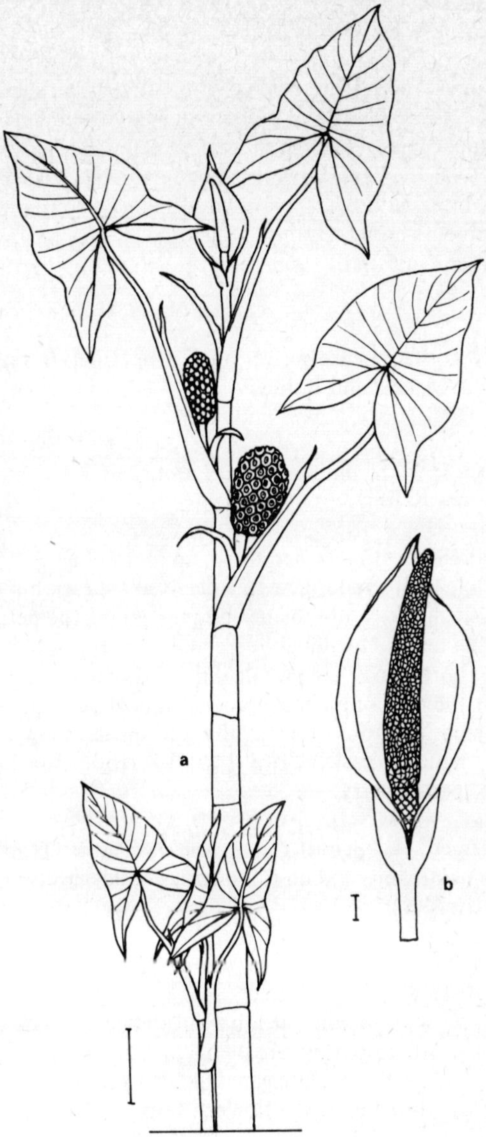

Fig. 56. *Montrichardia linifera* (Anuda) Schott: a, habit (10 cm); b, spathe and spadix (1 cm).

Lasia Lour., Fl. Cochinch. 81 (1790)
Fig. 59A.
Rhizome creeping or ascending, up to 75 cm long. Leaves with spiny petioles, up to 120 cm high; blades sagittate or pinnatifid, spiny on the main nerves beneath. Inflorescence about equalling the leaves; peduncle spiny. Spathe convolute, open only at the base, brownish, *c.* 35 cm long. Spadix *c.* 4 cm long, crowded with bisexual flowers; perianth segments 4 to 6, ovate-cuneate. Stamens 4 to 6, filaments broad. Ovary 1-locular; ovules solitary; stigma small, sessile; spadix in fruit up to 10 cm long, green, fleshy; berries warted on the exposed side.

c. 3 species: Indo-Malaysia; occurs in river mud, often tidal. *L. spinosa* (L.) Thwait. is frequent in Malaya and Borneo.

Montrichardia Crüger in Mohl and Schlechtdl., Bot. Zeit. 12: 25 (1854) nom. cons., [*Pleurospa* Rafin.]
Fig. 56.
Rhizome submerged. Stem stout, erect, leafy, up to 4 m high; internodes up to 30 cm long. Petiole *c.* 30 cm long, with a persistent sheath up to half its length, ending in a ligule-like point; blade about equalling the petiole, sagittate; lateral lobes about equalling the middle lobe. Inflorescences axillary in the leaf sheaths, shorter than the petiole. Spathe 17 to 20 cm long, whitish inside, convolute at the base only, deciduous. Spadix about equalling the spathe, dense-flowered, the upper ¾ male, the rest female. Flowers unisexual. Stamens 3 to 6, contiguous, sessile. Ovary 1-locular; ovules 1 or 2; stigmas sessile; fruit a large spongy berry.

2 species, *M. arborescens* (L.) Schott and *M. linifera* Schott: Tropical America. Common in lakes and streams, and on flood-plains beside rivers.

Orontium L., Sp. Pl. 324 (1753)
Fig. 51B.
Rhizome stout. Leaves with petioles sheathing at the base, up to 15 cm long; blades oblong elliptic, often floating. Spathe membranous forming a sheath at base of scape, or absent. Spadix elongated, golden-yellow above, green below, with a band of white below the flowers. Lower flowers bisexual, each with 6 perianth segments, 6 stamens and 3 carpels; upper becoming male only, with 2 to 6 perianth segments and 1 to 6 stamens. Staminodes sometimes present. Ovary 1-locular; ovules 5 to 9; fruit a few-seeded berry.

Grear, J. W. jr. Cytogeography of Orontium aquaticum. Rhodora 68: 25–34 (1966)

1 species, *O. aquaticum* L., occurs in shallow water in eastern N. America, frequently cultivated as a decorative plant elsewhere.

Fig. 57. *Peltandra virginica* L.: habit (1 cm).

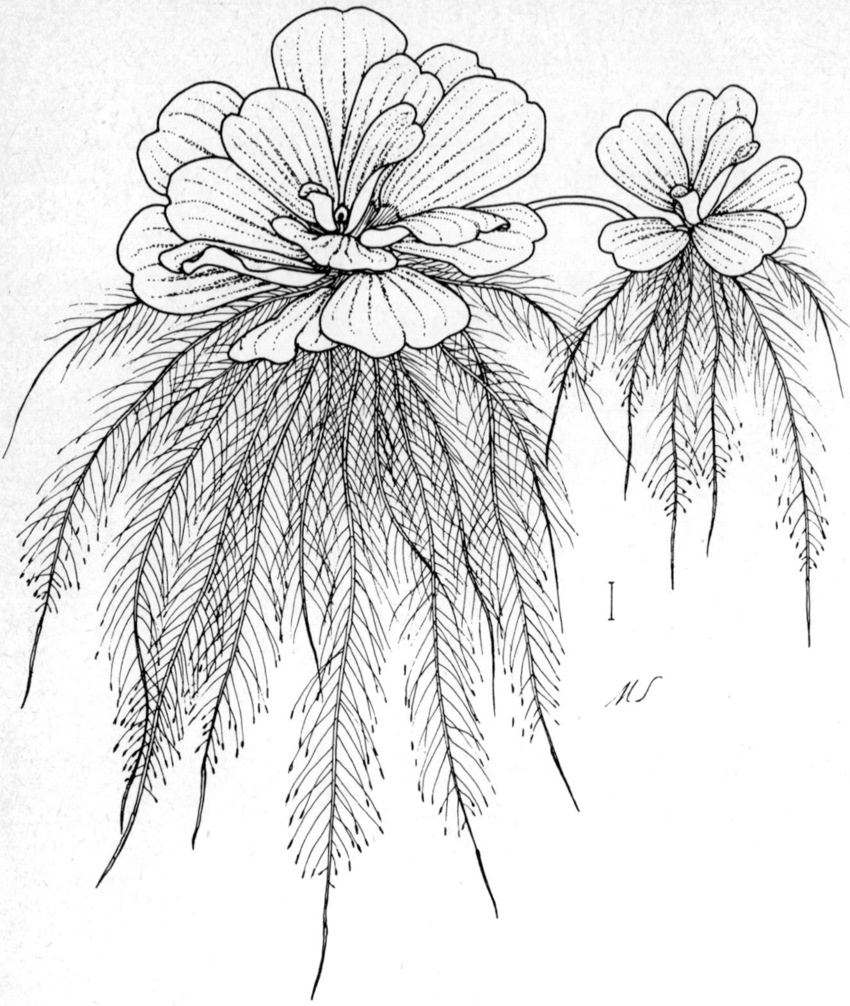

Fig. 58. *Pistia stratiotes* L.: habit (2 cm).

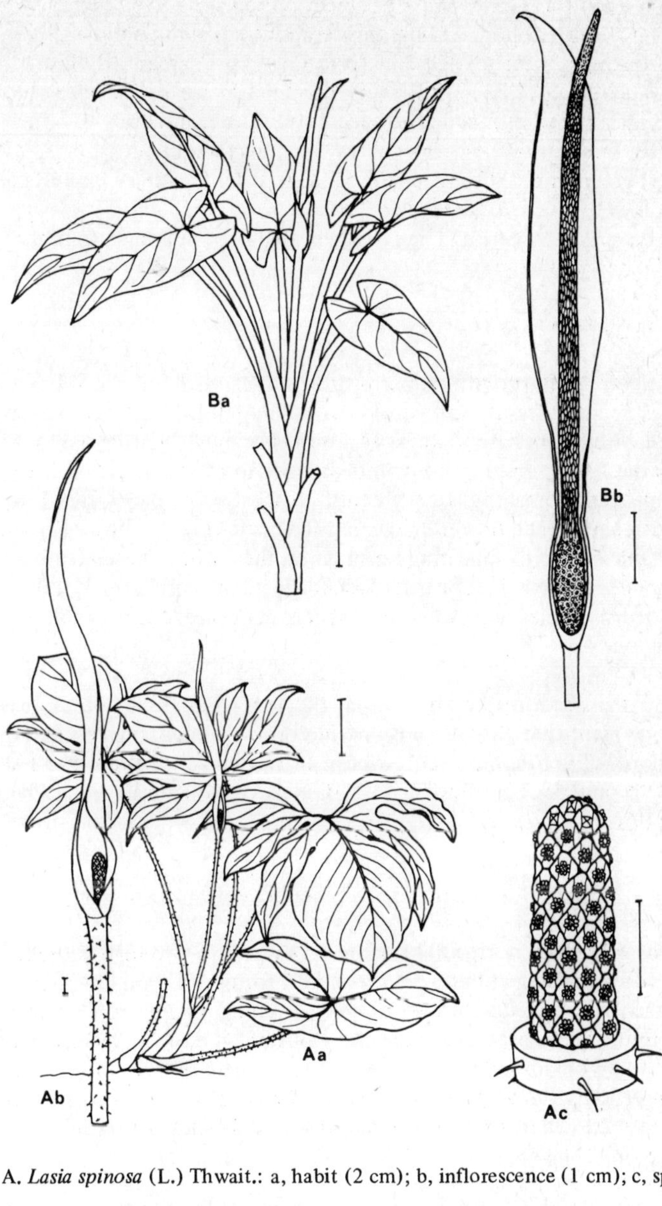

Fig. 59. A. *Lasia spinosa* (L.) Thwait.: a, habit (2 cm); b, inflorescence (1 cm); c, spadix (1 cm);
B. *Typhonodorum lindleyanum* Schott: a, habit (10 cm); b, spadix (10 cm).

Peltandra Rafin., Journ. Phys. Chem. 89: 103 (1819)
Fig. 57.
Rhizome short, stout. Leaves short petioled, up to 40 cm high; blades variable, sagittate, up to 25 cm long. Inflorescence about equalling petiole. Spathe convolute throughout or dilated and spreading above, green or white Spadix completely covered with flowers, or sometimes naked at apex. Flowers unisexual; perianth absent. Male flowers above, sessile; anthers 4 to 8 in a peltate head; female flowers below; ovary 1-locular; staminodes 4 or 5, white, fleshy; fruit green or red, fleshy, enclosed in persistent leathery base of spathe.

3 species: eastern N. America. They occur in shallow water and swamps.

Pistia L., Sp. Pl. 963 (1753)
Fig. 58.
Rhizome floating, short, with numerous roots, often stoloniferous. Leaves in a rosette, sessile, obovate to ovate-cuneate, densely pubescent, 13 to 15 cm long, with parallel ribs from base to apex. Inflorescence much shorter than the leaves; spathe 2 to 4 cm long, convolute at base, spreading above, constricted between male and female flowers, greenish, hairy. Spadix partly fused to spathe, with one female flower below and few male flowers above. Perianth absent. Stamens 2 to 8, appearing terminal on the spadix, subtended by a lobed membranous ring. Female flower flask-shaped; stigma with a thick stalk. Ovary 1-locular; ovules numerous; fruit green, many-seeded, crowned by the persistent style.

1 species, *Pistia stratiotes* L.: throughout the Tropics and Subtropics. It is a free-floating plant that grows in a very wide variety of aquatic habitats. In many regions it has become a serious pest. It is often called the Water Lettuce. For a full account see Sculthorpe, C. D. Biology of Aquatic Vascular Plants, London (1967).

Typhonodorum Schott, Oesterr. Bot. Wochenbl. 7: 69 (1857)
Fig. 59B.
Large herbs with stout horizontal rhizome. Leaves 3 to 4 m high, all radical, emergent and sheathing at base; petioles stout, forming a stem-like axis; blades lanceolate, ovate to cordate, up to 1.3 m long. Spathe 50 to 80 cm long, convolute below, constricted at the mouth of the tube, greenish-white. Spadix 35 to 55 cm long; female flowers at base, male flowers above, and sterile flowers between and above the males. Perianth absent. Anthers 4 to 8, united. Ovary 1-locular; ovule 1; stigma subsessile; berries large, each containing 1 nut-like seed.

1 species, *T. lindleyanum* Schott: Madagascar, Mauritius and Zanzibar. It occurs in large masses in rivers and ponds.

ASTERACEAE [COMPOSITAE]

c. 900 genera of which 5 contain aquatic species.*

Herbs. Flowers arranged on common receptacle in heads; heads surrounded by an involucre of bracts. Sepals absent, replaced by pappus of hairs or bristles or absent. Petals 5, fused, either actinomorphic (tubular florets) or zygomorphic and tongue-like (ligulate florets). Stamens 5; anthers loosely joined around style. Ovary inferior, 1-locular; ovule 1, erect on base of loculus; fruit an achene.

Hutchinson, J. Aquatic Compositae. Gard. Chron. 59: 305–306 (1916)

1A Green parts of plant clothed with whitish, short, appressed hairs
Eclipta

1B Green parts of plant not clothed with whitish, short appressed hairs.
- 2A Leaves simple, with entire margin; pappus bristles converging at tips
Hydropectis
- 2B Leaves not entire, serrate, pinnatisect or divided into linear segments; pappus bristles not converging at tips
  - 3A Pappus absent (widespread)
Cotula
  - 3B Pappus of bristles or awns
    - 4A Pappus of 3 to 6 spreading awns (Temperate N. America)
Megalodonta
    - 4B Pappus of numerous bristles (Mexico)
Erigeron

Cotula L., Sp. Pl. 891 (1753)
Fig. 60A, B.
Stems floating, creeping, ascending or erect. Leaves usually alternate, often heteromorphic, entire to pinnatisect with linear lobes (in *C. myriophylloides* leaves opposite, spuriously whorled; each leaf divided to the base into linear, filamentous segments); leaf bases usually sheathing. Flower heads hemispherical, solitary in the axils of leaves; involucral bracts in 2 or 3 rows, subequal, blunt. Ligulate florets much reduced or absent; tubular florets with flattened or winged tube. Achenes compressed, often wing-margined; pappus absent.

c. 75 species: almost cosmopolitan, chiefly S. Hemisphere. Most species are terrestrial. *C. myriophylloides* Harvey from the Cape Penninsula, S. Africa is the only obligate aquatic species. Some species such as the widely naturalised *C. coronopifolia* L. are essentially swamp species but may be found growing submerged.

* After going into print, our attention was drawn to an additional aquatic genus, *Cadiscus* E. H. F. Meyer. It is illustrated and described on page 549.

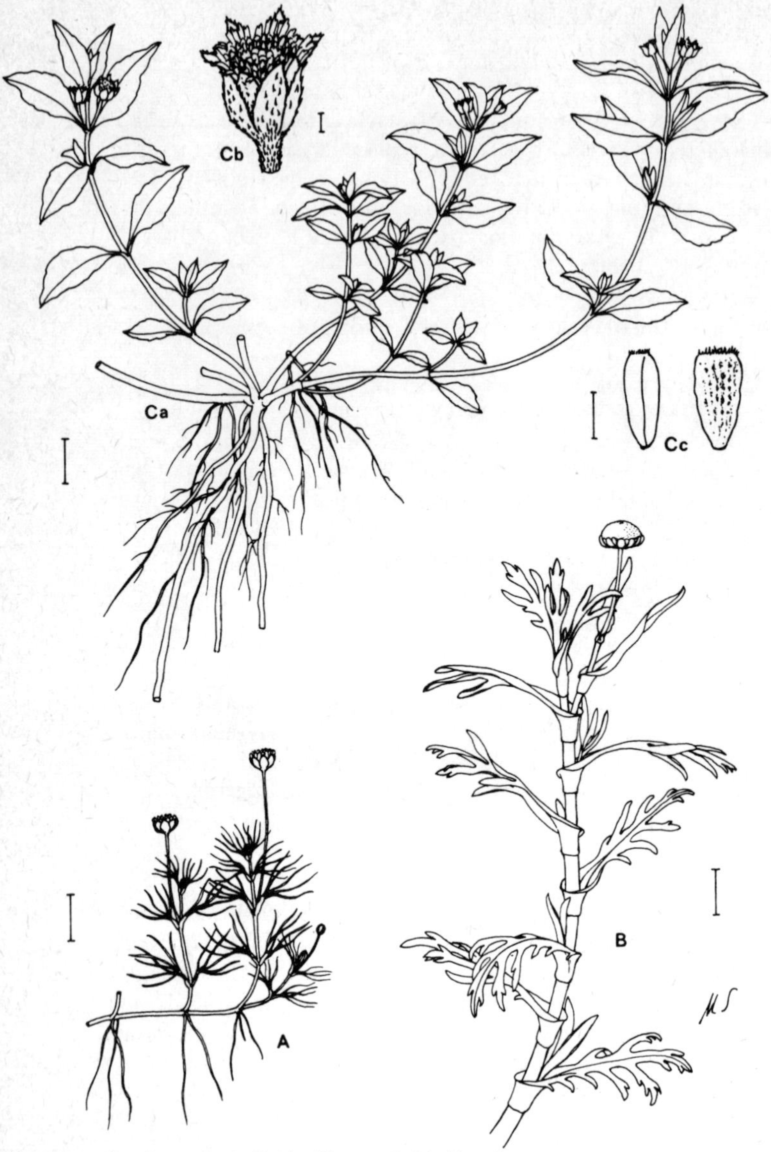

Fig. 60. A. *Cotula myriophylloides* Harvey: habit (1 cm);
B. *Cotula coronopifolia* L.: habit (1 cm);
C. *Eclipta alba* (L.) Hassk.: a, habit (1 cm); b, flowering head (1 mm); c, achenes (1 mm).

Eclipta L., Mant. 157, 286 (1771) nom. cons.
Fig. 60C.
Herbs clothed in white, short, appressed hairs. Stems creeping or erect, often swollen when submerged. Leaves opposite, lanceolate or narrowly ovate-lanceolate, 4 to 10 cm long, 1 to 3 cm wide; margin serrate. Flower heads axillary and terminal, about 1 cm diam.; involucral bracts few, overlapping, in 2 rows, ovate-lanceolate. Ligulate florets white, with tongue 2-lobed at apex; tubular florets 4-toothed at apex. Achenes dimorphic but alike in each head; most heads with straw-coloured, tuberculate, slightly flattened achenes; other heads with achenes dark grey to black, smooth, distinctly flattened; pappus reduced to a few hairs or teeth or absent.

3 to 4 species: cosmopolitan in warm regions. *E. prostrata* (L.) L. [*E. alba* (L.) Hassk.] although not strictly aquatic is commonly found in standing water in ricefields and irrigation ditches.

Erigeron L., Sp. Pl. 863 (1753)

c. 200 species: cosmopolitan, most in C. and N. America. *E. heteromorphus* is the only aquatic species.

E. heteromorphus Robinson, Proc. Amer. Acad. Arts. Sci.: 27: 173 (1893)
Fig. 61.
Stems smooth, simple or sparingly branched, floating, creeping or ascending, rooting at lower nodes. Leaves alternate or sub-opposite, dimorphic; terrestrial leaves sessile, up to 5 cm long, 3 cm wide, simple or irregularly lobed, with laciniate or cuneate bases; submerged leaves 10 cm or more long, divided into linear, filamentous segments. Flower heads *c.* 1.5 cm diam., borne singly; involucral bracts subequal, overlapping, smooth, green, scale-like, with narrow scarious margins. Ligulate florets up to 50, linear, white; tubular florets numerous, yellow. Pappus of bristles.

E. heteromorphus is known from the cascades of the Concepcion River, Micos, San Luis Potosi, Mexico where it grows on tufa more or less submerged.

Megalodonta Greene, Pittonia 4: 271 (1901), [*Bidens* L. pro parte]
Fig. 62B.
Perennial. Stem simple or sparingly branched, up to 2 m or more long. Leaves opposite, sessile, heteromorphic; lower or submerged leaves finely divided into forked, capillary segments; upper or emergent leaves somewhat reflexed, lanceolate, 2 to 4 cm long, with margin sharply and narrowly toothed; intermediate shaped leaves occur between the upper and lower ones. Flower heads usually terminal; outer involucral bracts usually 5 or 6, ovate or oblong, obtuse,

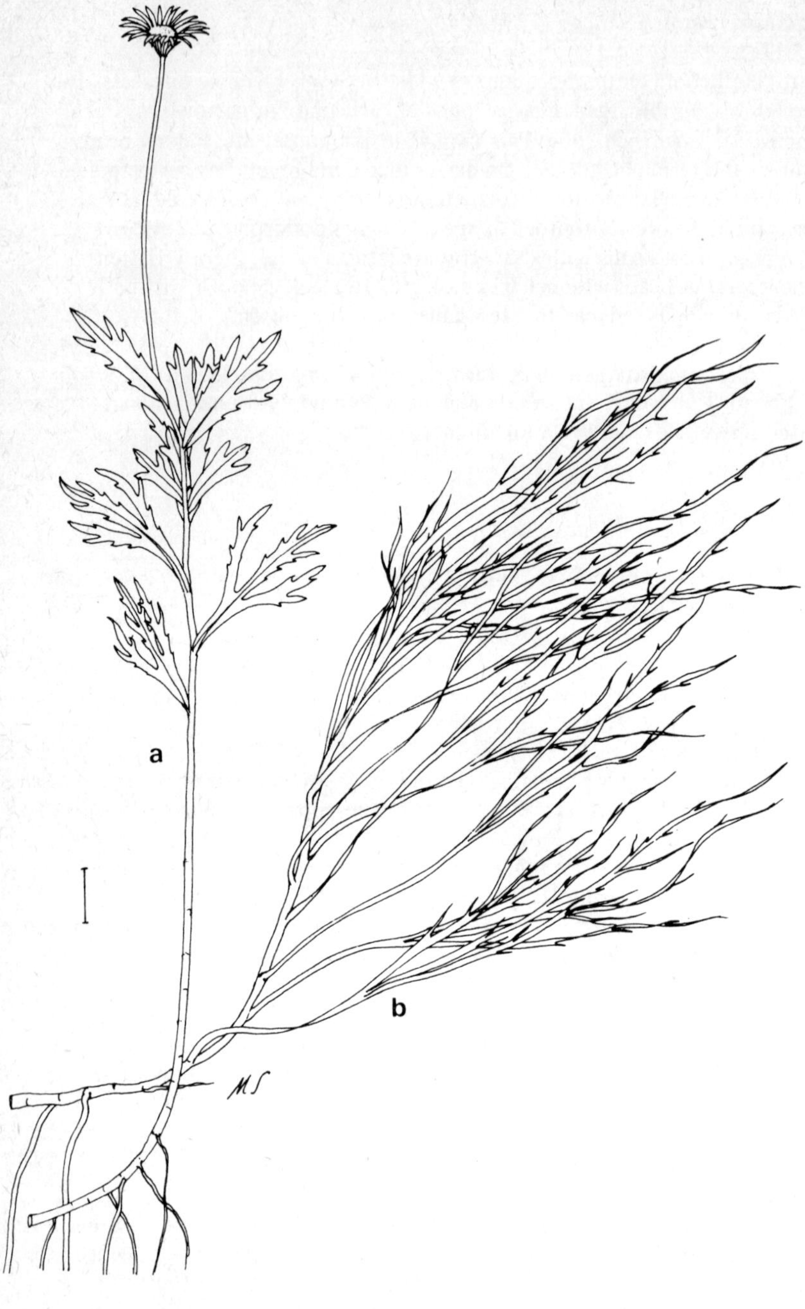

Fig. 61. *Erigeron heteromorphus* Robinson: a, flowering plant; b, submerged plant (1 cm).

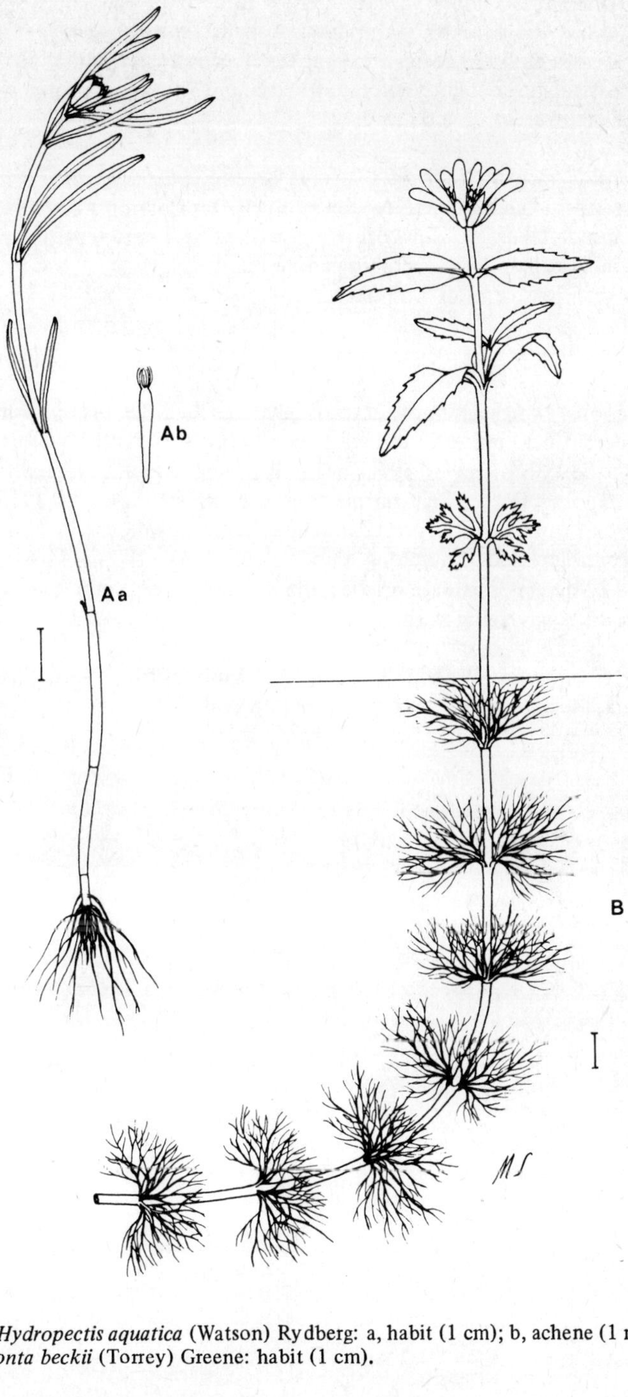

Fig. 62. A. *Hydropectis aquatica* (Watson) Rydberg: a, habit (1 cm); b, achene (1 mm); B. *Megalodonta beckii* (Torrey) Greene: habit (1 cm).

rarely toothed, 5 to 7 mm long, exceeded by inner ones. Ligulate florets usually 5 to 7, yellow, 12 to 18 mm long. Achenes linear-oblong, subterete, truncate at both ends, 10 to 15 mm long (excluding awns), with 3 to 6 apical awns; awns straight or slightly curved, spreading, 15 to 28 mm long, with numerous backwardly directed spines.

1 polymorphic species, *M. beckii* (Torrey) Greene [*Bidens beckii* Torrey]: N. America from Quebec and New Jersey to British Columbia and Oregon. Occurs in ponds, lakes and slowly flowing streams. It is very sensitive to water pollution and is apparently becoming a rare plant.

Hydropectis Rydberg, N. Amer. Fl. 34: 216 (1916), [*Pectis* L. pro parte] Fig. 62A.

Stems smooth, elongate, simple or sparingly branched. Leaves opposite, few, entire, linear, thick, blunt, 2 to 3 cm long, rarely glandular. Flower heads solitary in the axils of leaves; stalks becoming recurved with age; involucre of 5, overlapping, flat, obtuse, purple-tipped, *c.* 1 cm long bracts. Ligulate florets absent; tubular florets *c.* 12, included within involucre. Achenes somewhat compressed, 8 mm long, slender, with long, attenuate base; pappus of 10 to 12 unequal, serrulate bristles, the longest about 2 mm long, all ascending and converging at the tips.

1 species, *H. aquatica* (Watson) Rydberg: known only from Sierra Madra, Chihuahua, Mexico, where it grows in shallow water.

BALSAMINACEAE

4 genera; *c.* 500 species; Eurasia, Africa and N. America. *Hydrocera* is aquatic.

Hydrocera Blume, Bijdr. 241 (1825)
Fig. 63.
Glabrous, annual. Stem ascending or floating, branched, 5-angled; internodes swollen when submerged or floating, bearing numerous, long, fibrous roots. Leaves alternate, linear, up to 10 cm or more long; stipules absent; leaf-bases bearing paired glands; leaf margin toothed. Inflorescences axillary, cymose. Flowers bisexual, zygomorphic, borne in pairs. Sepals 5, free, petaloid, the 2 outer laterals flat, becoming adaxial due to twisting; the abaxial prolonged backwards into a short, hollow spur. Petals 5, free, the 2 outer abaxial ones longer, concave, reddish. Stamens 5; filaments broad, united towards the connective, anthers united round the style. Carpels 5, united, superior; ovules 2 to 3 in each carpel, axile; fruit a septicidally dehiscing capsule, purplish-red.

1 species: *Hydrocera triflora* (L.) Wt. and Arn. found in the Indo-Malayan region. It is an amphibious plant found in shallow water. In certain parts of Andhra Pradesh the flowers are used for dyeing finger-nails.

Fig. 63. *Hydrocera triflora* (L.) Wt. and Arn.: a, inflorescence; b, stem base (1 cm).

BRASSICACEAE [CRUCIFERAE]

c. 375 genera, of which 3 contain aquatics: *c.* 3200 species, of which *c.* 9 are aquatic.

Annual or perennial herbs. Leaves alternate, without stipules. Flowers actinomorphic, hypogynous. Sepals 4, free in 2 decussate pairs. Petals absent or 4, free, alternating with sepals. Stamens usually 6, rarely 4. Ovary of 2 united carpels, with 2 parietal placentas, 2-locular through the formation of a membranous false septum; fruit a dehiscent capsule opening by 2 valves.

1A Leaves subulate or linear, in a rosette
Subularia

1B Leaves not subulate or linear, or not in a rosette but scattered along the stem

2A Capsules flattened in transverse section; valves of capsule coiling spirally from the base at dehiscence
Cardamine

2B Capsules round in transverse section; valves of capsule not coiling spirally from the base at dehiscence
Rorippa

Cardamine L., Sp. Pl. 654 (1753)
Fig. 64A.
Annual or perennial herbs, glabrous or with unbranched hairs. Stems erect or creeping. Leaves borne along the stem, simple to pinnate. Petals white, pink or purple (rarely pale yellow). Fruit at least 3 times as long as wide, flattened in transverse section; valves of the fruit coiling spirally from the base at dehiscence; seeds in 1 row in each loculus.

Schultz, O. E. Monographie der Gattung Cardamine. Bot. Jahrb. 32: 280–623 (1930)

c. 160 species: cosmopolitan but chiefly from the temperate regions. Most species are marsh or bog plants but some species such as *C. lyrata* Bunge, *C. prorepens* Fischer and *C. variabilis* Philippi are frequently found growing submerged.

Rorippa Scop. Fl. Carn. 520 (1760), [*Nasturtium* R. Br. *Neobeckia* Greene, *Armoracea aquatica* (Eaton) Wiegland]
Fig. 64B.
Annual to perennial herbs; glabrous or with unbranched hairs. Stems erect or creeping. Leaves borne along the stem, simple to pinnate, occasionally divided into capillary segments. Petals white, pale blue, or yellow. Fruit ovate or

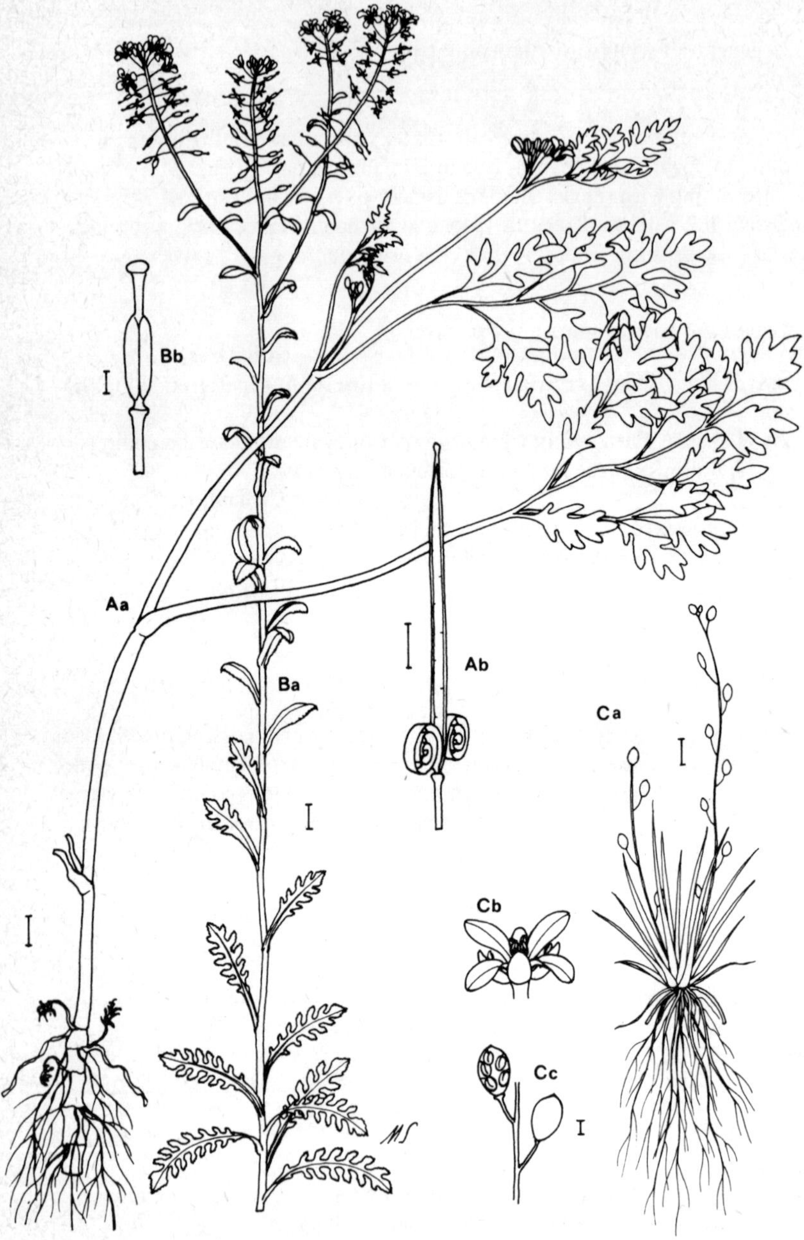

Fig. 64. A. *Cardamine geraniifolia* DC.: a, habit (1 cm); b, fruit (1 cm);
B. *Rorippa aquatica* (Eaton) Palmer and Steyermark: a, habit (1 cm); b, fruit (1 mm);
C. *Subularia aquatica* L.: a, habit (1 cm); b, flower; c, fruit (1 mm).

elongate, round in transverse section; valves of the fruit not coiling spirally from the base at dehiscence; seeds in 1 or 2 rows in each loculus.

Bleasdale, J. K. A. The flowering and growth of watercress (Nasturtium officinale R. Br.) Journ. Hort. Sci. 39: 227–233 (1964)
Green, P. S. Watercress in the New World. Rhodora 64: 32–43 (1962)
Howard, H. W. and Lyon, A. G. Biological Flora of the British Isles: Nasturtium officinale R. Br. Journ. Ecol. 40: 228–245 (1952)
Jonsell, B. Studies on the north-western European species of Rorippa s. str. Symb. Bot. Upsal. 19 (2): 1–222 (1968)

c. 70 species: almost cosmopolitan but chiefly from temperate regions and tropical mountains. About 4 species are aquatic, they are emergent herbs that grow in shallow, still or flowing, usually eutrophic water. *R. amphibia* (L.) Besser and *R. aquatica* (Eaton) Palmer and Steyermark develop finely divided submerged leaves. *R. nasturtium-aquaticum* (L.) Hayek [*Nasturtium officinale* R. Br.] and *R. microphylla* (Boenn) Hyl. [*N. microphyllum* (Boenn) Reichenb.] and their hybrid are known as watercress and are widely cultivated as salad plants.

Subularia L., Sp. Pl. 642 (1753)
Fig. 64C.
Annual, tufted, glabrous herbs. Stems reduced to corm-like structure. Leaves 1 to 7 cm long, numerous, in a rosette, subulate or linear, terete; apex acute. Flowers 2 to 12 (–18), when submerged usually cleistogamous. Sepals 0.7 to 1.3 mm long, erect. Petals slightly longer than sepals, white, sometimes absent. Ovary surrounded by a fleshy ring; capsule narrowly elliptic to broadly obovate, up to 5 mm long; seeds up to 6 in each loculus.

Mulligan, G. A. and Calder, J. A. The genus Subularia (Cruciferae). Rhodora 66: 127–135 (1964)

2 species: *S. aquatica* L. (leaves subulate) N. America, N. Europe and C. Asia but apparently absent in N. Asia from the Urals to Kamtschatka; *S. monticola* A. Br. ex Schweinf. (leaves linear, tapering above the middle) at high elevations on the mountains of Uganda, Congo, Keyna and Tanganyika. Both species are found submerged in pools and lakes and are rather characteristic for base-poor conditions.

BUTOMACEAE

1 genus, *Butomus* (the Limnocharitaceae is recognised as a separate family in this treatment).

Butomus L., Sp. Pl. 372 (1753)
Fig. 65.

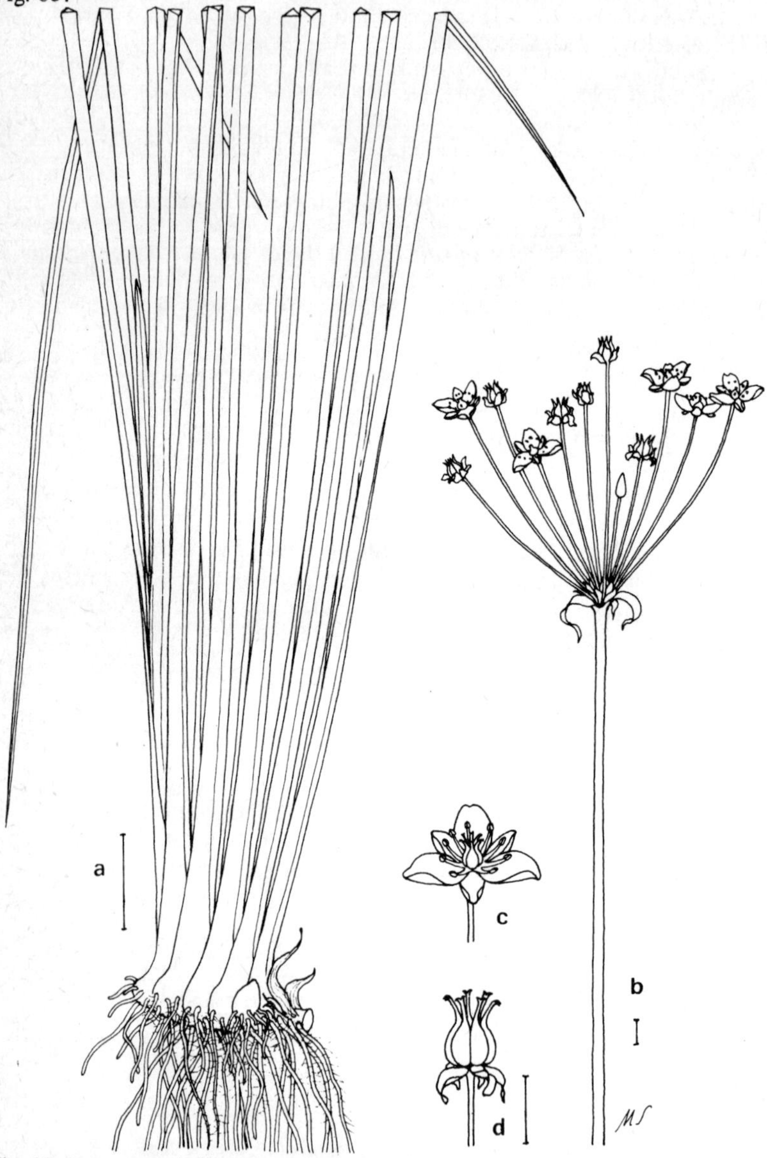

Fig. 65. *Butomus umbellatus* L.: a, habit (5 cm); b, inflorescence (1 cm); c, flower; d, fruit (1 cm).

Rhizomatous perennial. Leaves linear, 50 to 100 cm or more long, 0.3 to 1.0 cm wide, triangular in transverse section, arising in 2 ranks from rhizome; apex acuminate. Inflorescence 1 m or more long, exceeding the leaves, umbel-like (1 terminal flower surrounded by 3 cymes); pedicels 5 to 10 cm long. Flowers bisexual, actinomorphic. Sepals 3, petal-like, pink with darker veins, persistent, 1.0 to 1.5 cm long. Petals 3, like sepals but somewhat longer. Stamens 6 to 9; pollen 1-sulcate. Carpels superior, 6 to 9, slightly united at base; when ripe obovoid, crowned by persistent style; seeds numerous, scattered over the inner surface of the carpel wall, except on midrib and edges; embryos straight.

Buchenau, F. in Engler, A. Pflanzenreich 15 (IV.16) 1–12 (1903)
Stuckey, R. L. Distributional history of Butomus umbellatus in the Western Lake Erie and Lake St. Clair region. Michigan Bot. 7: 134–142 (1968)

1 species, *B. umbellatus* L.: Europe and temperate Asia, naturalised in N. America. Usually occurs in swamps, ditches, and along pools, lakes and rivers; it may grow in a sterile state completely submerged in flowing water. It is a decorative plant and is frequently cultivated.

CABOMBACEAE

2 genera: *Brasenia, Cabomba.*

Perennial aquatic herbs. Roots adventitious at nodes. Stems creeping and rhizomatous or erect and leafy. Leaves floating or submerged, palmately veined, either all peltate, undissected, alternate and floating, or submerged and palmately dissected and opposite or whorled. Flowers single on long axillary peduncles, all parts hypogynous and free. Sepals 3 or 4. Petals 3 or 4, white, yellow or purple. Stamens 3 to 36; pollen 1-sulcate. Carpels 1 to 18; stigmas terminal; ovules 1 to 3; fruit indehiscent.

Li, H.-L. Classification and phylogeny of Nymphaeaceae and allied families. Amer. Midl. Nat. 54: 33–41 (1955)

Wood, C. E. The genera of the Nymphaeaceae and Ceratophyllaceae in the Southeastern United States. Journ. Arnold Arb. 40: 94–112 (1959)

1A Dissected leaves present on submerged stems
Cabomba

1B Dissected leaves absent (all leaves peltate)
Brasenia

Brasenia Schreb., Gen. Pl. ed. 8, 1: 372 (1789)
Fig. 66A.
Submerged parts of plant heavily covered with mucilaginous jelly. Stem rhizomatous bearing leaves and axillary runners which creep on the surface of mud. Leaves alternate, peltate, oval, floating (seedling leaves often submerged). Sepals 3 or 4. Petals 3 or 4, linear, sessile, purple. Stamens 18 to 36. Carpels 4 to 18, oblong-ovate; ovules usually 2.

Adams, F. S. Winterbud production and function in Brasenia schreberi. Rhodora 71: 417–432 (1969)

Kanna, P. Morphological and embryological studies in Nymphaeaceae. II Brasenia schreberi Gmel. and Nelumbo nucifera Gaertn. Austral. Journ. Bot. 13: 379–387 (1965)

Raymond, M. and Dansereau, P. The geographical distribution of the bipolar Nymphaeaceae, Nymphaea tetragona and Brasenia schreberi. Proc. Pacific Sci. Congr. 7 (5): 122–131 (1953)

Richardson, F. C. Morphological Studies of the Nymphaeaceae. IV Structure and development of the flower of Brasenia schreberi Gmel. Univ. Calif. Publ. Bot. 47: 1–45 (1969)

1 species: *Brasenia schreberi* Gmel. widely but sporadically distributed in N. and C. America, West Indies, Africa, E. Asia and Australia. *Brasenia* is found in ponds and slow flowing rivers and streams. It is not known to be of any economic importance.

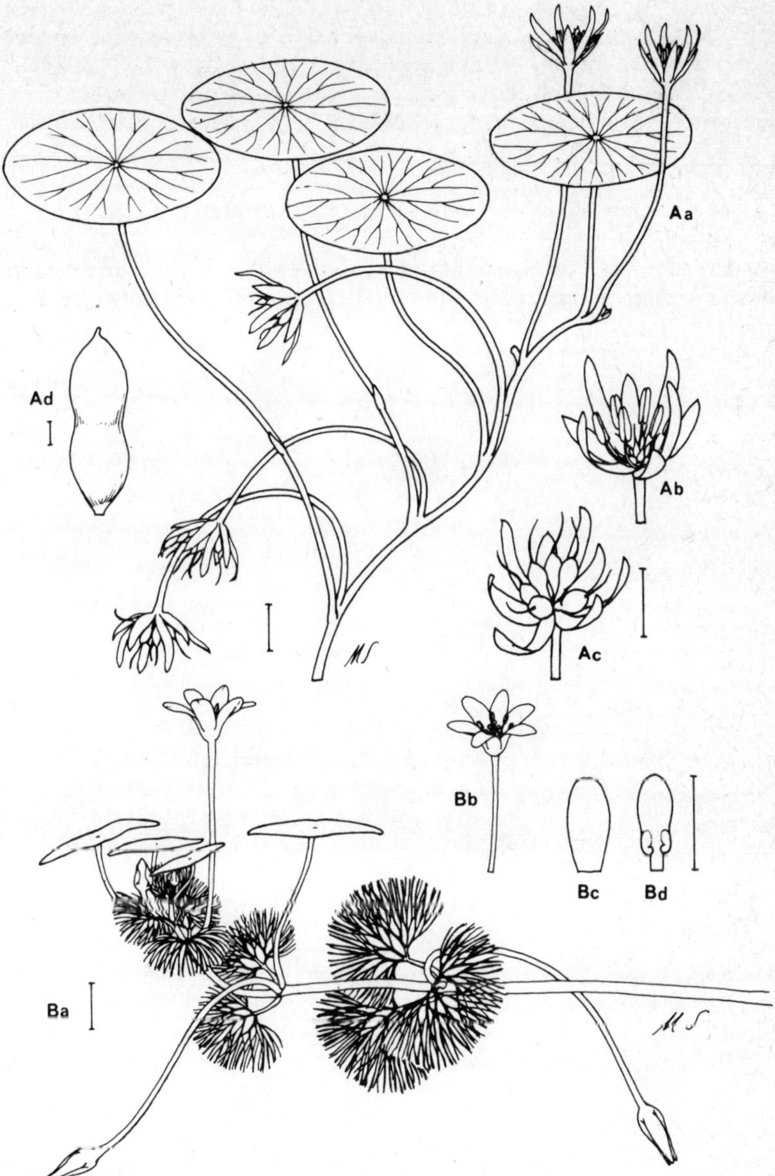

Fig. 66. A. *Brasenia schreberi* Gmel.: a, habit (1 cm); b, flower; c, fruiting head (1 cm); d, nutlet (1 mm);
B. *Cabomba piauhyensis* Gardner: a, habit (1 cm); b, flower; c, outer petal; d, inner petal (1 cm).

Cabomba Aublet, Hist. Pl. Guiane Franç. 1: 321, t.124 (1775)
Fig. 66B.
Strictly aquatic herbs. Leaves dimorphic; submerged leaves opposite or whorled, palmately dissected into many linear segments; floating leaves (when present) peltate and tapered to both ends or narrow and forked. Sepals 3. Petals 3, clawed, often with auricules, white, yellow or purple. Stamens 3 to 6. Carpels (1) 2 to 4, elongate; ovules usually 3.

Fassett, N. C. A monograph of *Cabomba.* Castanea 18: 116–128 (1953)

c. 7 species: warm and temperate regions of the New World. They are found in ponds and streams. Several species are very decorative and are widely used as aquarium plants.

CALLITRICHACEAE

1 genus.

Callitriche L., Sp. Pl. 969 (1753)
Fig. 69A.
Delicate annuals or perennials. Stems elongate when submerged, contracted when emersed. Leaves opposite (often rosette-like at tips of floating stems), linear, elliptic, oblong or spathulate, entire, often with forked apex; stipules absent. Flowers minute, unisexual, solitary or rarely 1 male and 1 female in the same leaf axil. Sepals and petals absent. Male flower consists of 1 stamen; filament slender, elongate; anther reniform, 2-locular, with longitudinal dehiscence; the dehiscence slits becoming continuous at the top. Female flower consists of 1 ovary; ovary 4-lobed, 4-locular; ovules 1 in each loculus, pendulous; styles 2, free, elongate, often papillose; fruit 4-lobed, with each lobe winged or keeled, at maturity splitting into 4 nutlets.

Fassett, N. C. Callitriche in the New World. Rhodora 53: 137–155, 161–182, 185–194, 209–222 (1951)
Mason, R. Callitriche in New Zealand and Australia. Austral. Journ. Bot. 7: 295–327 (1959)
Rubtzoff, R. Notes on Callitriche in western North America. Wasmann Journ. Biol. 27 (1): 103–114 (1969)
Schotsman, H. D. Les Callitriches – Espèces de France et Taxa nouveaux d'Europe, in Jovet, P. Flore de France 1: Paris (1967)
Záhradniková-Rošetzka, K. Rozširenie druhov rodu Callitriche na slovensku. Biológia (Bratislava) 23 (4): 257–266 (1968)

c. 17 species: almost cosmopolitan. Some species are obligate submerged aquatics with underwater pollination, others are amphibious and a few are terrestrial. All species are small slender plants but may grow gregariously. They occur in a wide variety of aquatic and semi-aquatic habitats and are most frequent in regions with temperate climates. Most species are very plastic in form and their identification at specific level is often difficult; before attempting any determinations it is recommended to consult the work of Schotsman (1967).

CANNACEAE

1 genus.

Canna L., Sp. Pl. 1 (1753)
Fig. 67.
Tall leafy perennials, up to 2 m high. Leaves pinnately nerved from the midrib; sheathing at the base, usually without a distinct petiole; nerves parallel. Flowers zygomorphic in racemes or panicles, irregular. Sepals 3, overlapping free. Petals 3, united at the base and fused to the staminal column. Stamens petaloid; the 3 outer sterile, 2 inner united, 1 free with a solitary 1-locular anther on its side. Ovary inferior, 3-locular; ovules numerous; fruit a capsule; seeds rounded, with a very hard endosperm.

Segeren, W. and Maas, P. J. M. The genus Canna in northern South America. Acta Bot. Neerl. 20 (6): 663–680 (1971)

c. 50 species: pantropical but most species in S. America. *C. glauca* L. in central S. America and *C. siamensis* Kränzl. in Thailand are reported to be aquatic.

Fig. 67. *Canna glauca* L.: a, leaf (2 cm); b, inflorescence (1 cm).

CENTROLEPIDACEAE

5 genera and about 40 species, mainly in the southern hemisphere. Several of the species are small submerged aquatics which flower when the habitat becomes dry.

Small tufted annuals or perennials. Leaves linear, mainly basal or distichous, without a ligule. Inflorescence a terminal head with few opposite bracts, and few to many flowers. Flowers unisexual or bisexual, subtended by 1 to 3 hyaline glumes. Sepals and petals absent. Stamens usually 1 or 2; filaments capillary. Styles capillary, 1 to many, free or united. Ovule solitary. Ovary 1- to 3-locular, or ovary of 2 or more carpels, superposed in rows.

There are two different interpretations of the flower and inflorescence of the Centrolepidaceae. Hieronymous (1873), followed by Bentham and Hooker (1883) and Hutchinson (1959) regarded the flowers of *Centrolepis* as bisexual, whereas Eichler (1875) followed largely by Hamann (1963), considered that all the flowers are unisexual, consisting either of a single anther or a single carpel, and the apparently bisexual flowers are reduced inflorescences or pseudanthia.

Moore, L. B. and Edgar, E. Centrolepidaceae in Flora of New Zealand 2: 79–85 (1970)

1A Flowers in a head, surrounded by usually 4 more or less equal bracts
- 2A Male and female flowers in separate heads **Hydatella**
- 2B Male and female flowers together in same head **Trithuria**

1B Flowers in a flattened spike, subtended by 2 large bracts
- 3A Lowest bracts enclosing the flowers; carpels 3 or more, united on the carpophore **Aphelia**
- 3B Lowest bract not enclosing the flowers; carpels solitary **Centrolepis**

Aphelia R. Br., Prodr. Fl. Nov. Holl. 251 (1810)
Fig. 68A.
Small tufted annuals up to 2 cm high. Leaves linear, capillary, somewhat sheathing at base. Flowers unisexual, rarely bisexual, solitary within distichous bracts, forming a short terminal spike. Male flowers each with a single stamen and a 1-locular anther, few at the base of the spike; the female above, each of a single style and 1 carpel.

c. 6 species: Australia. *A. gracilis* Sonder occurs in shallow water and wet places in South Australia.

Centrolepis Labill., Nov. Holl. Pl. 1: 7 (1840)
Fig. 68B.
Usually very small annuals (tufted perennials in a few non-aquatic species). Leaves linear, capillary, all basal, sheathing at base. Flowers bisexual, forming a terminal spike, enclosed by 2 leaf-like bracts. Stamen 1, with a 1-locular anther. Carpels 3 or more, on slender stalks of unequal length which are united into an erect carpophore, so that the carpels are arranged in 2 rows or spirally. Styles undivided, 1 per carpel. Seeds ovoid-oblong, enclosed in a pericarp or utricle.

c. 25 species: S. E. Asia and Australia. *C. banksii* (R. Br.) Roemer and Schultes occurs as a weed in ricefields in northern Australia and S. E. Asia.

Hydatella Diels in Diels & Pritzel., Bot. Jahrb. 35: 93(1904), [*Juncella* F. Mueller pro parte]
Fig. 68C.
Small annuals, often bright reddish, 1 to 4 cm high. Leaves all basal, linear, capillary, not sheathing at base. Flowers minute, crowded into a head, with 2 or 4 hyaline bracts. Male and female flowers on separate stems; the male with a 2-locular anther; the female with a short stalk and many unequal septate styles. Fruit apiculate, indehiscent.

2 species in Western Australia and New Zealand. *H. inconspicua* (Cheesem.) Cheesem. is found in shallow water in the edges of lakes in New Zealand.

Trithuria Hooker fil., Bot. Antarct. Voy. 3 (2): 78 (1858) [*Juncella* F. Mueller pro parte]
Fig. 68D.
Small annuals, 1 to 3 cm high, usually reddish. Leaves capillary, all basal. Flowers unisexual, mixed in a terminal head. Male flowers of 1 long filament and a 2-locular anther, female of one, 1-locular ovary and usually 3 simple or divided styles, more than the male. Ovules solitary in each loculus. Capsule triquetrous, opening from the base in three valves.

2 species in temperate Australia. *T. submersa* Hook. fil. is found in shallow water in South Australia and Tasmania. The genera *Trithuria* and *Hydatella* are sometimes united under the name *Juncella* F. Mueller., see Edgar, E. N. Z. Journ. Bot. 4: 157 (1966).

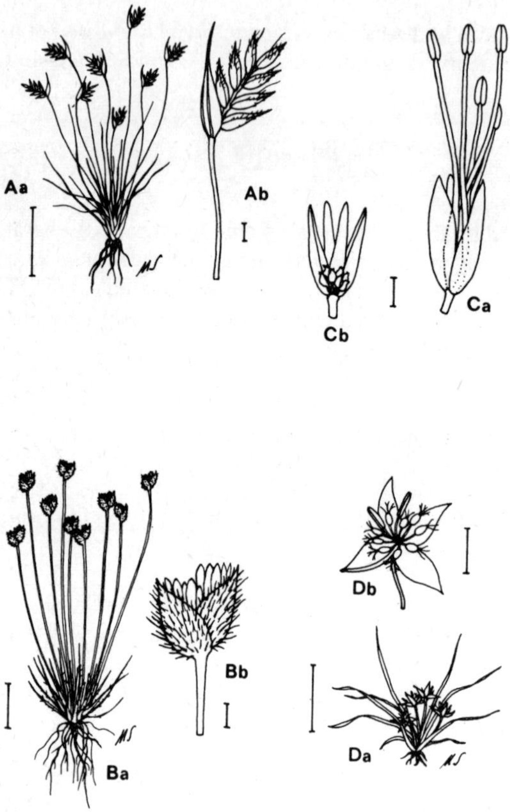

Fig. 68. A. *Aphelia gracilis* Sond.: a, habit (1 cm); b, inflorescence (1 mm);
B. *Centrolepis fascicularis* Labill.: a, habit (1 cm); b, inflorescence (1 mm);
C. *Hydatella inconspicua* (Cheesem.) Cheesem.: a, male inflorescence; b, female inflorescence (1 mm);
D. *Trithuria submersa* Hooker fil.: a, habit (1 cm); b, inflorescence (1 mm).

CERATOPHYLLACEAE

1 genus.

Ceratophyllum L., Sp. Pl. 2: 922 (1753)
Fig. 69B.
Entirely submerged herbs. Roots absent (even in embryo) but branches sometimes modified as "rhizoids". Stem branched, not more than one branch at a node. Leaves whorled, (3–) 6 to 8 (–10) at a node, rather rigid, often brittle, 1 to 4 times forked, with 2 rows of minute teeth along the ultimate segments; segments tipped by 2 bristles. Flowers unisexual, usually solitary in the axil of 1 leaf in a whorl. Perianth of 8 to 12 linear segments, united at base. Stamens numerous; filaments short or absent; connective prolonged apically. Ovary superior, 1-locular; fruit a 1-seeded nut, tipped by a persistent style and often with additional basal or marginal spines.

Carr, J. L. Primary productivity and physiology of Ceratophyllum demersum. Austral. Journ. Marine Freshw. Res. 20: 115–126, 127–142 (1969)
Wood, C. E. The genera of Nymphaeaceae and Ceratophyllaceae in the southeastern United States. Journ. Arnold Arb. 40: 94–112 (1959)

2 or *c.* 30 species: cosmopolitan. The species of *Ceratophyllum* are very variable and are taxonomically difficult. *Ceratophyllum demersum* L. and *C. submersum* L. are found throughout the world except in the colder regions. The other species are local and based mainly on fruit characteristics; more than 30 species have been described but most of them are probably no more than local variants of *C. demersum* and *C. submersum.* All species are obligately submerged aquatics and can not tolerate periods of emergence; they are found free-swimming or loosely anchored in silt in still and flowing water. Large populations may block or constrict water channels. For work on the economic importance of Ceratophyllum see: Hambric, R. N., Proc. 18th. Sth. Weed Conf. 458–463 (1965) and Little, E. C. S., Weed Res. 8: 79–105 (1958).

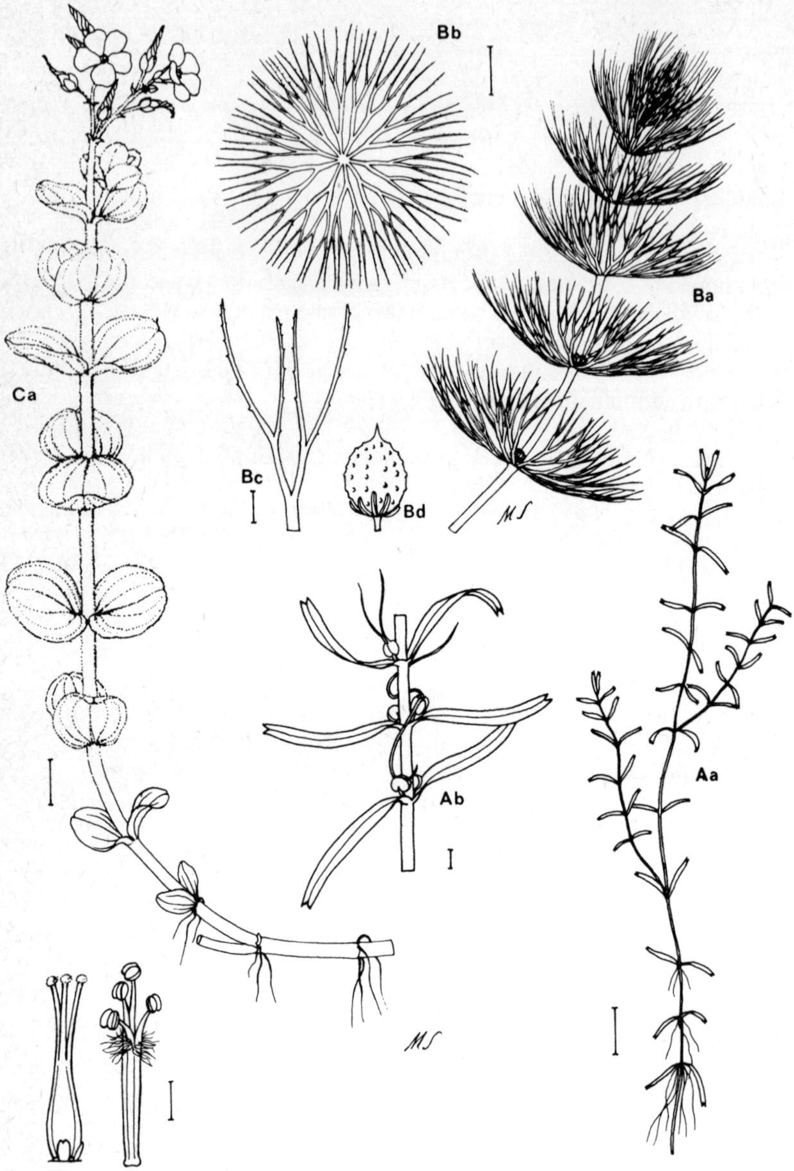

Fig. 69. A. *Callitriche truncata* Guss.: a, habit (1 cm); b, detail of flowering shoot (1 mm); B. *Ceratophyllum submersum* L.: a, habit; b, whorl of leaves (1 cm); c, detail of leaf (2 mm); d, fruit;
C. *Hypericum elodes* L.: a, habit (1 cm); b, styles; c, stamens (1 mm).

CLUSIACEAE [GUTTIFERAE]

c. 40 genera; *c.* 100 species; cosmopolitan. *Hypericum* is the only genus containing aquatics.

Hypericum L., Sp. Pl. 783 (1753)

Fig. 69C.

Annual, perennial or woody. Stems erect or creeping; internodes often swollen when submerged. Leaves opposite, entire, simple, with translucent, oil-filled glands. Flowers, actinomorphic. Sepals (4–) 5, free, often glandular. Petals (4–) 5, free, contorted in bud, yellow, sometimes tinged with red. Stamens numerous, grouped together in bunches. Ovary 1- to 5- locular, superior; ovules numerous; fruit a septicidal capsule, rarely fleshy and indehiscent.

c. 400 species: temperate regions and tropical mountains. Many species are found in bogs and marshes and can tolerate temporary flooding. *H. elodes* L., confined to W. Europe, is, perhaps, the only species habitually found in water.

COMMELINACEAE

The Commelinaceae has about 38 genera and 500 to 600 species. Several genera, such as *Aneilema* R. Br. sensu stricto, *Commelina* L., *Cyanotis* D. Don. and *Floscopa* Lour. have species that grow in wet places but only *Murdannia* Royle sensu stricto has habitually aquatic species.

Perennial herbs. Leaves alternate, parallel nerved, with basal, membranous, closed sheaths. Flowers solitary or in variously complex inflorescences, actinomorphic or zygomorphic, usually bisexual. Sepals 3, free, usually green. Petals 3, free, membranous, usually coloured. Stamens 5 or some absent or staminodial; anthers basifixed. Ovary superior, 3- or rarely 2-locular; style simple; stigma terminal, capitate or 3-lobed; ovules few or solitary in each loculus, axile; fruit a capsule or rarely fleshy and indehiscent.

Murdannia Royle, Ill. Bot. Himal. 403, t. 95, fig. 3 (1839), nom. cons., [*Aneilema* R. Br., pro parte]

Fig. 70A.

Stems creeping or floating in water. Leaves elongate. Flowers solitary or paired in leaf axils or in inflorescences. Sepals 3, subequal. Petals 3, subequal, blue, pink or white. Stamens usually 3, fertile; staminodes usually 3, alternating with stamens (occasionally staminodes reduced in number or absent, or 1 stamen transformed into a staminode or absent); apex of staminodes 3-lobed (like a pawnbroker's sign) or the lateral lobes sometimes so near the filament as to make the apex hastate.

Brenan, J. P. M. Notes on African Commelinaceae. Kew Bull. 1952: 179–190 (1952)

c. 50 species: pantropical. Some species are found in pools, streams and irrigation ditches. *M. blumei* (Hassk.) Brenan [*Aneilema hamiltonianum* Wall. ex Clarke] is frequently reported as a ricefield weed in Asia.

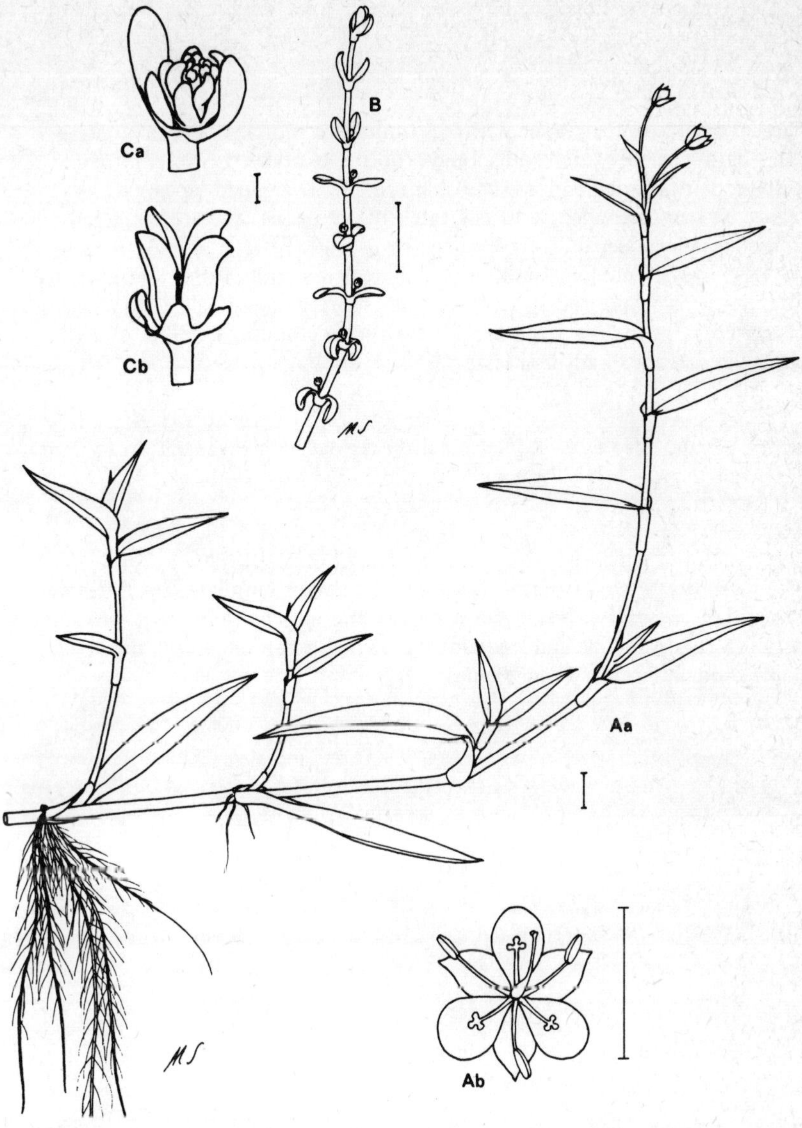

Fig. 70. A. *Murdannia blumei* (Hassk.) Brenan: a, habit (1 cm); b, flower (1 cm);
B. *Crassula granvikii* Mildbr.: habit (1 cm);
C. *Crassula aquatica* (L.) Schönl.: a, flower (1 cm); b, fruiting head.

CONVOLVULACEAE

c. 55 genera: *c.* 1650 species: cosmopolitan. A single aquatic species: *Ipomoea aquatica* Forsk.

Ipomoea L., Sp. Pl. 159 (1753)
Fig. 71.

Perennial or occasionally annual. Stems trailing on moist ground or floating, when floating usually thick and spongy, rooting at nodes. Leaves alternate, petiolate; blades variable in size and shape, ovate, triangular, lanceolate or linear; base truncate, cordate to sagittate. Inflorescences axillary, 1- to few-flowered, cymose. Sepals 5, free, persisting in fruit. Petals 5 united, funnel-shaped, 2.5 to 5.0 cm long, pink or pale lilac, often with purple centre, rarely white. Stamens 5, inserted at base of petals, shorter than petals. Ovary superior, of 2 united carpels; style 1, simple; fruit ovoid to globose, 4 valved, often splitting irregularly; seeds 4 or less.

Datta, S. C. and Biswas, K. K. Germination regulating mechanisms in aquatic angiosperms. I. Ipomoea aquatica Forsk. Ciências Naturais 39: 175–185 (1970)

c. 500 species of which 1 is aquatic, *I. aquatica* Forsk. [*I. reptans* Poir., *I. repens* Roth] : circumtropical. Occurs in marshy or inundated localities, in pools and ditches. Often forms dense masses and may be found as a weed in rice fields. It is palatable and frequently grown as a vegetable; it is also used as a pig food and locally in medicines.

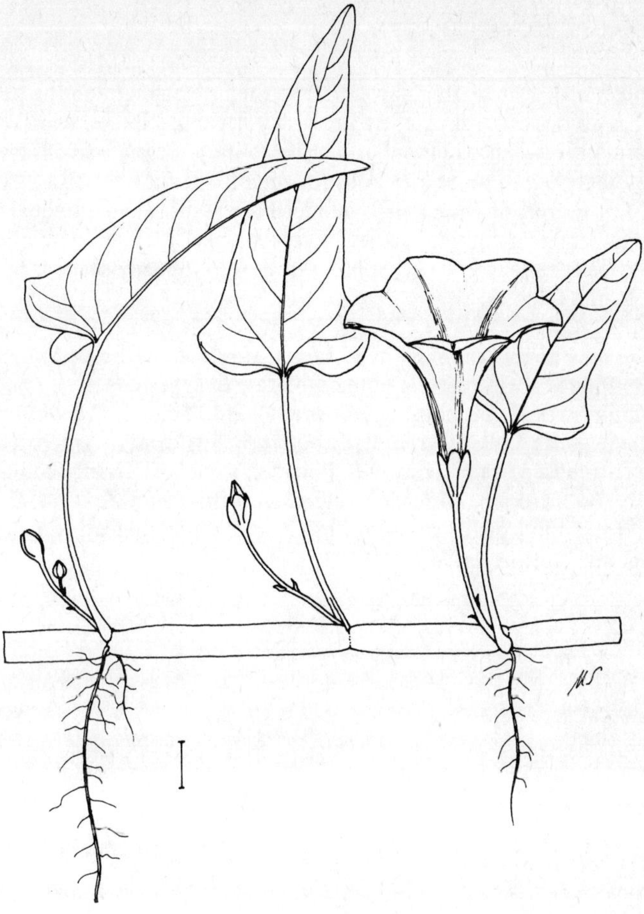

Fig. 71. *Ipomoea aquatica* Forsk.: habit (1 cm).

CRASSULACEAE

c. 35 genera: *c.* 1500 species: cosmopolitan. Only the genus *Crassula* contains aquatics.

Crassula L., Sp. Pl. 282 (1753), [*Tillaea* L.]
Fig. 70B, C.

Annual or perennial, succulent, glabrous herbs. Stems creeping floating or ascending. Leaves opposite, simple, entire, fleshy, lacking stipules, usually joined at base. Inflorescence cymose or usually solitary. Flowers small, bisexual, actinomorphic, 3-, 4- or 5-merous (in aquatic species). Sepals 3, 4 or 5. Petals 3, 4 or 5, free, or occasionally reduced or absent. Stamens equal in number to the petals (in aquatic species). Carpels superior, equal in number to the petals, free or slightly joined at base; fruit of 1- to many-seeded, free or partially joined follicles.

c. 300 species: cosmopolitan, most in S. Africa. Most species are xerophytic but a few are found in wet places and may tolerate long periods of submergence (i.e. *C. aquatica* (L.) Schönl. *C. vaillantii* (Willd.) Schönl. *C. bonariensis* (DC.) Camb. *C. tillaea* Lester-Garland); *C. bonariensis* is from S. America and has become naturalised in ricefields in S. Portugal. *C. natans* Thunb. and *C. inanis* Thunb., both from S. Africa, *C. helmsii* Kirk from New Zealand, *C. granvikii* Mildbr. from Uganda and a few other species are obligate aquatics which flower and fruit in water.

CYPERACEAE

Perennial or sometimes annual herbs, rarely shrubs, creeping, tufted, or floating in water. Flowering stems usually solid, triangular in transverse section, rarely terete or jointed. Leaves mostly basal, linear and grass-like, often reduced to sheaths, flat or inrolled, usually without a ligule. Inflorescence variable; from a single, terminal, few-flowered spikelet to a compound panicle with many spikelets, often with leaf-like involucral bracts. Spikelet few- to many-flowered, often with all the flowers except one sterile, unisexual or bisexual. Flowers unisexual or bisexual, solitary within a glume-like bract. Perianth reduced to hairs, bristles, scales or absent, sometimes forming a sac which surrounds the nut. Stamens usually 3. Ovary superior, with one erect ovule; style 1; stigmas 2 or 3; fruit a nut, either trigonous (if stigmas 3), or biconvex (if stigmas 2), often with a persistent beak.

Clarke, C. B. Illustrations of Cyperaceae, London (1909)
Hutchinson, J. The Families of Flowering Plants, vol II, Monocotyledons, Oxford (1959)
Koyama, T. Classification of the family Cyperaceae. Journ. Fac. Sci. Tokyo Univ. (Bot.): 37–148 (1961)
Kükenthal, G. Cyperaceae, Caricoideae in Engler, Pflanzenreich 38 (IV 20): 1–824 (1909); Cyperaceae, Cypereae in Engler, Pflanzenreich 101 (IV 20): 1–671 (1935–36)
Kukkonen, I. Gedanken und Probleme zur Systematik der Familie Cyperaceae. Eine Zusammenfassung. Aquilo, ser. Bot. 6: 18–42 (1967)
Lye, K. A. Studies in African Cyperaceae II. Bot. Not. 124: 280–286 (1971); Moderne oppfatving ar slekta Scirpus L. Blyttia 29: 141–147 (1971)
Mattfeld, J. Das morphologische Wesen und die phylogenetische Bedeutung der Blumenblätter. Ber. Deutsch. Bot. Ges. 56: 86–116 (1938)
Metcalfe, C. R. Anatomy of the Monocotyledons V. Cyperaceae. Oxford (1971)

There are many different interpretations of the Cyperoid flower. The "Synanthientheorie" of Mattfeld (1938) considers that the basic flower is unisexual and that bisexual flowers are reduced inflorescences. According to this theory, the hypogynous bristles or scales correspond not to perianth, but to bracts or parts of bracts. The opposing view is held by Hutchinson (1959). The scales and bristles are considered to be perianth, and the number of stamens in a single flower indefinite, usually 3, but up to 12 or more in the Hypolytreae.

In this account, Hutchinson's theory has been accepted as a convenient basis. The number of stamens in the flower is considered to be 3, in some cases reduced to one; many stamens arise by the grouping of a number of one-stamened flowers. The subject is discussed by Koyama (1961), Kukkonen (1967) and Metcalf (1971).

Several different arrangements of genera have been proposed, especially within *Cyperus* s.l. and *Scirpus* s.l. The treatment adopted here takes *Cyperus* in its broadest sense, and *Scirpus* in a narrow sense following Lye (1971).

1A All glumes arranged strictly in 2 rows on the axis of the spikelet; all flowers except lowest fertile
2A Inflorescence a terminal, simple or compound umbel or head; perianth bristles absent
Cyperus
2B Inflorescences axillary from the leaf sheaths; perianth bristles present with downward pointing hairs
Dulichium
1B Glumes arranged spirally on the axis, or the lowest glumes in two rows and sterile
3A Inflorescence a solitary terminal spikelet; bracts absent or very small
4A Flowers unisexual; nut enclosed in a sac
Carex
4B Flowers bisexual; nut not enclosed in a sac
5A Stems erect; plant tufted or with a creeping rhizome
Eleocharis
5B Stems creeping, or floating, or submerged, often rooting at the nodes
6A Plant with numerous sterile stems in successive whorls; perianth bristles present
Websteria
6B Plant without whorls of sterile stems; perianth bristles absent
Eleogiton
3B Inflorescence of rarely 1 apparently lateral, usually 2 or more spikelets; bracts present
7A Spikelets much reduced and crowded into spikelet-like spikes; glume-like bracts very hard and stiff
8A Inflorescence a compound panicle of many spikes
9A 7 or 8 stamens per spikelet
Diplasia
9B 2 stamens per spikelet
Hypolytrum
8B Inflorescence simple, of 1 to few lateral spikes
10A Spikes large, solitary
Lepironia
10B Spikes small, grouped in a lateral cluster
Chorisandra
7B Spikelets not reduced; glumes not hard and stiff
11A Flowers unisexual, ovary the stamens not enclosed in the same glume
12A Inflorescence of bisexual clusters of spikelets in the axils of leafy bracts
13A Clusters of spikelets sessile or shortly stalked, solitary or few in each leaf axil
Diplacrum
13B Clusters of spikelets long stalked; many in each leaf axil; stalks often recurved
Calyptocarya

12B Inflorescence paniculate, or if of axillary clusters, then spikes unisexual
14A Nut very large, often with a cupule or stipe at its base
Scleria
14B Nut small, without cupule
15A Female flowers enclosed in a sac
Carex
15B Female flowers enclosed by 2 glumes
Lagenocarpus
11B At least one flower of spikelet bisexual, with ovary and stamens in the same glume
16A Spikelet with many perfect flowers
17A Perianth bristles very long and silky, more than 6
Eriophorum
17B Perianth bristles short, or absent, or if silky then only 6
18A Inflorescence with several flat or leaf-like bracts at the base
19A Spikelets 1 to 6 in a terminal head; flower enclosed by small scales
Ascolepis
19B Spikelets not in a terminal head; perianth of bristles or absent
20A Inflorescence of sessile and shortly stalked heads; stigmas 2; bristles absent
Oxycaryum
20B Inflorescence several times compound; stigmas 3; bristles usu. 6. often silky
Scirpus
18B Inflorescence with one leafy or several non-leafy bracts at the base
21A Style swollen at the base, caducous or persistent
22A Spikelets in terminal and axillary cymes; style almost wholly persistent; beak large
Psilocarya
22B Spikelets in a terminal umbel; style base only persistent; beak small or absent
Fimbristylis
21B Style not swollen at the base, wholly caducous
23A Glumes very hairy; perianth of well-developed scales
Fuirena
23B Glumes not hairy; perianth, if present, of bristles or slender scales
24A Plant robust; perianth bristles usually present
Schoenoplectus

24B Plant very slender; perianth bristles always absent
Isolepis

16B Spikelet with few (1 to 4) perfect flowers
25A Stigmas 2; beak large
26A One fertile flower per spikelet; perianth bristles usually present
Rhynchospora
26B Usually 3 fertile flowers per spikelet; perianth bristles absent
Dichronema
25B Stigmas 3; beak small, or absent
27A Axis of spikelet thickened and flexuose between the flowers
Schoenus
27B Axis of spikelet not thickened and flexuose
28A Nut without beak; inflorescence corymbose
Cladium
28B Nut with beak; inflorescence not corymbose
29A Perianth bristles well developed; glumes 2-ranked at base of spikelet
Asterochaete
29B Bristles absent or very small; glumes all spiral
30A Stamens 3 to 6; filaments lengthening in fruit, persistent; only upper flower of spikelet fertile
Gahnia
30B Stamens 2 to 3; filaments not lengthening; usually 2 or 3 flowers of spikelet fertile
Baumea

Ascolepis Nees ex Steud., Syn. Pl. Cyp. 105 (1855), nom. cons.
Fig. 72.
Glabrous, usually tufted perennials, or annuals. Stems terete or triangular in transverse section, up to 80 cm high. Leaves shorter than the stem. Inflorescence a globose or hemispheric head of densely packed, many-flowered spikelets, occasionally of 2 or 3 distinct spikelets. Bracts leafy, strongly deflexed. Flowers bisexual. Glumes glabrous, lanceolate, spirally arranged. Perianth of 1 hypogynous scale, usually encircling the flower, longer than the glume and forming an elongated conical beak. Stigmas 2 or 3. Stamens 3 or 1. Nut very small, smooth, trigonous or biconvex.

c. 15 species: Tropical Africa and America. Most species occur in wet places. *A. pusilla* Ridley is found in rice-fields in W. Africa.

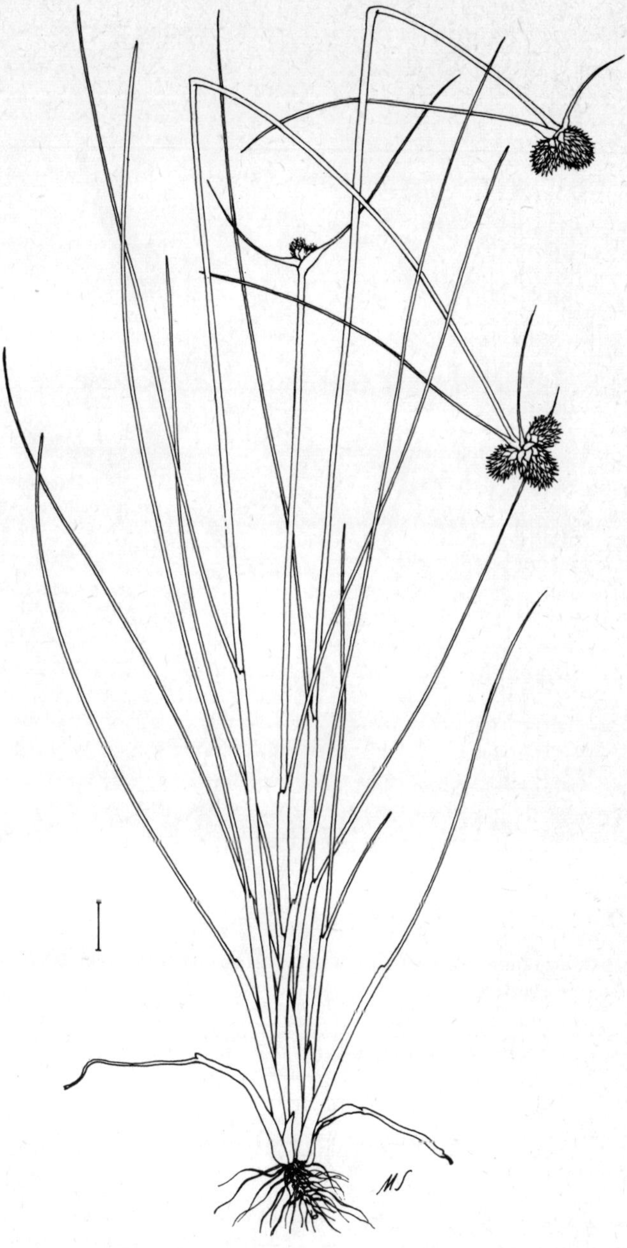

Fig. 72. *Ascolepis brasiliensis* (Kunth) Benth. ex C.B.Cl.: habit (1 cm).

Fig. 73. *Asterochaete glomerata* (Thunb.) Nees: a, habit (3 cm); b, spikelet (1 mm); c, nut (1 mm).

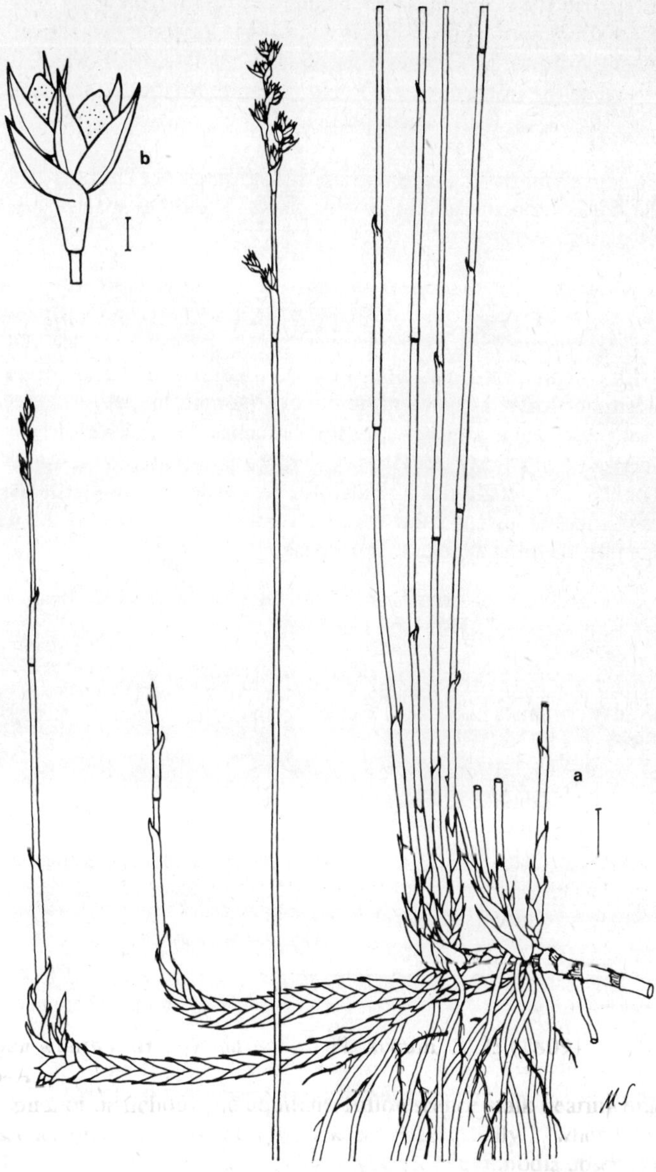

Fig. 74. *Baumea juncea* (R.Br.) Palla: a, habit (1 cm); b, spikelet (1 mm).

Asterochaete Nees, Linnaea 9: 300 (1834)
Fig. 73.
Perennials. Stems trigonous, up to 1 m high. Leaves both basal and cauline, usually shorter than the flowering stem; blades flat, up to 2 cm wide. Inflorescence a condensed panicle. Bracts wide at the base, narrow above. Spikelets usually 5-flowered. Glumes distichous, lanceolate, the lower 3 sterile, the rest bisexual or the uppermost male only. Perianth bristles 6. Stamens 3. Stigmas 3; nut trigonous, with a small beak.

4 species: S. Africa and the Mascarene Islands. *A. glomerata* (Thunb.) Nees occurs at the sides of streams and in marshy places. The other species usually occur in wet places.

Baumea Gaudich. in Freyc., Voy. Bot. 416 (1826), [*Machaerina* Vahl.]
Fig. 74.
Perennials with creeping rhizomes. Stems terete, trigonous or flattened, up to 200 cm high, not leafy. Leaves with terete or trigonous blades or reduced to sheaths. Inflorescence a panicle with short branches, the spikelets in usually sessile clusters, 4- to 11-flowered. Glumes spiral to imbricate, the lowest 1 to 4 empty, the next 1 to 3 fertile and bisexual, the rest male only or sterile. Perianth bristles 6, small or absent. Stamens 2 or 3. Stigmas 3; style base dilated, adnate to the nut, forming a large or small beak.

Kükental, G. Vorarbeiten zu einer Monographie der Rhynchosporoideae. Feddes Rep. 51: 1–7, 139–183 (1942)

c. 70 species: mainly in the tropics but absent from Africa. Several species occur at the edges of lakes in habitats similar to the closely related genus *Cladium* P. Br.

Calyptrocarya Nees, Linnaea 9: 304 (1834)
Fig. 75.
Annuals or perennials, rhizome tufted. Stem up to 40 cm high, trigonous, leafy. Lowest leaf blades long, exceeding the inflorescence. Inflorescence of groups of sessile and stalked, spherical clusters of spikelets in the leaf axils; some of the stems of the clusters arcuate-recurved. Spikelets bisexual with 1 terminal female flower with 2, distichous glumes, and at its base, 2 to 4 male spikelets each with 1 to 4 flowers. Perianth bristles absent. Stamens 1. Stigmas 3; nut small, biconvex.

5 species: tropical America. Found in wet forests and swamps.

Carex L., Sp. Pl. 972 (1753)
Fig. 76.
Rhizomatous or tufted perennials. Stems solid, usually leafy and trigonous, up to 200 cm high. Leaf blades flat, keeled or involute. Inflorescence various, from a much-branched panicle to a simple spike. Spikes unisexual, the upper

Fig. 75. *Calyptrocarya angustifolia* Nees: habit (1 cm).

Fig. 76. *Carex vesicaria* L.: habit (2 cm).

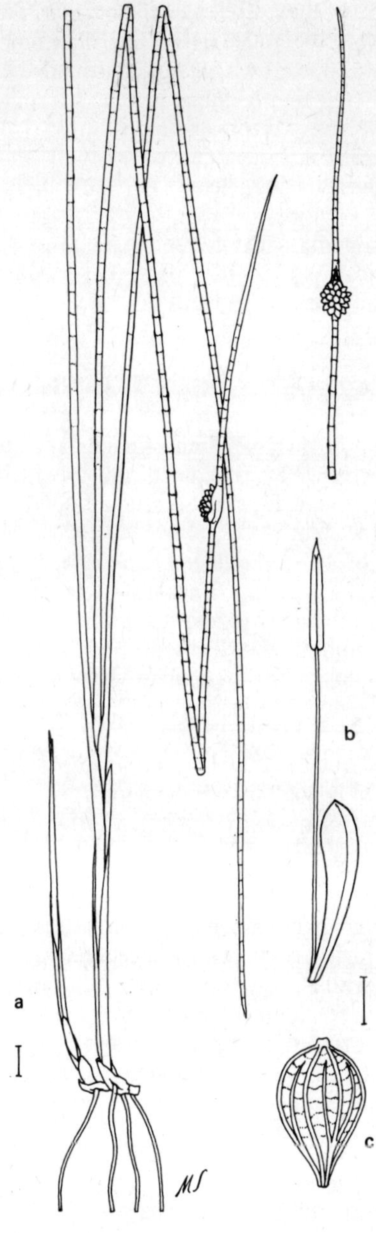

Fig. 77. *Chorisandra cymbaria* R. Br.: a, habit (1 cm); b, male flower (1 mm); c, nut (1 mm).

usually male, or bisexual with the male flowers usually uppermost; sometimes terminal and solitary, and then often plants dioecious. Flowers unisexual, in 1-flowered spikelets, each subtended by a glume. Male flowers with 2 or 3 stamens; perianth absent. Female flowers with 1 nut, surrounded by a globular, trigonous or compressed perianth ending in a beak from which the stigmas project. Stigmas 2 or 3; ovary biconvex or trigonous.

Kükenthal, G. in Engler, Pflanzenreich 38 (IV. 20): 1–824 (1909)

c. 1000 species: cosmopolitan, mainly temperate. Most species occur in damp places. Most of the aquatic species have unisexual spikes (subgenus *Carex*); e.g. *C. riparia* Curt. and *C. pseudocyperus* L.

Chorisandra R. Br., Prodr. Fl. Nov. Holl. 221 (1810)
Fig. 77.
Perennials with a creeping rhizome. Stems slender, stiff, terete, transversely septate, up to 100 cm high. Leaves all basal, the upper with stem-like blades, the lower reduced to sheaths. Bract exceeding the spike by *c.* 10 cm, appearing as a continuation of the stem. Spike solitary, 1 to 2 cm diam., globular, composed of many crowded and reduced bisexual spikelets. Bracteoles variable in shape, rigid. Spikelets with 10 to 12 male flowers, each with 1 glume and 1 stamen; and 1 female flower with about 3 glumes (2 empty male and 1 fertile female), and 1 nut. Perianth absent. Style linear, deeply divided with 2 or 3 stigmas; nut biconvex, with reticulate surface, and a very short truncate beak.

4 species: Australasia. All the species occur in wet places; *C. cymbaria* R. Br. and *C. gigantea* Raynal in lakes in New Caledonia.

Cladium P. Br., Hist. Jamaic. 114 (1756)
Fig. 78.
Stout perennials with creeping rhizomes. Stems hollow, terete, or bluntly trigonous, up to 3 m high. Leaves basal and cauline, blades up to 2 cm wide, flat, keeled upwards, with a triquetrous point. Inflorescence much branched, of 2 to several corymbs, each branch ending in a dense head of 3 to 10 spikelets. Lower bracts leaf-like. Glumes imbricate, the lower 2 or 3 sterile, the upper 1 (to 3) fertile, bisexual, the uppermost sometimes male only. Perianth bristles absent. Stamens 2. Stigmas 3; style base enlarged, deciduous; nut ovoid, acuminate, truncate at the base.

Conway, V. M. Biological Flora of the British Isles: Cladium mariscus (L.) R. Br. Journ. Ecol. 30: 211–216 (1942)

c. 3 species, *C. mariscus* (L.) Pohl is cosmopolitan, the other species are from America and Australasia. *C. mariscus* grows in swamps and shallow water; it has been extensively used for thatching.

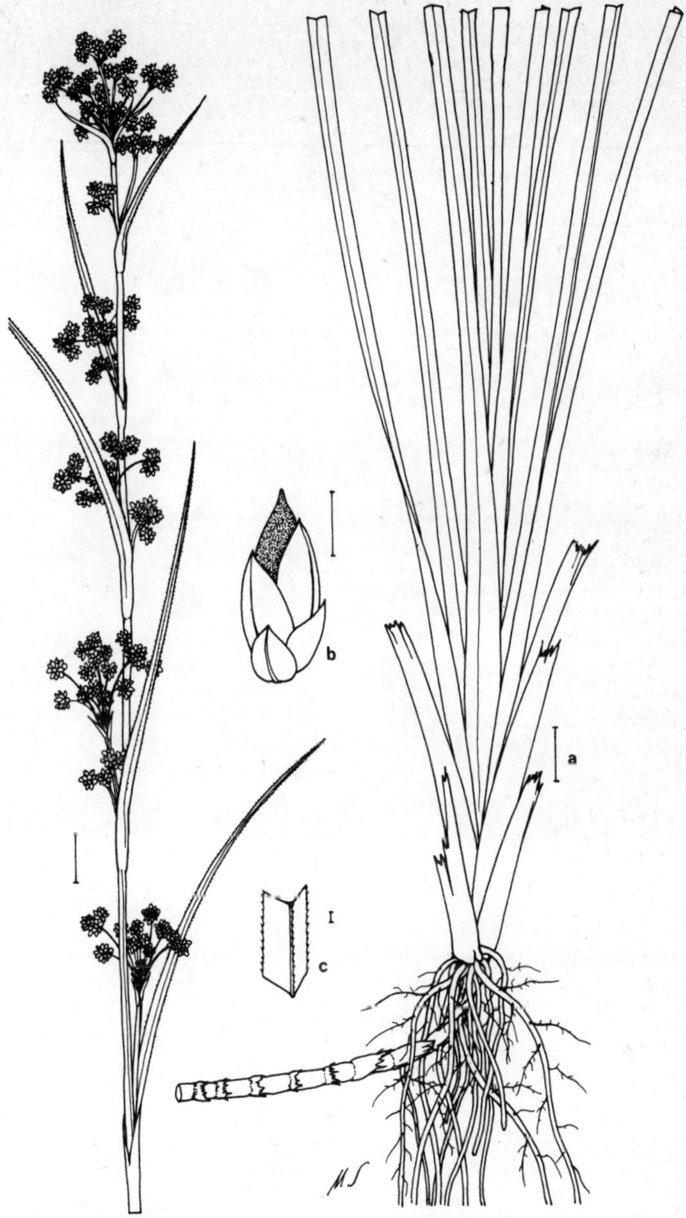

Fig. 78. *Cladium mariscus* (L.) Pohl: a, habit (2 cm); b, spikelet (1 mm); c, detail of leaf (1 mm).

Fig. 79. *Cyperus papyrus* L. (Subgen. *Cyperus*): a, habit (10 cm); b, inflorescence (10 cm).

Fig. 80. *Cyperus ferrugineus* Poir. (Subgen. *Pycreus* (P. Beauv.) C.B.Cl.): a, habit (1 cm); b, spikelet (1 mm).

Cyperus L., Sp. Pl. 44 (1753), [*Juncellus* Griseb., *Kyllinga* Rottb., *Mariscus* Vahl, *Pycreus* Beauv.]
Fig. 79, 80.

Annuals or perennials, sometimes with creeping or tuberous rhizomes. Stems usually trigonous, usually leafy at the base, up to 5 m tall. Leaves with flat blades or sometimes reduced to scale-like sheaths. Inflorescence of 1 to many heads in a simple or compound umbel, with branches bearing clusters of spikelets. Bracts leafy, usually long. Spikelets 1- to many-flowered, caducous or persistent, or breaking up into 1-seeded pieces. Glumes in 2, opposite rows. Flowers bisexual. Perianth bristles absent. Stamens 1 to 3. Stigmas 2 or 3; style not dilated at base, deciduous, leaving a short slender beak; nut trigonous or biconvex.

Carter, G. S. The papyrus swamps of Uganda. Heffer, Cambridge (1955)
Kükenthal, G. in Engler, Pflanzenreich 101 (IV 20): 1–671 (1935, 1936)

c. 900 species: cosmopolitan, mostly tropical. Most species are found in wet or regularly inundated regions. Many species are serious weeds on irrigated land and particularly the perennial species are a menace in irrigation ditches. However, some species, such as *C. esculentus* L., are widely cultivated for their edible tubers which are sold often but erroneously, under the name "water chestnut" – see *Trapa.* Among the more aquatic species the papyrus (*C. papyrus* L.) is particularly notable; it has all its leaves reduced to sheaths and bare stems up to 5 m tall; it is the main constituent of the Sudd on the Nile.

Dichronema Michx., Fl. Bor. Am. 1: 37 (1803)
Fig. 81.

Tufted or stoloniferous perennials, rarely annuals. Stems leafy, trigonous, up to 1 m high. Leaves flat, numerous, especially at the base of the stem. Inflorescence capitate with 1 to many spikelets; bracts 3 to 5, leaf-like, often with a white area at their base. Spikelets whitish, many flowered. Glumes ovate-lanceolate, distichous at the base, then spirally arranged, the 3 or more lowest sterile, the next 1 to 4 (usually 3) bisexual, the uppermost sterile. Perianth bristles absent. Stamens 2 to 3. Stigmas 2; style branches long; nut sessile, obovoid, compressed, topped by the persistent style base, often wrinkled or papillate.

c. 60 species: tropical and temperate America. Many species occur in marshes and shallow water. This genus is very similar to *Rhynchospora* Vahl.

Diplacrum R. Br., Prodr. Fl. Nov. Holl. 240 (1810)
Fig. 82.

Annuals or perennials, often with creeping rhizomes. Stem up to 1 m high, trigonous, leafy. Lowest leaf blades short, the rest long, flat, much exceeding the stem. Inflorescence of subsessile or shortly pedunculate clusters of

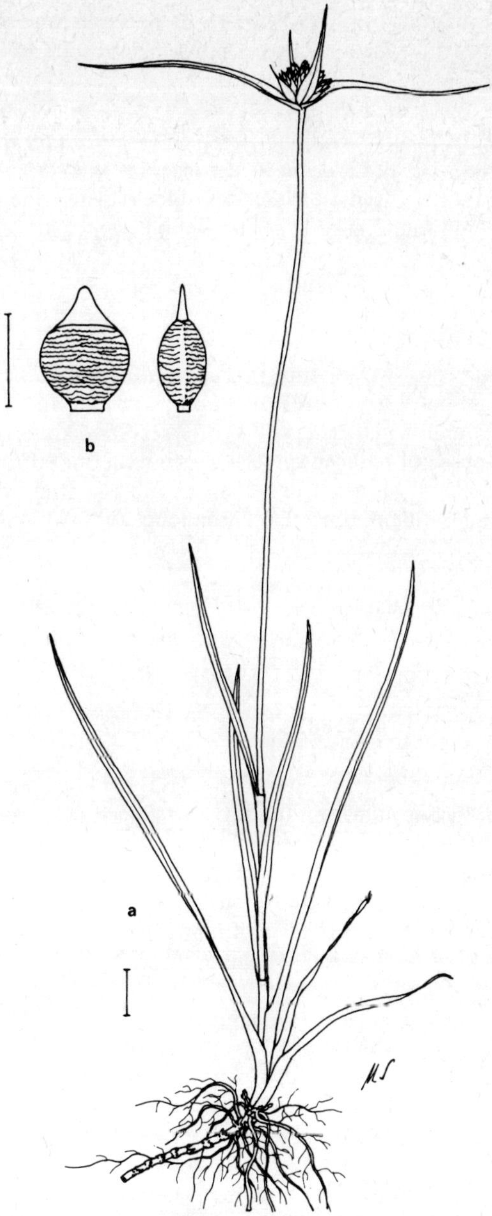

Fig. 81. *Dichronema cephalotes* (L.) A. S. Hitch.: a, habit (1 cm); b, nut (1 mm).

spikelets in the leaf axils. Glumes spirally arranged, the two subtending the female flower often 3-lobed. Flowers unisexual; spikelets bisexual, with one terminal female flower and 1 or 2 male flowers. Perianth bristles absent. Stamens 3. Stigmas 3; nut small, less than 1 mm long, subspherical, ribbed or reticulated.

c. 6 species: tropical Africa, E. Asia, and America. *D. africanum* C. B. Cl. and *D. longifolium* (Griseb.) C. B. Cl. occur in wet places in west tropical Africa; *D. africanum* is an annual, often associated with rice growing, the other species is perennial and found in streams and wet places.

Diplasia L. C. Richard in Pers., Syn. Pl. 1: 70 (1805)
Fig. 83.
Large perennials with thick woody rhizomes covered by brown scales. Stems up to 3 m high, trigonous, leafy. Leaf blades up to 4 m long and 6 cm wide. Inflorescence compound, with clusters of 3 to 12 sessile spikes on long peduncles. Spikes composed of reduced spikelets, each with one stiff, glume-like bracteole, 4 small pellucid glumes, 7 or 8 stamens and 1 central ovary. Perianth bristles absent. Stigmas 2; nut *c.* 6 mm long, with swollen persistent style base.

3 species, 2 in tropical S. America, 1 in Indo-China. *D. keratifolia* L. C. Richard grows at the edges of rivers in the Amazon basin and Trinidad.

Dulichium Pers., Syn. Pl. 1: 65 (1805)
Fig. 84.
Perennials with a creeping rhizome. Stems up to 100 cm high, terete, hollow between the nodes. Leaves numerous; basal ones reduced to sheaths; the cauline ones flat, linear, up to 8 mm wide. Inflorescence of axillary racemes, each with *c.* 8, distichous spikelets each 1.5 to 2 cm long. Bracts like the leaves, exceeding the racemes. Spikelets with 6 to 12 bisexual distichous flowers. Glumes lanceolate, carinate. Perianth bristles 6 to 9, with backward pointing barbs. Stamens 3. Stigmas 2; nut smooth, with a long beak.

Kükenthal, G. Vorarbeiten zu einer Monographie der Rhynchosporoideae. Bot. Jahrb. 75: 485–488 (1952)

1 species, *D. arundinaceum* (L.) Britt. occurs in swamps and shallow water throughout temperate N. America, and has been recorded as a fossil in Europe.

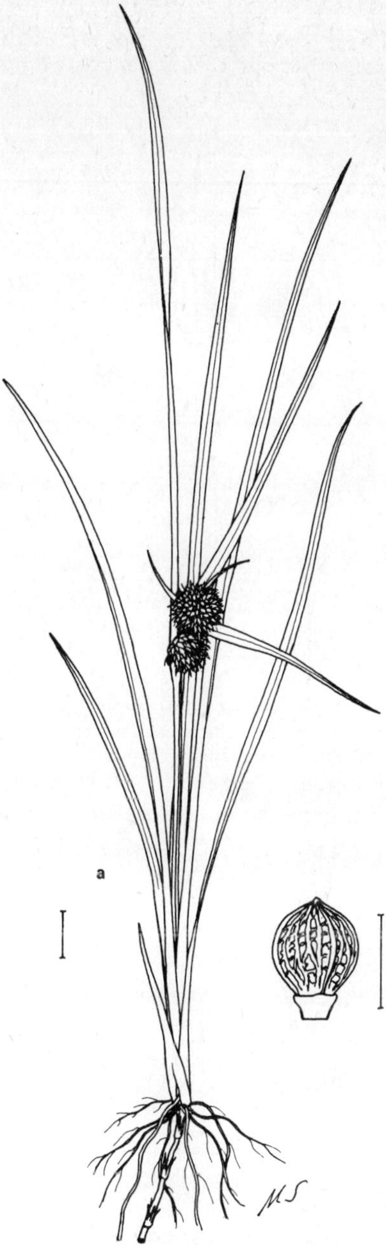

Fig. 82. *Diplacrum longifolium* (Griseb.) C.B.Cl.: a, habit (1 cm); b, nut (1 mm).

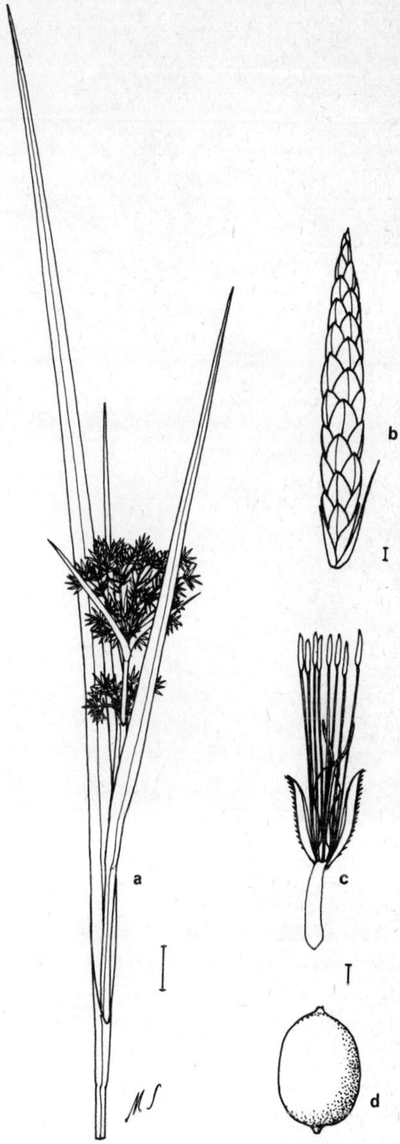

Fig. 83. *Diplasia keratifolia* L. C. Richard: a, inflorescence (5 cm); b, spike (1 mm); c, flower (1 mm); d, nut (1 mm).

Fig. 84. *Dulichium arundinaceum* (L.) Britt.: a, habit (1 cm); b, spikelet (1 mm).

Eleocharis R. Br., Prod. Fl. Nov. Holl. 224 (1810)
Fig. 85.
Perennials, or annuals, often with a creeping rhizome. Stems terete or rarely 4-angled, up to 2 m high. Transverse section of stem with numerous approximately equal air canals. Leaf blades where present, similar to the stems. At least the upper leaf sheaths without blade. Bracts absent. Inflorescence a solitary terminal spikelet, few- to many-flowered. Glumes all spirally arranged, the lowest usually sterile, or at least different in shape from the others. Flowers bisexual. Perianth bristles present, usually 6, rarely 0 or 12, shorter than or not much longer than the nut. Stamens 3 or 2. Stigmas 3 or 2; style usually with a swollen persistent base; nut trigonous or biconvex.

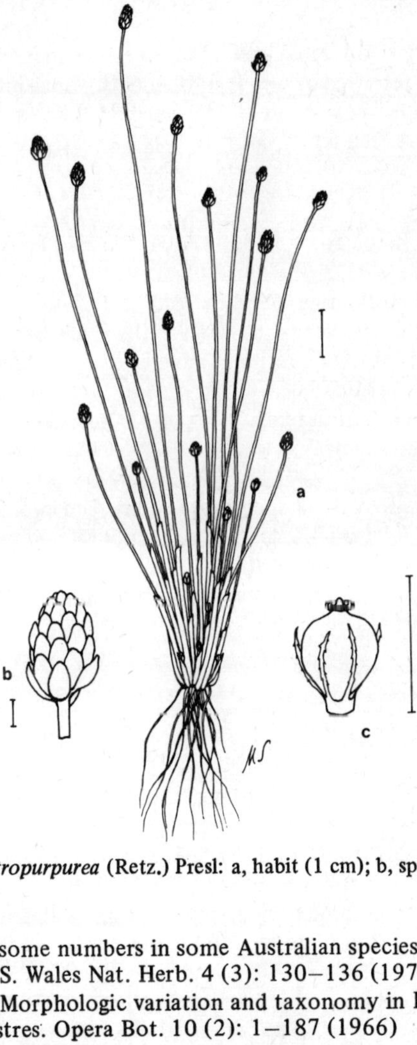

Fig. 85. *Eleocharis atropurpurea* (Retz.) Presl: a, habit (1 cm); b, spikelet (1 mm); c, nut (1 mm).

Briggs, B. Chromosome numbers in some Australian species of Eleocharis. Contr. New S. Wales Nat. Herb. 4 (3): 130–136 (1970)
Strandhede, S. O. Morphologic variation and taxonomy in European Eleocharis, subser. Palustres. Opera Bot. 10 (2): 1–187 (1966)

Strandhede, S. O. Eleocharis, subser. Eleocharis in North America. Bot. Notiser 120: 355–368 (1967)

Svenson, H. K. Monographic studies in the genus Eleocharis. Rhodora 31: 121–135, 152–163, 167–191, 199–219, 224–242 (1929), loc. cit. 34: 193–203, 215–227 (1932), loc. cit. 36: 377–389 (1934), loc. cit. 39: 210–231, 236–273 (1937), loc. cit. 41: 1–19, 43–77, 90–110 (1939)

Thiébaud, M.-A. Contribution à l'étude ecologique du genre Eleocharis R. Br. en Suisse. Bull. Soc. Neuchâtel Sci. Nat. 94: 55–65 (1971)

Walters, S. M. Eleocharis R. Br. Biological Flora of the British Isles. Journ. Ecol. 192–206 (1949)

Zukowski, W. Rodzaj Eleocharis R. Br. w. Polsce. Prace Kom. Biol. (Poznań) 30 (2): 1–113 (1965)

c. 200 species: cosmopolitan. Most species occur in marshes and shallow water and several have been reported as weeds in ricefields. When submerged in deep water *Eleocharis* remains sterile. *E. acicularis* (L.) Roem. and Schultes and *E. vivipara* Link are frequently cultivated as decorative aquarium plants.

Eleogiton Link, Hort. Berol. Alt. 1: 293 (1827), [*Scirpus* pro parte]
Fig. 86.

Slender perennials, often floating or submerged, leafy. Stems elongated, usually much branched, with bladeless leaf sheaths at the base of the peduncles. Spikelets 3- to 5-flowered, small, terminal, longer than the bracts. Flowers bisexual, without scales or bristles. Glumes broadly ovate, spirally arranged. Stamens 3. Stigmas 2 or 3; nutlet trigonous, or biconvex and flattened, obovate, smooth, without a beak.

c. 10 species: cosmopolitan. Most species are submerged aquatics with the spikelets emersed. *E. fluitans* (L.) Link, the commonest and most widespread species, occurs in Africa, Europe, southeast Asia, Australia and New Zealand. For bibliography see *Scirpus.* Distinguished from *Scirpus* by its habit and single terminal spikelet.

Eriophorum L., Sp. Pl. 52 (1753)
Fig. 87.

Perennials with tufted or creeping rhizomes. Stems leafy, up to 90 cm high, terete or triquetrous. Leaves green in winter, but mostly dead at flowering time, usually with a triquetrous point. Spikelets many-flowered, forming an umbellate inflorescence, or solitary. Flowers bisexual. Bristles numerous, elongating greatly and becoming silky after flowering. Stamens 2. Stigmas 3; style base not swollen; nut compressed, trigonous.

Faegri, K. Zur Hybridbildung in der Gattung Eriophorum, in Festschrift Werner Lüdi, Bot. Inst. Rübel, Zürich 50–58 (1958)

Fig. 86. *Eleogiton striata* Nees: a, habit (1 cm); b, spikelet (1 mm).

c. 20 species: N. Temperate regions and 1 species in S. Africa. Most species are characteristic of bogs and marshy places. *E. angustifolium* Honck. and *E. gracile* Roth often occur in shallow water in lakes and bog pools.

Fig. 87. *Eriophorum angustifolium* Honck: a, habit (1 cm); b, nut with bristles (1 cm).

Fig. 88. *Fimbristylis littoralis* Gaud.: a, habit (1 cm); b, ovary and style (1 mm).

Fimbristylis Vahl, Enum. Pl. 2: 285 (1805)
Fig. 88.

Annuals or perennials, usually tufted. Stems terete, erect, up to 1 m high, sometimes hairy, usually leafy at the base. Leaf blades flat. Inflorescence a loose umbel or capitate cluster of many-flowered spikelets, or sometimes a solitary terminal spikelet. Bracts leafy or setaceous, 1 to several. Glumes spirally arranged, often ciliate, the lowest 1 to 4 empty. Perianth bristles absent. Stamens 1 to 3. Styles 2 or 3, fimbriate, pubescent or glabrous, their bases swollen, usually caducous with the style; nut trigonous or biconvex.

Gordon-Gray, K. D. Fimbristylis and Bulbostylis: generic limits as seen by a student of southern African species. Mitt. Bot. Staatsamml. München 10: 549 (1971)

c. 300 species: tropics and subtropics. Several species are aquatic, most occur in wet places. *F. littoralis* Gaud. [*F. miliacea* (L.) Vahl] is a frequent weed in rice-fields, many other species are potential aquatic weeds.

Fuirena Rottb., Desc. et Ic. Pl. 70: t.19 (1773)
Fig. 89.

Perennials with creeping rhizome. Stems up to 150 cm high, triquetrous, nodose. Leaves flat, pilose, or sheath-like. Inflorescence capitate or corymbose, with heads of many-flowered spikelets. Bracts short, subequal to the spikelets. Glumes spirally arranged, rough or ciliate, the 2 lower sterile, the rest fertile. Flowers bisexual. Perianth of 3 bristles and 3 scales, often clawed and very variable in shape; sometimes of 6 bristles or rarely absent. Stamens 3. Stigmas 3; style not swollen at the base; nut trigonous.

Govindarajalu, E. The systematic anatomy of the S. Indian Cyperaceae: Fuirena Rottb. Journ. Linn. Soc. Bot. London 62: 27–40 (1969)
Lye, K. A. New taxa and combinations in Fuirena Rottb. Bot. Notis. 127 (1): 109–112 (1974)

c. 40 species: tropics and subtropics. *F. umbellata* Rottb., the largest species, is a common aquatic in all regions. *F. ciliaris* (L) Roxb. is recorded from rice-fields in India, and *F. robusta* Kunth along canals in Guiana.

Gahnia Forster, Char. Gen. 51 (1776)
Fig. 90.

Perennials, often stout, usually tufted. Stem terete, solid or hollow, leafy, up to 1 m high. Leaves with a ligule; blades without a distinct midrib, involute when dry. Inflorescence paniculate or spike-like. Bracts sheathing at the base, often exceeding the inflorescence. Spikelets 1- to 2-flowered with only the upper flower fertile. Glumes spirally arranged, 1-nerved, about 3 empty below the fertile flower. Perianth bristles absent. Stamens 3 to 6, the filaments often lengthened and persistent, so that the nut remains suspended by them after it has ripened. Stigmas usually 3; nut trigonous with a small conical beak.

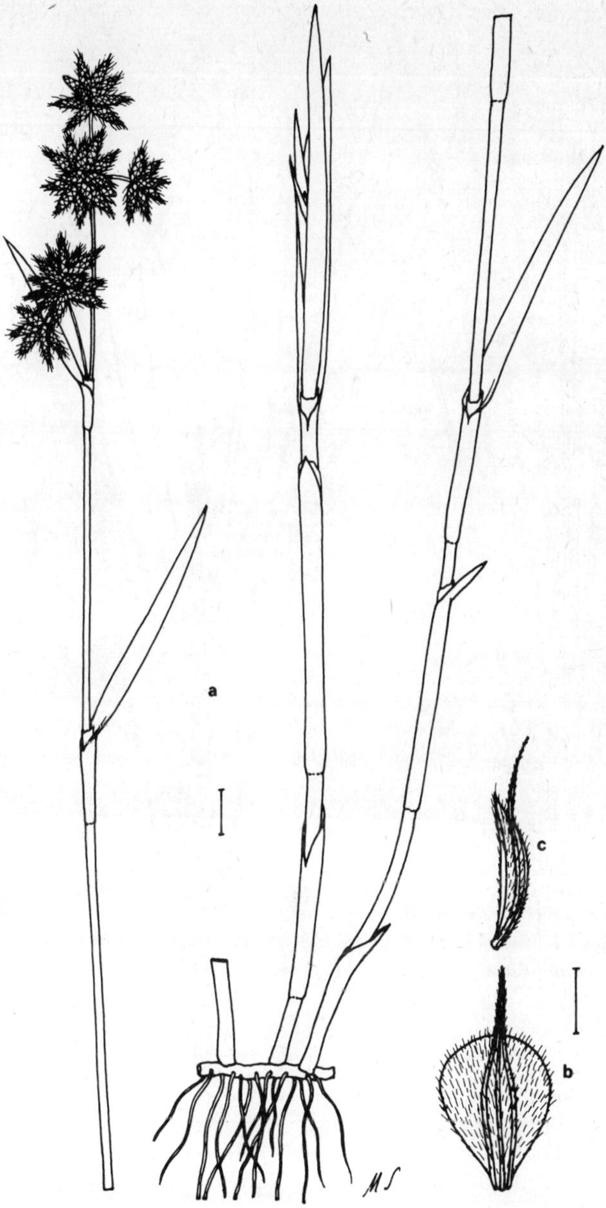

Fig. 89. *Fuirena ciliaris* (L.) Roxb.: a, habit (1 cm); b, glume from the back (1 mm); c, glume from the side.

Fig. 90. *Gahnia javanica* Mor.: a, habit (1 cm); b, spikelets (1 mm).

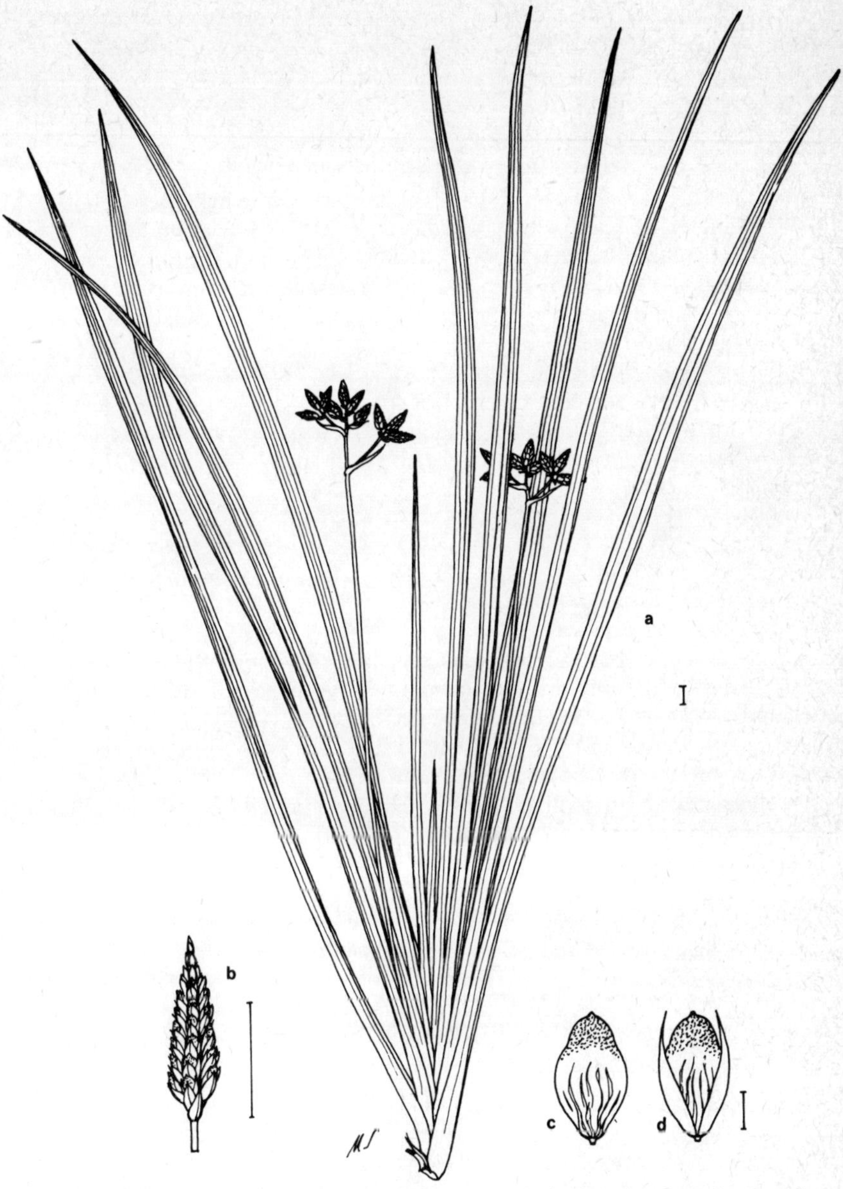

Fig. 91. *Hypolytrum africanum* Nees: a, habit (1 cm); b, spike (1 cm); c, nut; d, nut with glumes (1 mm).

Kükenthal, G. Vorarbeiten zu einer Monographie der Rhynchosporoideae. Feddes Rep. 52: 52–111 (1943)

c. 40 species: mainly in Australia. Most species are found in wet places that are regularly inundated.

Hypolytrum Richard in Pers., Syn. Pl. 1: 70 (1805)
Fig. 91.
Perennials with a short rhizome. Stem trigonous, up to 1 m high. Leaves flat, 3-nerved, lanceolate. Inflorescence paniculate; bracts long, leaf-like. Spikes solitary or in groups. Spikelets reduced, densely crowded on the spikes, each with 2 male and 1 female flower and a single bract; the male flowers each with 1 stamen and 1 bracteole. Style deeply divided, with 2 stigmas; nut biconvex with a short beak.

c. 80 species: tropics and subtropics. Most species are characteristic of forests, but a few are found in swamps and shallow water, e.g. *H. purpurascens* Cherm. in West Africa, *H. nudicaule* Cherm. in Madagascar and *H. longifolium* Nees in tropical S. America.

Isolepis R. Br., Prodr. Fl. Nov. Holl. 221 (1810), [*Scirpus* L. pro parte, *Lipocarpha* R. Br., *Hemicarpha* Nees]
Fig. 92.
Annuals or perennials up to 30 cm high, usually slender and tufted. Stems terete. Leaves basal, capillary, channelled, often few or absent. Bracts 1 or 2, often appearing as a continuation of the stem. Spikelets small, usually less than 5 mm long, 1 to 15 together, usually sessile, many-flowered. Glumes spirally arranged. Flowers bisexual. Perianth absent or of scales (always absent in *Isolepis* sensu stricto; with one thin hyaline scale in *Hemicarpha*, and 2 thin hyaline scales in *Lipocarpha*). Stamens 3. Stigmas 2 or 3; nut trigonous, smooth, longitudinally striate or dotted.

Haines, R. W. and Lye, K. A., Studies in African Cyperaceae IV, Bot. Not. 124: 287–291, 473–482 (1971)

c. 60 species: cosmopolitan. Most species occur in shallow water and wet mud. Distinguished from *Schoenoplectus* mainly by size.

Lagenocarpus Nees, Linnaea 9: 304 (1834)
Fig. 93.
Robust perennials. Stems trigonous. Leaves flat, up to 3.5 cm wide. Inflorescence a large, loose, panicle with heads of 6 to 15 spikelets. Spikelets unisexual; the female with 3 to 7 glumes, of which 1 only fertile; the male similar but with several fertile glumes; lower glumes mucronate, all pubescent. Perianth bristles absent. Stamens 2. Stigmas 3; nut 3 mm long, ovate, with an obtuse beak.

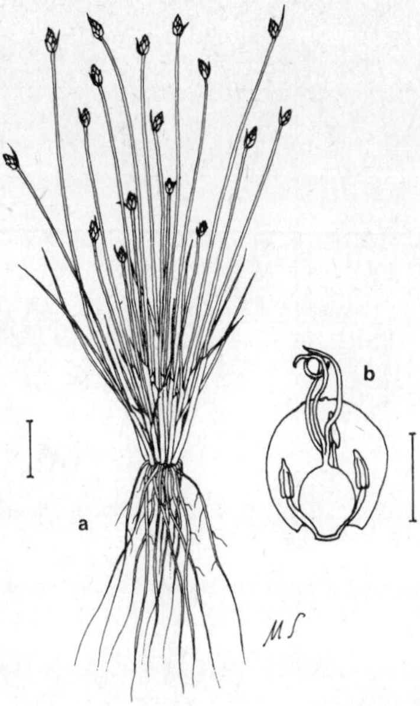

Fig. 92. *Isolepis cernua* (Vahl) Roem & Schultes: a, habit (1 cm); b, flower (1 mm).

Pfeiffer, H. Additamenta ad cognitionem generis Lagenocarpus. Feddes Rep. 21: 34–36 (1925)

c. 34 species: tropical America. *L. guianensis* Nees occurs in swamps.

Lepironia L.C. Richard in Pers., Syn. Pl. 1: 70 (1805)
Fig. 94.
Perennials with creeping rhizome. Stems 5 to 100 cm high, simple, leafless, terete, transversely septate, with a few sheaths at the base. Bract exceeding the spike, appearing as a continuation of the stem. Spike solitary, *c.* 2.5 cm long, oblong-ellipsoid, composed of many spirally arranged crowded and reduced bisexual spikelets. Bracteoles ovate, obtuse, rigid. Spikelet composed of 2 kinds of male flowers; one with 1, broad, boat-shaped glume and 1 stamen; the other with 6 to 12 flowers each with 1 linear glume and 1 stamen; and 1 female flower with 8 glumes and 1 ovary. Style rather short; stigmas 2; nut biconvex, compressed.

2 species, *L. articulata* (Retz.) Domin [*L. mucronata* Richard] occurs in shallow lakes in Madagascar, Australasia and India; it is cultivated in China for a fibre used for making the sails of Junks. *L. compressa* Boek., the other species, was described from Hong Kong.

Fig. 93. *Lagenocarpus martii* Nees: a, habit (1 cm); b, spikelet (1 mm); c, nut with glumes.

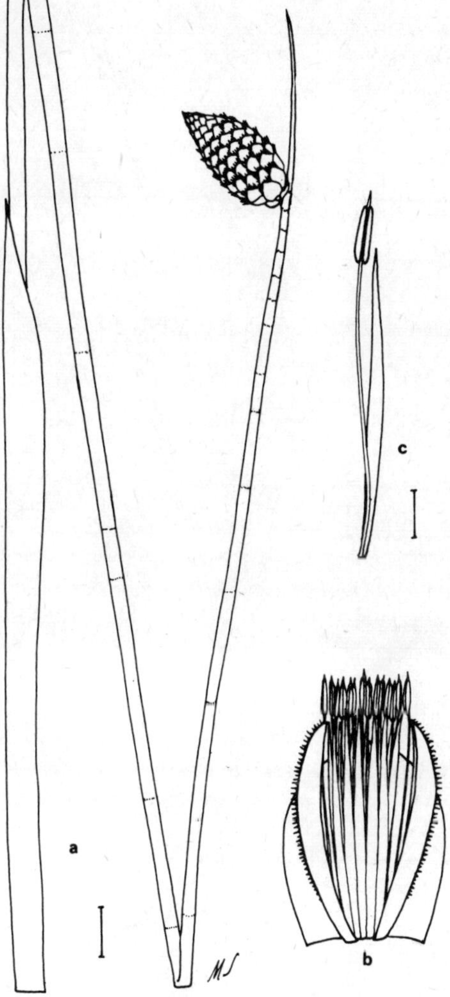

Fig. 94. *Lepironia articulata* (Retz.) Domin: a, habit (1 cm); b, spikelet (1 mm); c, male flower (1 mm).

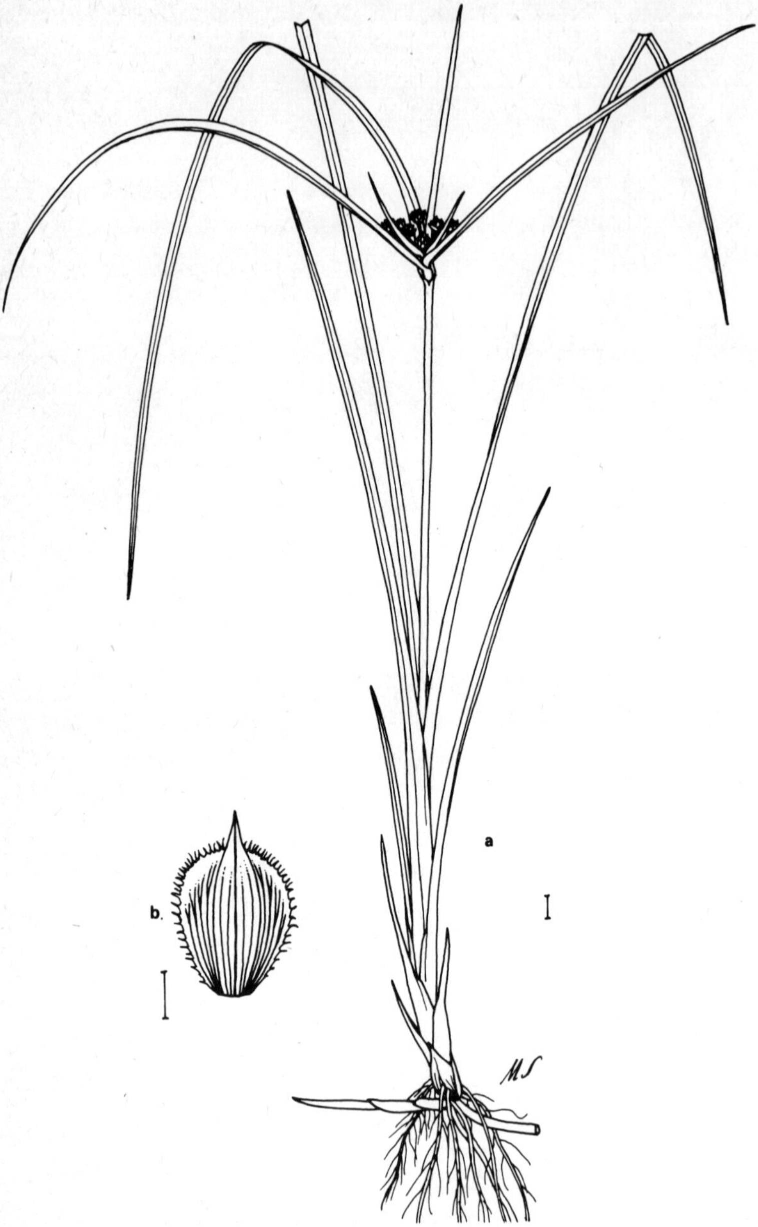

Fig. 95. *Oxycaryum cubense* (Poep. & Kunth) K. Lye: a, habit (1 cm); b, glume (1 mm).

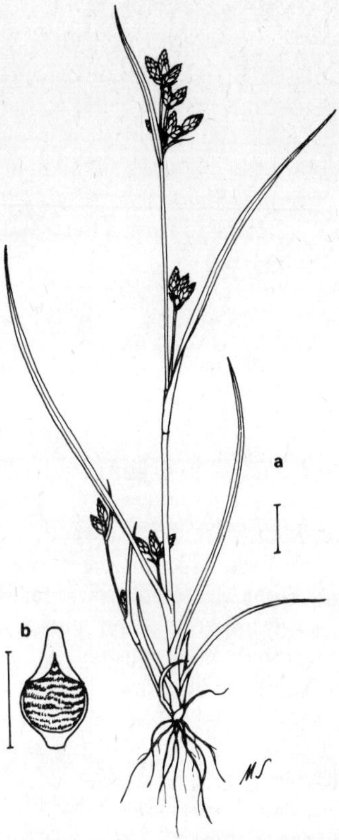

Fig. 96. *Psilocarya scirpoides* Torr.: a, habit (1 cm); b, nut (1 mm).

Oxycaryum Nees in Mart., Fl. Brasil. 2, 1: 90 (1842), [*Scirpus cubensis* Poep. and Kunth]
Fig. 95.
Robust, stoloniferous, perennial with long, scaly stolons rooting at the nodes. Stem up to 100 cm high, sharply trigonous, smooth. Leaves all basal or nearly so; leaf blades flat, much longer than the stem. Inflorescence of 3 to 10 sub-umbellately arranged globose heads, 5 to 15 mm across, with numerous densely packed, many-flowered spikelets. Bracts very long. Glumes spirally arranged, stiff, with ciliate margins and often with spiny teeth on the keel. Flowers bisexual. Perianth bristles and scales absent. Stamens 3. Stigmas 2; nut fusiform, with a long beak.

Lye, K. A. The genus Oxycaryum Nees. Bot. Not. 124: 280–286 (1971)

1 species, *O. cubense* (Poep. and Kunth) K. Lye [*Scirpus cubensis* Poep. and Kunth] occurs in tropical America and Africa. It has become important as an early coloniser of *Salvinia* mats on Lake Kariba, and so contributes to the formation of Sudd. It is also a persistent weed in irrigation ditches.

Psilocarya Torrey, Ann. Lyc. N.Y. 3: 360 (1836)
Fig. 96.
Usually annuals up to 1 m high. Stems trigonous or terete. Leaves mainly basal; blades flat, often pubescent. Bracts similar to the leaves, usually exceeding the inflorescence. Spikelets ovoid, nearly sessile, in terminal or axillary cymes, 20- to 30-flowered. Glumes all alike, imbricate. Flowers all bisexual. Perianth bristles absent. Stamens usually 2. Stigmas 2; style almost wholly persistent, its base enlarging to form a hard and large beak; nut biconvex, transversely wrinkled.

c. 10 species: mainly in tropical S. America, with a few species in the West Indies, Australia and N. America. *P. scirpoides* Torrey is frequent in wet places in N. America:

Rhynchospora Vahl, Enum. Pl. 2: 229 (1806)
Fig. 97.
Usually tufted perennials. Stems terete or trigonous, up to 2 m high. Leaves flat, basal and cauline, up to 1 cm wide. Inflorescence of spikelets in 1 or more heads, or paniculate, often clustered. Bracts leaf-like, usually exceeding the head. Spikelets 7 or 8 flowered; the lower glumes distichous, the upper spirally arranged; the three lowest glumes usually sterile, the fourth bisexual, the rest male only or sterile, or the uppermost rarely bisexual. Perianth bristles usually 6, rarely more, or absent. Stamens 1 to 3. Stigmas 2; style entire, 2-lobed at apex or branched, dilated at the base; nut oblong or ovoid, beaked by the persistent style-base.

c. 200 species: cosmopolitan, but mainly American and tropical. Most species occur in wet places. *R. corymbosa* (L.) Britton is widely distributed in the tropics in marshes and shallow water.

Schoenoplectus (Reichenb. fil.) Palla, Bot. Jahrb. 10: 298 (1888), [*Scirpus* L. pro parte]
Fig. 98.
Stout perennials or more slender annuals. Stems triquetrous or terete, up to 4 m high, sometimes with many transverse septa and numerous approximately equal air canals, often nearly or quite leafless; ribbon-like submerged leaves sometimes present. Inflorescence apparently lateral, sessile, capitate or with short branches; lowest bract appearing as a continuation of the stem, exceeding the inflorescence. Spikelets many-flowered. Flowers bisexual. Glumes spirally arranged. Perianth bristles 6, sometimes fewer or absent. Stamens 3. Stigmas 3 or 2; nut trigonous or biconvex.

Bakker, D. Miscellaneous notes on Scirpus lacustris L. sensu lato in the Netherlands. Acta Bot. Neerl. 3: 425–445 (1954)
Koyama, T. Taxonomic study of the genus Scirpus Linné. Journ. Fac. Sci. Tokyo Univ. Bot. 7: 271–366 (1958)

Fig. 97. *Rhynchospora corymbosa* (L.) Britt.: a, habit (1 cm); b, spikelet (1 mm).

Lye, K. A. A new species of Schoenoplectus and some new combinations. Bot. Notiser 124 (2): 287–291 (1971)

c. 50 species: cosmopolitan. Most species are aquatic, rooting on the bottom with erect emergent stems. Several species (such as *S. lacustris* (L.) Palla) are used for weaving. *S. tatora* (Kunth) Palla is used for making boats and sails on Lake Titicaca. *S. juncoides* (Roxb.) Krecz is reported to be a troublesome weed in ricefields in Madagascar.

Schoenus L., Sp. Pl. 42 (1753)
Fig. 99.
Usually tufted, sometimes creeping annuals or perennials. Stems terete, up to 50 cm high. Leaves longer or shorter than the stem, all or the lower often reduced to sheaths; blades usually subterete. Inflorescence of compressed or lax bracteate terminal heads of spikelets or paniculate. Rachilla prominently flexuose between the flowers. Spikelets 1 to 5 flowered; the lower flowers sterile, the upper bisexual; glumes distichous. Perianth of 1 to 6 bristles, or glabrous or plumose scales, or absent. Stamens 3 or 6. Stigmas 3; nut trigonous, often 3-ribbed; beak small or absent.

Kükenthal, G. Vorarbeiten zu einer Monographie der Rhynchosporoideae. Feddes Rep. 44: 1–32, 65–101, 161–195 (1938)

c. 100 species: mainly Australian. Many species occur in bogs and marshes. *S. natans* Bentham from Western Australia, and *S. fluitans* Hooker fil. from South Australia, Tasmania and New Zealand usually grow floating in water.

Scirpus L., Sp. Pl. 49 (1753)
Fig. 100.
Stout usually creeping perennials. Stems trigonous, leafy, up to 2 m high. Leaves flat or keeled with well-developed blades. Inflorescence terminal, much branched, corymbose. Bracts several, leaf-like, equalling or exceeding the inflorescence. Spikelets many-flowered. Flowers bisexual, spirally arranged; glumes sometimes awned. Perianth of 1 to 6 bristles, sometimes long and silky, occasionally absent. Stamens 3. Stigmas 3; nut variable.

Lye, K. A. Moderne oppfatning avslekta Scirpus L. Blyttia 29: 141–147 (1971)

c. 30 species: cosmopolitan. The genus is best represented in N. America where several species are aquatic, e.g. *S. fluviatilis* (Torr.) Gray which occurs in ponds and lakes throughout northern N. America.

Scirpus L. as interpreted here, includes only sect. *Phyllantheli* Beurl., sect. *Taphrogeton* Reichenb., and sect. *Trichophorum* (Pers.) Darl.; the genera *Schoenoplectus, Isolepis, Oxycaryum, Eleogiton*, and *Websteria* are treated separately, see the references cited with *Schoenoplectus.*

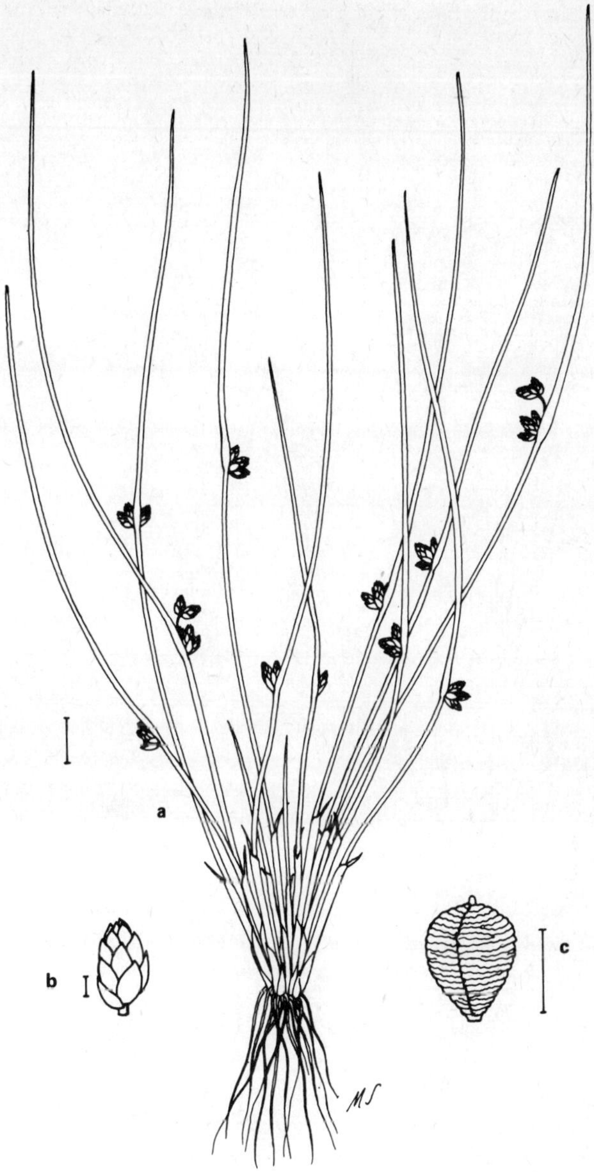

Fig. 98. *Schoenoplectus supinus* (L.) Palla: a, habit (1 cm); b, spikelet (1 mm); c, nut (1 mm).

Fig. 99. *Schoenus fluitans* Hooker fil.: a, habit (1 cm); b, spikelet (1 mm).

Fig. 100. *Scirpus longii* Fernald: habit (1 cm).

Fig. 101. *Scleria tessellata* Willd.: a, habit (1 cm); b, spikelet (1 mm); c, nut (1 mm).

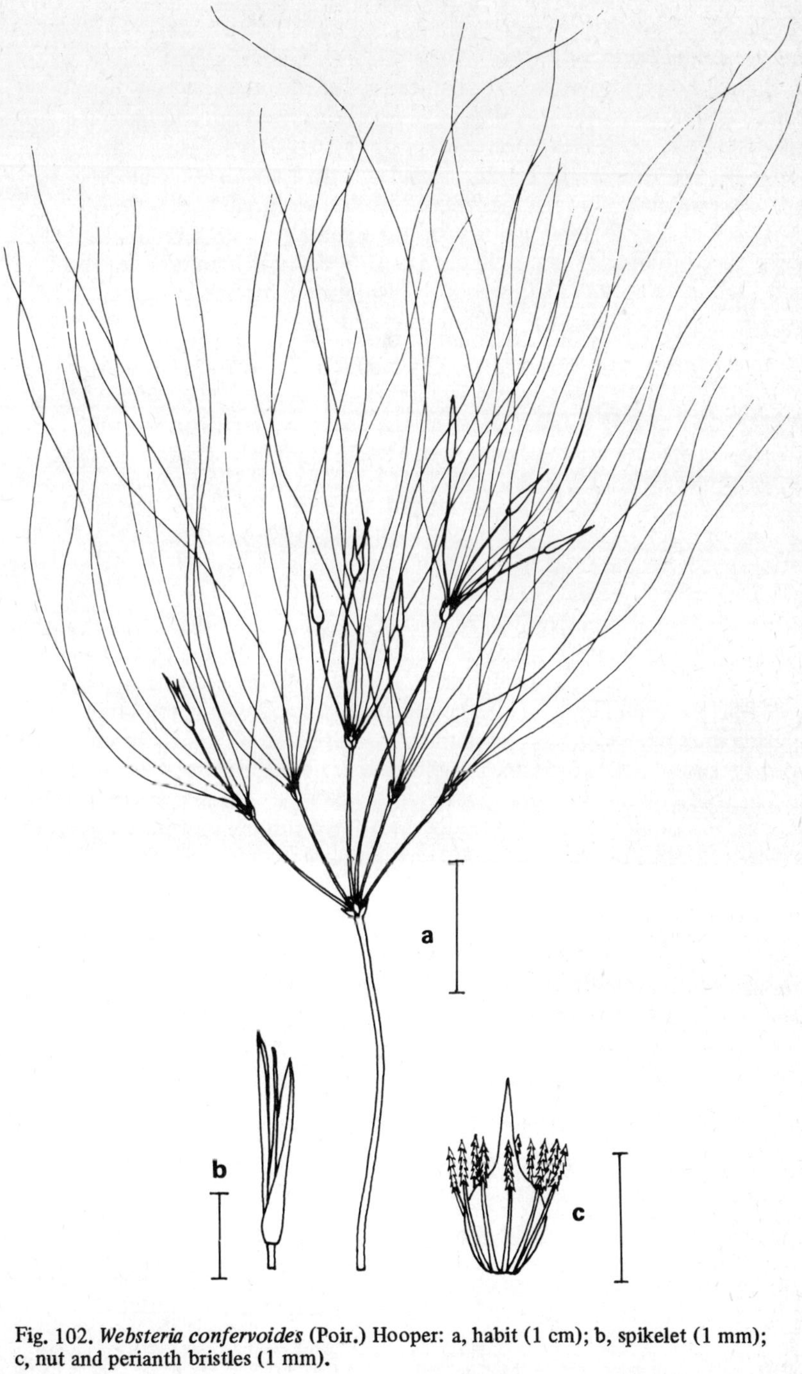

Fig. 102. *Websteria confervoides* (Poir.) Hooper: a, habit (1 cm); b, spikelet (1 mm); c, nut and perianth bristles (1 mm).

Scleria Bergius, Vet. Acad. Handl. Stokh. 26: 142 (1765)
Fig. 101.
Usually perennial with a tough or woody rhizome. Stem trigonous, leafy, up to 2 m high. Lowest leaves short or reduced to sheaths, the rest long, usually flat. Inflorescence usually a leafy panicle, occasionally an interrupted spike. Spikelets solitary or in groups, bisexual or unisexual. Glumes spirally arranged, the lowest 1 to 4 sterile; the bisexual spikelets with 1 terminal female flower and 1 or several male flowers; the female spikelets with one female flower surmounted by 1 or 2 empty glumes; the male spikelets with several male flowers. Perianth bristles absent. Stamens 1 to 3. Stigmas 3; nuts large, ovoid or subglobose, on a short stalk, smooth or sculptured, often white.

Robinson, E. A. A provisional account of the genus Scleria Berg (Cyperaceae) in the 'Flora Zambesiaca' area. Kew Bull. 18 (3): 487–551 (1963)

c. 200 species: mainly tropical. Most species occur in swamps and wet places, and many in shallow water. *S. oryzoides* Presl is found as a weed in ricefields in E. Africa, India, and S. E. Asia.

Websteria Hart Wright, Bull. Torrey Bot. Club 14: 135 (1887)
Fig. 102.
Slender, floating or submerged leafy perennial. Stems leafy, the branches arising in clusters, 20 to 40 cm long. Leaves capillary, in clusters at the ends of the branches. Inflorescence compound, with spikelets on long pedicels among the leaves. Spikelet 5 mm long, oblong-lanceolate, the two lower glumes thin enclosing the spikelet, the lowest sterile, the next bisexual, the upper male or sterile. Perianth bristles 6. Stamens 3. Stigmas 2; nutlet trigonous, obovate.

1 species, *W. confervoides* (Poir.) Hooper; scattered throughout the Tropics. It is a green, alga-like plant found in still or running water.

DROSERACEAE

4 genera; *Aldrovanda* is the only genus containing aquatics.

Aldrovanda L., Sp. Pl. 281 (1753)
Fig. 103A.
Submerged perennial, perennation by persistent terminal buds. Stems floating; roots absent. Leaves 10 to 15 mm long, in crowded whorls of 6 to 9; each with a cuneate basal part and terminating in 4 to 6 bristles and an orbicular lobe, hinged along the midrib, which can close rapidly and trap small animals. Flowers axillary, solitary, shortly stalked, rarely produced, usually cleistogamous. Sepals 5, united below. Petals 5, free, 4 to 5 mm long, greenish-white. Stamens 5. Ovary 1-locular, superior; styles 5; fruit a capsule.

Lloyd, F. E. The carnivorous plants. Waltham, Mass. USA. 194–212 (1942)

1 species, *A. vesiculosa* L.; C. and E. Europe, Caucasus, E. and S. E. Asia, Timor and N. E. Australia; usually found in shallow, still, oligotrophic water.

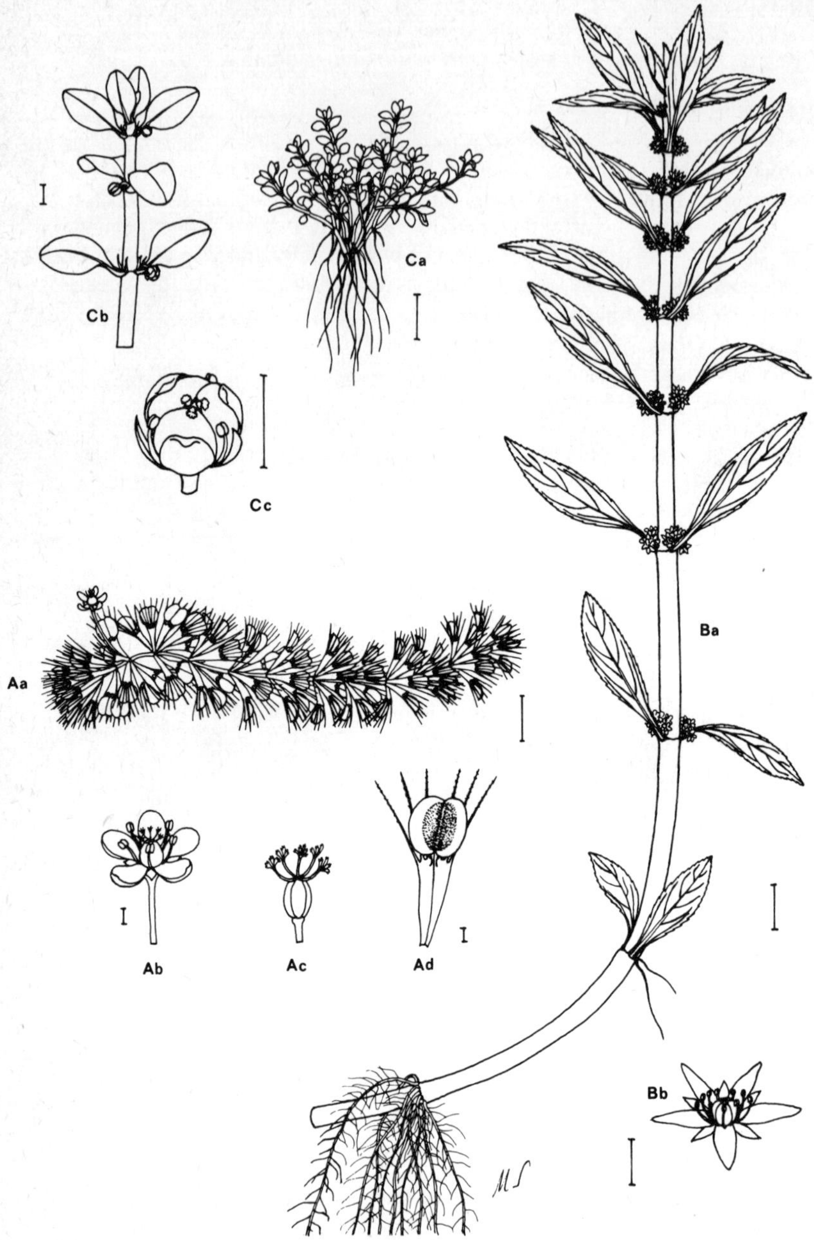

Fig. 103. A. *Aldrovanda vesiculosa* L.: a, habit (1 cm); b, flower (1 mm); c, ovary and styles; d, leaf (1 mm);
B. *Bergia capensis* L.: a, habit (1 cm); b, flower (1 mm);
C. *Elatine ambigua* Wight: a, habit (1 cm); b, shoot apex (1 mm); c, flower (1 mm).

ELATINACEAE

2 genera; *c.* 32 species; cosmopolitan

Annuals, perennials or somewhat woody. Leaves opposite, rarely whorled, simple; margin entire or toothed; stipules present. Flowers axillary, solitary or in cymes, actinomorphic, often inconspicuous. Sepals 2 to 5, free or united at base. Petals free, as many as sepals. Stamens as many as or twice as many as sepals. Ovary 2- to 5-locular, superior; ovules numerous; placentation axile; fruit a septicidal capsule.

1A Sepals 5, free, acute, with distinct nerve; plants usually robust and conspicuous; flowers usually in dense clusters
Bergia

1B Sepals 2, 3 or 4, united at base, obtuse, without distinct nerve; plants usually small and inconspicuous; flowers not in dense clusters
Elatine

Bergia L., Mantissa Alt. 2: 152, no. 1309 (1771)
Fig. 103B.
Annuals, perennials or woody, usually robust and conspicuous. Stems when floating often swollen. Leaves opposite, not whorled; margin frequently toothed. Flowers often in dense, axillary clusters. Sepals 5 and acute (in aquatic species), free, with distinct nerve, often glandular. Ovary 5-locular.

c. 20 species; throughout the warmer regions of the world. About half the species are aquatic or semi-aquatic. *B. capensis* L. and other species are found as weeds in ricefields.

Elatine L., Sp. Pl. 367 (1753)
Fig. 103C.
Annuals or short-lived perennials, usually small and inconspicuous. Leaves opposite (or in *E. alsinastrum* L. whorled); margin not toothed. Flowers not in dense clusters, usually solitary. Sepals 2, 3 or 4, obtuse, united below, without a distinct vein, not glandular. Ovary 2- to 4-locular.

Bacigalupo, N. M. Obscrvaciones sobra el género Elatine en la Argentina. Darwiniana 16 (1–2): 106–115 (1970)
Gauthier, R. and Raymond, M. Le genre Elatine dans le Québec. Contr. Inst. Bot. Univ. Montréal 64: 29–35 (1949)
Schotsman, H. Notes sur Elatine alsinastrum L. Bull. Cent. Étud. Rech. Sci. Biarritz 6 (4): 739–747 (1967)
Schotsman, H. and Bosserdet, P. Notes sur Elatine brochoni Clav. Bull. Cent. Étud. Rech. Sci. Biarritz 7 (1): 81–86 (1968)

c. 12 species; cosmopolitan. Most species are adapted to fluctuating water levels and are found in shallow water that seasonally dries out. They are frequent in fish ponds that are regularly drained. *Elatine* is often regarded as a beneficial plant as it effectively consolidates mud.

ERIOCAULACEAE

13 genera, of which 6 contain aquatic species.

Perennials or annuals. Stems corm-like or elongate. Leaves in basal rosettes or cauline, usually linear or lanceolate. Flowers unisexual, borne in dense heads, subtended by an involucre of bracts; heads solitary or in umbels; peduncle usually exceeding the leaves, sheathed at base; within each head male and female flowers mixed or male in the middle and female outside or plants rarely dioecious. Perianth in 2 series, not clearly differentiated into sepals and petals; outer series 2- or 3-merous, with segments free, united or partly united; inner series 2- or 3-merous, usually united, at least at base, or absent. Stamens as many or twice as many as outer perianth segments, inserted on inner perianth segments when present; male flowers bear a rudimentary ovary. Ovary superior, 2- or 3-locular; style 1, terminal; stigmas 2 or 3, free, elongate; ovules solitary and pendulous in each loculus; fruit a membranous capsule, loculicidally dehiscent.

Hess, H. Zur Kenntnis der Eriocaulaceae von Angola und dem unteren Belgischen Kongo. Ber. Schweiz. Bot. Ges. 65: 115–204 (1955), op. cit. 67: 83–90 (1957)

Meikle, R. D. and Baldwin, J. R. Eriocaulaceae and Xyridaceae in Liberia. Amer. Journ. Bot. 39: 44–51 (1952)

Meikle, R. D. Notes on the Eriocaulaceae of West Tropical Africa. Kew Bull. 22: 141–144 (1968)

Moldenke, H. N. A long series of papers published in Phytologia from 1939 onwards with privately printed summaries appearing in 1946, 1949, 1959 and 1971

Ruhland, W. in Engler, A. Pflanzenreich 13 (IV. 30): 30–117 (1903)

1A Heads spiny; stamens 2-locular; inner perianth segments of female flowers minute, free, elongate, with long hairs (stems elongate and leafy) **Tonina**

1B Heads not spiny; stamens 4-locular; inner perianth segments of female flowers free, united or partly united, or very rarely absent, if minute and free then without long hairs

- 2A Stamens 4 or 6
  - 3A Perianth segments of male flowers united into a funnel-shaped tube; perianth segments of female flowers united above, free at base **Mesanthemum**
  - 3B Perianth segments of male and female flowers free; inner segments rarely absent **Eriocaulon**
- 2B Stamens 2 or 3
  - 4A Inner perianth segments of female flowers partly united, at apex and base free **Syngonanthus**

4B Inner perianth segments of female flowers free
5A Involucral bracts usually hairy; hairs on floral bracts club-shaped; stigmas usually bifid; stylar appendages when present exserted at same height as stigmas

Paepalanthus

5B Involucral bracts glabrous; hairs on floral bracts acute, not club-shaped; stigmas simple; stylar appendages exserted below the stigmas

Leiothrix

Eriocaulon L., Sp. Pl. 87 (1753)
Fig. 104A, B.
Stems simple, usually corm-like, occasionally elongate. Leaves more or less linear, membranous, often translucent and windowed. Perianth 2-or 3-merous; outer segments free or united into a tube split down one side; inner segments free, bearing an apical black or rarely red gland, occasionally absent in female flowers. Stamens 4 or 6, in 2 whorls; anthers 4-locular, mostly black or brown. Ovary 2- or 3-locular; stylar appendages absent; stigmas simple, 2 or 3.

c. 400 species: chiefly in Tropical and Subtropical regions, some species in the N. Temperate zone; largest number of species in the New World. The majority of species are found in marshes, swamps or seasonally inundated regions and many are ricefield weeds. Some species are submerged aquatics (for example, *E. melanocephalum* Kunth from S. America, *E. setaceum* L. from S. E. Asia and *E. bifistulosum* v. Heurck and Mueller from Africa).

Leiothrix Ruhl., in Engler, A. Pflanzenreich 13 (IV. 30): 225 (1903)
Fig. 105A.
Stems simple, elongate. Leaves in rosettes or cauline, in aquatic species dense, hairlike. Involucral bracts glabrous. Hairs on floral bracts and perianth smooth and acute. Perianth 3-merous; segments free. Stamens 3; anthers 4-locular. Ovary 3-locular; style long; stylar appendages exserted below the stigmas, not papillose; stigmas 3, simple.

c. 65 species: Tropical S. America. Most species are found in seasonal marshes; one species, *L. fluitans* (Mart.) Ruhl from Brazil occurs submerged in streams.

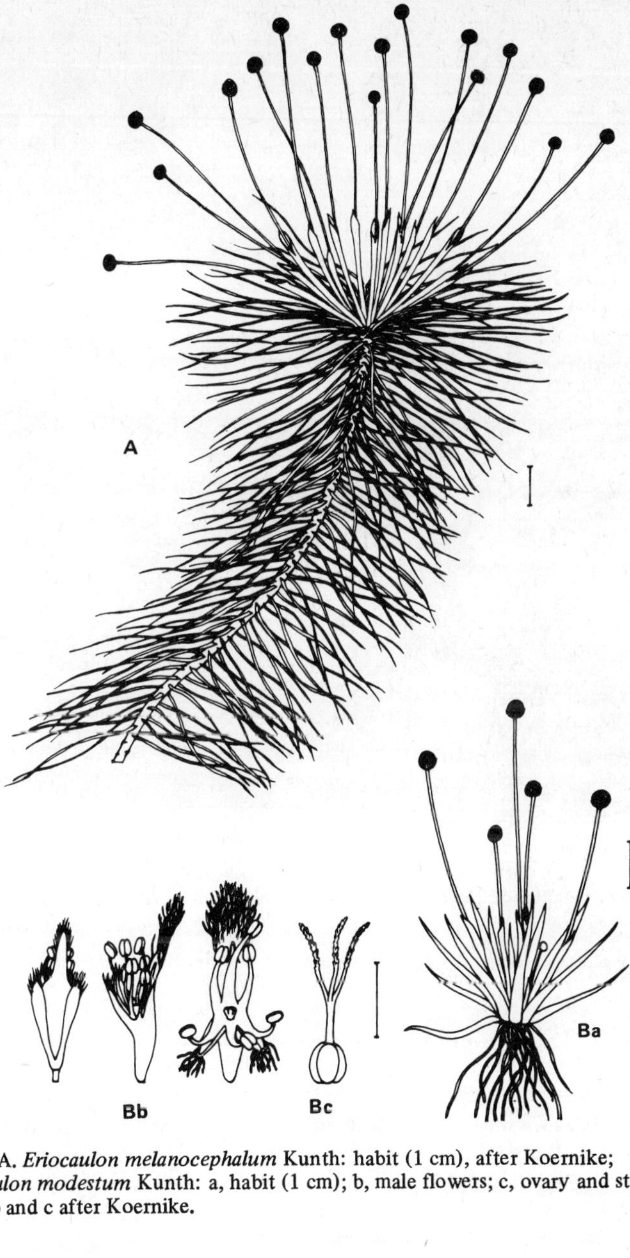

Fig. 104. A. *Eriocaulon melanocephalum* Kunth: habit (1 cm), after Koernike; B. *Eriocaulon modestum* Kunth: a, habit (1 cm); b, male flowers; c, ovary and styles (1 mm); b and c after Koernike.

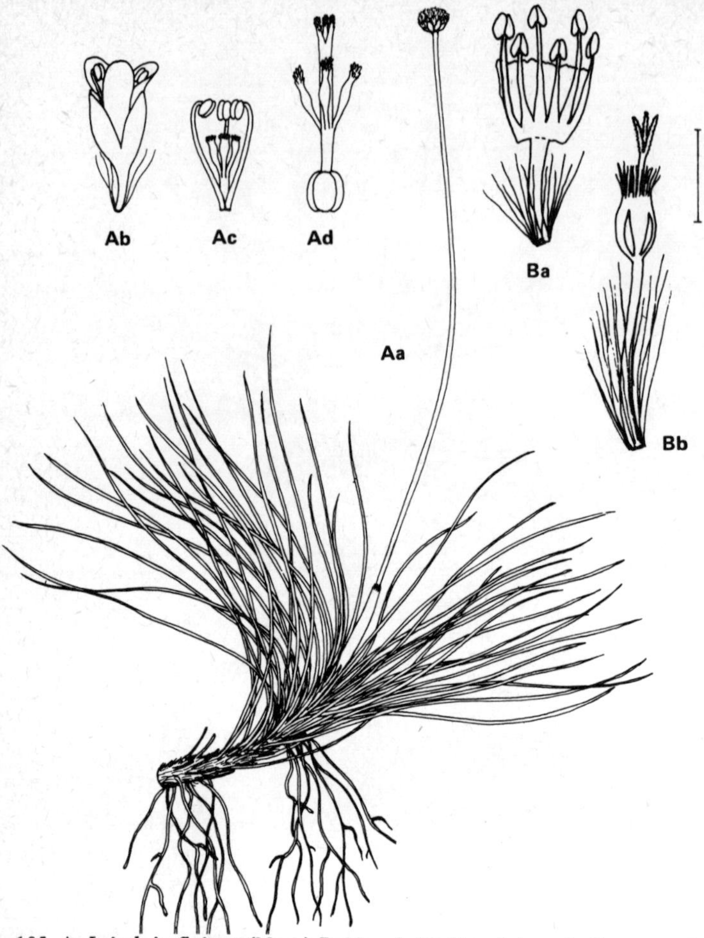

Fig. 105. A. *Leiothrix fluitans* (Mart.) Ruhl: a, habit (1 cm); b, male flower; c, stamens and abortive styles; d, ovary and styles (1 mm), after Koernike;
B. *Mesanthemum reductum* Hess: a, male flower with perianth opened; b, female flower (1 mm), after Hess.

Mesanthemum Körn., Linnaea 27: 572 (1856)
Fig. 105B.
Stems corm-like. Leaves in rosettes, not translucent, not windowed. Involucral bracts hairy; hairs simple, smooth, acute. Perianth 3-merous; outer segments in male flowers united at base, in female free, caducous; inner segments in male united to a tube, in female at base free, middle and upper parts united. Stamens 6, in 2 whorls; anthers 4-locular, yellowish. Ovary 3-locular; style long, without appendages; stigmas 3, simple.

c. 12 species: Tropical Africa and Madagascar.Most species occur in marshes, *M. reductum* Hess from the Quiriri River in Angola, is a totally submerged aquatic.

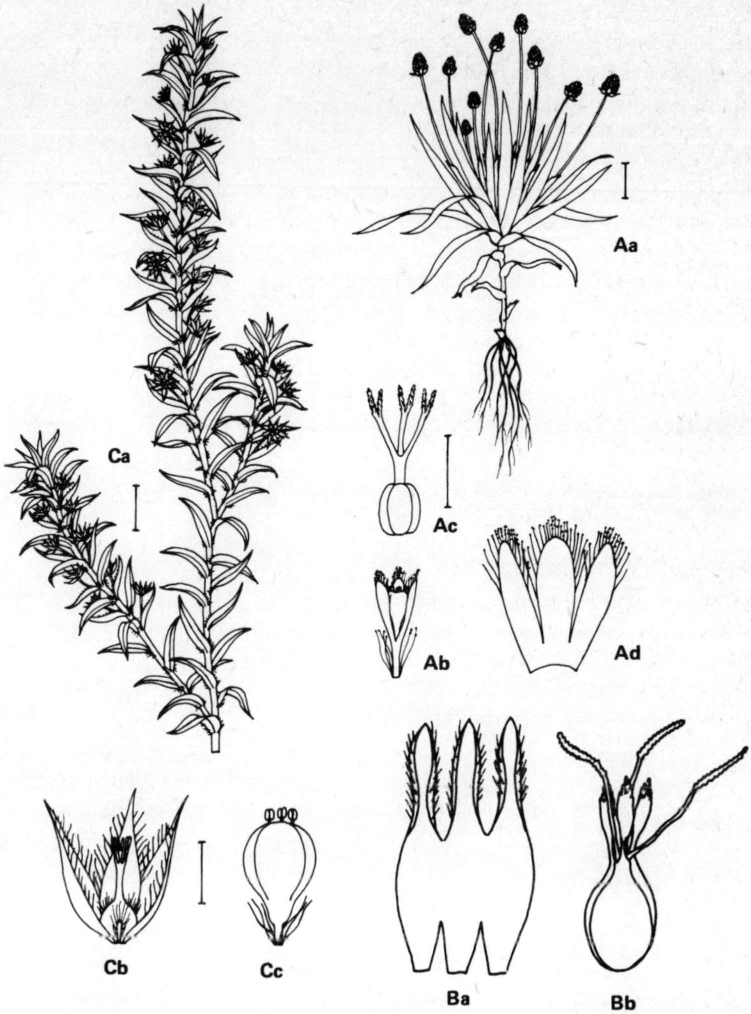

Fig. 106. A. *Paepalanthus myocephalus* Mart. ex Koern.: a, habit (1 cm); b, male flower; c, ovary and styles; d, perianth opened (1 mm), after Koernike;
B. *Syngonanthus anthemidiflorus* Ruhl: a, perianth opened; b, female flower (1 mm), after Koernike;
C. *Tonina fluviatilis* Aubl.: a, habit (1 cm); b, female flower; c, male flower (1 mm), after Koernike.

Paepalanthus Kunth, Enum. Pl. 3: 498 (1841), nom. cons.
Fig. 106A.
Stems simple or branched, corm-like or elongate. Leaves membranous to leathery, not windowed. Involucral bracts hairy and/or glandular. Hairs on floral bracts and perianth granular, usually club-shaped, often tuberculate. Perianth 2- or 3-merous; outer segments united at base; inner segments of male flowers united into a membranous, funnel-shaped tube, involute at maturity, in female free. Stamens 2 or 3, in 1 whorl, a second whorl may appear as small rudiments; anthers 4-locular. Ovary 2- or 3-locular; stylar appendages when present, exserted at same height as stigmas, papillose at apex; stigmas usually bifid, occasionally simple.

c. 485 species: Tropical America, most species in Brazil. The majority of species are found in seasonally wet places; a few species grow submerged in temporary pools.

Syngonanthus Runl. in Urban, Symb. Ant. Ill. 1: 487 (1900)
Fig. 106B.
Stems simple or branched, corm-like or elongate. Perianth 3-merous; outer segments more or less free; inner segments united into a tube in the male, in female free at base and apex, with apical lobes involute. Stamens 3; anthers 4-locular. Ovary 3-locular; style short, with appendages exserted at same height as stigmas; stigmas 3, simple elongate (probably only apical 1/3 of stigma pollen-receptive).

c. 200 species: most in S. America, some in Tropical Africa and Madagascar. The majority of species occur in marshes and swamps and may be found temporarily submerged. *S. hygrotrichus* Ruhl., from Brazil, is a submerged aquatic.

Tonina Aubl., Pl. Guiane 856 (1775)
Fig. 106C.
Stems elongate, branched, slender, 50 cm or more long. Leaves cauline, 8 to 15 mm long, 1 to 2.5 mm wide lanceolate or oblong, often somewhat recurved; margin and base hairy; apex acute. Heads spiny, shortly stalked, borne in leaf axils. Involucral and floral bracts with long cuspidate apex, hairy. Perianth 3-merous; outer segments united at base; inner segments in male united, shortly tubular, in female free, linear, minute, with long hairs. Stamens 3; anthers 2-locular. Ovary 3-locular; style elongate with 3, non-papillose appendages inserted at apex; stigmas 3, about equalling stylar appendages, with bifid apex.

1 species, *T. fluviatilis* Aubl., Tropical America. Occurs in swamps, rivers and lakes; stems often straggling or floating in water.

EUPHORBIACEAE

c. 300 genera: *c.* 6000 species: cosmopolitan but not found in tundra regions. *Caperonia* and *Phyllanthus* contain aquatics.

1A Leaves in 2 rows, orbicular, not exceeding 2 cm long, margin not toothed (free floating)

Phyllanthus fluitans

1B Leaves not in 2 rows, ovate to linear, exceeding 2 cm long, margin toothed (bottom rooted)

Caperonia

Caperonia St.-Hil., Hist. Pl. Remarq. Brésil 244 (1824)
Fig. 107A.
Annual or perennial, hispid to pubescent (rarely glabrous) herbs. Stem erect, when submerged usually with swollen internodes. Leaves alternate, entire, ovate to linear, shortly petiolate; margin toothed; stipules paired, small. Inflorescence axillary racemes or spikes. Flowers unisexual; male flowers with sepals 5, petals 5, stamens 10 borne on an elongated receptacular stalk; female flowers with sepals 5, petals 5 or absent; ovary of 3 united carpels, superior; each carpel containing 1 seed.

Pax, F. in Engler, A. Pflanzenreich 6 (IV. 147): 27–49 (1912)

30 to 60 species: tropical regions of Africa and America. *Caperonia* is taxonomically somewhat confused and in need of revision. Most species are found in wet regions that are regularly subjected to flooding. When submerged, the stems usually become swollen. *C. palustris* (L.) St.-Hill. is reported to be a pest in drainage canals in British Honduras.

Phyllanthus L., Sp. Pl. 981 (1753)

c. 600 species: tropical and sub-tropical regions of the world. *P. fluitans* is the only aquatic species and is vegetatively unlike all other species of *Phyllanthus*.

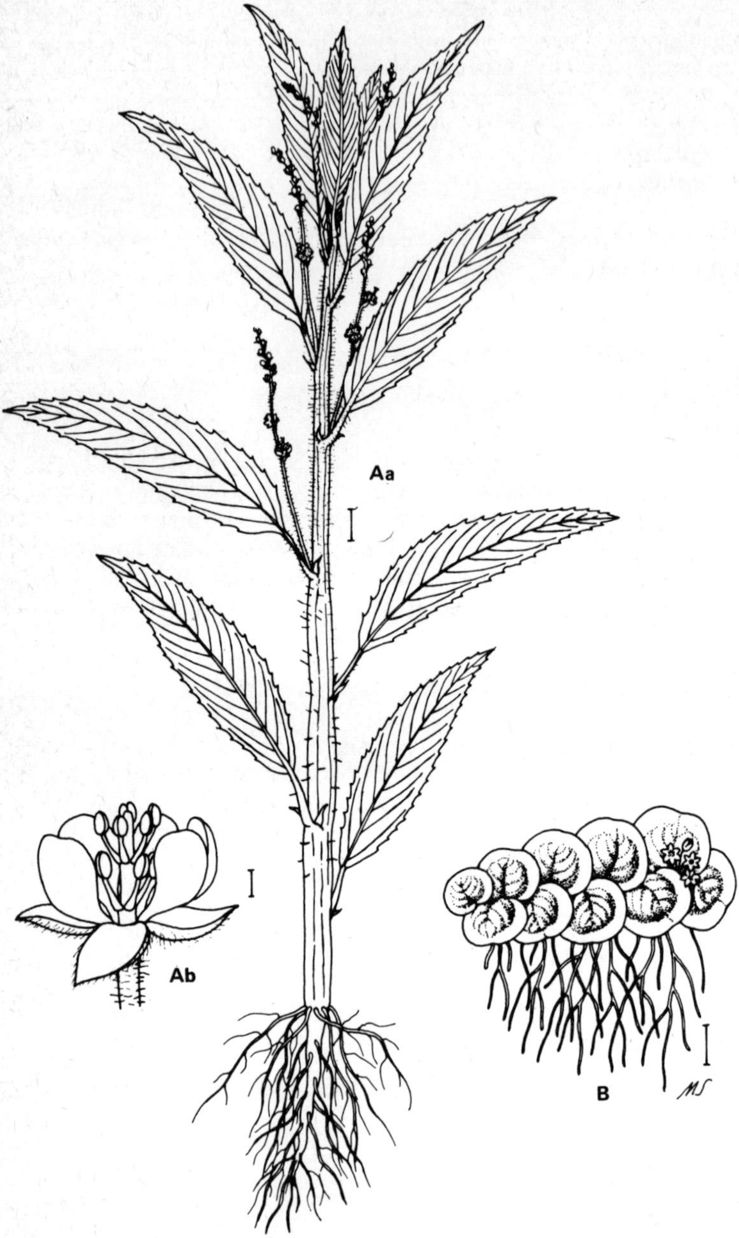

Fig. 107. A. *Caperonia palustris* (L.) St.-Hill.: a, habit (1 cm); b, flower (1 mm); B. *Phyllanthus fluitans* Müll.: habit (1 cm).

Phyllanthus fluitans Müll. Arg.
Fig. 107B.
Free-floating, annual or short-lived perennial, green when young, becoming red at maturity. Stems floating, horizontal, up to 8 cm long, sparsely branched, rooting at internodes. Leaves floating, alternate, in two rows, overlapping, almost sessile, entire, orbicular, *c.* 15 mm diameter, cordate at base; 2 blister-like swellings containing aerenchyma found in the centre of the blade, one each side of the midrib (presumably functioning as floats); upper leaf surface with small translucid spots; stipules paired, small, membranous. Inflorescence axillary, 3- or 4-flowered cymes. Flowers somewhat insignificant, *c.* 1.5 mm diameter, unisexual; disc. 5- or 6-lobed; male flowers with 3, free stamens; female flowers with 3, fused, superior carpels; each carpel containing 2 seeds.

Chodat, R. Observations sur le macroplancton des étangs du Paraguay. Bull. Herb. Boiss. sér. 2, 6: 143–147 (1906)

P. fluitans is a poorly known plant recorded from Brazil, Ecuador and Paraguay It is free-floating and grows with *Salvinia, Azolla* and *Lemna.* Its superficial resemblance to *Salvinia* is remarkable.

FABACEAE [LEGUMINOSAE, including CAESALPINIACEAE, MIMOSACEAE and PAPILIONACEAE]

500 to 600 genera: 12 000 to 13 000 species: cosmopolitan. Three genera (*Aeschynomene, Neptunia* and *Sesbania*) contain aquatics.

1A Leaves 2-pinnate; flowers actinomorphic, lower flowers of inflorescence sterile; fruit without transverse septae
Neptunia

1B Leaves 1-pinnate; flowers zygomorphic, all fertile; fruit with transverse septae

2A Fruit indehiscent, breaking into regularly-shaped portions
Aeschynomene

2B Fruit marginally dehiscent, not breaking into portions
Sesbania

Aeschynomene L., Sp. Pl. 713 (1753), [*Herminiera* Guill. and Perr.]
Fig. 108A.
Shrubs or herbs. Stems erect, creeping or floating, internodes often swollen when submerged or floating. Stipules usually persistent. Leaves alternate, 1-pinnate. Inflorescence simple or branched. Flowers usually in axillary racemes, rarely terminal, bisexual, zygomorphic. Calyx tubular, either 5-lobed or 2-lipped. Petals yellow, often lined with purple; posterior petal orbicular; spreading; lateral petals obliquely obovate or oblong; keel either obovate and slightly incurved, or often narrow and much incurved. Stamens 10 usually united, rarely 9 united, and 1 free; stamen tube occasionally split below or on both sides. Fruit a jointed pod, transversely septate between seeds; joints separating, indehiscent.

Léonard, J. Notulae Systematicae. XV. Aeschynomene. Bull. Jard. Bot. Et. Bruxelles 24: 63–84 (1954)
Rudd, V. The American species of Aeschynomene. Contr. U.S. Nat. Herb. 32: 1–172 (1955), Supplementary studies in Aeschynomene. Journ. Wash. Acad. Aci. 49: 45--52 (1959)

c. 150 species: warmer regions of the world. Approximately half of the species are aquatic or semi-aquatic and some contribute to the development of sudd. Many species are troublesome weeds of ricefields and irrigation canals. *Aeschynomene* is a poor forage plant but can be used for manure. The swollen floating stems of *A. aspera* L. and other species supplies a soft pith-like material (secondary xylem) which is employed in the making of such varied items as artificial flowers, toys, sunhats, life-belts and swimming jackets.

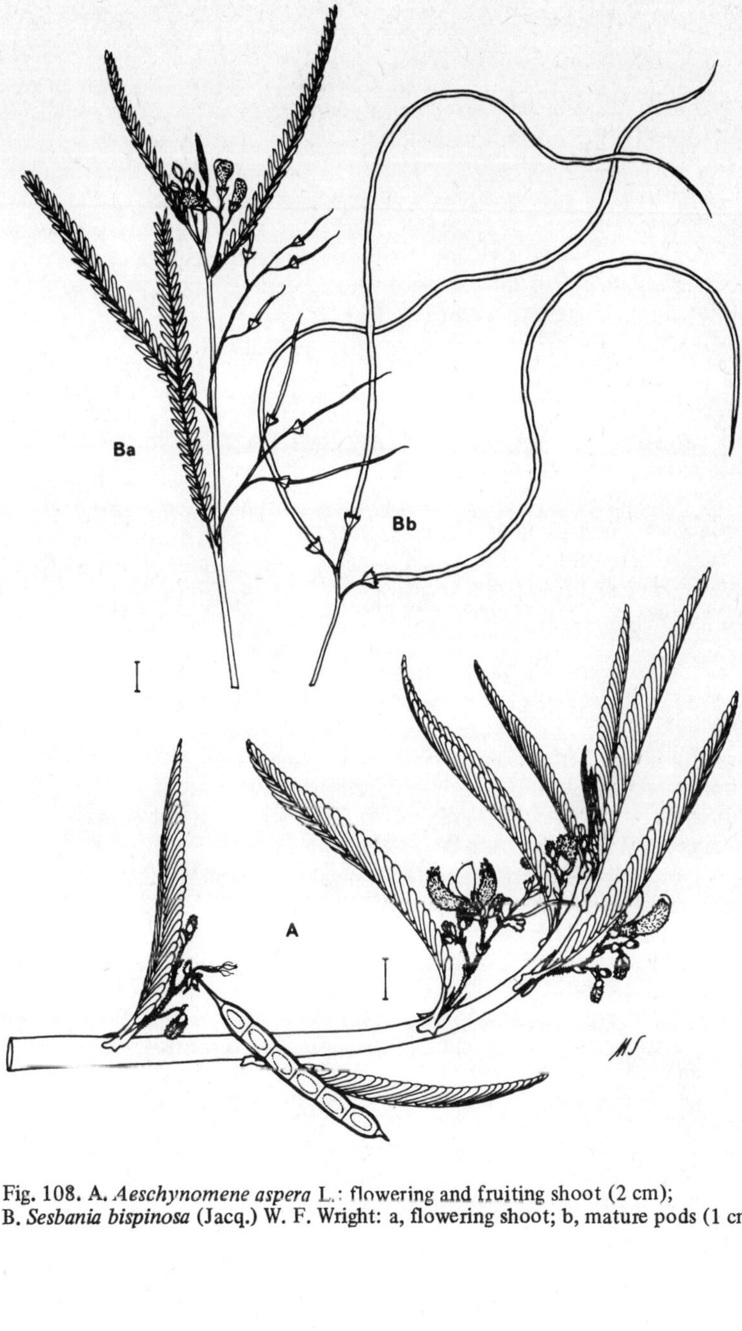

Fig. 108. A. *Aeschynomene aspera* L.: flowering and fruiting shoot (2 cm); B. *Sesbania bispinosa* (Jacq.) W. F. Wright: a, flowering shoot; b, mature pods (1 cm).

Neptunia Lour., Fl. Cochinch. 654 (1790)
Fig. 109.
Perennial herbs. Tap root often becoming woody. Stems prostrate, ascending or floating, terete, occasionally angled when young, unbranched or few-branched; internodes often swollen when floating. Stipules present, often not evident on floating stems. Leaves alternate, 2-pinnate and usually sensitive. Inflorescence a congested, pedunculate spike. Flowers dimorphic, upper bisexual and lower male, the stamens of lower flowers usually sterile or occasionally anther-bearing. Sepals campanulate, 5-lobed, green or rarely yellow. Petals 5, green or rarely yellow. Stamens free, 10 (in aquatic species) or 5; sterile stamens petaloid, yellow. Fruit a pod, flattened, marginally dehiscent.

Windler, D. R. A revision of the genus Neptunia (Leguminosae). Austral. Journ. Bot. 14: 379–420 (1966)

11 species warmer regions of the world. Most species are found in dry conditions. *N. triquetra* (Vahl.) Benth. is found in wet regions and withstands flooding. *N. plena* (L.) Benth. from the warmer regions of America and Asia is semi-aquatic. *N. oleracea* Lour. [*N. prostrata* (Lam.) Baill.] is found in tropical Asia, Africa and C. and S. America where it floats on the surface of still or slow-moving and frequently stagnant waters; it is often found as a weed in irrigation channels.

Sesbania Scop., Introd. 308 (1777), nom. cons.
Fig. 108B.
Small trees, shrubs or herbs. Stems usually erect occasionally floating; internodes often swollen when submerged or floating. Stipules frequently caducous. Leaves alternate, 1-pinnate. Inflorescence of lax, axillary racemes. Flowers bisexual, zygomorphic. Calyx tubular, broad, 5-lobed or 5-toothed. Petals yellow, purplish, spotted or white; posterior petal orbicular or ovate, spreading or reflexed; lateral petals oblong, falcate, keel incurved with long claws. Stamens 10, 9 united, the upper 1 free. Fruit a straight or slightly curved pod, transversely septate between seeds but not breaking into joints, marginally dehiscent.

Gillet, J. B. Sesbania in Africa (excluding Madagascar) and southern Arabia. Kew. Bull. 17: 91–153 (1963)
Schenk, H. Ueber das Aerenchym, ein dem Kork homologes Gewebe bei Sumpfpflanzen. Jahrb. Wiss. Bot. 20: 526–574 (1889)
Uppal, H. L. Green manuring with special reference to Sesbania aculeata for treatment of alkaline soil. Indian Journ. Agr. Sci. 25: 211–235 (1955)
Varada Chari, K. Agricultural development in Madras State, World Crops 9: 33–37 (1957)

c. 70 species; warmer regions of the world. Most species grow in wet land that is liable to be flooded but some are semi-aquatic or aquatic. Some species (such as *S. bispinosa* (Jacq.) W. F. Wright and *S. hirtistyla* Gillett) are troublesome weeds of ricefields. *S. sesban* (L.) Merrill is used as a fodder plant on land subject to flooding and is especially valuable in saline areas. *S. bispinosa* [*S. aculeata* (Willd.) Pers.] is used in India as a green manure in ricefields and can be effectively used for reclaiming alkaline land.

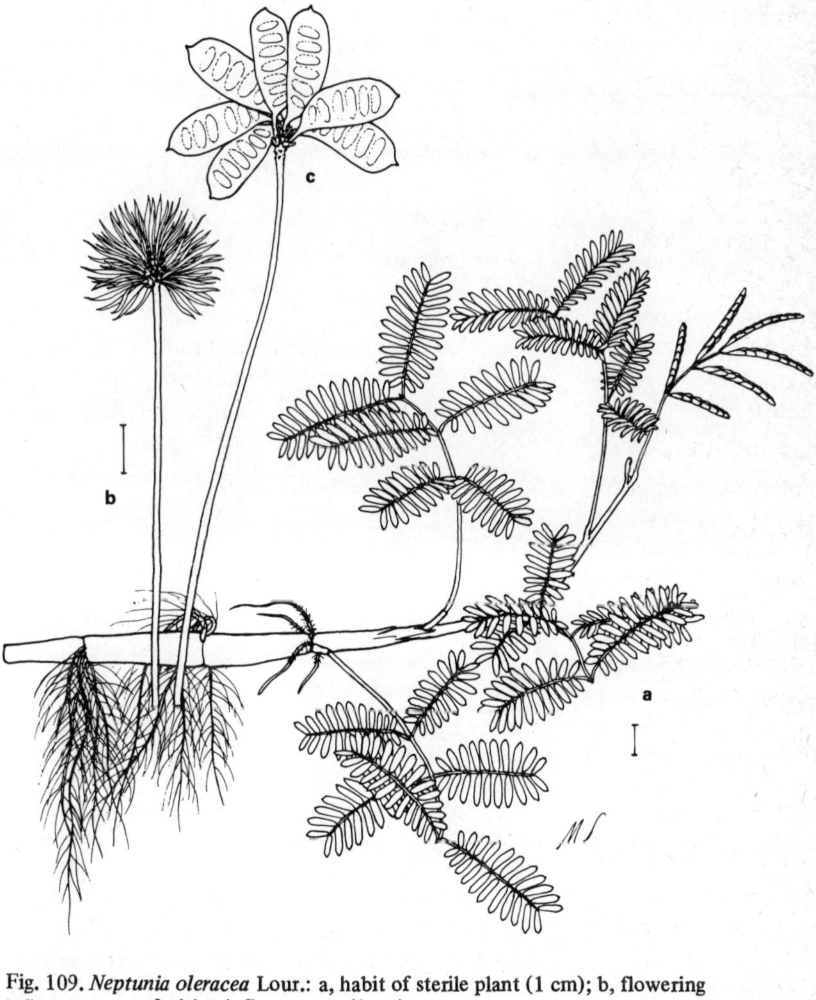

Fig. 109. *Neptunia oleracea* Lour.: a, habit of sterile plant (1 cm); b, flowering inflorescence; c, fruiting inflorescence (1 cm).

HALORAGACEAE [HALORAGIDACEAE, HALORRHAGACEAE]

5 genera (4 of which contain aquatics or semi-aquatics); *c.* 120 species: cosmopolitan (most species in Australia). The circumscription of the family in this text excludes the Callitrichaceae, Gunneraceae and Hippuridaceae.

Annual or perennial herbs or woody undershrubs. Leaves alternate, opposite or whorled, entire or pinnately divided (aquatics often heterophyllous), pinnately nerved; stipule-like appendages often present. Inflorescence variable, flowers occasionally solitary. Flowers bisexual or unisexual, actinomorphic, usually inconspicuous. Sepals 2 to 4 (in female flowers sometimes absent), united at base. Petals 2 to 4 or absent, boat-shaped. Stamens as many or twice as many as sepals. Ovary inferior, of 1 to 4 united carpels; each carpel with 1 pendulous ovule; fruit nut-like, or a schizocarp separating into 1-seeded nutlets.

Schindler, A. K. in Engler, A. Pflanzenreich 23 (IV. 225): 1–133 (1905). This treatment is today unacceptable and the Haloragaceae is in need of a new revision

1A Fruit splitting up into 1-seeded nutlets
Myriophyllum
1B Fruit not splitting up into 1-seeded nutlets
 2A Floral parts 3-merous; anthers ellipsoidal (confined to E. N. America)
Proserpinaca
 2B Floral parts 2- or 4-merous; anthers linear (S. Hemisphere and E. Asia)
  3A All flowers bisexual, sessile or shortly-stalked
Haloragis
  3B Some flowers unisexual, males distinctly stalked
Laurembergia

Haloragis J. R. and G. Forst., Gen. Pl. 31 (1776), [*Meionectes* R. Br., *Meziella* Schind.]
Fig. 110A.
Herbs or woody undershrubs. Stems ascending or creeping. Leaves usually opposite below and alternate above, simple or palmately divided. Flowers bisexual, sessile or shortly-stalked. Sepals 2 or 4. Petals 2 or 4. Stamens as many or twice as many as sepals; anthers linear. Ovary 1-, 2- or 4-locular; fruit indehiscent, hard or spongy.

c. 60 species (Schindler op. cit. recognised 70 species); most S. Hemisphere (centre in Australia) also known in the Himalayas, Japan and Korea. Most species are terrestrial but some (*H. brownii* (Hooker) Schind.) are sub-aquatic while some others (*H. halconensis* Merr., *H. micrantha* (Thunb.) R. Br.) are amphibious.

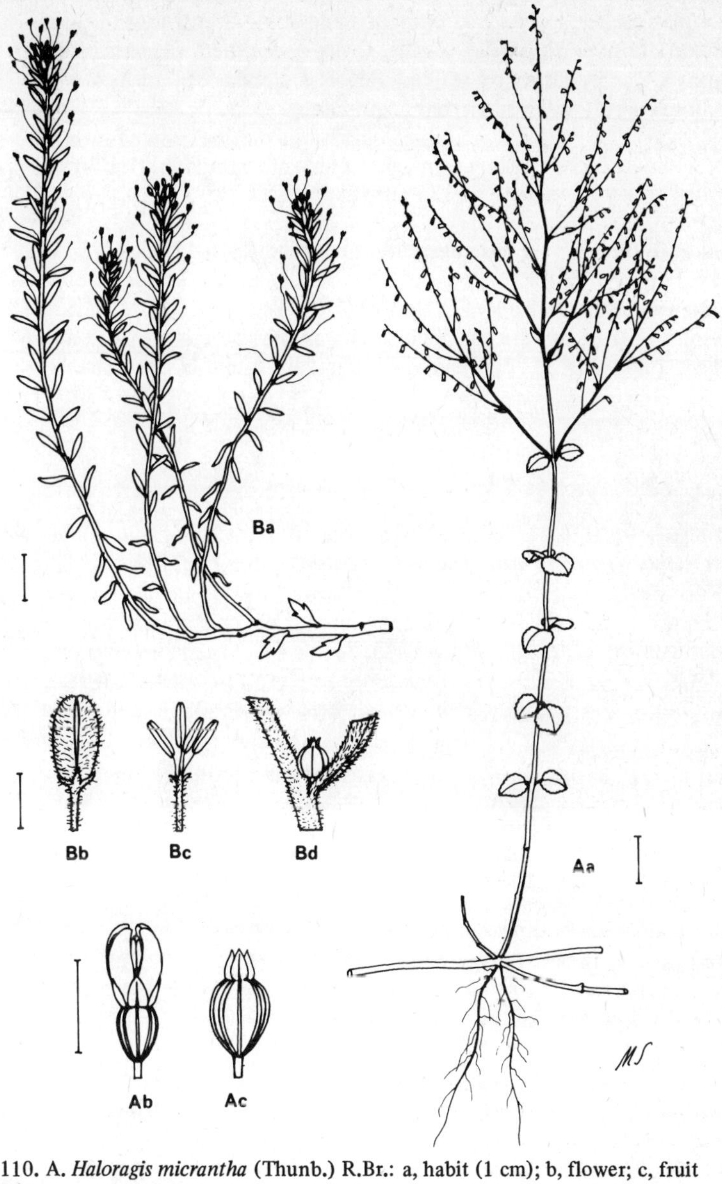

Fig. 110. A. *Haloragis micrantha* (Thunb.) R.Br.: a, habit (1 cm); b, flower; c, fruit (1 mm);
B. *Laurembergia repens* Berg.: a, habit (1 cm); b, male flower in bud; c, male flower after petals have fallen; d, female flower (1 mm).

Laurembergia Berg., Descr. Pl. Cap. 350 (1767), [*Serpicula* L.]
Fig. 110B.
Perennial herbs, occasionally woody at the base. Stems ascending or prostrate and rhizomatous, the lower parts branched and somewhat woody. Leaves opposite or alternate, subsessile, simple, entire or toothed. Flowers bisexual or unisexual, usually distinctly stalked. Sepals 4. Petals 4. Stamens 4 or 8; anthers linear. Fruit indehiscent, hard, nut-like.

Raynal, A. Les espèces africaines du genre Laurembergia Berg. (Halorrhagaceae), et leur répartition. Webbia 119 (2): 683–695 (1965)

c. 4 species (more than 20 have been described, see Raynal op. cit.); *L. coccinea* (Bl.) Kan. S. and S. E. Asia, *L. repens* Berg. S. Africa, *L. tetrandera* (Schott) Kan. Africa and S. America and *L. veronicifolia* (Bory) Schind. Madagascar and Réunion. All species are amphibious and are characteristically found in regularly flooded areas. *L. coccinea* is occasionally found in ricefields.

Myriophyllum L., Sp. Pl. 992 (1753)
Fig. 111A.
Annual or perennial herbs. Stems rhizomatous, floating or ascending. Leaves whorled, opposite or alternate, divided or entire (submerged leaves usually divided), or rarely reduced to scales, often flanked by stipule-like out-growths. Inflorescences mostly terminal, spike-like, or flowers in leaf axils. Flowers bisexual or unisexual, sessile or nearly so. Sepals 4 or absent, usually very small. Petals 2, 4 or absent (usually reduced or absent in female flowers). Stamens 4 or 8. Ovary 2- or 4-locular; fruit breaking up into 1-seeded nutlets.

Blackburn, R. D. and Weldon, L. W. Eurasian watermilfoil–Florida's new underwater menace. Hyacinth Control Journ. 1967 (6): 15–18 (1967)
Meijden, R. van der. An annotated key to the South-East Asiatic, Malesian, Mascarene, and African species of Myriophyllum (Haloragaceae). Blumea 17 (2): 303–311 (1969)
Patten, B. C. The status of some American species of Myriophyllum as revealed by the discovery of intergrade material between M. exalbescens Fern. and M. spicatum L. in New Jersey, Rhodora 56: 213–225 (1954)
Steenis, J. H. Eurasian watermilfoil, a rapidly spreading water plant. Hyacinth Control Journ. 1967 (6): 18–19 (1967)

c. 40 species: almost cosmopolitan (lacking in the Arctic and rare in Africa). Most species are aquatic or amphibious, found in a variety of habitats. *M. aquaticum* (Velloso) Verdc. [*M. brasiliense* Cambess.], originally from S. America has become naturalised in many countries where it spreads vegetatively and is often considered a serious nuisance; in Java, however, it is cultivated and the tips of the shoots are eaten as a vegetable. *M. spicatum* is very widespread and is often regarded as a menace because it sometimes may reduce the flow of water or cause blockages in water channels.

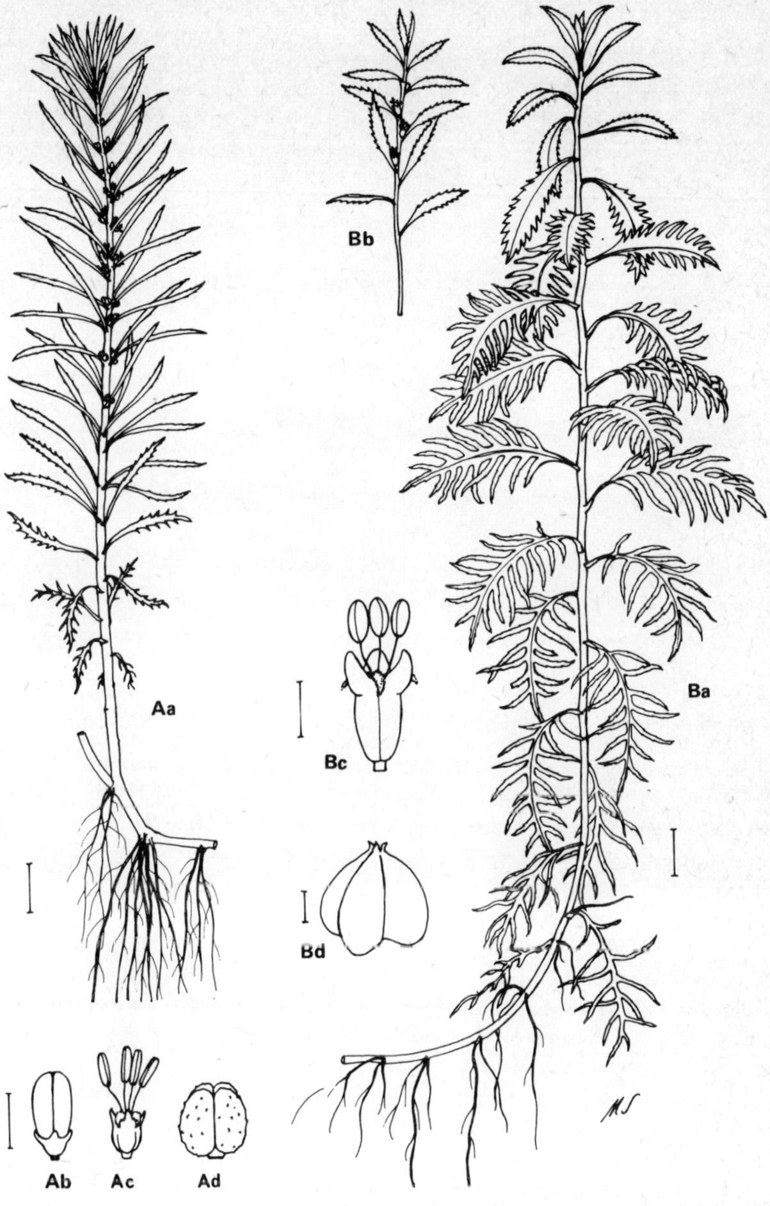

Fig. 111. A. *Myriophyllum oliganthum* (Wr. and Arn.) F.v.M.: a, habit (1 cm); b, bud; c, flower; d, fruit (1 mm);
B. *Proserpinaca palustris* L.: a, habit; b, flowering shoot (1 cm); c, flower (1 mm); d, fruit (1 mm).

Proserpinaca L., Sp. Pl. 88 (1753)
Fig. 111B.
Perennial herbs. Stems ascending or prostrate and rhizomatous, the lower parts branched and somewhat woody. Leaves alternate, sub-sessile, pinnately divided or the upper occasionally entire but distinctly toothed. Flowers bisexual, sessile, solitary in leaf axils. Sepals 3. Petals absent. Stamens 3; anthers ellipsoidal. Fruit nut-like, 3-angled, 3-seeded.

Fernald, M. L. and Griscom, L. Proserpinaca palustris and its varieties. Rhodora 37: 177–178 (1935)
Burns, G. P. Heterophylly in Proserpinaca palustris. Ann. Bot. 18: 579–587 (1904)

2 or 3 species: E. N. America. Found in marshes, shallow ponds and temporary pools.

HANGUANACEAE

Monotypic.

Hanguana Blume, Enum Pl. Java 15 (1827), [*Susum* Blume]
Fig. 112A.
Robust, erect, dioecious, 50 to 200 cm tall herbs. Stems erect, ascending or floating, often with long creeping or floating stolons. Leaves stiff, erect, mostly basal, the lower petiolate, the upper sessile, variable in size, 20 to 120 cm long, 1.5 to 15 cm wide, usually lanceolate, longitudinally nerved, with many fine crossnerves; leaf bases sheathing. Inflorescence a panicle, 10 to 120 cm long, 1 to 3 times branched; branches subtended by large bracts. Flowers unisexual, solitary or in small clusters, sessile. Perianth 2-seriate; segments united below, green or yellowish, persistent; outer segments 3, 2.0 to 2.5 mm long; inner segments 2.5 to 3 mm long, often red-dotted. Male flower with stamens 6, inserted on perianth, about as long as inner perianth segments. Female flower with staminodia 6; the 3 outer small, triangular, the 3 inner larger, rounded, dorsally compressed; ovary 3-locular, with 1, axile ovule in each loculus; stigma sessile, 3-lobed; fruit an oblong, 2 cm long, red, fleshy drupe.

Airy Shaw, H. K. Hanguanaceae. Kew Bull. 18: 260–261 (1965)
Smithson, E. The comparative anatomy of the Flagellariaceae. Kew. Bull. 1956: 491–501 (1956)

1 species, *H. malayanum* (Jack.) Merr.: Ceylon, S. E. Asia, Malaysia, Palau Islands. It occurs in swamps, pools, lakes, slowly flowing freshwater, along river and lake shores, and in humid forests. It often forms floating islands. This species is very unusual in many respects; it shows a very wide ecological amplitude but recognition of the aquatic form as a distinct species (*Susum anthelminthicum* Blume or *H. aquatica* Kanehira) does not seem justified. For a floating aquatic it is exceptionally large. Its taxonomic history is somewhat chequered and it has appeared in the following families: Liliaceae, Smilacaceae, Juncaceae and Flagellariaceae.

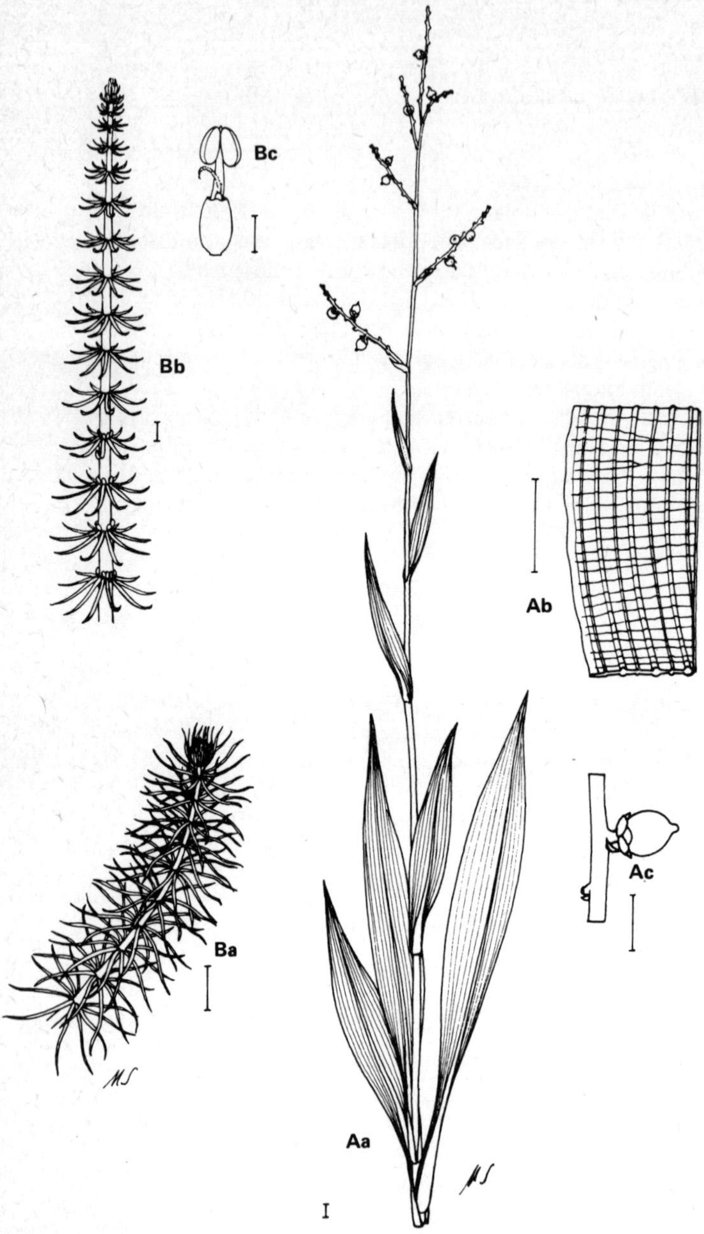

Fig. 112. A. *Hanguana malayanum* (Jack.) Merr.: a, habit (1 cm); b, detail of leaf (1 cm); c, fruit (1 cm);
B. *Hippuris vulgaris* L. a, submerged shoot; b, aerial shoot (1 cm); c, flower (1 mm).

HIPPURIDACEAE

1 genus: *Hippuris.*

Hippuris L., Sp. Pl. 4 (1753)
Fig. 112B.
Perennial. Stem a creeping rhizome from which arise erect, leafy stems. Leaves whorled, simple, obovate to linear, entire; stipules absent. Flowers bisexual, unisexual or sterile, borne in leaf-axils, the upper sessile, the lower usually shortly stalked. Perianth reduced to a rim around the top of the ovary. Stamen 1. Ovary inferior, 1-locular; ovule solitary, pendulous; fruit a small, ovoid, smooth nut.

McCully, M. E. and Dale, H. M. Variations in leaf number in Hippuris. Canad. Journ. Bot. 39: 611–625 (1961)
McCully, M. E. and Dale, H. M. Heterophylly in Hippuris. Canad. Journ. Bot. 39: 1099–1116 (1961)

c. 1 species: temperate and cold regions of N. Hemisphere. *H. vulgaris* L. is found in still or slowly flowing calcareous water. It is very variable and some variants have been given specific status.

HYDROCHARITACEAE

15 genera; cosmopolitan. All aquatic, 3 genera (*Enhalus, Thalassia* and *Halophila*) are marine and thus excluded from this account.

Annual or perennial, dioecious or monoecious herbs. Leaves various, usually submerged, rarely floating or partly emergent. Flowers arranged in a bifid spathaceous bract or between 2 opposite bracts, bisexual or unisexual; when unisexual then males usually more than 1, females solitary. Perianth segments free, 1- or 2-seriate, 3 or rarely 2 in each series; inner series when present usually showy and petal-like. Stamens 1 to numerous, in 1 or more whorls, the inner ones sometimes staminodial. Ovary inferior, of 2 to 15 united carpels; placentas parietal, occasionally protruding nearly to centre of ovary; fruit globose to linear, dry or pulpy, indehiscent or rupturing irregularly; seeds numerous, smooth, warty or spiny.

Dandy, J. E. Notes on Hydrocharitaceae. Journ. Bot. (London) 72: 132–139 (1934); 73: 209–211 (1935)

Ernst-Schwarzenbach, M. Zur Blütenbiologie einiger Hydrocharitaceen. Ber. Schweiz. Bot. Ges. 55: 33–69 (1945)

Ernst-Schwarzenbach, M. Zur Kompatabilität von Art- und Gattungsbastardierung bei Hydrocharitaceen. Oesterr. Bot. Zeit. 100: 403–425 (1953)

Hartog, C. Den. The sea-grasses of the world. London and Amsterdam (1970)

Kaul, R. B. Floral morphology and phylogeny in the Hydrocharitaceae. Phytomorphology. 18 (I): 13–35 (1968)

Kaul, R. B. Evolution and adaption of inflorescences in the Hydrocharitaceae. Amer. Journ. Bot. 57 (6): 708–715 (1970)

1A Leaves differentiated into blade and petiole
  2A Petioles with 1 or 2, transparent stipules near base; stolons present
    3A Petals absent or when present up to 1½ times as long as sepals
Limnobium
    3B Petals present, more than 1½ times as long as sepals
Hydrocharis
  2B Petioles without stipules: stolons absent
Ottelia
1B Leaves not differentiated into blade and petiole
  4A Leaves in a basal rosette
    5A Leaves rigid, with conspicuous, spinous-serrate margins; entire plant rising to water surface to flower
Stratiotes
    5B Leaves not rigid, without conspicuous, spinous-serrate margins (margins occasionally minutely serrate or leaves irregularly spiny); entire plant not rising to water surface to flower
      6A Male flowers minute, breaking off and floating on the water surface; petals minute; greater part of leaf parallel-sided
Vallisneria

6B Male flowers not minute, not breaking off and floating on the water surface; petals large and conspicuous or absent; greater part of leaf not parallel-sided, or when parallel-sided then bearing spines

7A Styles 3 to 15, each bifid; spathes inflated and distinctly winged or spiny; mature leaves lanceolate, broad-ovate to suborbicular
Ottelia

7B Styles 3, each entire; spathes not inflated, not distinctly winged but occasionally ribbed, never spiny; mature leaves linear, tapering gradually to a fine point
Blyxa

4B Leaves arranged along the stem

8A Leaf-bases more or less clasping the stem

9A Midrib prominent; male flowers not free-floating
Blyxa

9B Midrib not prominent; male flowers free-floating
Nechamandra

8B Leaf-bases not clasping the stem

10A At least the middle and upper leaves opposite or whorled

11A Petals about 3-times as long and as broad as sepals; flowers with nectaries
Egeria

11B Petals rarely exceeding sepals or absent; flowers without nectaries

12A Male spathes containing numerous flowers
Lagarosiphon

12B Male spathes each containing 1 flower

13A Male spathe globose, spiny; stamens 3; all flowers unisexual
Hydrilla

13B Male spathe ovoid or tubular, smooth; stamens rarely 3, usually 6 or 9; some flowers often bisexual
Elodea

10B All leaves spirally arranged

14A Male spathes of 2 united bracts, containing numerous pedicelled flowers; female spathes containing 1 to 3 flowers; female flowers with 2-seriate perianth (through introductions widespread)
Lagarosiphon

14B Male spathes of 2 distinct bracts, containing numerous stamens (male flower represented by single, naked stamen); female spathes containing 1 flower; female flowers with 1-seriate perianth (N. W. Australia)
Maidenia

Blyxa Noronha ex Thouars, Gen. Nov. Madag. 4 (1806)
Fig. 113.
Submerged monoecious or dioecious, annuals or perennials. Stem corm-like or up to 60 cm long. Leaves linear, spirally arranged, radical or along the stem; base sheathing the stem; margins entire or minutely serrate; apex attenuate; nerves parallel; midrib prominent. Spathes sessile or stalked, tubular, with

Fig. 113. *Blyxa aubertii* Rich.: a, habit (1 cm); b, flower.

6 longitudinal ribs, bifid at the apex, 1-flowered or in male spathes of dioecious plants with up to 10 flowers. Flowers unisexual or bisexual, female and bisexual ones sessile, male ones stalked. Sepals 3, linear or linear-lanceolate, green, persistent. Petals 3, linear, longer than the sepals, white, flaccid, fringed or reduced or absent. Stamens 3, 6, or 9; filaments capillary; anthers linear or lanceolate, 2-locular. Ovary of 3 carpels, linear, with long, capillary beak; styles 3, entire, linear, united at the base; fruit linear or linear-lanceolate, membranous; seeds 10 to numerous, elliptic or fusiform, 1 to 2 mm long, smooth or spiny.

Govindappa, D. A. and Naidu, T. R. B. The embryosac and endosperm of Blyxa oryzetorum Hooker fil. Journ. Indian Bot. Soc. 35: 417–422 (1956)
Lakshmanan, K. K. Embryological studies in the Hydrocharitaceae I. Blyxa octandra Planch. Journ. Madras Univ. ser. B.31: 133–142 (1961)
Rangaswamy, K. A. A morphological study of the flower of Blyxa echinospora Hook. Journ. Indian Bot. Soc. 20: 123–133 (1941)

c. 10 species: warmer regions of Old World, introduced in N. America. In a variety of aquatic habitats, frequently found as weeds in ricefields and irrigation ditches.

Egeria Planchon, Ann. Sci. Nat. Bot. sér. 3, 11: 79 (1849)
Fig. 114A.
Like *Elodea* but male spathe 2 to 4 flowered; female spathe split halfway down one side; petals about 3 times larger than sepals; stamens 9 (–10), distinct; anthers loculicidally dehiscent; flowers with nectaries.

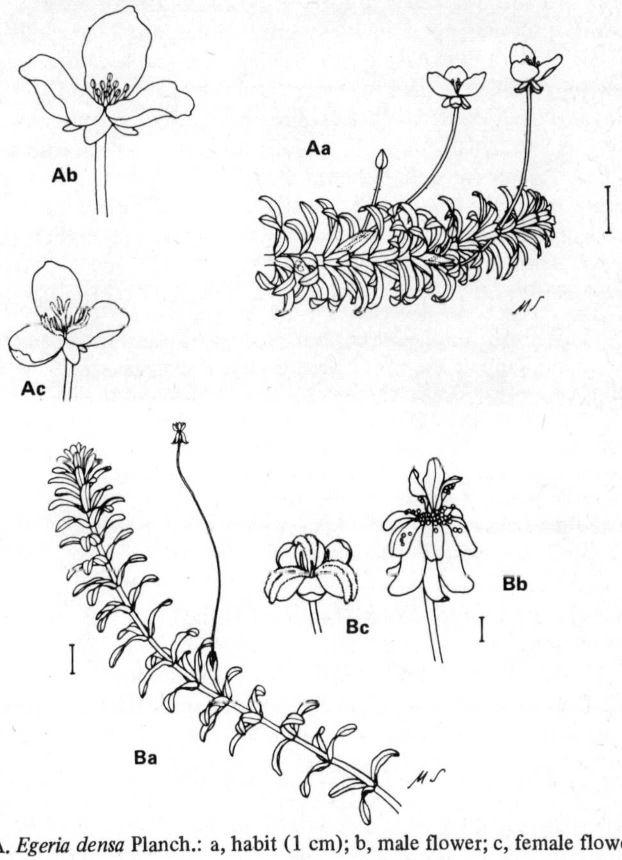

Fig. 114. A. *Egeria densa* Planch.: a, habit (1 cm); b, male flower; c, female flower (1 mm);
B. *Elodea canadensis* Rich.: a, habit (1 cm); b, male flower; c, female flower (1 mm).

St. John, H. Monograph of the genus Egeria Planchon. Darwiniana 12: 293–307, 523 (1961)

2 species, *E. densa* Planch. [*Elodea densa* (Planch.) Casp.] and *E. najas* Planch.: originally in warm-temperate S. America. *E. densa* has been introduced to N. America, Africa, Europe and Japan. It usually occurs in lakes, streams and ponds. It is frequently grown in biological laboratories and in aquaria. In some regions it is considered a menace to navigation.

Elodea Michx., Fl. Bor. Amer. 1: 20 (1803) [*Anacharis* Rich.]
Fig. 114B.
Dioecious or monoecious, often stoloniferous perennials. Stems slender, simple or sparsely branched. Leaves sessile; lowest ones alternate, opposite, or in whorls of 3, middle and upper ones opposite or in whorls of 3 to 7, linear to oblong, acute or obtuse, sharply serrulate, 1-nerved; leaf-axils with minute squamulae intravaginales. Inflorescences unisexual, or bisexual. Spathes axillary, ovoid or tubular, apex shallowly bifid; containing 1 (or rarely 3) male or 1 female, or rarely 1 bisexual flower. Male spathes sessile or nearly so. Male flowers long-stalked, separating early from the mother plant and floating on the water surface. Female spathes sessile. Female flowers long-stalked. Bisexual flowers like female ones, but with 3 stamens. Sepals 3, elliptic, paper-like. Petals 3 (absent in *E. richardii* St. John), elliptic to linear membranous, white to purple smaller than or slightly exceeding the sepals. Stamens 3 to 9, usually 6 in a lower, outer ring, and 3 in an inner united filament column; anthers 2-locular, septifragally dehiscent. Ovary 1-locular; styles 3, bifid or rarely entire; fruit a capsule; seeds 1 to 5, cylindrical to fusiform.

Kent, D. H. Plant notes: Elodea. Proc. Bot. Soc. Brit. Isles, 5: 232 (1964)
St. John, H. Monograph of the genus Elodea (Hydrocharitaceae). I. The species found in the Great Plains, the Rocky Mountains, and the Pacific States and Provinces of North America. Res. Stud. Wash. State Univ., 30: 19–44 (1962); II. The species found in the Andes and western South America. Caldasia, 9: 95–113 (1964); III. The species found in northern and eastern South America. Darwiniana, 12: 639–652 (1963); IV. The species of eastern and central North America, and Summary. Rhodora, 67: 1–35, 155–180 (1965)

17 species: originally from N. and S. (but not Central) America: occurring gregariously in lakes, ponds, canals and slowly flowing water. Several species have become established in the Old World. *E. canadensis* Rich. was introduced to Ireland in 1836 and aggressively invaded Europe soon after; it is now established in Asia, Africa, Australia and New Zealand. In many regions it is considered a pest.

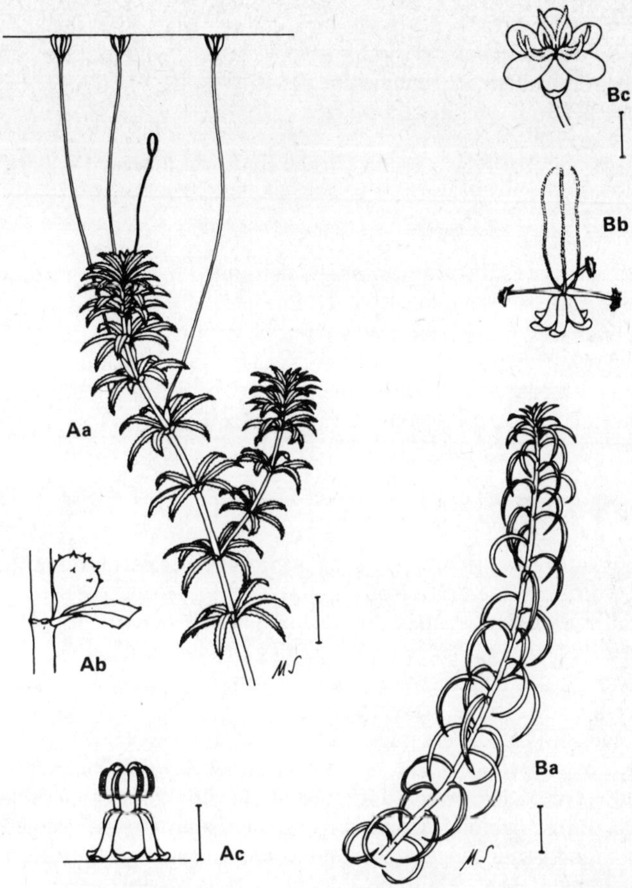

Fig. 115. A. *Hydrilla verticillata* (L. fil.) Royle: a, habit (1 cm); b, male inflorescence (1 mm); c, male flower;
B. *Lagarosiphon muscoides* Harvey. a, habit (1 cm); b, male flower; c, female flower (1 mm).

Hydrilla L. C. Rich., Mém. Cl. Sci. Math. Phys. Inst. France 12 (2): 9, 61, 76 (1814)

Fig. 115A.

Submerged, monoecious or dioecious, perennials. Stems branched, about 1 mm thick and up to 3 m long; internodes 3 to 50 mm long. Leaves in whorls of 3 to 8, sessile, 7 to 40 mm long, linear to lanceolate, rarely elliptic, midrib conspicuous, margin toothed; leaf-axils with minute (up to 0.5 mm long) squamulae intravaginales. Inflorescences unisexual. Male spathe about 1.5 mm long, subsessile, solitary in the leaf-axils, flattened, globose, somewhat spiny. Male flower pedicelled (1 to 2 mm long) small, solitary. Female spathes about 5 mm long, sessile, solitary in the leaf-axils, cylindrical, apex bifid, membranous, containing a solitary female flower. Sepals 3, imbricate, white, sometimes with reddish dots, oblong to obovate, scarious, up to 3 mm long and 1.0 mm wide.

Petals 3, imbricate, white, up to 3 mm long and 0.5 mm wide. Stamens 3. Ovary cylindric to narrowly conical; beak capillary; styles 3, alternating with the petals; fruit cylindrical or narrowly conical, softly spiny, 7 mm long and 1.5 mm wide; seeds 2 to 6, oblong-ellipsoidal.

Lakshmanan, K. K. Embryological studies in the Hydrocharitaceae IV. Post-fertilization development in Hydrilla verticillata Royle. Phyton (B. Aires) 22: 45–50 (1969)

1 species, *H. verticillata* (L. fil.) Royle: widely distributed in warmer regions of the Old World; it is also found locally in N. Europe and introduced in N. America. In ricefields, pools, lakes, slow streams, even in tidal waters. Often gregarious. When occurring in large quantities it can be used as manure. Often considered a serious menace as it reduces the rate of flow of water in irrigation canals. It has recently become a serious pest in Florida, USA.

Hydrocharis L., Sp. Pl. 1036 (1753)
Fig. 116A.
Floating, monoecious, stoloniferous, perennials. Stolons originating from the leaf-axils. Leaves differentiated into blade and petiole; blade floating or emergent, ovate to suborbicular, apex rounded or acute; base more or less cordate or reniform; nerves curved, parallel, joining the marginal nerve, connected by straight parallel cross-veins, in their turn connected again by very fine veinlets, parallel to the nerves; base of petiole with 1 or 2 ligulate, transparent stipules. Inflorescences unisexual. Spathe segments lanceolate, membranous; the male spathe stalked, containing 1 to 4 flowers, the female one sessile, 1-flowered. Male flowers short stalked, female flowers long stalked. Sepals 3, elliptic, obtuse, white or greenish white, persistent. Petals 3, larger than the sepals, broadly-obovate, with broadly rounded apex and cuneate persistent base. Stamens 9 to 12, anthers 2-locular. Ovary elliptic, nearly 6-locular; styles 6, flat, bifid; fruit berry-like, elliptic to globose, with 6 ribs, bursting irregularly at the apex; seeds numerous, ellipsoidal.

Cutter, E. G. and Feldman, L. J. Trichoblasts in Hydrocharis. Amer. Journ. Bot. 57 (2): 190–201 (1970)
Dore, W. G. Progress of the European Frog-bit in Canada. Canad. Field-nat. 82: 76–84 (1968)

3 to 6 species: widespread in the Old World, *H. morsus-ranae* L., introduced in N. America. Free-floating or rooting in shallow water, occurring in pools, lakes and the wetter parts of marshes, often locally gregarious.

Lagarosiphon Harvey, in Hooker's Journ. Bot. 4: 230 (1841)
Fig. 115B.
Submerged, dioecious, rhizomatous, perennials. Leaves widely spaced below, crowded above, alternate, subopposite or rarely whorled, each with 2 minute,

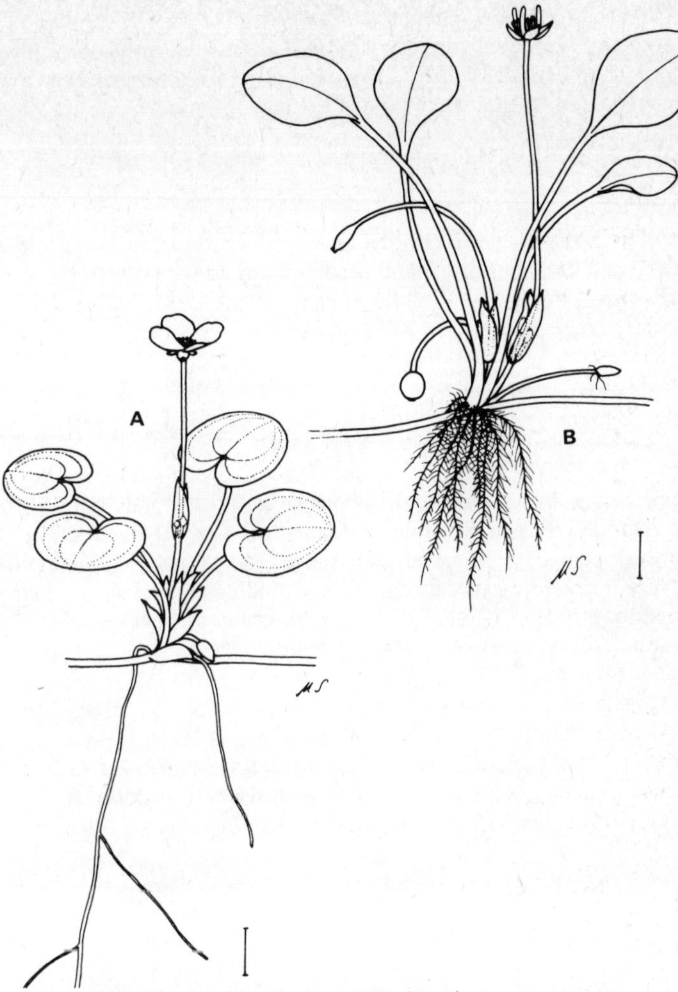

Fig. 116. A. *Hydrocharis morsus-ranae* L. habit (1 cm);
B. *Limnobium laevigatum* (Humb. & Bonpl. ex Willd.) Heine: habit (1 cm).

narrowly ovoid intravaginal stipules; margins toothed or entire. Male spathes axillary, consisting of 2 united bracts, obovate or ovate to lanceolate, compressed or cup-shaped, dentate; containing numerous stalked flowers (up to 50 in *L. major*); the buds usually becoming detached and rising to the surface where they expand. Sepals 3. Petals 3, slightly broader than sepals. Stamens 3; staminodes 3, longer than the stamens, papillate and usually coloured above, joined at the top, acting as a sail. Female spathes axillary, consisting of 2 united, narrowly oblong, ovoid or cylindrical bracts, entire or toothed, containing 1, 2 or 3 flowers; perianth as in male flowers, perianth-tube exserted laterally near the apex of the spathes; staminodes 3, small, capillary; ovary 1-locular with 3 parietal placentas, styles 3, adnate to perianth-tube, each divided above into 2, long, papillate, often brightly coloured stigmas; ovules 6 to 30; fruit a capsule ovoid or cylindrical, attenuate at the apex into a beak (the persistent base of the perianth), protruding from torn spathe-valves; seeds cylindrical, with a short stipe at the base, pointed at the apex.

Obermeyer, A. The South African species of Lagarosiphon. Bothalia 8: 139–146 (1964)

c. 16 species: Africa and Madagascar, introduced elsewhere. *L. major* (Ridley) Moss has established itself in Europe and New Zealand; in the latter it is considered a pest.

Limnobium L. C. Rich., Mém. Cl. Sci. Math. Phys. Inst. France 12 (2): 66 (1814), [*Hydromystria* G. F. W. Mey, *Jalambicea* Cerv.]
Fig. 116B.
Stoloniferous herbs. Leaves differentiated into blade and petiole; blade floating or emergent, elliptic or ovate to suborbicular, apex rounded or obtuse; base cuneate, rounded or cordate, under surface of the floating leaves often covered by a thick aerenchymatous cushion; nerves curved, parallel, joining the marginal nerve, connected by straight, parallel, ascending cross-veins, in their turn connected by very fine, parallel veinlets; petiole near the base with 1, transparent, stipule. Inflorescences unisexual. Spathe segments linear-lanceolate, membranous, containing 3 to 9 male flowers or 1 female one; male and female flowers pedicelled. Sepals 3, oblong or lanceolate. Petals 3, linear-lanceolate, narrower than the sepals and only up to about 1½ times as long, absent or reduced in the female flowers of some species. Stamens 6 to 12, anthers 2-locular, laterally dehiscent. Ovary elliptic; styles 6, each bifid; fruit berry-like, elliptic; seeds numerous, subglobose.

Heine, H. Limnobium laevigatum correct name for Hydromystria stolonifera, Jalambicea repens, Trianea bogotenses, Limnobium stoloniferum, Hydrocharis stolonifera. Adansonia 8 (3): 314–316 (1968)

c. 3 species: warmer parts of America, introduced elsewhere. Free-floating or rooting in shallow water. *L. laevigatum* (Humb. and Bonpl. ex Willd.) Heine has become established in Europe and Java.

Maidenia Rendle, Journ. Bot. (London) 54: 313–316 (1916)
Fig. 117A.
Dioecious, submerged, stoloniferous, perennial. Stem erect, simple, 50 to 60 mm long, about 2 mm in diameter, leafy. Leaves sessile, alternate, up to 40 mm long and 4 mm wide, setaceous, margins sparsely toothed. Male spathes axillary, subsessile, 2.5 mm long, consisting of 2 distinct bracts, containing numerous stamens, spirally arranged on a central axis. Each stamen is considered to be morphologically equivalent to 1 male flower; filaments short;

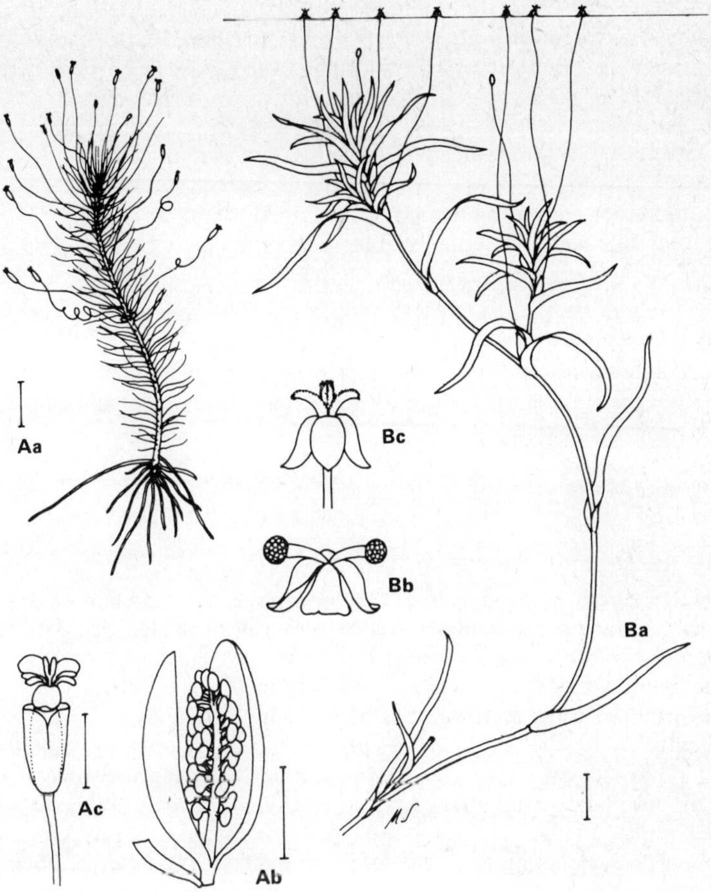

Fig. 117. A. *Maidenia rubra* Rendle: a, habit (1 cm); b, male inflorescence (1 mm); c, female flower (1 mm); after Highley;
B. *Nechamandra alternifolia* (Roxb.) Thwaites: a, habit (1 cm); b, male flower (1 mm); c, female flower (1 mm).

anthers broadly-ellipsoidal. Female spathe axillary, 1-flowered, bifid, with a 20 to 30 mm long stalk. Perianth 1-seriate, segments broad-elliptic, about 1 mm long. Ovary subcylindrical; stigmas 3, bifid; fruit fusiform, obscurely 3-carinate; seeds numerous, minute.

1 species, *M. rubra* Rendle: N. W. Australia. In ponds and ditches. This species has not been well studied.

Nechamandra Planchon, Ann. Sci. Nat. Bot. sér. 3, 11: 78 (1849)
Fig. 117B.
Submerged, dioecious, perennial or annual. Stems 60 to 90 cm long, filiform, ascending. Leaves sessile, alternate, linear-lanceolate, 3 to 8 cm long and up to

0.6 cm wide, acute, often twisted; margins toothed; bases clasping the stem. Male spathes axillary, consisting of 2 bracts, containing numerous minute flowers, the buds becoming detached and rising to the surface before anthesis. Sepals 3. Petals 2, small. Stamens 2, divergent, anthers ovate, transversely dehiscent. Female spathes axillary, 1-flowered tubular, with about 40 mm long bifid apex. Perianth of 3 spreading, orbicular, concave sepals. Perianth-tube filiformly attenuated above into a neck-like limb. Ovary ovate-lanceolate, gradually attenuated above; stigmas 3, cuneate, bilobed; fruit an ovoid, indehiscent utricule, included within spathe; seeds numerous, oblong, pitted.

Lakshmanan, K. K.: Embryological studies in the Hydrocharitaceae. III. Nechamandra alternifolia (Roxb.) Thw. Phyton (B. Aires) 20: 49–58 (1963)

1 species, *N. alternifolia* (Roxb.) Thw.: India and S. E. Asia. In ponds, tanks and lakes. In its native range it has not been reported to be a pest but it is a potentially dangerous weed.

Ottelia Pers., Syn. Pl. 1: 400 (1805); [*Bootia* Wall., *Oligolobos* Gagnep., *Xystrolobos* Gagnep.]

Fig. 118.

Monoecious or dioecious, perennials. Leaves linear, lanceolate, broadly-ovate, suborbicular, cordate or reniform, smooth or spiny; apex rounded, acute or apiculate; base cuneate, truncate or cordate; nerves 3 to 11, parallel, straight or curved, connected by fine parallel crossveins, often forming a characteristic pattern, midrib sometimes prominent; petiole sheathing at the base, often merging gradually into the blade. Spathes peduncled, elliptic or ovate, with 6 more or less prominent ribs or 2 to 10 wings, occasionally spiny; female and bisexual ones 1-flowered, male ones many-flowered. Flowers unisexual or bisexual, female and bisexual ones sessile, male ones stalked. Sepals 3, linear, oblong or ovate, green with scarious margins, persistent. Petals 3, oblong, broad-obovate to orbicular, 2 to 3 times as long as the sepals, white or coloured. Stamens 6 to 15; filaments filiform, flattened, anthers linear or oblong, laterally dehiscent. Ovary oblong, narrowed at the top, incompletely divided by the parietal placentas into 6 loculae; styles 6 to 15, bifid (in male flowers 3 stylodia are present); fruit oblong, attenuate at apex, pericarp thickened; seeds numerous, minute, oblong or fusiform, with rather thick testa.

Dandy, J. E. Notes on Hydrocharitaceae. Journ. Bot. (London) 72: 132–139 (1934), 73: 209–211 (1935)

Kaul, R. B. Morphology and development of the flowers of Bootia cordata, Ottelia alismoides and their synthetic hybrid. Amer. Journ. Bot. 56 (8): 951–959 (1969)

Ernst-Schwarzenbach, M. Kleistogamie and Antherenbau in der Hydrocharitaceen-Gattung Ottelia. Phytomorphology 6: 296–311 (1956)

c. 40 species: warmer parts of the world, 1 species in Brazil and the rest in the Old World with centres of speciation in C. Africa and S. China. This genus is much in need of revision. *O. alismoides* (L.) Pers. is naturalised in Europe.

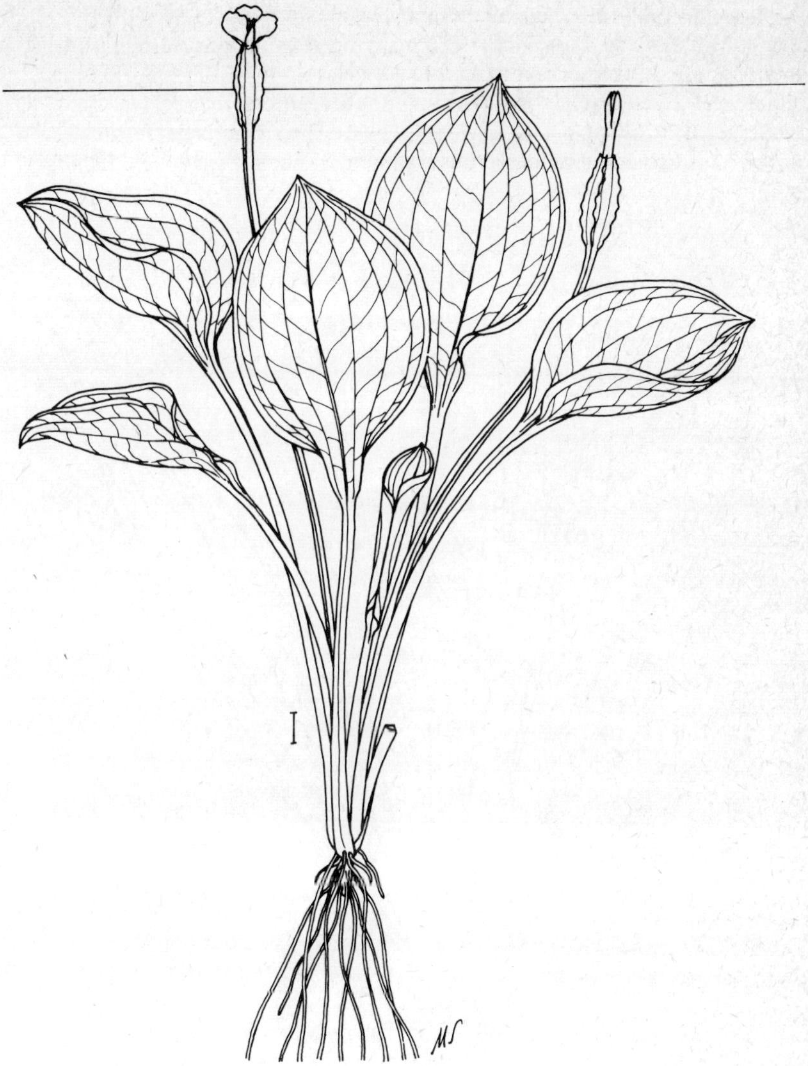

Fig. 118. *Ottelia alismoides* (L.) Pers.: habit (1 cm).

Stratiotes L., Sp. Pl. 535 (1735)
Fig. 119.

Dioecious, stoloniferous, perennial. Leaves sessile, in large rosettes, 15 to 50 cm long and up to 4 cm wide, rigid, many-nerved; margins distinctly spinous-serrate. Inflorescences axillary; peduncles up to 30 cm long. Spathes composed of 2 bracts; male spathe containing several bracteolate, stalked flowers; sepals 3, ovate; petals 3, white, obovate, 2 to 3 cm wide; fertile stamens 12, surrounded by numerous staminodes. Female spathe containing

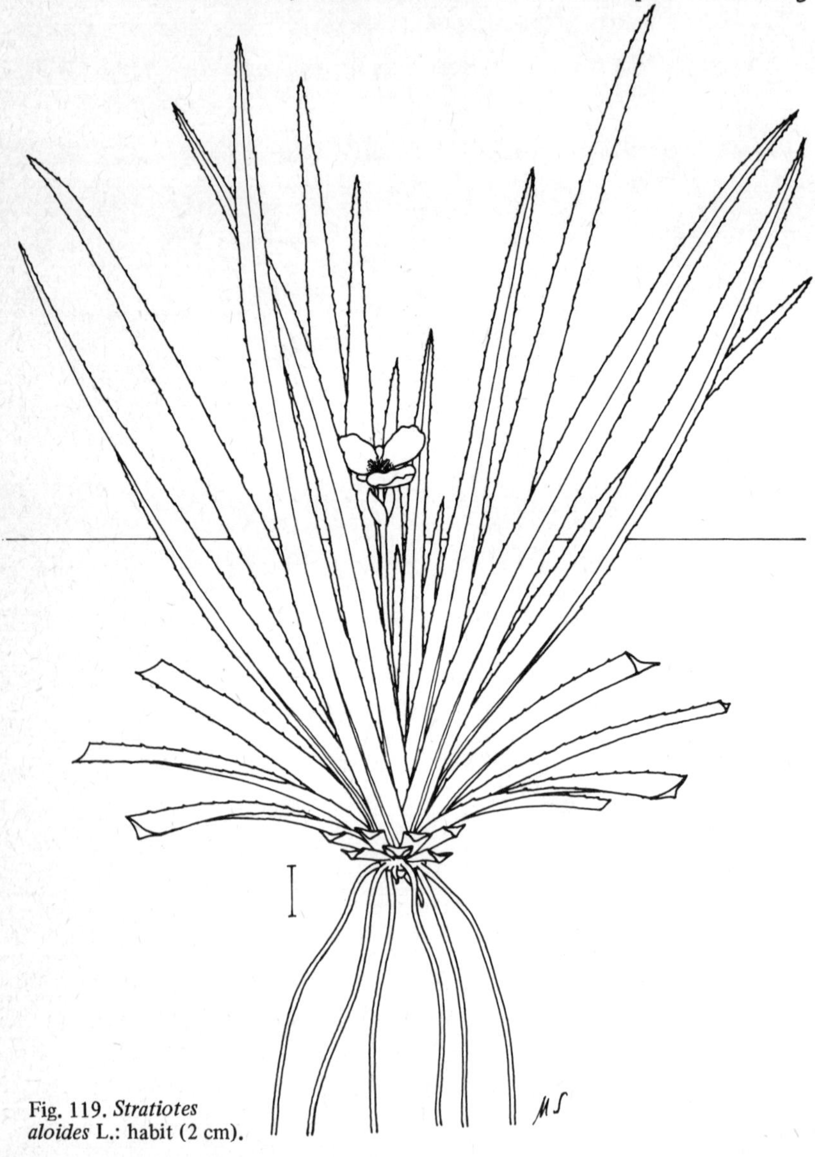

Fig. 119. *Stratiotes aloides* L.: habit (2 cm).

1 sessile flower; female flower smaller than male one, hardly exserted, staminodes longer than in the male flower. Ovary 6-locular; styles 6, bifid; fruit a fleshy capsule; seeds not numerous, about 9 mm long, brown.

Arber, A. On root development in Stratiotes aloides L. Proc. Camb. Phil. Soc. Biol. Sci. 17: 369–379 (1914)

Bande, E. Die Embryoentwicklung von Stratiotes aloides L. Planta 46: 649–671 (1956)

Chandler, M. E. J. The geological history of the genus Stratiotes. Quart. Journ. Geol. Soc. London 79: 117–138 (1923)

1 species, *S. aloides* L.: parts of Europe and N. W. Asia, introduced elsewhere. At least 8 fossil species have been described. *S. aloides* occurs in lakes, pools and canals, at the moment it is extending its range in Europe apparently by vegetative propagation because most new populations are entirely female. In habit it resembles an *Aloe* and is remarkable as the whole plant rises to the surface of the water in summer to bloom and then sinks during autumn; only the seedling stage is bottom rooted.

Vallisneria L., Sp. Pl. 1017 (1753)
Fig. 120.
Submerged, dioecious, stoloniferous, perennials. Leaves radical, linear, sheathing at the base, with longitudinal air-channels; apex obtuse; margins minutely toothed or entire; nerves 3 to 9, parallel, connected by cross veins, only the midrib reaching the apex, the other nerves gradually joining together near the apex. Male spathe shortly stalked, smaller than female one, containing many stalked flowers, which break off and rise to the water surface. Sepals 3, ovate or oblong-ovate, convex. Petals 3, minute. Stamens 1 to 3. Female spathe united, tubular with bifid apex; stalk very long, spirally contracted after fertilisation. Sepals 3, oblong or oblong-ovate. Petals 3, minute, scarious. Ovary linear, nodding; styles 3, each split into 2 lobes; fruit linear, often very long; seeds numerous, oblong to fusiform, testa membranous.

Ant, H.: Zur Ausbreitung der Sumpfschraube, Vallisneria spiralis (Hydrocharitaceae) im Norden ihres Areals. Decheniana 122 (2) 195–197 (1970)

Choudhuri, G. N.: Seed germination and flowering in Vallisneria spiralis. North-west Sci. 40: 31–35 (1966)

Marie-Victorin, Fr.: Les Vallisnéries américaines. Contr. Inst. Bot. Montréal 46: 1–38 (1943)

Miki, S.: On fresh water plants new to Japan. Bot. Mag. Tokyo 48: 326–337 (1934)

6 to 10 species: almost cosmopolitan, absent from the cold regions. Found in still or flowing water up to 1 m or more deep in a variety of habitats. Most species spread vegetatively by stolons and often build up large populations which may seriously hinder the flow of water in irrigation canals. *Vallisneria* species are frequently cultivated as decorative aquarium plants. Several species have become naturalised outside their normal range. *V. spiralis* L. is actively extending its range particularly in Europe.

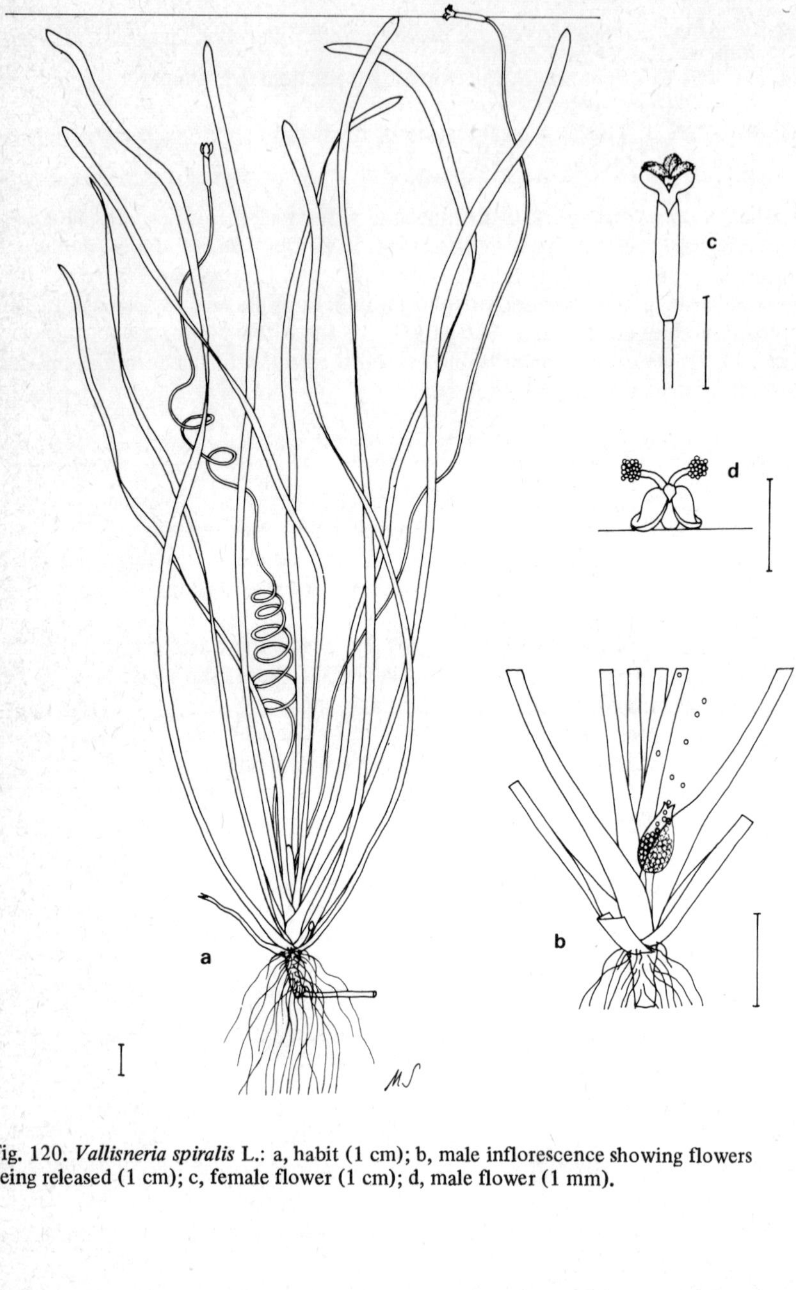

Fig. 120. *Vallisneria spiralis* L.: a, habit (1 cm); b, male inflorescence showing flowers being released (1 cm); c, female flower (1 cm); d, male flower (1 mm).

HYDROPHYLLACEAE

c. 18 genera. *Hydrolea* is the only genus containing aquatics.

Hydrolea L., Sp. Pl. ed. 2, 328 (1763), nom. cons.
Fig. 121A.
Annual or perennial, unarmed or spinous herbs. Stems terete, often swollen and spongy. Leaves alternate, simple, entire, usually narrow; stipules absent. Flowers bisexual, actinomorphic, solitary or in terminal or axillary cymes. Sepals 5, almost free to base, persistent. Petals 5, united at base, blue. Stamens 5, free, inserted at base of petals, alternating with petal lobes. Ovary superior 2- (rarely 3- or more-) locular, with parietal placentas; styles 2 (rarely 3 or more); stigmas capitate; ovules numerous; fruit a capsule splitting by valves or bursting irregularly.

c. 20 species: pantropical. Mostly species of swamps and seasonally inundated regions. Often gregarious and occasionally reported as a weed in ricefields and irrigation ditches. When in water the stems are usually swollen and spongy.

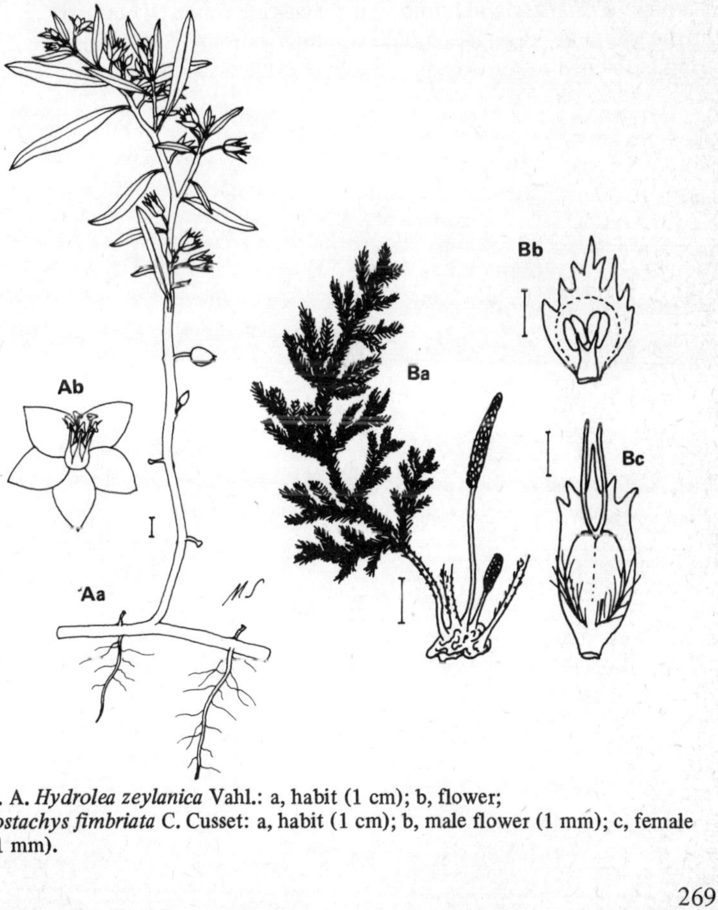

Fig. 121. A. *Hydrolea zeylanica* Vahl.: a, habit (1 cm); b, flower;
B. *Hydrostachys fimbriata* C. Cusset: a, habit (1 cm); b, male flower (1 mm); c, female flower (1 mm).

HYDROSTACHYACEAE

1 genus, *Hydrostachys.*

Hydrostachys Du Petit-Thouars, Gen. Nov. Madag 2 (1806)
Fig. 121B.
Roots creeping, closely attached to rocks; stems, thick, tuber-like. Leaves usually in rosettes, simple to 3 times pinnate; ultimate segments usually filamentous or occasionally scale-like. Plants unisexual. Flowers borne in dense spikes; each flower borne in axil of a bract. Sepals absent. Petals absent. Male flowers reduced to 1 stamen. Female flower reduced to 1 ovary; styles 2, divergent; seeds numerous, parietal; fruit a capsule opening by 2 equal valves.

Cusset, C. Revision des Hydrostachyaceae. Adansonia sér. 2, 13 (1): 75–119 (1973)
Rau, W. and Jäger-Zürn. I. Zur Kenntnis der Hydrostachyaceae. Sitz. Ber. Heidelberger Akad. Wiss. Math.-Nat. Kl. 1966 (I): 3–117 (1966)

22 species: Madagascar and Africa from the equator southwards. The greatest morphological diversity and number of species are found in Madagascar. *Hydrostachys* in the vegetative phase is similar to some Podostemaceae. They are found attached to rocks and stones in flowing water.

HYPOXIDACEAE

c. 7 genera; *Hypoxis,* has one aquatic species in S. Africa.

Hypoxis, L., Syst. ed. 10: 986 (1759), [*Spiloxene* Salisb.]
Fig. 122.
Herbs with thick rootstock covered in loose, fibrous scales. Leaves mostly basal, erect, emergent, white and spongy below. Inflorescence as long or slightly shorter than the leaves, with a few-flowered, long pedicellate umbel. Bracts membranous, greenish white, one at the base of each pedicel. Perianth segments 6, white, the outer pale green below, free or united only at the base. Stamens 6, inserted at the base of each perianth segment. Ovary inferior, 3-locular, ovules numerous; seeds small.

c. 20 species: *H. aquatica* L. fil. occurs in shallow pools and ditches in S. Africa. Some other species occur in marshes.

Fig. 122. *Hypoxis aquatica* L. fil.: habit (1 cm).

IRIDACEAE

c. 60 genera; cosmopolitan. *Iris* has several aquatic representatives.

Iris L., Sp. Pl. 38 (1753)
Fig. 123.

Fig. 123. *Iris pseudacorus* L.: a, habit (10 cm); b, stem base; c, inflorescence (1 cm).

Perennials with thick creeping rhizome. Stems erect, up to 2 m tall, leafy. Leaves equitant, monofacial, glabrous, sheathing at base. Flowers up to 15 cm in diameter, yellow, white or bluish, subtended by a spathe-like bract. Perianth in 2 petaloid series, the outer 3 deflexed, the inner 3 usually ascending. Stamens 3, inserted at the base of the outer perianth segments. Ovary 3-locular; style branches bifid at the apex, bent over the anthers, petaloid; ovules axile, numerous.

Foster, R. C. A cytotaxonomic survey of the N. American species of Iris. Contr. Gray Herb. 119: 1–82 (1936)

c. 300 species of which about ten are aquatic and many others occur in wet places: throughout the temperate regions of the northern hemisphere. Most aquatic species belong to section *Apogon* Baker.

JUNCACEAE

9 genera; *Prionium* and *Juncus* have aquatic representatives.

Glabrous perennials or annuals. Stems up to 2 m long, floating or emergent. Leaves terete, capillary or flat, occasionally sheathing at base, sometimes absent. Inflorescence a cluster of terminal or apparently lateral cymes, sometimes reduced to a 1- or 2-flowered head. Bracts glume-like. Flowers bisexual. Perianth segments 6, in two whorls, reduced, scale-like, greenish or brownish. Stamens 3 or 6. Ovary 1- or 3-locular; stigmas 3, brush-like; fruit a capsule.

Buchenau, F. in Engler, A. Pflanzenreich 36(IV. 36): 1–284 (1906)

1A Subshrubby; leaves crowded at the top of the stem; perianth segments rigid; capsule few-seeded
Prionium

1B Herbaceous; leaves arising from base of stem or cauline; perianth segments thin; capsule many-seeded
Juncus

Juncus L., Sp. Pl. 325 (1753)
Fig. 124.
Glabrous perennials or annuals. Stems up to 1 m long, usually annual, floating or emergent. Leaves terete, capillary or sometimes absent, sheathing at the base. Inflorescence a cluster of terminal or apparently lateral cymes, sometimes reduced to a 1- or 2-flowered head. Flowers bisexual. Perianth segments 6, thin, scarious. Stamens 3 or 6, shorter than perianth. Ovary 1- or 3-locular; capsule with numerous small seeds.

Adamson, M. A. A revision of the South African species of Juncus. Journ. Linn. Soc. London (Bot.) 50: 2–38 (1935)
Barros, M. Las Juncaceas de la Argentina, Chile y Uruguay. Darwiniana 10: 358–458 (1953)
Edgar, E. The leafless species of Juncus in New Zealand. New Zealand Journ. Bot. 2: 177–204 (1964)
Snogerup, S. Studies in the genus Juncus III. Observations on the diversity of chromosome numbers. Bot. Not. 116: 142–156 (1963)

c. 225 species: cosmopolitan. Most species occur in wet places. *J. heterophyllus* Dufour has capillary submerged leaves in winter and terete emergent leaves in summer, it occurs in S. Europe. *J. bulbosus* L. often grows submerged throughout Europe and *J. subtilis* E. Mey has the same habit in N. America.

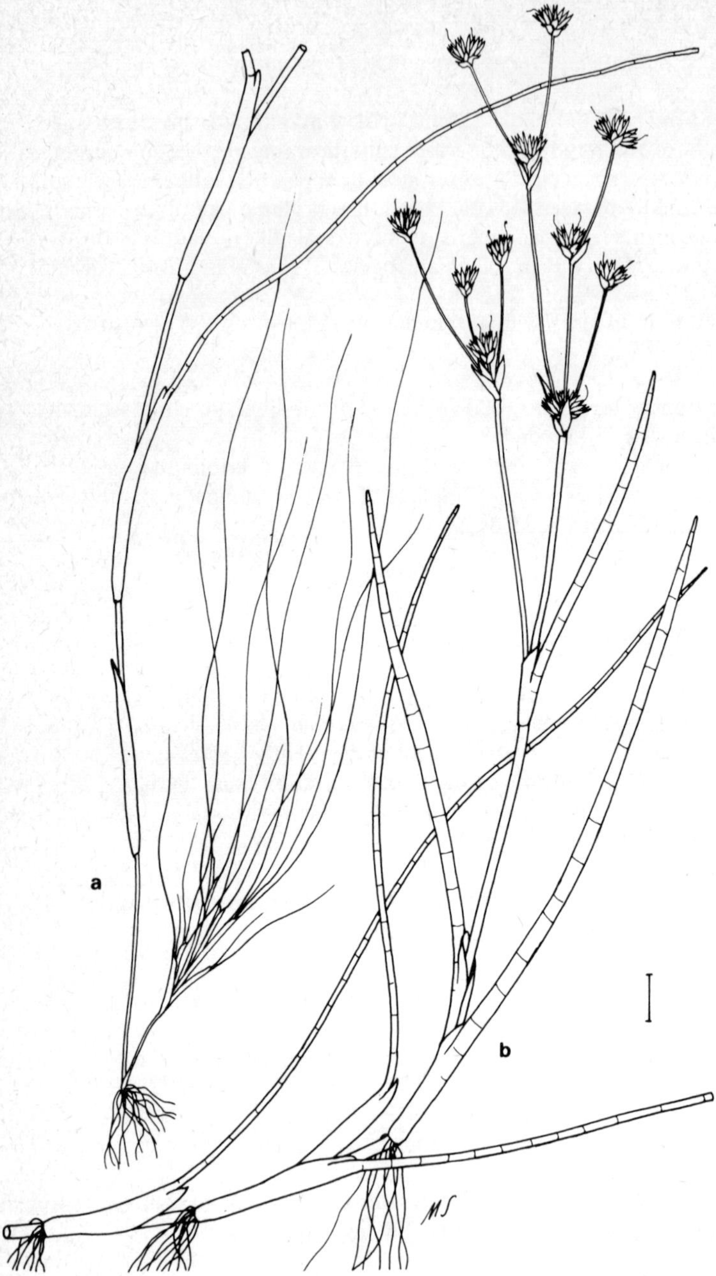

Fig. 124. *Juncus heterophyllus* Dufour: a, submerged stem showing hair-like leaves; b, emersed flowering stem (1 cm).

Prionium E. Mey, Linnaea 7: 131 (1832)
Fig. 125.

Stout erect perennial. Stem persistent, little branched, covered with the black fibrous remains of old leaves, up to 2 m high, 5 to 10 cm in diameter. Leaves flat, rigid, serrate, densely crowded at the stem apex. Inflorescence much branched, leafless, many flowered, up to 50 cm high. Perianth segments rigid, acute, bright brown. Stamens 6, longer than the perianth. Capsule few-seeded.

1 species, *P. serratum* (L. fil.) Drège, occurs in mountain streams in S. Africa, see Curtis's Bot. Mag. ser. 3, 24: t.5722 (1868).

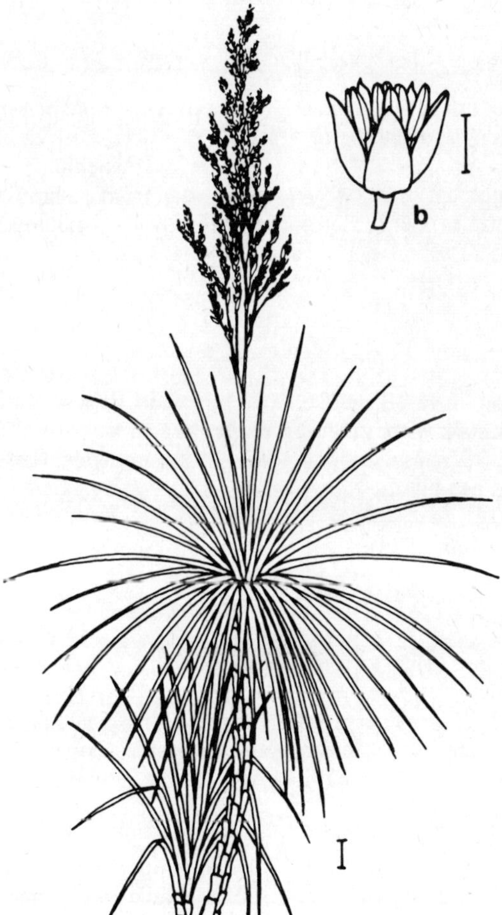

Fig. 125. ***Prionium serratum*** **(L. fil.) Drège: a, habit (10 cm); b, flower (1 cm).**

JUNCAGINACEAE nom. cons.

3 genera: *Triglochin* and *Maundia* contain aquatics.

Annuals or perennials. Stems reduced, corm-like or rhizomatous. Leaves mostly radical, linear, sheathing at base. Inflorescence racemose or spike-like; flowers bisexual (in aquatic genera), actinomorphic; bracts absent. Perianth 2-seriate; segments free, herbaceous, 2, 4 or 6. Stamens 6 (in aquatic genera); anthers subsessile, 2-locular, opening by longitudinal slits. Ovary of 3, 4 or 6, superior, free or united carpels; style short and stout or absent; stigma plumose or papillose; ovule solitary in each carpel; fruits dry, adaxially dehiscent.

Buchenau, F. Scheuchzeriaceae in Engler, A. Pflanzenreich (IV. 14): 1–20 (1903)

1A Carpels 6, united, separating from a central axis when ripe; perianth segments 6, caducous; ovules basal, erect
Triglochin

1B Carpels 3 or 4; united at base only, not separating from a central axis when ripe; perianth segments 2 or 4, persistent; ovules pendulous
Maundia

Maundia F. Mueller, Fragm. 1: 22 (1858)
Fig. 126A.
Perennial. Leaves basal, linear, flaccid, 6 to 15 mm wide, floating. Inflorescence spicate; inflorescence stalk without leaves. Perianth segments 2 or 4, persistent; apex curved inwards. Stamens 6; anthers sessile, *c.* 1.5 mm long. Carpels 3 or 4, united at base, ovules pendulous.

1 species, *M. triglochinoides* F. Mueller: E. Australia. Apparently grows in flowing water.

Triglochin L., Sp. Pl. 338 (1753), [*Cycnogeton* Endl.]
Fig. 126B.
Annuals or perennials. Leaves linear, basal. Inflorescence spicate or racemose. Perianth segments 6, caducous. Stamens 6; anthers sessile. Carpels 6, occasionally 3 sterile, united for most of their length, parting from a central axis when ripe; ovules basal, erect.

c. 15 species: cosmopolitan, most in Australia and temperate S. America. Mostly marsh plants but *T. procera* R. Br. from Australia and Tasmania is aquatic.

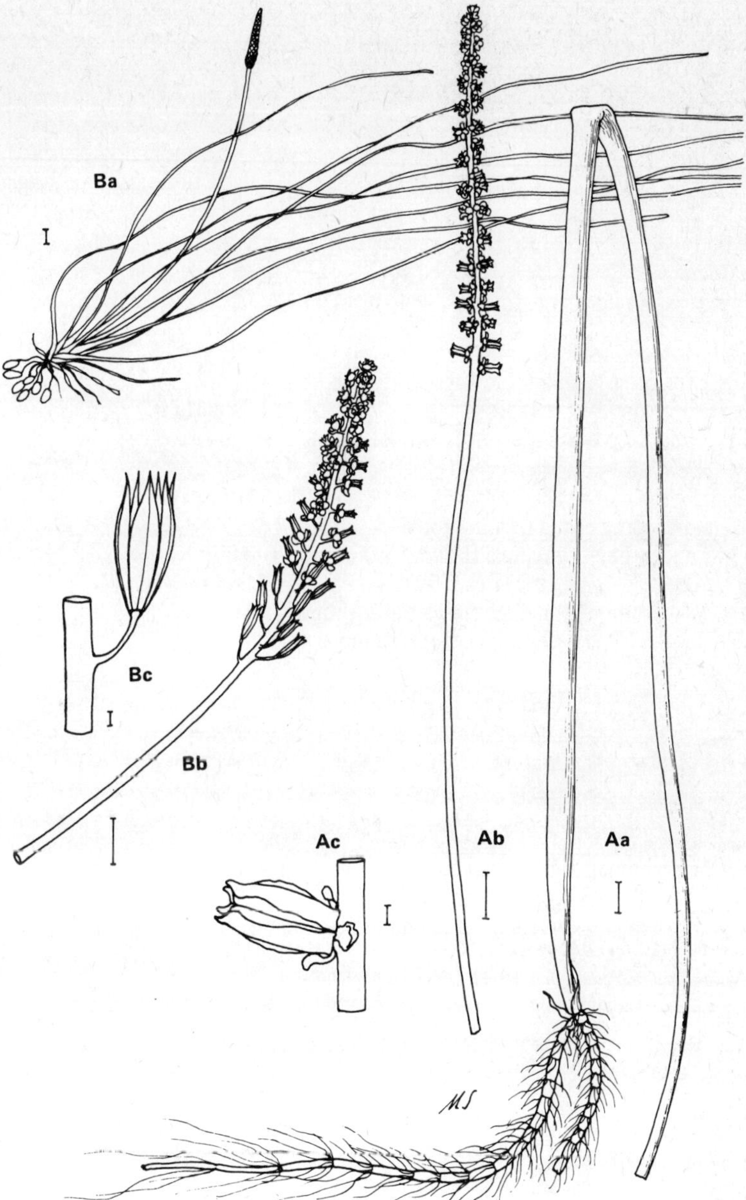

Fig. 126. A. *Maundia triglochinoides* F. Mueller: a, habit (1 cm); b, inflorescence (1 cm); c, ripe fruit (1 cm);
B. *Triglochin procera* R.Br.: a, habit (1 cm); b, inflorescence (1 cm); c, ripe fruit (1 mm).

LAMIACEAE [LABIATAE]

c. 180 genera of which 5 contain aquatics.

Annual or perennial herbs, often aromatic. Stems usually square in transverse section. Leaves opposite. Inflorescence whorl-like; the whorls often crowded at the stem apex, forming a short spike. Sepals 5 (or 4), united below, often 2-lipped, with 3 adaxial lobes and 2 abaxial ones. Petals 5, united into a long tube; the 2 adaxial lobes forming an often simple lip, the abaxial 3 forming a 3-lobed lower lip, or rarely (in *Teucrium*) all 5 petals forming the abaxial lip. Stamens usually 4, rarely reduced to 2. Ovary of 2 carpels, each with 2 ovules, finally appearing 4-lobed; style bifid above; fruit of 4 nutlets.

1A Petals forming 1, abaxial, 5-lobed lip
Teucrium

1B Petal tube 2-lipped; lobes more or less equal
- 2A Sepal lobes unequal
Pogogyne
- 2B Sepal lobes equal or subequal
  - 3A Anthers 1-locular; flowers minute, crowded in long spikes
Dysophylla
  - 3B Anthers 2-locular; flowers not minute, mainly axillary
    - 4A Plant not aromatic; stamens 2
Lycopus
    - 4B Plant aromatic; stamens 4
Mentha

Dysophylla Blume, Bijdr. 826 (1826)
Fig. 127.
Annuals or perennials. Stem usually erect. Leaves linear to lanceolate, opposite or many in a whorl. Bracts much smaller than leaves. Flowers minute, crowded into long spikes. Sepals 5, equal. Petals equal, or the abaxial lip 2-lobed. Stamens 4, equal, longer than petals; anthers 1-locular. Nutlets ovoid or oblong.

c. 25 species: in temperate and tropical Asia and Australia. Many species occur in swamps and as weeds in ricefields.

Lycopus L., Sp. Pl. 21 (1753)
Fig. 128A.
Odourless perennials with a creeping rhizome, and stolons or tubers. Stems erect. Submerged leaves (produced in spring) various but occasionally pinnate with linear segments; emersed leaves ovate to lanceolate, toothed to deeply pinnately lobed. Bracts not differentiated from the leaves. Flowers small, white to purple, clustered in the leaf axils. Sepals equal, 4 or 5, united to form

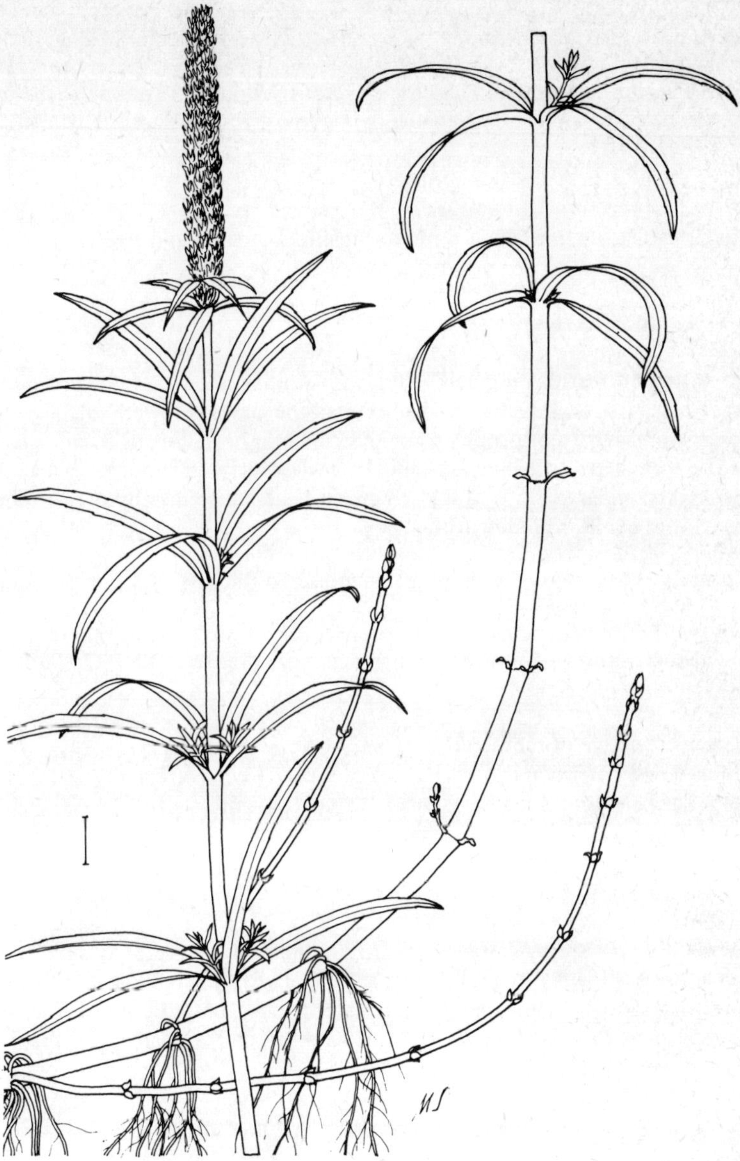

Fig. 127. *Dysophylla yatabeana* Makino: habit (1 cm).

a 13-nerved tube. Petals subequal, the adaxial lip entire or emarginate. Stamens 2, longer than petals. Nutlets tetrahedral.

Henderson, N. C. A taxonomic revision of the genus Lycopus L. (Labiatae) Amer. Midl. Nat. 68: 95–138 (1962)
Skalický, V. Studie o evropských druzích rodu Lycopus L. (Studie über die europäischen Arten der Gattung Lycopus L.) Sborn. Nàrod. Mus. Praha 24 (5): 185–216 (1968)

c. 14 species: throughout N. America, Europe and temperate Asia; 1 species in Australia. Occurs in shallow water at the edges of lakes and rivers.

Mentha L., Sp. Pl. 576 (1753)
Fig. 128B.
Rhizomatous aromatic perennials, often with stolons. Stems usually erect, emergent. Leaves toothed; bracts smaller than the leaves. Flowers small, in axillary whorls, often forming a terminal spike or head. Sepals 5, subequal, united to form a 10- to 13-nerved tube. Petals 4, usually pale purple, subequal, the adaxial often larger, emarginate. Stamens 4, subequal, usually longer than petals. Nutlet ovoid, rounded at the apex.

Graham, R. A. Mint Notes V. Mentha aquatica and the British Mints. Watsonia 3: 109–121 (1954)
Murray, M. J. The genetic basis for the conversion of methone to menthol in Japanese mint (Mentha arvensis piperascens). Genetics 45: 931–937 (1960)

c. 25 species: native of temperate Europe, Asia and Australia, probably introduced in S. Africa and America. Many species occur in shallow water at the edges of lakes and rivers.

Pogogyne Bentham, Lab. Gen. et Sp. 414 (1834)
Fig. 129A.
Annuals with erect stems. Leaves ovate to spathulate. Bracts mostly smaller than the leaves. Flowers small, in dense spikes, or whorled at the base of the inflorescence. Sepals 5, unequal, the adaxial 3 shorter, united into a 15-nerved tube. Petals blue or purple, equal, the adaxial erect, entire. Stamens 4, the upper sometimes sterile. Nutlets obovoid.

5 species: western N. America. Characteristically found in vernal pools.

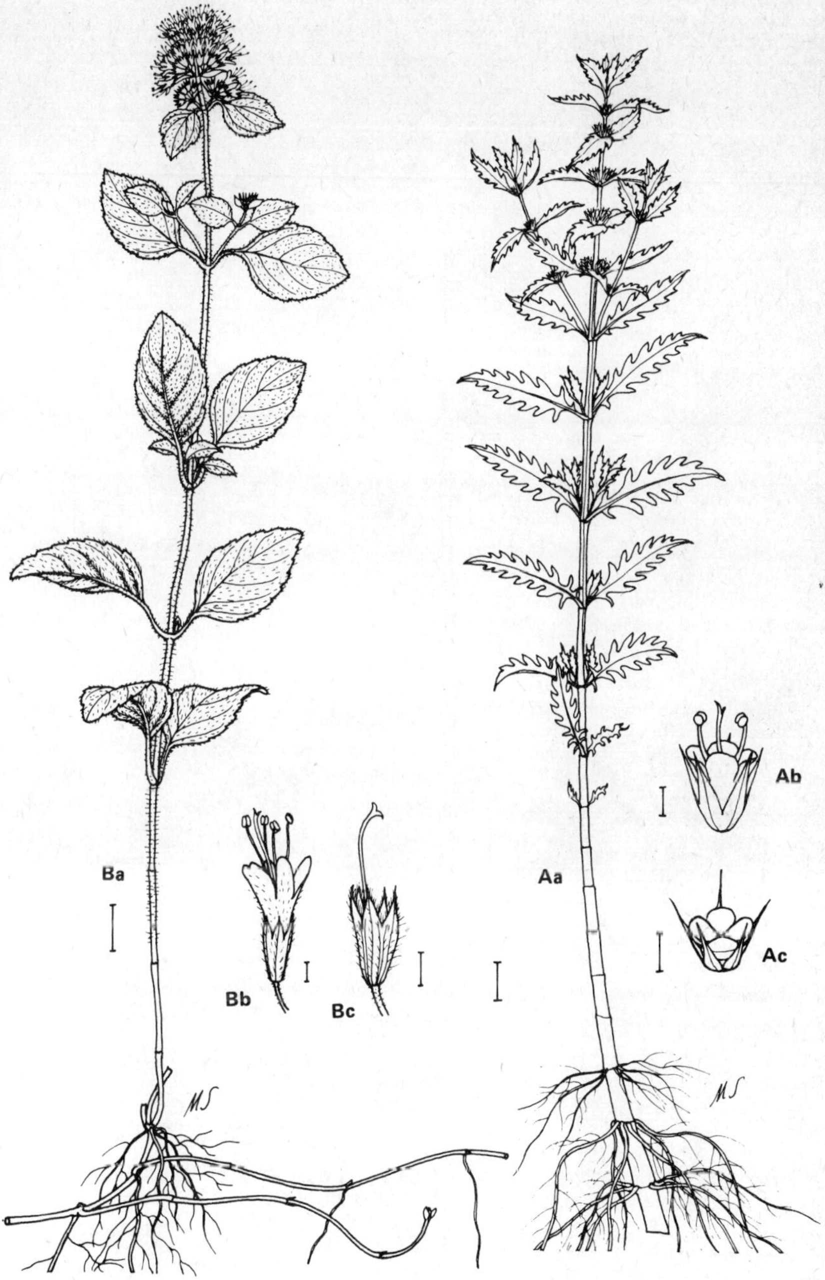

Fig. 128. A. *Lycopus europaeus* L.: a, habit (1 cm); b, flower (1 mm); c, fruit (1 mm); B. *Mentha aquatica* L.: a, habit (1 cm); b, flower (1 mm); c, sepals (1 mm).

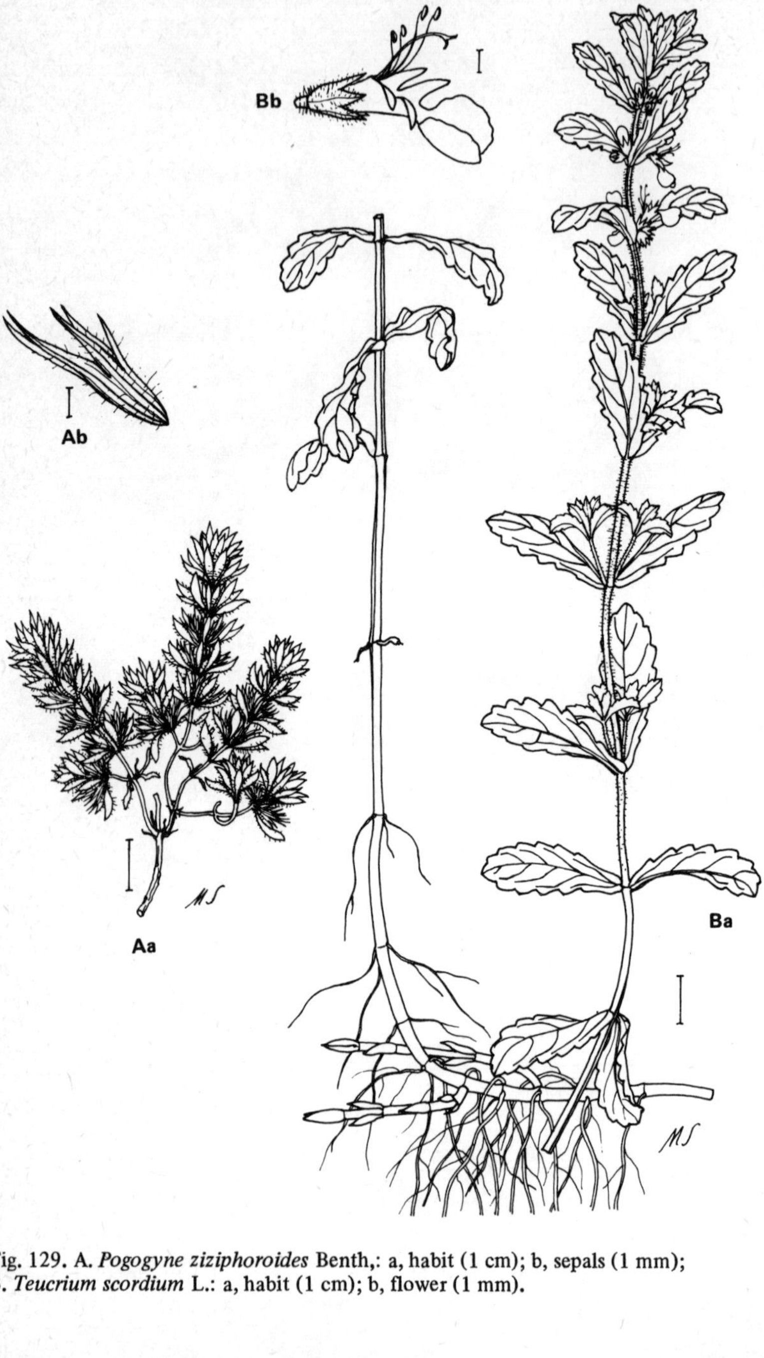

Fig. 129. A. *Pogogyne ziziphoroides* Benth,: a, habit (1 cm); b, sepals (1 mm); B. *Teucrium scordium* L.: a, habit (1 cm); b, flower (1 mm).

Teucrium L., Sp. Pl. 564 (1753)
Fig. 129B.
Perennials with creeping rhizomes and leafy stolons. Stems ascending. Leaves oblong; margin serrate. Bracts similar to the leaves. Flowers in distant whorls; pedicels short. Sepals 5, equal, or adaxial sepal larger. Petals united to form one 5-lobed lip; the middle lobe largest. Stamens 4, longer than petals; the outer pair longer than the inner. Nutlets obovoid.

c. 100 species: cosmopolitan. *T. scordium* L. occurs in shallow pools and the edges of rivers in Europe and N. W. Asia.

LEMNACEAE

6 genera: *c.* 29 species: all aquatic.

Small thalloid, entirely or partly floating or submerged herbs, solitary or connected in groups. Roots simple, several, 1 or absent. Flowers unisexual, naked or enclosed in a sheath. Perianth absent. Male flowers consisting of 1 stamen with 1- or 2-locular anthers. Female flowers consisting of 1, sessile, 1-locular ovary; fruit a 1- to 4-seeded utricle.

The thallus has variously been interpreted as a modified stem, a leaf or partly stem and partly leaf.

Blazey, E. and McClure, J. The distribution and taxonomic significance of lignin in the Lemnaceae. Amer. Journ. Bot. 55: 1240–1245 (1968)

Clark, H. L. and Thieret, J. W. The duckweeds of Minnesota. Michigan Bot. 7: 67–76 (1968)

Daubs, E. H. Monograph of Lemnaceae. Illinois Biological Monographs 34: 1–118 (1965) – for criticism of this work see: Clark, H. L. and Thieret, J. W. Sida 2 (6): 437–438 (1966) and Hartog, C. den. Blumea 15 (2): 575–576 (1967)

Evans, O. D. Some observations on the Lemnaceae or "duckweeds" of New South Wales. Contr. New S. Wales Nat. Herb. 4 (3): 87–94 (1970)

Harrison, D. E. and Beal, E. O. The Lemnaceae of North Carolina. Journ. Elisha Mitchell Sci. Soc. 80 (I): 12–18 (1964)

Hartog, C. den and Plas. F. van der. A synopsis of the Lemnaceae. Blumea 18 (2): 355–368 (1970)

Hegelmaier, F. Die Lemnaceen: eine monographische Untersuchung. 1–169. Leipzig (1868)

Hepper, F. The duckweeds of W. Africa, Nigerian Field 31: 18–21 (1966)

Hillman, W. S. The Lemnaceae or duckweeds. A reivew of the descriptive and experimental literature. Bot. Rev. 27: 221–287 (1961)

Jovet-Ast, S. Contribution à l'étude des eaux douces de l'Ennedi 2. Lemnaceae. Bull. Inst. Fond. Afr. Noire, sér. A, 30 (3): 830–847 (1968)

Schulz, B. Wasserlinsen. Wittenberg Lütherstadt (1962)

1A Thalli with roots; 2 budding pouches; inflorescence of 1 female and 2 male flowers, enclosed by a membranous sheath; anther 2-locular, transversely dehiscent

- 2A Roots solitary on each thallus segment; thalli without dorsal and ventral scales, usually 1 to 5 mm long, under-surface green or streaked with brown; nerves 1 to 3, often indistinct

  Lemna

- 2B Roots several from each thallus segment; thalli with 1 dorsal and 1 ventral scale, usually 4 to 10 mm long, undersurface usually reddish; nerves 3 to 15

  Spirodela

1B Thalli without roots; 1 budding pouch; inflorescence of 1 female and 1 male flower, not enclosed by a membranous sheath; anther 1-locular, apically dehiscent

3A Thalli globose, solitary or usually only mother and daughter thalli connected; budding pouch funnel-shaped with a circular opening **Wolffia**

3B Thalli flat, usually connected in groups; budding pouch triangular

4A Thalli floating on the water surface, slightly swollen; without brown pigment cells in the epidermis (thallus symmetric, round or nearly so; budding pouch symmetric; inflorescence 1) **Pseudowolffia**

4B Thalli submerged (base only may be floating), thin, membranous; with brown pigment cells in the epidermis

5A Thalli asymmetric, linear-oblong to almost strap-like, often curved; budding pouch asymmetric; inflorescence 1 **Wolffiella**

5B Thalli symmetric, broad-elliptic to ovate, slightly curved; budding pouch symmetric; inflorescences 2 **Wolffiopsis**

Lemna L., Sp. Pl. 970 (1753)
Fig. 130A, B.
Thalli free-floating upon or beneath the water surface, mostly small and rarely exceeding 5 mm long (except *L. trisulca*), solitary or remaining attached in groups of 2 to 10 or sometimes more, mostly orbicular to ovate in outline (much elongated in *L. trisulca*), flat or swollen; nerves 1 to 3 rarely 5, often indistinct; dorsal and ventral scales absent; roots solitary on each thallus segment, peltately attached, Inflorescences and daughter thalli borne in 2, marginal budding pouches, 1 on each side, at base of thallus. Inflorescence of 1 female and 2 male flowers, enclosed in a membranous sheath. Anthers 2-locular. Carpel with 1 ovule; seed ovoid, longitudinally ribbed with transverse striations or rarely smooth.

9 species: almost cosmopolitan. Usually gregarious and often forms green floating mats on the surface of still or slowly flowing water. Most species grow best in eutrophic conditions. Populations often reach pest proportions, for control methods see: Little, E. C. S. Weed Res. 8: 79–105 (1968).

Pseudowolffia Hartog and Plas, Blumea 18 (2): 365 (1970)
Fig. 131A.
Thalli floating on the water surface, symmetric, slightly swollen, round or nearly so, without pigment cells in the epidermis; margin entire or toothed; daughter thalli borne in 1, triangular pouch at base of thallus. Inflorescence 1, of 1 female and 1 male flower, without a membranous sheath. Anther 1-locular.

3 species: N. and C. Africa.

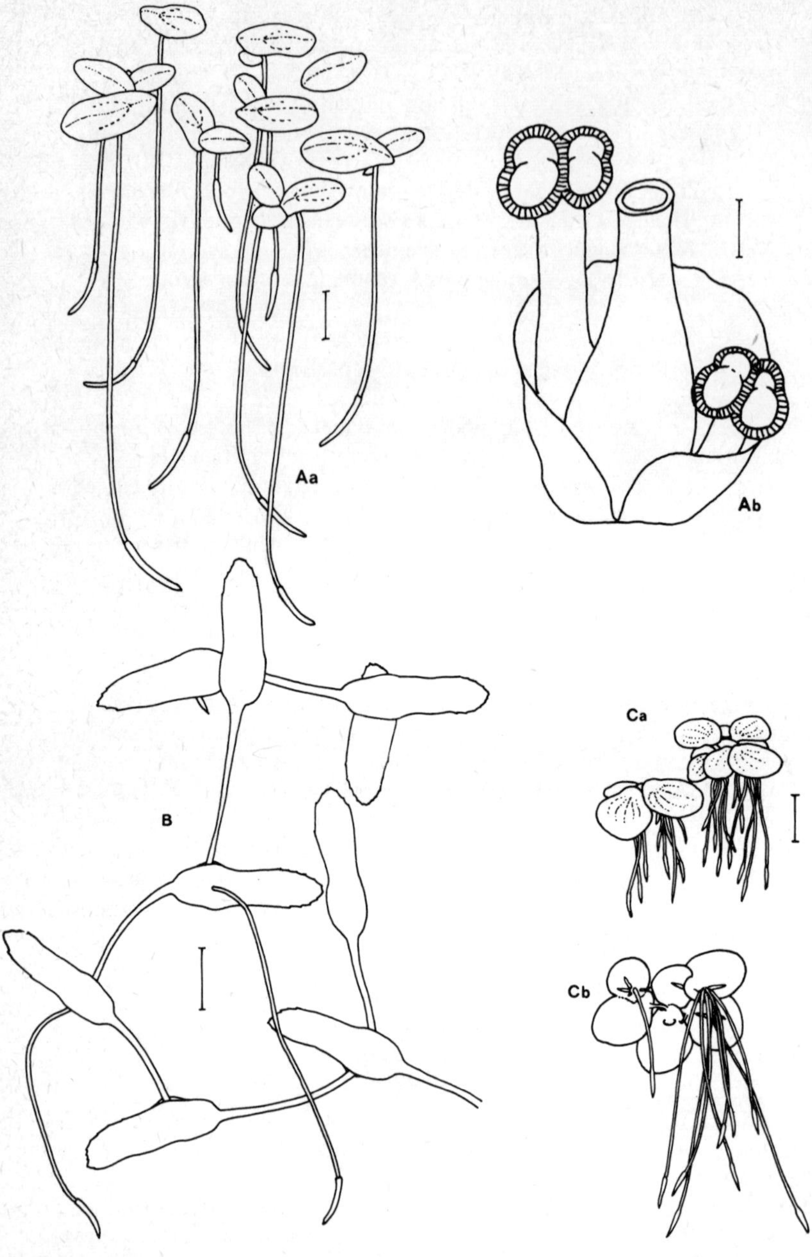

Fig. 130. A. *Lemna perpusilla* Torrey: a, habit (1 mm); b, inflorescence (0.1 mm);
B. *Lemna trisulca* L.: habit (5 mm);
C. *Spirodela polyrhiza* (L.) Schleid.: a, habit, from above (5 mm); b, habit, from below.

Spirodela Scheiden, Linnaea 13: 391 (1839)
Fig. 130C.
Thalli free-floating upon the water surface, 3 to 10 mm long, orbicular, obovate or slightly reniform, 2 to 5 or more remaining connected by elongate stipes,

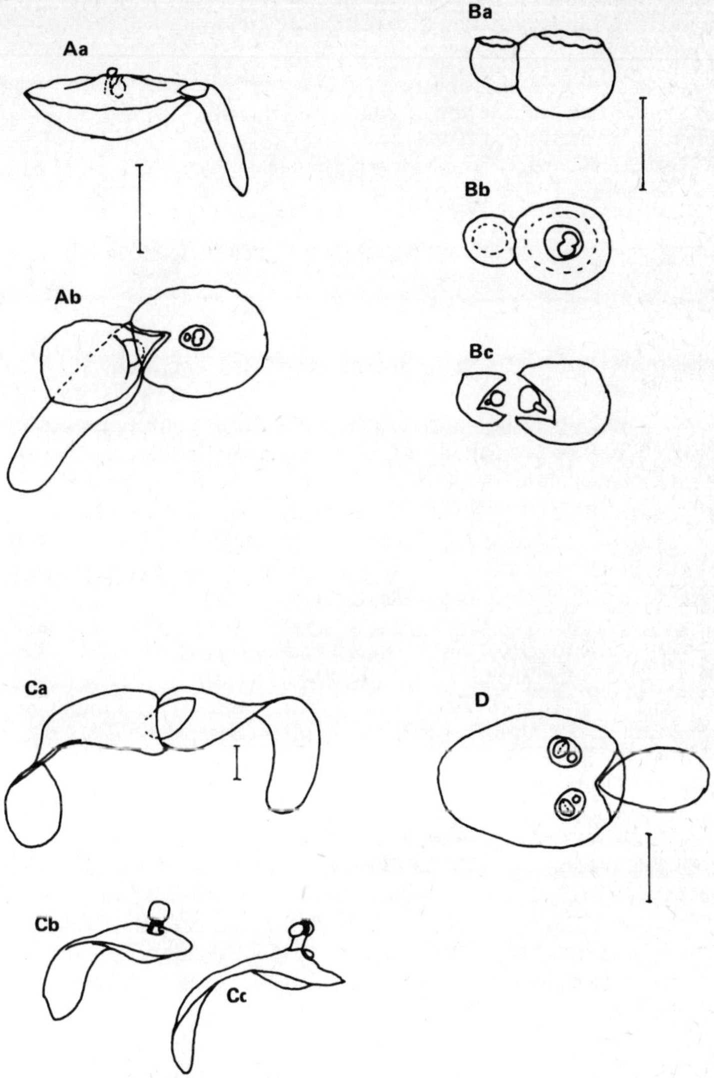

Fig. 131. A. *Pseudowolffia hyalina* (Delile) Hartog and Plas: a, habit, from side; b, habit, from above (1 mm);
B. *Wolffia arrhiza* (L.) Horkel ex Wimm.: a, habit, from side; b, habit, from above; c, longitudinal section (1 mm);
C. *Wolffiella lingulata* (Hegelm.) Hegelm.: a, habit, from above; b, habit, from side; showing "pollination droplet"; c, habit, from side, showing male flower (1 mm);
D. *Wolffiopsis welwitschii* (Hegelm.) Hartog and Plas: a, habit, from above (1 mm).

flat or inflated, underside usually red; brown pigment cells in the epidermis; margin entire; roots 2 to 18 or more, rarely 1, on each thallus segment; dorsal and ventral scales present. Inflorescences and daughter thalli borne in 2, ventral budding pouches. Inflorescence of 1 female and 2 male flowers, enclosed in a membranous sheath. Anthers 2-locular. Carpel with 1 ovule; seed ovoid, longitudinally ribbed with transverse striations or smooth with spongy outer layer.

Henssen, A. Die Dauerorgane von Spirodela polyrhiza (L.) Schleid. Flora (Jena) 141: 523–566 (1954)
Jacobs, D. L. An ecological life-history of Spirodela polyrhiza. Ecol. Monog. 17: 437–469 (1947)

c. 4 species: almost cosmopolitan. Forms floating mats often together with *Lemna.*

Wolffia Horkel ex Schleiden, Beitr. Bot. 1: 233 (1844), nom. cons. prop. Fig. 131B.
Thalli free-floating upon the water surface; rarely exceeding 1 mm in any dimension, solitary or 2 remaining attached, symmetric, globose to ovoid, sometimes flattened above, with or without brown pigment cells in the epidermis; roots absent. Daughter thalli borne in 1, funnel-shaped basal pouch, with circular opening. Inflorescence 1, of 1 female and 1 male flower, without membranous sheath. Anthers 1-locular. Seed globose or slightly compressed, smooth, with spongy outer layer.

Countryman, W. D. Wolffia in New Hampshire. Rhodora 70: 491 (1968)
Dore, W. G. Wolffia in Canada. Canad. Field-Nat. 71: 10–16 (1957)
Jäger, E. Zur Deutung des Arealbildes von Wolffia arrhiza (L.) Wimm. und einiger anderer ornithochorer Wasserpflanzen. Ber. Deutsch. Bot. Ges. 77: 101–111 (1964)
Rimon, D. and Galun, E. Morphogenesis of Wolffia microscopica: frond and flower development. Phytomorph. 18 (3): 364–372 (1968)
Toth, L. On some chemical properties of Wolffia arrhiza (L.) Wimm. Ann. Inst. Biol. Tihany 20: 275–282 (1962)

c. 7 species: almost cosmopolitan, absent from the colder regions. *Wolffia* species are among the smallest flowering plants, the thalli are scarcely visible as individuals but are frequently numerous enough to form a solid cover of the surface of water.

Wolfiella (Hegelm.) Hegelm., Bot. Jarhb. 21: 303 (1896)
Fig. 131C.
Thalli usually submerged in water except for a small area at the base, solitary or remaining connected and forming extensive colonies, asymmetric, flat, thin, membranous, linear-oblong to strap-like, frequently curved, with brown pigment cells in the epidermis; margin entire or slightly toothed; roots absent. Daughter thalli borne in 1, asymmetric, triangular budding pouch. Inflorescence 1, of 1 female and 1 male flower, without membranous sheath. Anthers 1-locular. Seed globose or ovoid, smooth.

5 species: 4 in warmer areas of America, 1 (*W. denticulata* (Hegelm.) Hegelm.) in S. Africa. Usually found mixed with other free-floating plants such as: *Lemna, Salvinia, Eichhornia,* etc.

Wolffiopsis Hartog and Plas, Blumea 18 (2): 366 (1970)
Fig. 131D.
Thalli usually submerged in water except for a small area at base, symmetric, flat, thin, membranous, broad-elliptic to ovate, slightly curved, with brown pigment cells in the epidermis; margin entire; base and apex obtuse; roots absent. Daughter thalli borne in 1, symmetric, triangular budding pouch. Inflorescences 2, each with 1 female and 1 male flower, without membranous sheath. Anthers 1-locular.

1 species, *W. welwitschii* (Hegelm.) Hartog and Plas: tropical Africa and tropical America. It is not certain that the American and African material belong to the same species.

LENTIBULARIACEAE

4 genera of which three are primarily aquatic.

Carnivorous perennials, rarely annuals. Leaves entire or much divided, often bearing traps. Flowers bisexual, zygomorphic, in bracteate racemes. Sepals 2, 4, or 5. Petals united at the base, forming a 2-lipped corolla; abaxial lip spurred or saccate. Stamens 2, attached to the base of the corolla. Anthers dorsifixed. Ovary superior; fruit a few to many seeded capsule.

Lloyd, F. E. The Carnivorous Plants: Waltham, Mass. (1942)

1A Traps with long tubular mouth and two spirally twisted arms
Genlisea

1B Traps globose

2A Traps solitary, terminal on subterranean leaves; emersed leaves linear, ascending, without traps; sepals 4
Polypompholyx

2B Traps usually several and lateral; emersed leaves absent or if linear usually trap-bearing; sepals 2
Utricularia

Genlisea, A. St. Hil., Voy. Distr. Diam. 2: 428 (1833)
Fig. 132A.
Rootless annuals with a slender rhizome and root-like appendages. Foliage leaves in a basal rosette, ascending, spathulate to linear; trap leaves reduced, descending into the substrate, ending in a single swollen flask-shaped trap with a long tubular mouth and 2 spirally twisted lobes. Inflorescence glandular, few- to many-flowered. Sepals 5. Petals united below forming a 2-lipped corolla; abaxial lip spurred. Capsule many-seeded.

c. 15 species: America, Tropical and S. Africa, Madagascar. Occurs in wet places, though rarely totally submerged.

Polypompholyx Lehmann, Nov. Stirp. Pugill. 8: 48 (1844)
Fig. 132B.
Rootless perennials with a swollen rhizome and root-like appendages. Foliage leaves in a basal rosette, ascending, linear: trap leaves reduced, descending into the substrate, ending in a single, globose, bladder-like trap. Inflorescence usually few-flowered. Sepals 4, 2 long and 2 short. Petal-tube usually pink. Capsule many-seeded.

2 species: endemic to W. Australia. Occurs in wet bogs.

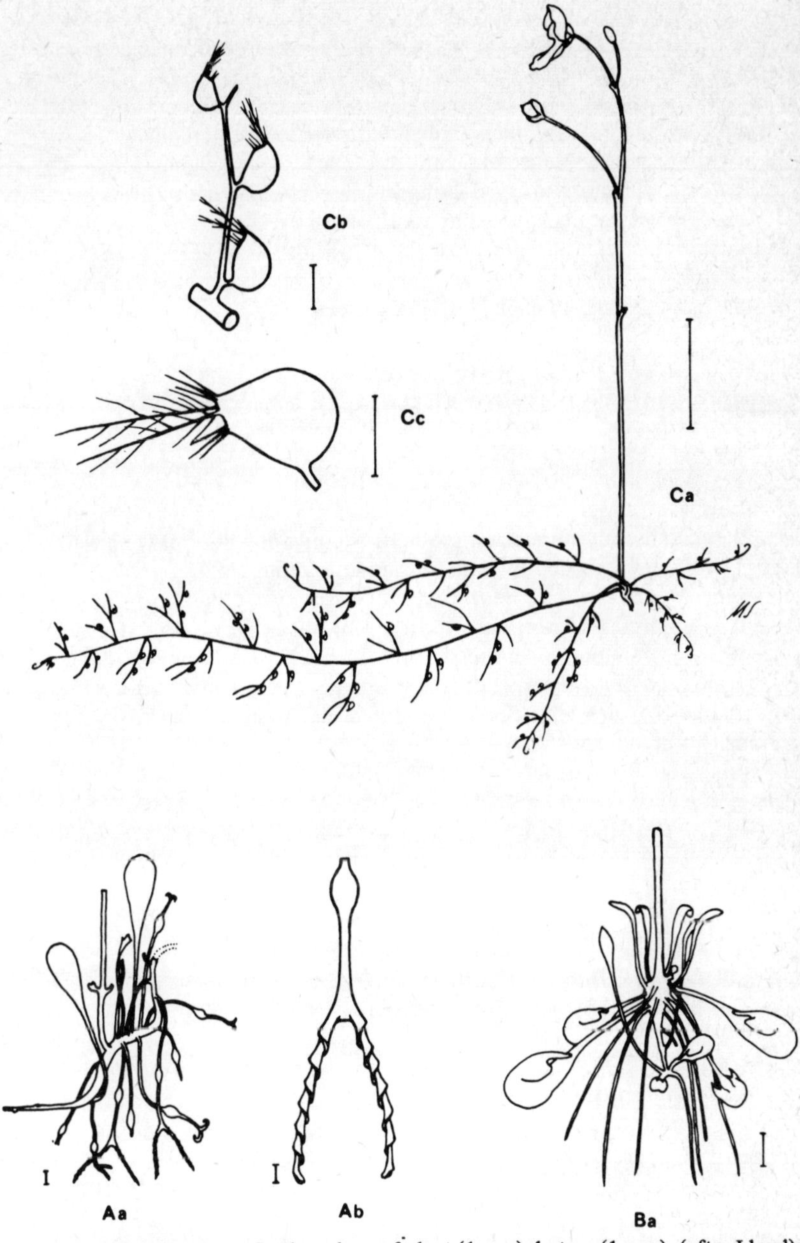

Fig. 132. A. *Genlisea repens* Benj.: a, base of plant (1 mm); b, trap (1 mm), (after Lloyd);
B. *Polypompholyx tenella* Lehm.: base of plant (1 mm), (after Lloyd);
C. *Utricularia gibba* Le Conte: a, habit (1 cm); b, traps (1 mm); c, trap (1 mm).

Utricularia L., Sp. Pl.: 18 (1753), [*Biovularia* Kamienski]
Fig. 132C.
Rootless perennials, floating in water or fixed by root-like appendages; sometimes forming tubers or turions. Emergent leaves entire, peltate, kidney-shaped to linear and thalloid, often bearing traps near their base; submerged leaves capillary and usually much branched, sometimes very reduced, bearing ovoid or globose bladder-like traps. Inflorescence emergent, sometimes supported by a whorl of spongy floating leaves. Sepals 2. Petals united below forming a 2-lipped corolla; adaxial lip more or less erect, abaxial lip with spreading or deflexed limb, spurred or saccate. Flowers sometimes cleistogamous. Capsule with 1 to many seeds; seeds very variable in size and shape, sometimes with glochidia (small hook-like processes).

Sorenson, D. R. and Jackson, W. T. The utilisation of Paramecia by the carnivorous plant Utricularia. Planta 83 (2): 166–270 (1968)
Taylor, P. The genus Utricularia in Africa (south of the Sahara) and Madagascar. Kew Bull. 18: 1–245 (1964)

c. 150 species: cosmopolitan. 25 to 30 species are aquatic, with only capillary leaves: they are usually found submerged or floating in still or moving water. Other species grow attached to stones in running water, e.g. *U. rigida* Benjamin in W. Africa and *U. neottioides* in S. America. Others grow in wet bogs and as epiphytes. *U. olivacea* Wright ex Griseb. a minute white-flowered species from shallow pools in eastern tropical America has been considered as a separate genus *Biovularia*. It differs from most species of *Utricularia* in having only 2 ovules, all the leaves modified into traps and in the sepals becoming toothed in fruit.

LILAEACEAE

1 genus.

Lilaea Humb. and Bonpl., Pl. Aequin. 1: 221 (1808)
Fig. 133.

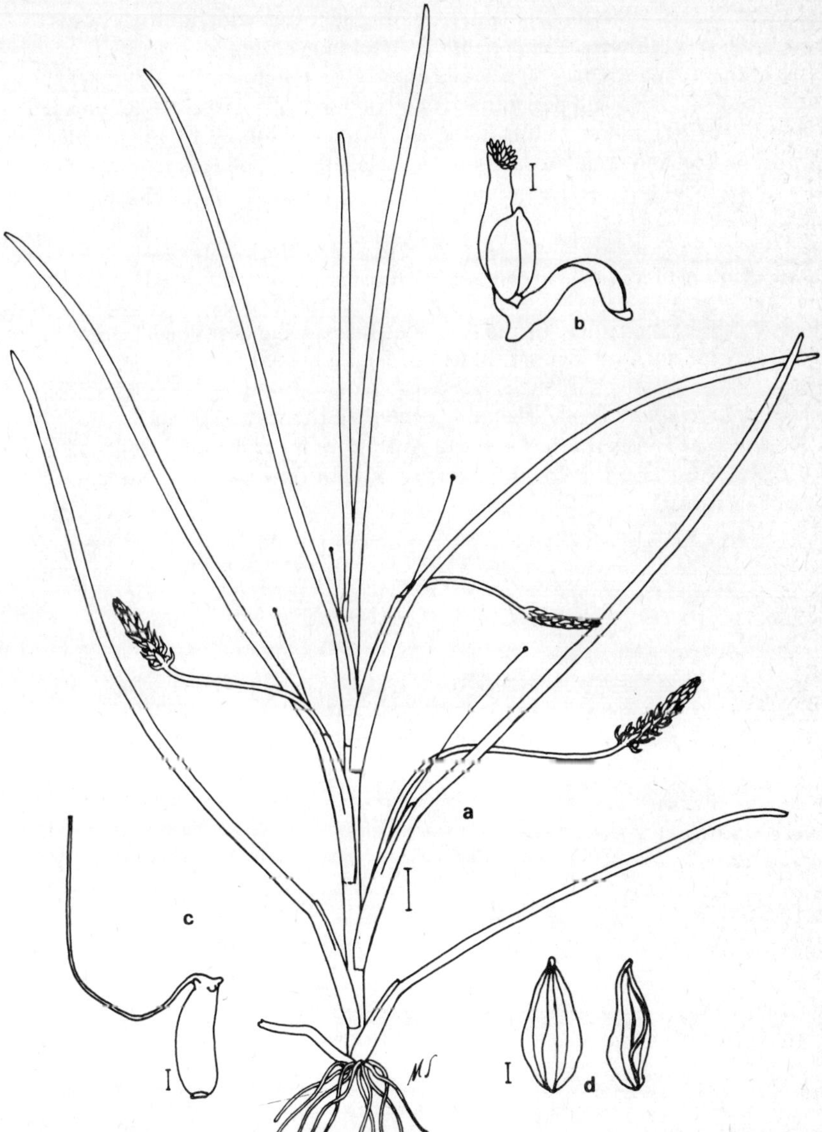

Fig. 133. *Lilaea scilloides* (Poir.) Haum.: a, habit (1 cm); b, bisexual flower from spike (1 mm); c, axillary female flower (1 mm); d, seeds (1 mm).

Annual. Stem reduced. Leaves basal, simple, linear, terete, 3 to 35 cm long, entire; stipules membranous, sheathing and converging across the top to form a short ligule. Inflorescence complex; each leaf axil gives rise to 2, basal, female flowers and 1, stalked spike bearing bisexual and male flowers. Female flowers enclosed in sheathing leaf base; perianth lacking; carpel 1; style thread-like, 20 to 300 mm long, arising laterally at top of carpel but at length ascending; stigma capitate, papillate; fruits angled, often with hooks or horns at apex. Bisexual flowers with perianth 1, bract-like, arising from base of sessile anther (perianth may represent a connective outgrowth); stamen 1; anther sessile (stamen and perianth caducous); ovary of 1 carpel; styles unequal in length 1 to 20 mm long; fruits flattened, 4 to 6 mm long, 2 to 3 mm wide, with dorsal keel and undulate lateral wings. Male flowers of 1, bract-like perianth segment (connective outgrowth) arising from base of the single, sessile anther.

Mason, H. L. A Flora of the marshes of California. Univ. Calif. Press. 100–102 (1957)

Singh, V. Vascular anatomy of the flower of Lilaea scilloides (Poir.) Haum. Proc. Indian Acad. Sci. ser. B, 61: 316–325 (1965)

1 species: *L. scilloides* (Poir.) Haum. [*L. subulata* Humb. and Bonpl.]: W. N. America, S. America to Chile and Argentine and probably introduced in Victoria, Australia. It is found in shallow, still or flowing water that may dry out each year.

LIMNOCHARITACEAE

4 genera: all aquatic.

Annuals or perennials, with latex. Juvenile leaves linear; adult leaves differentiated into petiole and blade; blades ovate to cordate; nerves parallel and curved. Inflorescence umbel-like or flowers solitary. Flowers bisexual, actinomorphic. Sepals green, persistent, with latex tubes. Petals white or yellow, delicate, not persistent. Stamens 6 to 9 or numerous; pollen with 4 or more pores. Carpels superior, 3 to numerous, free, in 1 or rarely 2 whorls, when ripe opening by the adaxial (vental) suture; seeds numerous, scattered over the inner surface of the carpel wall; embryos curved or folded.

Buchenau, F. in Engler, A. Pflanzenreich 15 (IV. 16): 1–12 (1903)
Takhtajan, A. Die Evolution der Angiospermen. Jena (1959)

1A Leaf-blades obtuse or cordate at base, usually floating; petioles with transverse septae
Hydrocleis

1B Leaf-blades cuneate at base, usually erect above the water; petioles without transverse septae

- 2A Leaf-blades ovate to suborbiculate, apex rounded or emarginate; petals yellow; stamens more than 9, surrounded by a whorl of staminodes
Limnocharis
- 2B Leaf-blades lanceolate to oblanceolate, apex tipped by a hard, blunt mucro; petals white; stamens about 9; staminodes absent
Tenagocharis

Hydrocleis L. C. Rich., Mém. Mus. Hist. Nat. 1: 368, 373 (1815), [*Ostenia* Buchenau]
Fig. 134.
Stoloniferous perennials or annuals. Leaves radical or in groups along the stem; petioles with transverse septae; blades ovate to suborbiculate, with cordate bases. Flowers in axillary fascicles or terminal and umbel-like or solitary. Petals yellow, delicate, not persistent. Stamens 6 to numerous; staminodes absent or present between stamens and petals, equalling or exceeding the stamens in number. Carpels usually 3 or 6, linear-lanceolate, united at base; style beak-like, persistent; seeds small, horseshoe-shaped.

Johri, B. M. The embryo sac of Hydrocleis nymphoides Buchen. Beih. Bot. Zbl. 48: 165–172 (1938)
Pedersen, T. M. New species of Hydrocleis, Scirpus and Stellaria. Bot. Tidsskr. 57: 38–46 (1961)

c. 9 species: S. America. Usually partly floating, and partly rooted in still water. *H. nymphoides* (H. and B. ex Willd.) Buchenau is very decorative and is widely cultivated, it has been grown in heated greenhouses in Europe since 1830.

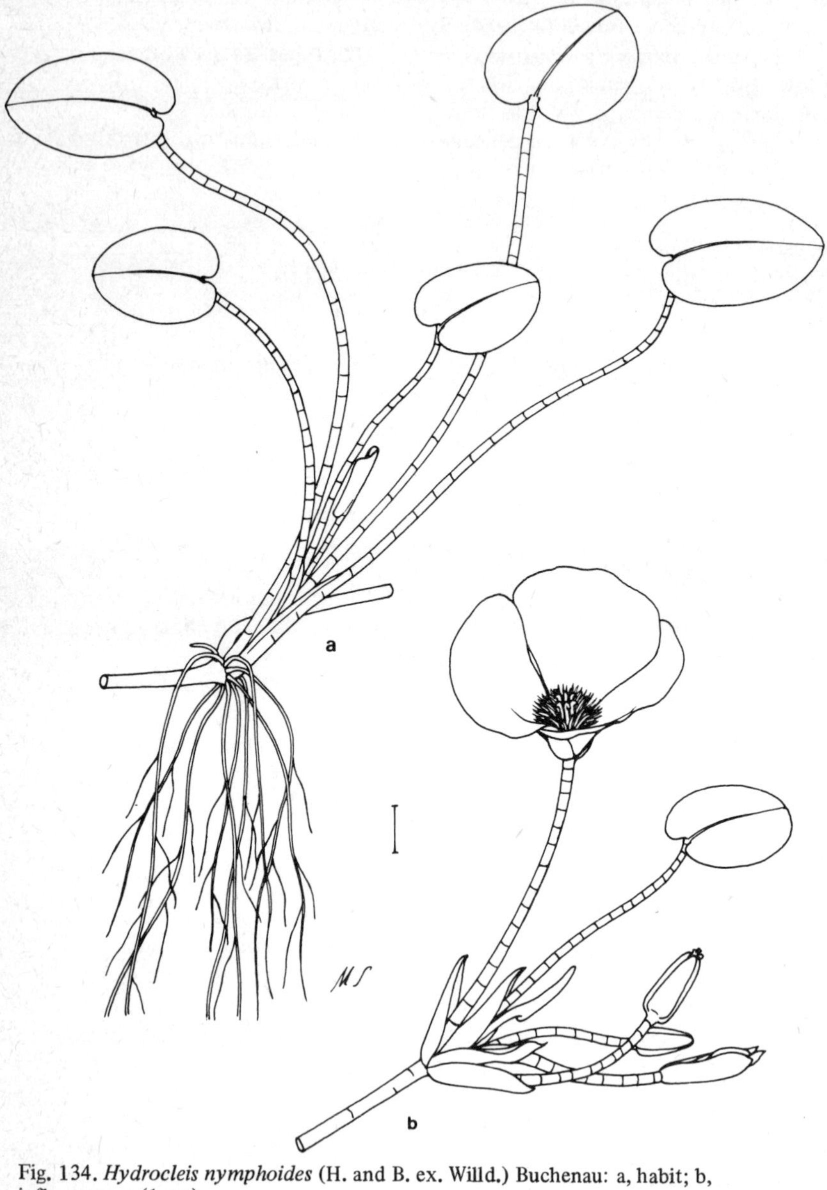

Fig. 134. *Hydrocleis nymphoides* (H. and B. ex. Willd.) Buchenau: a, habit; b, inflorescence (1 cm).

Limnocharis Humb. and Bonpl., Pl. Aequin. 1: 116 (1808)
Fig. 135.
Perennials or annuals. Leaves in rosettes; petioles thick, triangular in transverse section; blades ovate to suborbicular, with cuneate bases, inrolled when young, with rounded or emarginate apices. Inflorescence umbel-like, 2- to 15-flowered. Petals pale yellow with darker base, ovate to orbicular, not persistent. Stamens numerous, surrounded by a whorl of staminodes. Carpels numerous, laterally flattened and firmly pressed together, apparently forming one ovary.

Fig. 135. *Limnocharis flava* (L.) Buchenau: habit (5 cm).

Fig. 136. *Tenagocharis latifolia* (D. Don) Buchenau: a, habit (1 cm); b, flower (2 mm).

Bahadur, K. N. and Raizada, M. B. Limnocharis flava (L.) Buch. – a new record for India. Indian For. 94: 641–644 (1968)
Kaul, R. B. Ontogeny and anatomy of the flower of Limnocharis flava. Amer. Journ. Bot. 54 (10): 1223–1230 (1967)

2 species, *L. flava* (L.) Buchenau, originally from the warmer regions of America and naturalised in India and S. E. Asia, and *L. mattogrossensis* O. Kuntze from the Mattogrosso in Brazil. In swamps, ditches and pools and also found as a weed in ricefields. In S. E. Asia *L. flava* is eaten as an alternative for spinach or endive or used as fodder for pigs.

Tenagocharis Hochst., Flora 24: 369 (1841), [*Butomopsis* Kunth]
Fig. 136.
Annuals or perennials. Leaves erect; petioles up to 20 cm or more long; blades lanceolate to oblanceolate, 3 to 12 (–15) cm long, 1.5 to 3.5 (–5) cm wide, with cuneate base, and an acute or obtuse apex, tipped by a hard blunt mucro. Inflorescence exceeding the leaves, umbel-like, or 1 whorl of flowers (sometimes 2 in larger plants). Petals white, very delicate, withering and disintegrating after anthesis. Stamens *c.* 9. Carpels 4 to 9, united at base, in fruit 9 to 12 mm long, exceeding the sepals; seeds horseshoe-shaped, less than 0.5 mm long, minutely warted.

Johri, B. M. The life history of Butomopsis lanceolata Kunth. Proc. Indian Acad. Sci. ser. B., 4: 139–162 (1936)

1 species, *T. latifolia* (D. Don) Buchenau: tropical W. Africa, one locality in Uganda, India, Malaysia and N. Australia. In marshes and swamps, and found as a weed in ricefields.

LOBELIACEAE [CAMPANULACEAE pro parte]

c. 30 genera: 10 genera with aquatic species.

Annuals or perennials. Leaves usually alternate, often reduced and bract-like; stipules absent. Flowers bisexual, rarely unisexual, solitary in axils of leaves, or in loose terminal racemes, zygomorphic. Sepals 5, persistent in fruit, united at the base to form a tube which is adnate to ovary. Petals 5, united at least at the base, forming a 2-lipped corolla; abaxial lip 3-lobed, adaxial lip 2-lobed. Stamens 5, alternating with petal-lobes, united into a tubular column; filaments united or free only at the base; anthers united, introrse. Ovary 1- or 2-locular, usually appearing inferior; style 1, simple; stigma 2- to 4-lobed; fruit a capsule or berry.

Wimmer, F. E. in Engler, Pflanzenreich 106 to 108 (IV. 276): (1943, 1953, 1968)

1A Sinus between adaxial petals shorter than sinuses between lateral petals; flowers not cleistogamous
- 2A Fruit a juicy berry **Hypsela**
- 2B Fruit a capsule
  - 3A Petal-tube weakly 2-lipped, tube about equalling the lobes **Isotoma**
  - 3B Petal-tube strongly 2-lipped, tube much shorter than the lobes
    - 4A Capsule obconical, stalked **Porterella**
    - 4B Capsule linear, sessile (resembling a thickened pedicel) **Downingia**

1B Sinus between adaxial petals much longer than that between lateral petals; flowers occasionally cleistogamous
- 5A Fruit a juicy berry **Pratia**
- 5B Fruit a capsule
  - 6A Capsule 2-locular
    - 7A Capsule sessile or shortly stalked, linear, curved **Grammatotheca**
    - 7B Capsule stalked, obconical to ovate, not curved
      - 8A Petal-lobes very unequal in length, usually more than 5 mm long **Lobelia**
      - 8B Petal-lobes almost equal in length, less than 3 mm long **Mezleria**
  - 6B Capsule 1-locular
    - 9A Leaves linear to oblanceolate; seeds numerous; emergent annual **Legenere**
    - 9B Leaves capillary; 2 seeds per capsule; flaccid aquatic **Howellia**

Downingia Torrey, Explorat. Railroad Rout from Mississippi to Pacific Ocean Report Bot. 4: 116 (1856)

Fig. 137A.

Annuals 10 to 30 cm tall. Leaves often sessile, alternate, dimorphic; submerged leaves linear to capillary; emersed leaves lanceolate to oblanceolate. Flowers

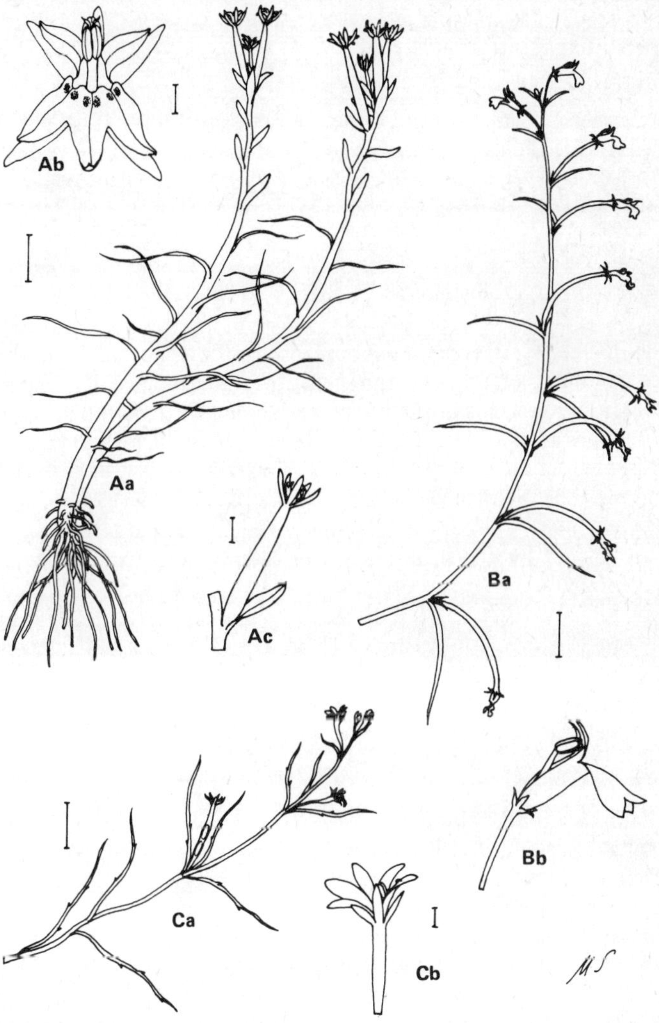

Fig. 137. A. *Downingia laeta* (Greene) Greene: a, habit (1 cm); b, flower (1 mm); c, fruit (3 mm); (b and c after Mason);
B. *Grammatotheca bergiana* (Cham.) Presl: a, habit (1 cm); b, flower (1 mm);
C. *Howellia aquatilis* A. Gray: a, habit (1 cm); b, flower (1 mm).

solitary and sessile. Sepal-lobes linear, subequal. Petal-tube strongly 2-lipped, not split adaxially, with a short tube; abaxial lip spreading or reflexed, usually with a white patch and yellow or purple spots or ridges; adaxial lip narrower, erect to recurved. Stamens united throughout their length; anthers papillose at the apices, the 2 abaxial each with a bristle. Ovary linear, much elongating in fruit, giving the appearance of a thick pedicel; capsule 1- or 2-locular, dehiscing by longitudinal slits; seeds numerous, 0.5 to 1.0 mm.

McVaugh, R. A monograph of the genus Downingia. Mem. Torrey Bot. Club 79 (4): 1–57 (1941)

13 species: western N. America and Chile. Centre of distribution is California with 12 species; the Chilean species is also found in N. America. In vernal pools and other seasonally wet places. Flowering occurs when the plant reaches the surface or the water level drops.

Grammatotheca Presl, Prodr. Monogr. Lobel.: 43 (1836)
Fig. 137B.
Perennial with creeping rhizome. Stems ascending, rooting at the base. Lower leaves ovate-oblong to oblong, the upper narrower, linear-lanceolate acute. Flowers solitary, in the axils of the leaves, sessile or with a short pedicel, with 2 linear bracteoles at their base. Petal-tube blue-violet with a white throat, strongly 2-lipped, split adaxially to the base; abaxial lip longer than the adaxial, subequal to the tube. Stamens united into a tube; anthers hairy at the apex. Capsule narrow, linear, cylindrical, 2-locular, often somewhat reflexed, dehiscing at the apex by two small valves, and later by longitudinal slits.

1 species, *G. bergiana* (Cham.) Presl divided by Wimmer, op. cit. (1953) into 4 varieties, 3 in S. Africa, 1 in Australia. In pools and streams which dry out in summer, and in permanent swamps.

Howellia A. Gray, Proc. Amer. Acad. 14: 25 (1879)
Fig. 137C.
A flaccid glabrous annual. Submerged leaves alternate or whorled, capillary, 1 to 5 cm long; emersed leaves linear-lanceolate, shorter, sessile, toothed towards the apex. Flowers either submerged and cleistogamous or emersed and chasmogamous, with short pedicels. Sepal-lobes 1 to 2 mm long, linear, acute or obtuse, spreading. Petal-tube split adaxially to base, about equalling the sepal-lobes; petal-lobes whitish, oblong, almost equal, *c.* 1 mm long. Stamens free from the petals, the 2 abaxial anthers with 3 small apical bristles. Capsule 1-locular, linear-oblong, curved, tapering towards base and apex, finally 6 to 10 mm long, usually 2-seeded; seeds 3 mm long.

1 species: *H. aquatilis* A. Gray occurs in ponds and lakes in N. W. North America (Oregon to British Columbia).

Hypsela Presl, Prodr. Monogr. Lobel: 45 (1836)
Fig. 138A.
Small creeping or ascending perennials. Leaves alternate, ovate to oblong, petiolate or sessile, Flowers on slender pedicels or sessile. Sepal-lobes 1 to 2 mm long, narrowly triangular. Petal-tube weakly 2-lipped, not split adaxially; petal-lobes almost equal, shorter than the tube, whitish. Filaments free, inserted about halfway up the petal-tube; anthers united, each of the 2 abaxial ones with 2 bristles at their apex. Fruit a berry, globose, *c.* 3 mm diam., 2-locular; seeds numerous.

4 species: *H. reniformis* (HBK.) Presl in S. America, *H. rivalis* E. Wimm. in New Zealand, *H. tridens* E. Wimm. and *H. sessiliflora* E. Wimm. in Australia. In streams and the edges of lakes. Differs from *Pratia* in its entire petal-tube and in the position of insertion of the stamens.

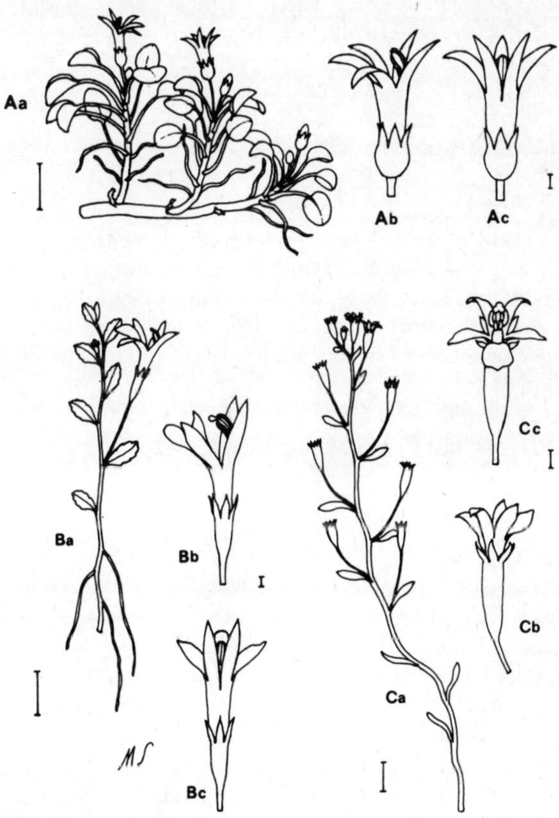

Fig. 138. A. *Hypsela reniformis* (HBK.) Presl: a, habit (1 cm); b, flower, from side; c, flower, from above (1 mm);
B. *Isotoma fluviatilis* (R.Br.) Mueller: a, habit (1 cm); b, flower, from side; c, flower, from above (1 mm);
C. *Legenere limosa* (Green) McVaugh: a, habit (1 cm); b, flower, from side; c, flower, from below (1 mm), (after Mason).

Isotoma Lindley, Bot. Reg. 12: 964 (1826), [*Laurentia* sect. *Isotoma* (R. Br.) Endl.]

Fig. 138B.

Slender annuals. Stems ascending. Leaves sessile or shortly petiolate, lanceolate to obovate; the upper usually narrower. Flowers sometimes unisexual; pedicellate. Sepal-lobes linear, *c.* 2 mm long. Petal-tube blue to whitish, weakly 2-lipped, not split adaxially; petal-lobes more or less equal, spreading, as long as the tube. Filaments inserted on the corolla; the 2 abaxial anthers hairy at the apex, each with 1, long bristle. Capsule obconical to cylindrical, 2-locular, dehiscing by 2 apical valves.

McComb, J. A. A revision of the species Isotoma fluviatilis. Contr. New S. Wales Nat. Herb. 4 (3): 106–111 (1970)

10 species: Australia: *c.* 2 aquatic. McComb (l.c.) cites 3 subspecies of *I. fluviatilis* (R. Br.) Mueller, differing in chromosome number and sexuality. In wet places around lakes and streams.

Legenere McVaugh, North Amer. Fl. 32A: 13 (1943)

Fig. 138C.

Ascending annuals. Submerged leaves linear, 1 to 3 cm long; emersed leaves oblong-lanceolate to subcuneate. Flowers on long slender pedicels. Sepal-lobes narrowly triangular, *c.* 1 mm long. Petal-tube sometimes absent, whitish, weakly 2-lipped, split adaxially to base; petal-lobes more or less equal. Filaments united into a tube; anthers united, the abaxial pair each with 2 minute bristles at its apex. Capsule 1-locular, 6 to 10 mm long, obconical, dehiscing at the apex by two small valves.

2 species: *L. valdeviana* (Phil.) E. Wimm. in Chile; *L. limosa* (Green) McVaugh in California: occurs in the beds of vernal pools.

Lobelia L., Sp. Pl. 929 (1753)

Fig. 139A, B.

Annuals or perennials. Stems reduced and corm-like, or elongate, erect, or floating, often somewhat fleshy. Leaves sessile, petiolate, or in basal rosettes, linear to ovate. Flowers solitary in the axils of the leaves or in terminal racemes, sometimes cleistogamous when submerged, on slender pedicels. Sepals 5, acute, subequal. Petal-tube 2-lipped, split adaxially to base; the three abaxial lobes spreading or deflexed, the 2 adaxial erect. Stamens not inserted on the petal-tube; filaments free below; anthers united into a tube, all or the abaxial pair with hairs at their apices. Capsule subglobose to clavate, 2-locular, dehiscing by 2 apical valves.

McVaugh, R. Studies in the taxonomy and distribution of the eastern North American species of Lobelia. Rhodora 38: 241–263, 276–298, 346–362 (1936)

c. 365 species: cosmopolitan. Many species are found in wet or seasonally inundated regions. On the basis of their habit the aquatic species may be divided into three groups. 1. *L. dortmanna* L. occurs in acid lakes in N. America and subarctic Europe south to the Gironde; it has a rosette of basal submerged leaves and an emersed inflorescence. 2. *L. paludosa* Nutt. and related species occur in the wetlands of southeastern N. America; they are erect perennials and the lower parts of the plant are submerged, the inflorescence and cauline leaves emersed. 3. *L. aquatica* Cham. from S. America and *L. alsinoides* Lam. from S. E. Asia are flaccid annuals and often occur floating in shallow water.

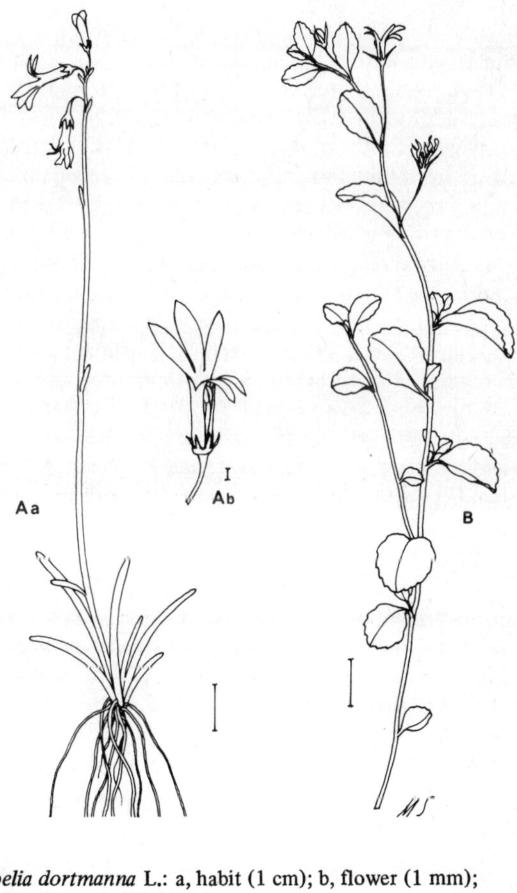

Fig. 139. A. *Lobelia dortmanna* L.: a, habit (1 cm); b, flower (1 mm); B. *Lobelia alsinoides* Lam.: habit (1 cm).

Mezleria Presl, Prodr. Monogr. Lobel. 7 (1836), [*Lobelia* subgen. *Mezleria* (Presl) E. Wimm.]
Fig. 140A.
Small creeping or ascending annuals or perennials. Leaves alternate, obovate to lanceolate, sessile or shortly petiolate. Flowers very small with slender pedicels. Sepal-lobes sometimes 4, usually 5, acute. Petal-tube weakly 2-lipped, with a deep adaxial split; petal-lobes almost equal, longer than the tube, white to pale mauve. Filaments united; anthers united, the 2 abaxial ones each with 2 bristles and a tuft of hairs at its apex. Capsule 2-locular, subglobose, dehiscing by 2 apical valves.

8 species: S. Africa. *M. quadrisepala* Good occurs in shallow pools and vleis in Cape Province. Adamson in Flora of the Cape Peninsula (1950) reports that it is "submerged in the young state but exposed at flowering time".

Porterella Torrey in Hayd., Rep. Geol. Survey Montana: 488 (1872), [*Laurentia* (Mich.) Adans. sect. *Solenopsis* Endl. pro parte]
Fig. 140B.
Erect, glabrous, somewhat fleshy annuals. Leaves linear to lanceolate, or scale-like below, narrowing from base towards apex, sessile. Flowers on slender spreading or ascending pedicels. Sepal-lobes linear-lanceolate, 3 to 10 mm long. Petal-tube short, strongly 2-lipped, without a deep adaxial slit, white to bluish; abaxial petal-lobes spreading, broadly orbicular to obovate, mucronate, with 2 yellow ridges in the throat and a wide white patch; adaxial petal-lobes oblong-lanceolate, erect or recurved. Stamens united throughout their length; anthers tufted at apex, the 2 abaxial ones each with a short apical bristle. Capsule 2-locular, obconical to turbinate, 8 to 14 mm long, dehiscing by two apical valves.

2 species: *P. carnulosa* (Hook. and Arn.) Torr. and *P. eximia* A. Nelson in western N. America. Characteristic of vernal pools and wet meadows.

Pratia Gaudich., in Freyc. Voy. Bot.: 456 (1826)
Fig. 140C.
Small creeping perennials, often dioecious. Leaves alternate, usually sessile, orbicular to linear-lanceolate. Flowers solitary, on slender pedicels or sessile. Sepal-lobes 1 to 2 mm long, narrowly triangular. Petal-tube weakly 2-lipped, deeply split adaxially, whitish. Filaments free below, united above, inserted at the base of the petal-tube; anthers united, the 2 abaxial ones each with 2 bristles at its apex. Fruit an ellipsoidal berry; seeds numerous, small.

c. 20 species: *c.* 10 in Australia, 3 in New Zealand, 3 in S. America, 3 in S.E. Himalayan region to New Guinea, and 1 in W. Africa. *P. prostrata* E. Wimm. in east Australia appears to be aquatic; it occurs in small streams. *P. concolor* (R. Br.) Druce also grows in water. *Pratia* is distinguished from *Lobelia* by its succulent, berry-like fruit. However in *P. angulata* (Forst.) Hooker in New Guinea and the Celebes this distinction breaks down; Moeliono [in Van Steenis, Fl. Malaysiana 6 (1): 135 (1960)] has united the two genera.

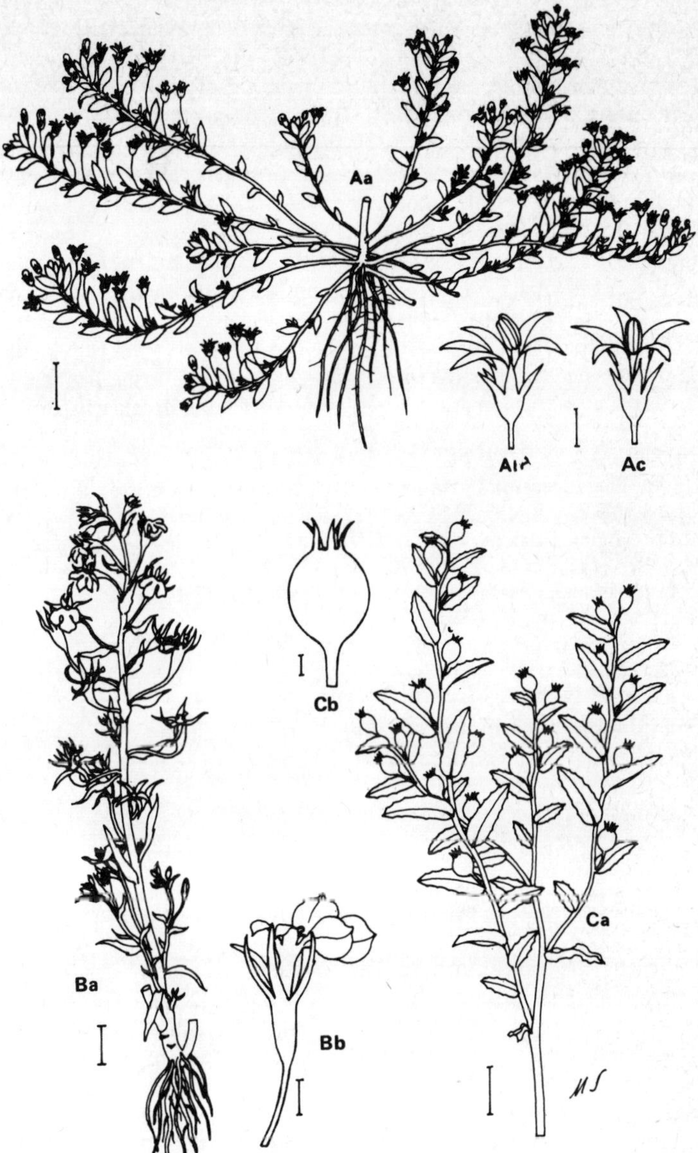

Fig. 140. A. *Mezleria depressa* Sond.: a, habit (1 cm); b, flower, from side; c, flower, from above (1 mm);
B. *Porterella carnulosa* (Hook. & Arn.) Torr.: a, habit (1 cm); b, flower (2 mm); (after Mason);
C. *Pratia concolor* (R.Br.) Druce: a, habit (1 cm); b, fruit (1 mm).

LYTHRACEAE

c. 25 genera; *c.* 550 species; cosmopolitan. 7 genera contain aquatics.

Herbs or occasionally shrubs. Stems floating, creeping, ascending, or erect often swollen when submerged. Leaves usually opposite, occasionally whorled, rarely alternate, simple, entire; stipules absent. Flowers bisexual, actinomorphic or zygomorphic, axillary or in terminal racemes or spikes, sometimes di- or trimorphic, occasionally cleistogamous. Sepals 4 to 6 (–7) united into a tube, valvate, often with appendages between the lobes. Petals 4 to 6 (–7) or absent, usually caducous, inserted towards the top of the sepals. Stamens as many or twice as many as the sepals, rarely more, often unequal in length, inserted on the inner surface of the sepal-tube below the petals; anthers frequently versatile, rarely basifixed. Ovary superior, free from the sepal-tube, (1–) 2- to 6-locular; septae separating the locules often incomplete; placentation axile, fruit a capsule, opening by a transverse slit, by valves, by splitting irregularly, or indehiscent.

Graham, S. A. The genera of Lythraceae in the southeastern United States. Journ. Arnold Arb. 45: 235–250 (1964)

Koehne, E. in Engler, Pflanzenreich 17 (IV. 216): 1–326 (1903), Nachträge. Bot. Jahrb. 41 (2): 74–110 (1907), 42 (3): 47–53 (1909)

Lourteig, A. Lythraceae Argentinae. Lilloa 9: 317–421 (1943), 10: 387–394 (1944), Darwiniana 8: 263–278 (1948)

Webb, D. A. Generic limits in European Lythraceae. Feddes Repert. 74 (1, 2): 10–13 (1967)

Generic delimitation in the Lythraceae is problematical and needs revision; the following key has been constructed on the basis of the aquatic species:

1A Flowers distinctly zygomorphic; disc at base of ovary present
Cuphea

1B Flowers actinomorphic or slightly zygomorphic; disc at base of ovary absent

- 2A At least some leaves in whorls, lanceolate, usually more than 5 cm long; stems distinctly woody at base, more than 1 m tall (N. America)
Decodon
- 2B Leaves not in whorls or when in whorls, linear, less than 5 cm long; stems herbaceous or slightly woody at base, usually less than 1 m tall
  - 3A Flowers solitary and more or less sessile in leaf axils
    - 4A Capsule indehiscent or splitting irregularly
      - 5A Stems creeping, much branched; ovary appearing 1-locular (incompletely 2-locular by reduction of septum) (N. America)
Didiplis
      - 5B Stems erect, branching only from base; ovary usually 4-locular (widespread in warm regions)
Ammannia

4B Capsule regularly dehiscent, opening by valves or slits
6A Flowers usually 4- or 5-merous; sepal tube campanulate or urn-shaped, usually without longitudinal ribs.
Rotala

6B Flowers usually 6-merous; sepal tube cylindrical below, with distinct longitudinal ribs
Lythrum

3B Flowers in cymes, umbels or crowded into bracteate heads
7A Septae of ovary complete, the placenta continuous with the style; petals corrugated in bud, or absent
Nesaea

7B Septae of ovary incomplete, the placenta not continuous with the style; petals not corrugated in bud, or absent
8A Flowers usually 6-merous; sepal tube cylindrical below, with distinct longitudinal ribs
Lythrum

8B Flowers usually 4-merous or 5-merous; sepal tube campanulate or urn-shaped, without or rarely with longitudinal ribs
9A Capsule dehiscing irregularly, without horizontal lines
Ammannia

9B Capsule dehiscing regularly by valves, with faint horizontal lines
Rotala

Ammannia L., Sp. Pl. 119 (1753) [*Hionanthera* A. Fernandes and Diniz.]
Fig. 141A.
Annuals or short-lived perennials. Stems 4-angled. Leaves opposite, linear to lanceolate or oblanceolate, sessile; base cordate to auriculate, rarely attenuate. Flowers actinomorphic, in sessile or stalked, axillary cymes, (1-) 3- to 15-flowered. Sepal tube campanulate to urn-shaped, becoming globose in fruit; sepal-lobes 4 (5); sepal appendages present or absent. Petals 0 to 4 (5), caducous. Stamens equal to or twice as many as sepals. Ovary 1- to 5-locular (usually 4); without disc at base; stigma capitate; style very short to longer than the ovary; capsule dehiscing irregularly.

c. 30 species: warmer parts of the world. Most species grow in wet regions, some species (*A. auriculata* Willd. *S. senegalensis* Lam.) are found as weeds in ricefields

Cuphea P. Brown, Civ. Nat. Hist. Jamaica 216 (1756)
Fig. 141B.
Annual or perennial herbs, or shrubs. Leaves opposite, whorled or rarely alternate. Flowers strongly zygomorphic, sessile or stalked, solitary or in complex inflorescences. Sepal tube cylindrical; the base occasionally spurred; sepal lobes usually 6. Petals 0 or 6 (rarely 2 or 4). Stamens (5–) 11 (–12). Ovary

Fig. 141. A. *Ammannia baccifera* L.: a, habit (1 cm); b, fruit (1 mm);
B. *Cuphea speciosa* (Anders.) O.Ktze.: a, habit (1 cm); b, apex of sepal tube, opened (1 mm).

incompletely 2-locular, often appearing 1-locular by reduction of septum; disc present at base on upper side only; style variable in length.

Bacigalupi, R. Taxonomic studies in Cuphea. Contr. Gray Herb 95: 1–26 (1931)
Barroso, G. M. Contribuicao ao estudo do genero Cuphea Adans. Rodriguesia 16, 17: 193–212 (1954)

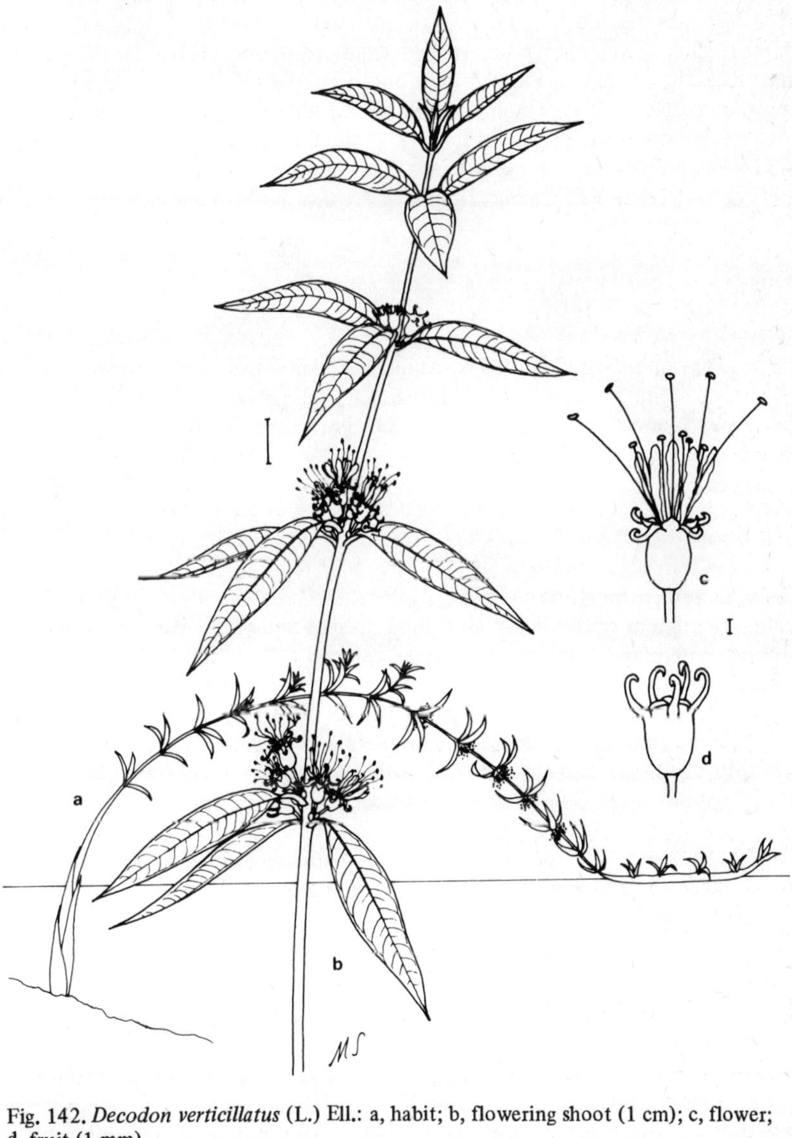

Fig. 142. *Decodon verticillatus* (L.) Ell.: a, habit; b, flowering shoot (1 cm); c, flower; d, fruit (1 mm).

c. 250 species: tropical and subtropical America. Most species are terrestrial, a few, such as *C. speciosa* (Anders.) O. Kuntze and *C. multicaulis* Koehne, are emergent aquatics in pools and slow-flowing rivers.

Decodon J. F. Gmelin in L., Syst. Nat. ed. 13, 2 (1): 677 (1791)
Fig. 142.
Perennial herbs woody at base or short-lived shrubs, spreading by means of arching branches rooting at the tips. Stems terete to 6-angled, the submerged portions thickened and spongy. Leaves whorled or opposite, lanceolate, shortly stalked; base acute. Flowers actinomorphic, 1 to 3, shortly stalked, arranged in axillary cymes at nodes; stalks partly adnate to petioles. Sepal tube campanulate; sepal lobes 4 or 5 (–7); sepal appendages present. Petals 4 or 5 (–7), about twice as long as sepals. Stamens 8 to 10, varying in length. Ovary 3- or rarely 4-locular; septa complete; style variable in length; capsule dehiscing by valves.

Schrenk, J. On the floating tissue of Nesaea verticillata (L.) HBK. Bull. Torrey Bot. Club 16: 315–323 (1889)

1 species, *D. verticillatus* (L.) Ell.: N. America, from Louisiana to Florida and northwards to Minnesota and Maine. Fossil material has been found in Asia and Europe. Extensive aerenchyma is produced in concentric layers on submerged stems.

Didiplis Rafin., Atl. Journ. 177 (1833), [*Peplis diandra* Nutt. ex DC.]
Fig. 143A.
Delicate, annual or short-lived perennial, glabrous herbs. Stems much branched, creeping, floating or occasionally ascending. Leaves opposite, sessile, narrowly linear when submerged, oblanceolate when emergent, 7 to 25 mm long, 2 to 4.5 mm wide. Flowers actinomorphic, axillary, solitary, sessile. Sepal tube 2 to 3 mm long, campanulate; sepal lobes 4; sepal appendages absent. Petals absent. Stamens 4. Ovary 1-locular (incompletely 2-locular by reduction of the septum); style short in chasmogamous flowers, absent in cleistogamous flowers; capsule indehiscent or splitting irregularly.

1 species, *D. diandra* (Nutt. ex DC.) Wood: N. America, Florida to E. Texas northwards to Minnesota and N. Carolina. Found along lake margins or submerged in shallow water.

Lythrum L., Sp. Pl. 446 (1753), [*Middendorfia* Trauv., *Peplis* L. pro parte]
Fig. 143B.
Annual or perennial herbs, or shrubs. Leaves opposite, whorled or alternate, linear to ovate or obovate, sessile or shortly petiolate; base attenuate to cordate. Flowers actinomorphic or slightly zygomorphic, axillary or terminal, solitary

to many-flowered. Sepal tube campanulate to tubular; sepal lobes (4–) 6; sepal appendages usually present. Petals (4–) 6 (rarely absent). Stamens as many or twice as many as sepal lobes. Ovary 2-locular, septum often incomplete; style present, variable in length; capsule dehiscing by slits or valves.

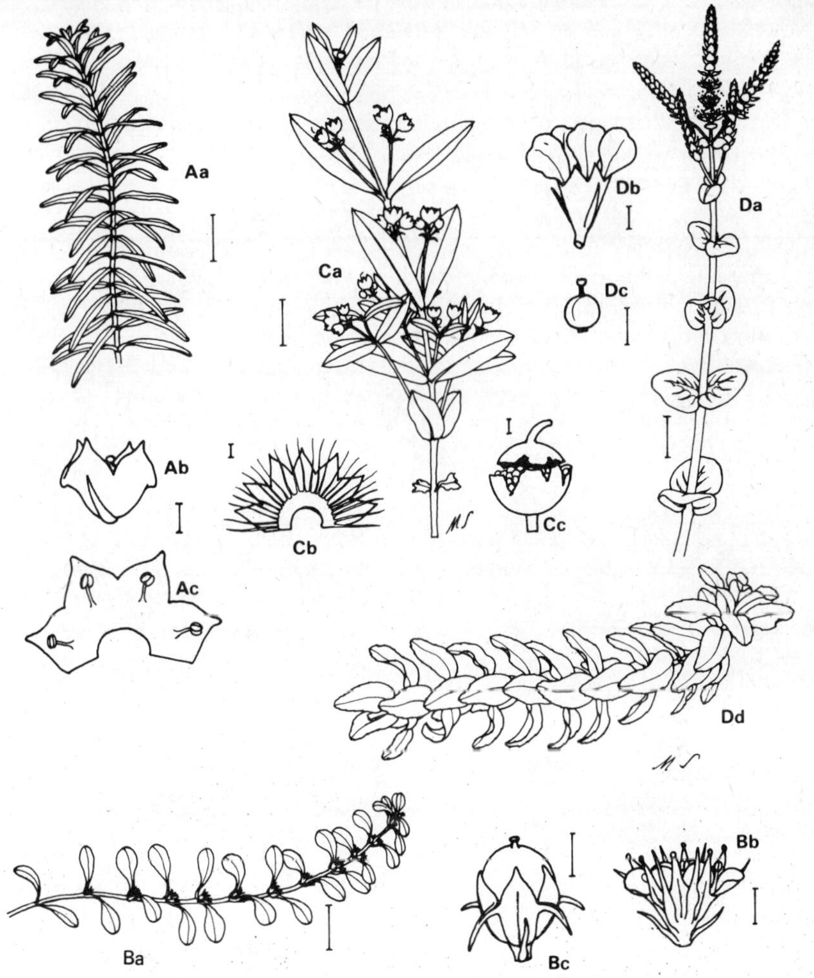

Fig. 143. A. *Didiplis diandra* (Nutt. ex DC.) Wood: a, habit (1 cm); b, flower; c, sepal-tube opened (1 mm);
B. *Lythrum portula* (L.) D. A. Webb: a, habit (1 cm); b, flower (1 mm); c, fruit (1 mm);
C. *Nesaea icosandra* Kotschy & Peyr.: a, habit (1 cm); b, sepal-tube opened (1 mm); c, fruit (1 mm);
D. *Rotala macrandra* Koehne: a, habit of aerial shoot (1 cm); b, flower (1 mm); c, ovary and style (1 mm); d, submerged shoot.

Allen, D. E. Variation in Peplis portula L. Watsonia 3: 85–91, Proc. Bot. Soc. Brit. Isles 5: 66–67 (1963)
Borja, Carbonell, J. Revision de las Especies Espanolas del Genero Lythrum L. Anal. Inst. Bot. Cavanilles 23: 145–162 (1968)

c. 38 species; cosmopolitan. Several species are found in wet regions that are regularly flooded. The aquatic or amphibious species have a more or less campanulate sepal tube and are occasionally placed in the genera *Middendorfia* (*L. borysthenicum* (Bieb. ex Schrank) Litwinov) or *Peplis* (*L. portula* (L.) D. A. Webb, *L. volgensis* D. A. Webb). They are found in and around lakes, pools and streams.

Nesaea Commerson ex. Humb., Bonpl. and Knuth. Nov. Gen. Sp. 6: ed. fol. 151, ed. qu. 191 (1823)
Fig. 143C.
Annual, perennial, or suffruticose herbs. Stems 4-angled. Leaves opposite, rarely alternate or whorled. Flowers actinomorphic, axillary, or terminal, in many-flowered, bracteate heads, or rarely, solitary, sessile or long-stalked. Sepal tube campanulate, urn-shaped or semiglobose; sepal lobes 4 to 8; sepal appendages absent or present. Petals 0 to 8, corrugated in bud. Stamens 4 to 23. Ovary 2- to 5-locular, septum complete, the placenta therefore continuous with the style; style variable in length; capsule dehiscing by splits or irregularly.

c. 50 species: warmer parts of the world. Most species are terrestrial, a few are found in regularly inundated regions. *N. cordata* Hiern and *N. crassicaulis* (Guill. and Perr.) Koehne are found as weeds in drainage ditches and ricefields.

Rotala L., Mantissa 143, 175 (1771)
Fig. 143D.
Annual or rarely perennial, glabrous herbs. Stems floating, creeping or ascending, terete. Leaves opposite, whorled, or rarely alternate, linear to ovate, sessile or very shortly stalked. Flowers actinomorphic, often inconspicuous, axillary, solitary, or in few-flowered clusters, or in terminal heads, sessile or nearly so. Sepal tube campanulate, globose, or urn-shaped; sepal lobes (3–) 4 (–6); sepal appendages present or absent. Petals 0 to 6, frequently small and inconspicuous. Stamens 4 (–6). Ovary (3–), 4- or (6–)-locular; style long, short or frequently absent; capsule dehiscing by valves.

Leeuwen, B. L. J. van. A preliminary revision of the genus Rotala in Malesia. Blumea 79 (1): 53–56 (1971)
Raynal, A. Les Rotala d'Afrique occidentale et centrale dans l'herbier de Paris. Adansonia 7: 535–545 (1967)

c. 50 species: in warmer parts of the world. All species are aquatic or amphibious. Many species are found as weeds in ricefields and irrigation channels.

MARANTACEAE

30 genera: members of the genera *Thalia, Phrynium* and *Donax*, occur in open water and swamps and at the edges of rivers.

Usually rhizomatous perennials. Stems sometimes woody. Leaves in 2 rows, differentiated into a sheath, and an often swollen short petiole, and a blade with several parallel nerves diverging obliquely from the midrib. Flowers bisexual, zygomorphic, in a terminal or subterminal bracteate spike or panicle. Sepals free, subequal, usually 3. Petals united at the base, usually 3. Only 1 stamen fertile; the rest petaloid, usually larger than the petals, united below; anther 1-locular. Ovary inferior, 1- or 3-locular; ovule solitary; fruit fleshy or a capsule.

Schumann, K. Marantaceae in Engler, A. Pflanzenreich 11, (IV.48): (1902)

1A Flowers in short, dense spikes **Phrynium**

1B Flowers in lax, spreading panicles
  2A Ovary 3-locular; petaloid staminodes 2 **Donax**

  2B Ovary 1-locular; petaloid staminode 1 **Thalia**

Donax Lour., Fl. Cochinch. 14 (1790)
Fig. 144A.
Stems woody, much branched, up to 3 m high. Leaves with ovate or oblong blades; sheaths long, closed throughout their length. Flowers in lax panicles; bracts persistent, narrowly lanceolate. Petaloid staminodes unequal, 2 per flower, small, white. Ovary 3-locular.

c. 6 species: Indo-Malaysia. *D. arundastrum* Lour. forms thickets in the edges of rivers from Burma and Siam, south to Borneo.

Phrynium Willd., Sp. Pl. I: 17 (1797), emend. K. Schum. in Engler, Pflanzenreich 11 (IV.48): 52 (1902)
Fig. 144B.
Stems not woody, up to 1.5 m high. Leaf-blades oblong-elliptic, 16 to 25 cm long; sheaths open to the base. Flowers in dense short spikes; spikes sessile in the leaf axils. Bracts lanceolate to obovate, reddish. Petaloid staminodes 2 per flower, pink or white. Ovary 3-locular.

c. 20 species: mainly in forests of S. E. Asia; 2 species occur in Tropical Africa, one of which, *P. confertum* (Bentham) K. Schum., is common in swamps in W. Africa.

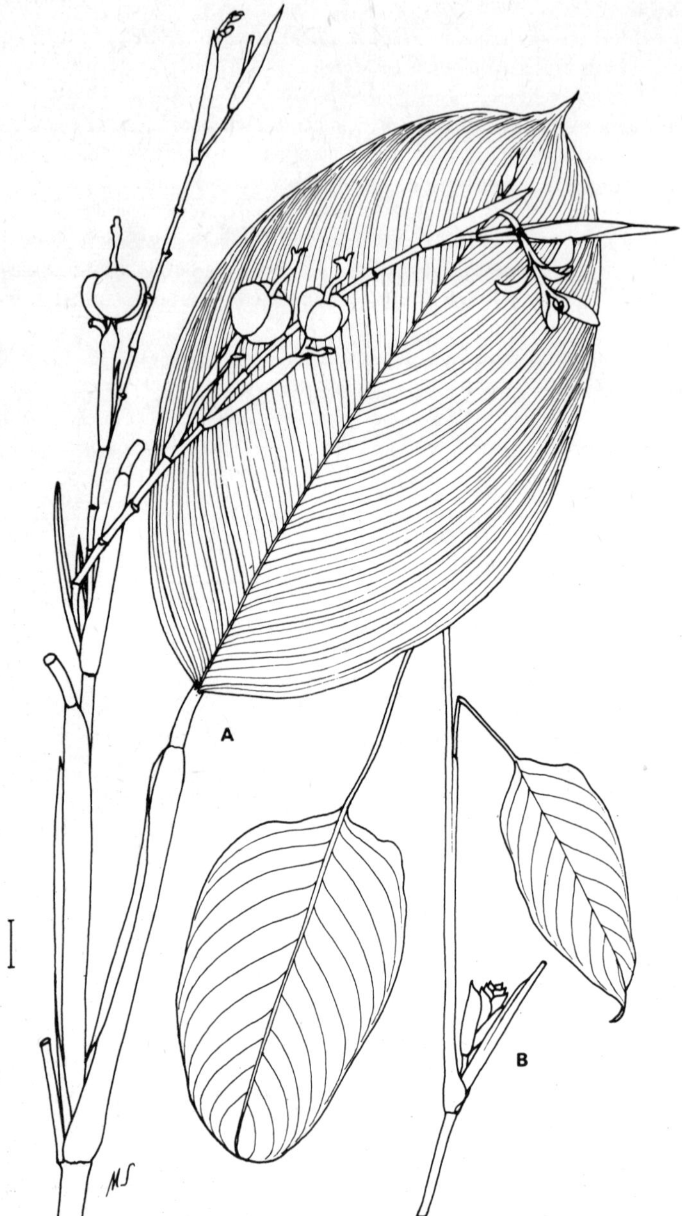

Fig. 144. A. *Donax arundastrum* Lour.: flowering stem (1 cm);
B. *Phrynium confertum* (Benth.) K. Schum.: flowering stem.

Fig. 145. *Thalia geniculata* L.: a, flowering stem (1 cm); b, flower (1 cm).

Thalia L., Sp. Pl. 3 (1753)
Fig. 145.
Stems not woody, up to 2 m high. Leaf blades ovate-lanceolate; sheaths closed throughout most of their length. Flowers in sessile pairs in lax panicles. Bracts usually deciduous. Petaloid staminodes 1 per flower, usually purplish. Ovary 1-locular.

c. 7 species: tropical America; one, *T. geniculata* L., also in Africa. Most species occur in swamps.

MAYACACEAE

1 genus.

Mayaca Aublet, Hist. Pl. Guiane Franç. 1: 42, pl. 15 (1775), [*Coletia* Vell.] Fig. 147A.
Herbs. Stems submerged, floating or creeping. Leaves spirally arranged, simple, linear, capillary, 1-nerved, notched at apex. Flowers bisexual, solitary or few in leaf axils, long stalked; flower stalk subtended by membranous bracts, usually reflexed after flowering. Perianth 3-merous, double; the outer sepal-like, of 3, lanceolate, persistent segments; the inner petal-like, of 3, ovate or obovate segments, white, pink or violet. Stamens 3, hypogynous, opposite sepal-like segments; filaments free, elongate; anthers 4-locular, basifixed, opening by an apical pore or pore-like slit. Ovary 1-locular; style elongate; fruit a 3-sided capsule, opening by 3-valves, each bearing a placenta in the middle; seeds numerous.

Bruggen, H. W. E. van, Mayaca spp. Het Aquarium (Den Haag) 28: 151–154 (1958)
Lourteig, A. Mayacaceae. Not. Syst. (Paris) 14: 234–248 (1952)
Lourteig, A. Sur Mayaca sellowiana (Mayacaceae). Taxon 17 (6): 742–743 (1968)

c. 10 species: America, from S. E. USA to Paraguay and Angola, Africa. Small mat-forming herbs in shallow water or on mud.

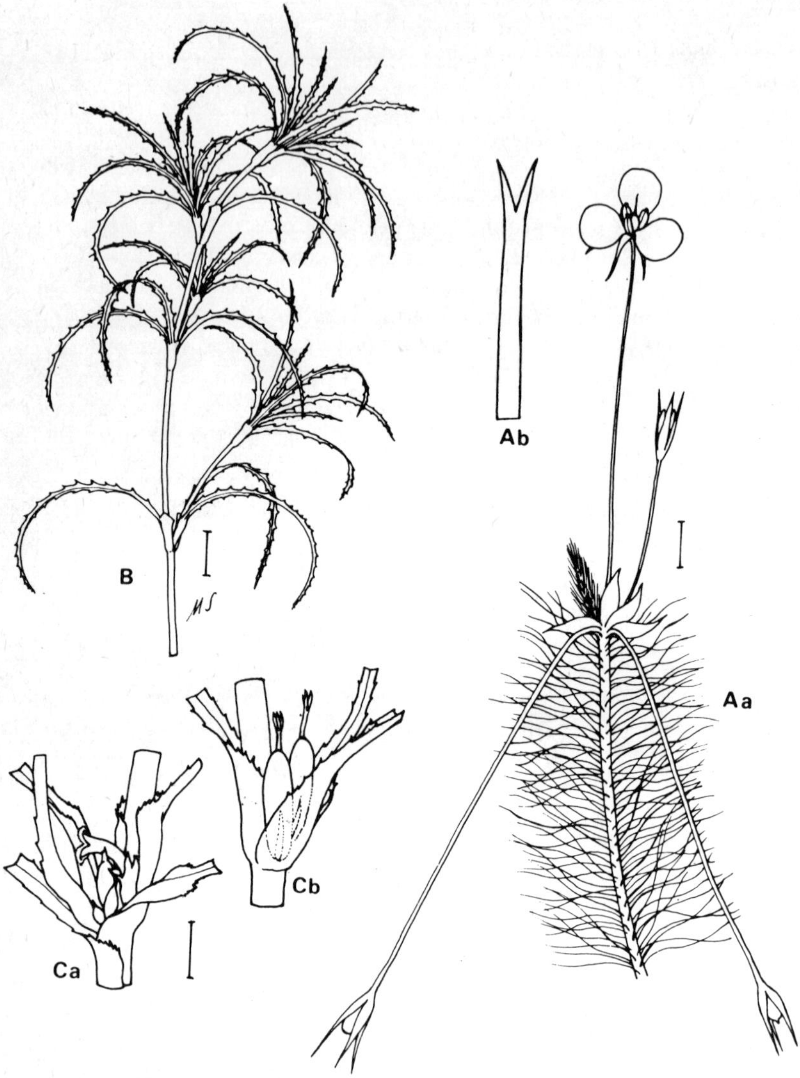

Fig. 147. A. *Mayaca sellowiana* Kunth: a, habit (1 cm); b, leaf apex;
B. *Najas indica* Cham.: habit (1 cm);
C. *Najas graminea* Del.: a, male flowers; b, female flowers (1 mm), after Mason.

MELASTOMACEAE

c. 200 genera; 2 contain aquatics, the family description is applicable to the aquatics.

Annual or perennial herbs. Stem erect, creeping or floating, simple or branched, often swollen below. Leaves opposite, simple, entire or toothed, stalked or sessile, 1-, 3-, 5-, or 7-nerved or bract-like and without nerves; stipules absent. Inflorescence cymose or flowers solitary. Flowers bisexual, actinomorphic, perigynous; hypanthium cup-shaped, bearing sepals, petals and stamens on the upper rim. Sepals 4 or 5, usually triangular. Petals 4 or 5, white, pink or purple, rolled in bud. Stamens usually twice as many as petals or fewer by abortion; anthers opening by terminal pores; connectives prolonged into 2-lobes. Ovary 2-, 3- or 4-locular, surrounded by hypanthium; style 1, elongate; stigma capitate; ovules numerous, axile; fruit a capsule, opening by 2, 3 or 4 valves.

1A Petals ovate, with acute to acuminate apex; anthers isomorphic; inflorescence freely branched and many-flowered
Nepsera

1B Petals obovate, with blunt apex; anthers dimorphic; inflorescence usually small and few-flowered
Acisanthera

Acisanthera P. Br., Hist. Jamaic. 217 (1756)
Fig. 146A.
Annuals or rarely perennials, usually glandular hairy above. Leaves entire or toothed, stalked or sessile, with 1 or 3 nerves or reduced and bract-like. Sepals 4 or 5, equalling or exceeding hypanthium in length. Petals small, usually obovate, white, pink or purple. Stamens somewhat dimorphic; the inner series smaller and in some species imperfect; the connectives of the outer anthers 2-lobed, rounded to linear, prolonged below the pollen sacs; the connectives of the inner anthers shortly 2-lobed, not or slightly prolonged. Ovary free from hypanthium; capsule 2-, 3- or 4-valved.

c. 35 species: tropical America and West Indies. Most species grow in wet places. Some species such as *A. bivalvis* (Aubl.) Cogn., *A. fluitans* Cogn., *A. inundata* (DC.) Triana, *A. limnobious* (DC.) Triana and *A. nana* Ule are found flowering in still or flowing water.

Nepsera Naud., Ann. Sci. Nat. Bot. III, 13: 28 (1850)
Fig. 146B.
Annual. Stems slender, erect, up to 1 m high, 4-angled, often swollen below, freely branched. Leaves stalked; stalk 5 to 10 mm long; leaf blades ovate to ovate lanceolate, 2 to 6 cm long, 5- to 7-nerved, acute at apex, rounded or

Fig. 146. A. *Acisanthera nana* Ule: a, habit (1 cm); b, flower (1 mm); c, different kinds of stamen;
B. *Nepsera aquatica* (Aubl.) Naud.: a, habit (2 cm); b, flower (1 mm); c, stamens.

cordate at base; margin finely toothed. Sepals 4, triangular, *c.* 3 mm long, about equalling hypanthium. Petals 4, 5 to 7 mm long, ovate to ovate-lanceolate, acute to acuminate. Stamens 8, isomorphic, differing slightly in size; connective with 2, erect, slender appendages, prolonged below the pollen sacs. Ovary free from hypanthium; capsule 3-valved.

1 species, *N. aquatica* (Aubl.) Naud.: tropical America and West Indies. It is a common and widespread species found in a variety of wet habitats and is frequently found in shallow water.

MENYANTHACEAE

5 genera: all aquatic.

Usually perennial, with tufted rootstocks or horizontal creeping rhizomes. Leaves alternate; blades orbicular to linear, or trifoliate; petioles sheathing at the base. Flowers bisexual, actinomorphic, often heterostylous, in simple or branched cymes or racemes, or in dense heads or clusters. Sepals 5, united at the base, persistent, sometimes adnate to the ovary below. Petals 5, united at the base, caducous, often heavily fimbriate inside. Ovary superior, 1-locular with 2 parietal placentas; fruit a capsule, dehiscent or somewhat fleshy and indehiscent.

1A Leaf-blade floating or submerged, on flaccid petiole
  2A Inflorescence a cluster of pedicellate flowers; fruit indehiscent or splitting irregularly
    Nymphoides
  2B Inflorescence an open panicle; fruit dehiscent by 4 valves
    Villarsia
1B Leaf-blade emersed, on erect petiole
  3A Leaf-blade trifoliate
    Menyanthes
  3B Leaf-blade simple entire or toothed
    4A Leaf-blade linear to spathulate
      Liparophyllum
    4B Leaf-blade elliptic to reniform
      5A Plant with creeping rhizome, leaf-blade retuse at apex, petals not fimbriate at base
        Nephrophyllidium
      5B Plant tufted, sometimes stoloniferous, leaf-blade not retuse at apex, petals fimbriate at base
        Villarsia

Liparophyllum Hooker fil., in Hook. London Jour. Bot. 6: 472 (1847)
Fig. 148A.
A small creeping perennial with thick rhizome and tufted leaves. Leaf blades emersed, fleshy, linear to linear-spathulate, obtuse. Inflorescence terminal, 1-flowered, shorter than the leaves; flowering stalk leafless. Sepals linear, united at the base, about equalling the petals. Petals white, with a broad, thin margin, undulate, united at the base for one third of their length, *c.* 5 mm long. Stamen filaments short and broad. Style very short; capsule globose, fleshy or succulent, indehiscent.

1 species, *L. gunnii* Hooker fil. in Tasmania and New Zealand. Occurs in mountain bogs.

Fig. 148. A. *Liparophyllum gunnii* Hooker fil.: a, habit (1 cm); b, flower; c, petal-tube opened (1 mm);
B. *Villarsia ovata* (L. fil.) Vent.: habit (1 cm).

Fig. 149. A. *Menyanthes trifoliata* L.: habit (1 cm);
B. *Nephrophyllidium crista-galli* (Menz.) Gilg.: habit (1 cm).

Menyanthes L., Sp. Pl. 145 (1753)
Fig. 149A.
Perennial with thick creeping rhizome. Leaf blades emersed, trifoliate; leaflets obovate to elliptic, entire. Inflorescence a 10- to 20-flowered raceme on a leafless stalk. Pedicels 5 to 10 mm. Flowers heterostylous. Sepals ovate, somewhat recurved, united at the base. Petals united at the base, much fimbriate inside, outside pink, inside paler or white, about 10 mm long. Stamen filaments short and broad. Styles long or short; capsule subglobose, opening by 2 valves.

1 species, *M. trifoliata* L.: circumpolar, S. to Marocco. Occurs in bogs and the shallow water of lakes and canals.

Nephrophyllidium Gilg. in Engler & Prantl, Nat. Pflanzenfam. 4, (2): 105 (1895), [*Fauria* Fernald]
Fig. 149B.
Perennial with thick creeping rhizome. Leaf blades emersed, reniform, crenate, often retuse at apex. Inflorescence a few-(about 3-)flowered raceme; flowering stalk leafless; pedicels 3 to 10 mm. Sepals narrowly triangular, united and adnate to ovary at the base. Petals united at the base forming a tube slightly shorter than the spreading lobes; petal-lobes white with a wavy crest along the median nerve and margin. Stamen filaments thin and elongate. Style short, about 4 mm long; capsule with free part narrowly conic, splitting into four at the apex, much exceeding the sepals.

1 species, *N. crista-galli* (Menz.) Gilg., occurs in bogs around the N. Pacific from Honshu (Japan) to Washington (N. W. America).

Nymphoides Hill, Brit. Herb.: 77 (1756), [*Limnanthemum* S. S. Gmelin]
Fig. 150.
Rhizomatous perennials, sometimes annuals. Leaf blades floating, broadly ovate to orbicular, entire or crenate, greenish above and often dark below. Inflorescence on a lax scape, with flowers pedicellate in clusters, supported by floating cauline leaves, so that the pedicels may appear to arise from the petiole (or base of the leaf blade). Adventitious root tubers sometimes present at the base of the inflorescence. Flowers bisexual or unisexual, often heterostylous, emersed at anthesis, usually submerged in bud and fruit. Sepals united at the base, narrowly triangular, appressed. Petals yellow or white, united at the base, often winged at the margins and much fimbriate inside, 10 to 30 mm long. Stamen filaments very short; capsule indehiscent.

Ornduff, R. Neotropical Nymphoides, Meso-american and W. Indian species. Brittonia 21: 346–352 (1969)
Raynal, A. Répartition géographique des Nymphoides (Menyanthaceae) Africains et Malgaches. Mitt. Bot. Staats. München 10: 122–134 (1971)

c. 20 species: cosmopolitan. In still or slow flowing water. *Nymphoides* and *Villarsia* are distinguished largely on habit and inflorescence type and in Australia at least these distinctions are somewhat unsatisfactory. *Nymphoides indica* (L.) O. Kuntze and *N. cristata* (Roxb.) O. Kuntze are common rice-field weeds.

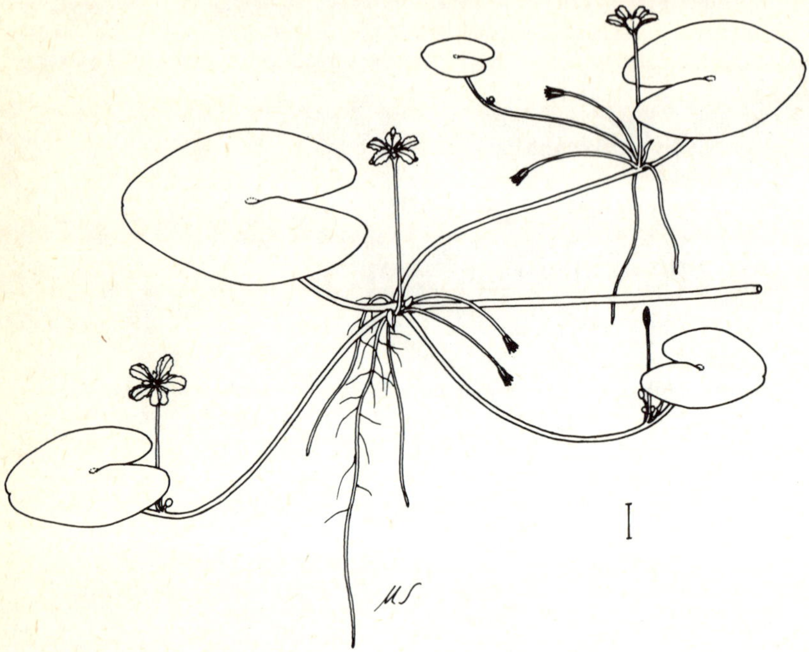

Fig. 150. *Nymphoides cristata* (Roxb.) O. Kuntze: habit (1 cm).

Villarsia Vent., Choix Plant. 9 (1803), nom cons.
Fig. 148B.
Tufted annuals or perennials, sometimes stoloniferous. Leaf-blades submerged, emersed or floating, elliptic to reniform, rounded to cordate at the base; margin sometimes toothed; petioles winged. Inflorescence usually an open panicle (flowers occasionally in dense heads), with leafy bracts at the base. Flowers homostylous or heterostylous. Sepals united at the base, sometimes enlarging in fruit. Petals united at the base conspicuously fimbriate in the throat, wavy on the edges, yellow or white, up to 25 mm long. Stamen filaments very short or almost lacking. Capsule splitting into four or more valves, not or scarcely exceeding sepals.

Aston, H. I. The genus Villarsia (Menyanthaceae) in Australia. Muelleria 2, 1: 3 (1969)

c. 16 species: all aquatic: Australia to S. E. Asia; 1 species *V. ovata* (L.fil.) Vent. in S. Africa.

NAJADACEAE

1 genus.

Najas L., Sp. Pl. 1015 (1753)
Fig. 147B, C. (see page 322)
Monoecious or dioecious, obligately submerged annuals or perennials. Stems slender, either sparsely branched and diffuse or much branched and condensed. Leaves appearing opposite, pseudo-whorled or in bunches (crowded in leaf axils), simple, linear; margin usually toothed; base sheathing, occasionally auriculate. Flowers unisexual, small, axillary, solitary or in small groups. Male flowers enclosed in a perianth-like, membranous sheath (spathe), ending in 2 thickened lips; stamen 1; anthers subsessile, 1- or 4-locular. Female flowers naked or surrounded by a membranous sheath (spathe); ovary of 1 carpel, tapering into a short style; stigmas 2 to 4, linear; ovules solitary; fruit a nutlet, enclosed in a loose, separable, membranous coat.

Backman, A. L. Najas minor All. in Europa einst und jetzt. Acta Bot. Fenn. 48: 1–32 (1951)
Clausen, R. T. Studies in the genus Najas in the northern United States. Rhodora 38: 333–345 (1936)
Rantzien, H. Horn Af. Notes on some tropical African species of Najas in the Kew Herbarium. Kew Bull. 1952: 29–40 (1952)
Rendle, A. B. in Engler, A. Pflanzenreich 7 (IV. 12): 1–19 (1901)

c. 50 species: cosmopolitan. It is found in a large variety of aquatic habitats, often growing at depths of 5 m or more. Many species are found in ricefields. On the whole *Najas* is not considered to be of great economic importance although it may bc a nuisancc in irrigation ditchcs. Many spccics of *Najas* are readily eaten by *Tilapia melanopleura*, an economically important fish.

NELUMBONACEAE

1 genus.

Nelumbo Adanson, Fam. Pl. 2: 76 (1763)
Fig. 151.
Large perennial, aquatic herbs with milky latex. Roots borne at nodes of stem. Stem dimorphic, with slender horizontal vegetative rhizomes and thickened storage rhizomes. Leaves peltate, when immature floating, when mature raised high above the water surface; petiole up to 2 m or more long, bearing numerous spines; blade reniform to orbicular, 10 to 60 (100) cm diameter, glaucous, nerves prominent but without spines. Flowers 10 to 20 (35) cm diameter; perianth segments 14 to 30, the outer 2 to 5 persistent and sepal-like, the inner caducous, white, yellow or red. Stamens up to 200, with conspicuous fleshy terminal appendage; pollen tricolpate. Carpels (9) 12 to 30 (40), borne singly in cavities in the pithy obconical receptacle; fruit a very hard-walled nut.

Godwin, H. and Willis, E. H. The viability of lotus seeds (Nelumbium nucifera Gaertn.). New Phytol. 63: 410–412 (1964)
Hall, T. F. and Penfound, W. T. The biology of the American lotus (Nelumbo lutea (Willd.) Pers. Amer. Midl. Nat. 31: 744–758 (1944)
Kanna, P. Morphological and embryological studies in the Nymphaeaceae. II. Austral. Journ. Bot. 13: 379–387 (1965)
Wood, C. E. The genera of Nymphaeaceae and Ceratophyllaceae in the south-eastern United States. Journ. Arnold. Arb. 40: 94–112 (1959)

2 species: *N. pentapetala* (Walt) Fernald [*N. lutea* (Willd.) Pers.] (perianth white to yellow, fruits nearly spherical) is native in the West Indies, C. America and the southern and eastern parts of N. America; *N. nucifera* Gaertn. [*N. speciosa* Willd.] (perianth white to red, fruits ellipsoidal) is native in E. and S. E. Asia, Australia, India and Persia. Both species have been introduced into many different countries and have become established in many of the warmer parts of the world. *Nelumbo* grows best in organic soil in water up to 2.5 m deep. It spreads as much as 20 m radially each year by slender runners with internodes up to 2 m long. In favourable conditions it may crowd out other aquatic vegetation and become a pest. The storage rhizomes and fruits are edible and in parts of Asia *N. nucifera* is cultivated as a crop. For the culture and economics see Malik, H. C. Indian Farming 11 (8): 23–24 (1961).
N. nucifera is considered sacred by Hindus.

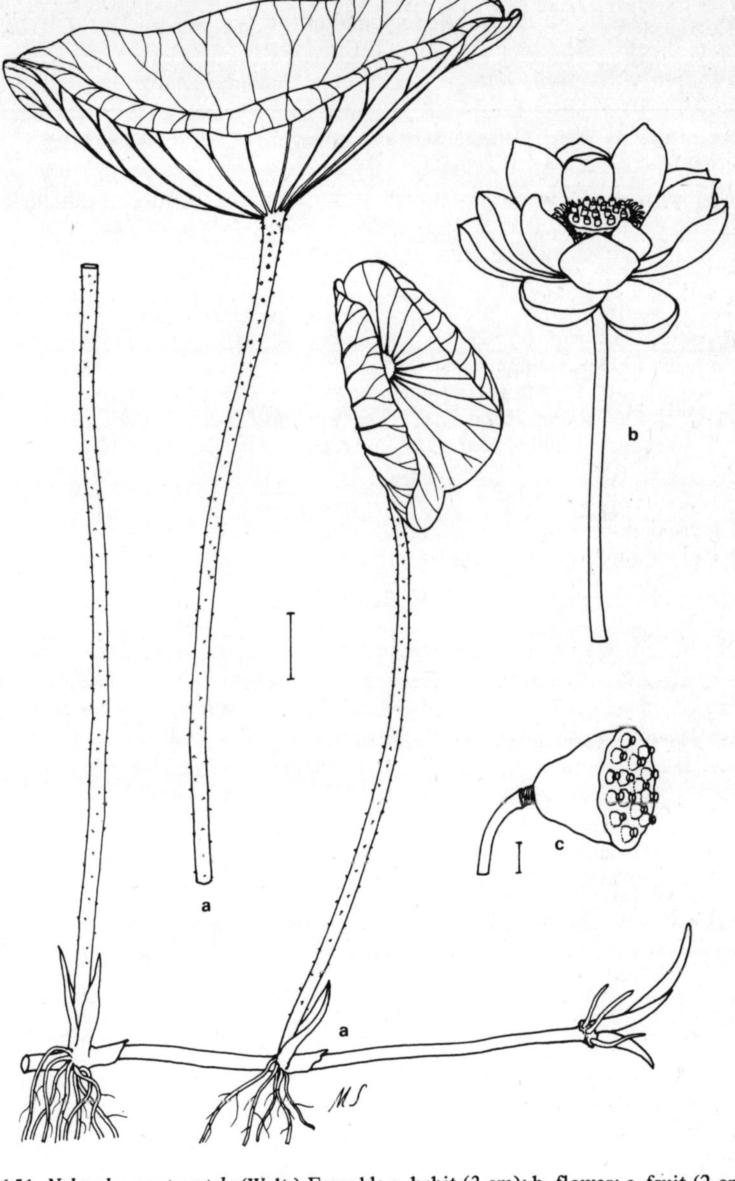

Fig. 151. *Nelumbo pentapetala* (Walt.) Fernald: a, habit (3 cm); b, flower; c, fruit (2 cm).

NYMPHAEACEAE

6 genera: *c.* 65 species: cosmopolitan. The genera *Brasenia, Cabomba* and *Nelumbo* in this treatment are not included in the Nymphaeaceae.

Acaulescent herbs. Stem a submerged corm or rhizome. Leaves alternate, cordate or peltate, linear to orbicular. Flowers solitary, axillary. Sepals 4 to 6, free or adnate to ovary. Petals absent or numerous, in many series, often passing into staminodes or stamens. Stamens numerous, free or adnate to petals. Ovary 3 or more carpels, united or partially united; stigmas united into a disc with radiating rays; fruit a many seeded, irregularly dehiscent, berry; seeds with or without aril.

Li, H. L. Classification and phylogeny of Nymphaeaceae and allied families. Amer. Midl. Nat. 54: 33–41 (1955)

Vinogradov, I. The systematics of the family Nymphaeaceae. Zap. Tsent-Kavkaz Otd Vses Bot. Obshchest 2: 5–1 (1967) – in Russian

Wood, C. E. The genera of the Nymphaeaceae and Ceratophyllaceae in the Southeastern United States. Journ. Arnold Arb. 40: 94–112 (1959)

1A Leaves peltate when mature; petiole with spines
- 2A Leaf-blade bearing spines on both surfaces; leaf margin not turned upwards; staminodes absent
  Euryale
- 2B Leaf-blade bearing spines on lower surface only; leaf margin turned upwards; staminodes replacing inner stamens
  Victoria

1B Leaves not peltate; petiole without spines
- 3A Petals absent
  Ondinea
- 3B Petals present
  - 4A Stamens adnate to petals; staminodes absent; leaves usually entirely submerged
    Barclaya
  - 4B Stamens not adnate to petals; staminodes present; leaves mostly floating or emersed
    - 5A Sepals 5 or 6; ovary superior; petals not spreading, scale- or stamen-like; seeds without aril
      Nuphar
    - 5B Sepals 4: ovary semi-inferior; petals spreading, showy; seeds with bell-shaped aril
      Nymphaea

Barclaya Wallich, Trans. Linn. Soc. London 15: t.18, 443 (1827)
Fig. 152.
Stem dimorphic corm-like and with slender horizontal rhizomes. Leaf blades linear to orbicular, cordate at base, usually entirely submerged but occasionally emersed for short periods. Sepals 4 or 5, hypogynous. Petals 8 to numerous, perigynous, united into a tube below and free above; inner petals pink or red. Stamens numerous, adnate to petals; staminodes absent. Ovary up to 10 carpels, united, inferior; stigmatic disc conical; seeds covered in fine spines.

Heine, H. Barclaya longifolia Wallich, eine neueingeführte wertvolle Aquarienpflanze. Die Aquarien- und Terrarien-Zeitschr. 11 (10): 314–317 (1958)
Hu, S.-Y., Studies in the Flora of Thailand. 48. The genus Barclaya (Nymphaeaceae). Dansk. Bot. Arkiv. 23: 533–540 (1968)

c. 4 species: tropical S. E. Asia. In pools and streams, usually in forested regions. Frequently cultivated as a decorative aquarium plant.

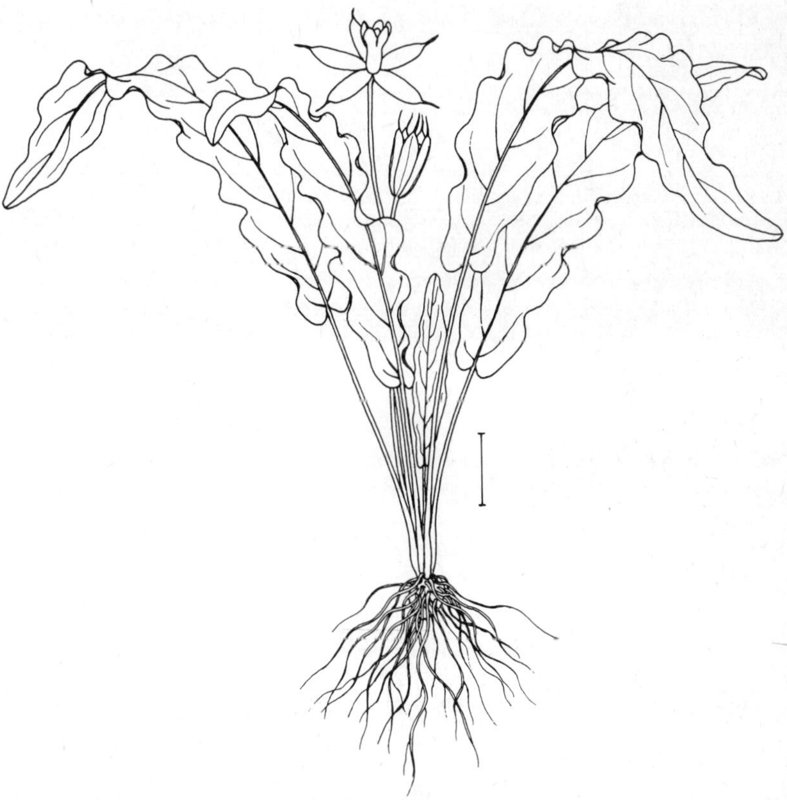

Fig. 152. *Barclaya longifolia* Wallich: habit (5 cm).

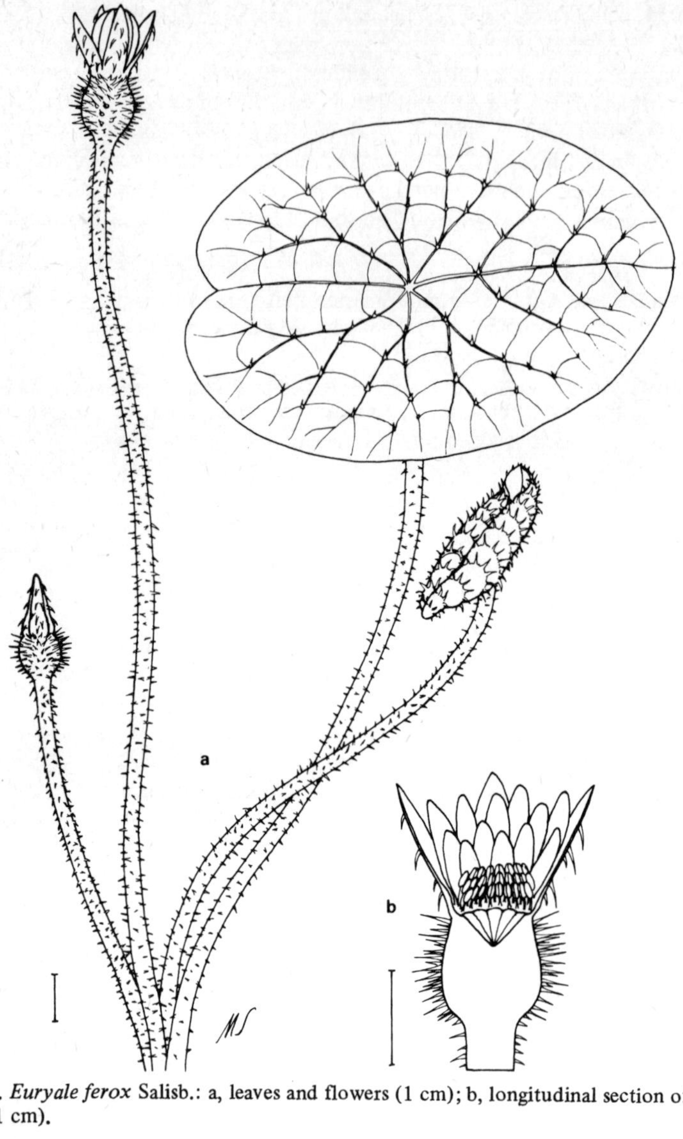

Fig. 153. *Euryale ferox* Salisb.: a, leaves and flowers (1 cm); b, longitudinal section of flower (1 cm).

Euryale Salisb., Ann. Bot. (Koenig and Sims) 2: 73 (1805)
Fig. 153.
Annual or perennial, densely prickly herb. Stem a short, thick rhizome. Petiole with spines; mature leaf blades floating, peltate, orbicular, up to 1 m or more diam.; both surfaces of leaf blade with spines; margin not upturned. Flowers usually cleistogamous partially or entirely submerged, 5 to 6 cm diam. Sepals 4, epigynous. Petals numerous, epigynous, shorter than sepals, blue, violet or

red. Stamens numerous, grouped in fascicles of 8; staminodes absent. Ovary of 8 carpels, united; stigmatic disc obconical; fruit crowned by persistent sepals and covered with stout prickles; seeds with pulpy aril.

Khanna, P. Morphological and embryological studies in Nymphaeaceae. I. Eurale ferox Salisb. Proc. Indian Acad. Sci. ser. B, 59: 237–243 (1964)

Sobolewska, M. Eurale ferox Salisb. in the pleistocene of Poland. Acta Palaeobotanica 11 (1): 13–20 (1970)

1 species: *Euryale ferox* Salisb. in N. India, Manchuria, China, Japan, Hainan and Taiwan. In pools and slowly flowing water rarely more than 1 to 5 m deep. Fruits and seeds are eaten in S. Asia. *E. ferox* is decorative and widely cultivated. Fossil material is known from Europe.

Nuphar, J. E. Smith in Sibth. and Smith, Fl. Graec. Prodr. 1: 361 (1809), nom. cons.

Fig. 154.

Stem a stout, creeping rhizome. Leaf blades when mature ovate to orbicular, with deep sinus at base; floating and emersed leaves leathery; submerged leaves thin and delicate. Sepals *c.* 6, hypogynous, the outer green, the inner yellow, tinged red or green. Petals numerous, hypogynous, linear, to oblong, scale- or stamen-like, bearing a nectary on abaxial surface. Stamens numerous, inserted with petals on the receptacle below the ovary. Ovary 5 to 20 (30) carpels, united; stigmatic disc concave; fruit ripening above water; seeds without aril.

Beal, E. O. Taxonomic revision of the genus Nuphar Sm. of N. America and Europe. Journ. Elisha Mitchell Scient. Soc. 72: 317–346 (1956)

De Poe, C. E. and Beal, E. O. Origin and maintenance of clinal variation in Nuphar (Nymphaeaceae). Brittonia 21 (1): 15–28 (1969)

Heslop-Harrison, Y. Biological Flora of the British Isles: Nuphar Sm. Journ. Ecol. 43: 342–364 (1955)

7 to 20 species: N. Temperate Zone. *Nuphar* is taxonomically difficult and is not well understood. It is found in still and slowly flowing water up to about 1.5 m deep. The seeds are eaten by water birds.

Nymphaea L., Sp. Pl. 510 (1753) partim. emend. J. E. Smith in Sibth. and Smith, Fl. Graec. Prodr. 1: 360 (1809) nom cons.

Fig. 155.

Stem a slender or stout, creeping rhizome or corm-like. Leaf blades when mature usually floating, ovate to orbicular, cleft at base; submerged leaves often present, filmy and delicate; aerial leaves exceptional, similar to floating leaves in shape. Sepals 4, green or streaked with red. Petals numerous, showy, in many series passing into staminodes or stamens. Stamens numerous, with or without appendaged connective; staminodes frequent. Ovary of 3 to 35 carpels, united or partially united; stigmatic disc obconical, concave or almost flat; fruit ripening underwater; seeds with bell-shaped aril.

Conard, H. S. The Waterlilies. A monograph of the genus Nymphaea. Carnegie Inst. Washington, Publ. No. 4, 1–279 (1905)

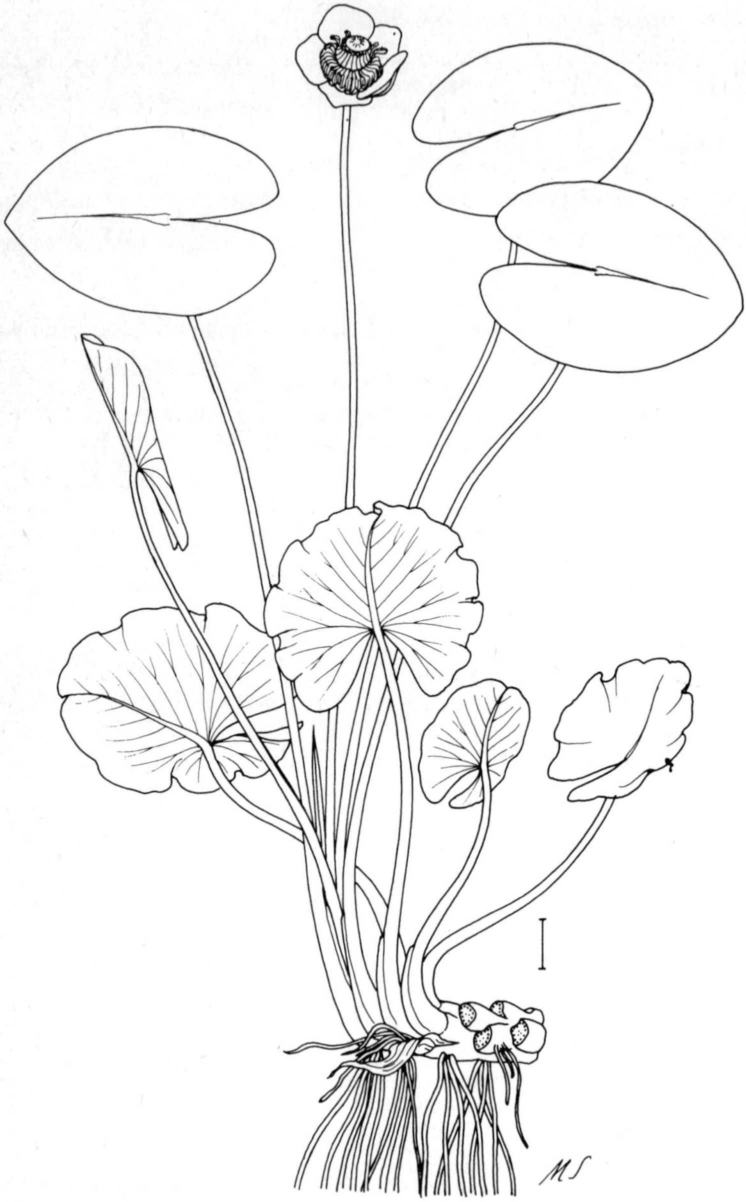

Fig. 154. *Nuphar lutea* (L.) Sibth. and Smith: habit (2 cm).

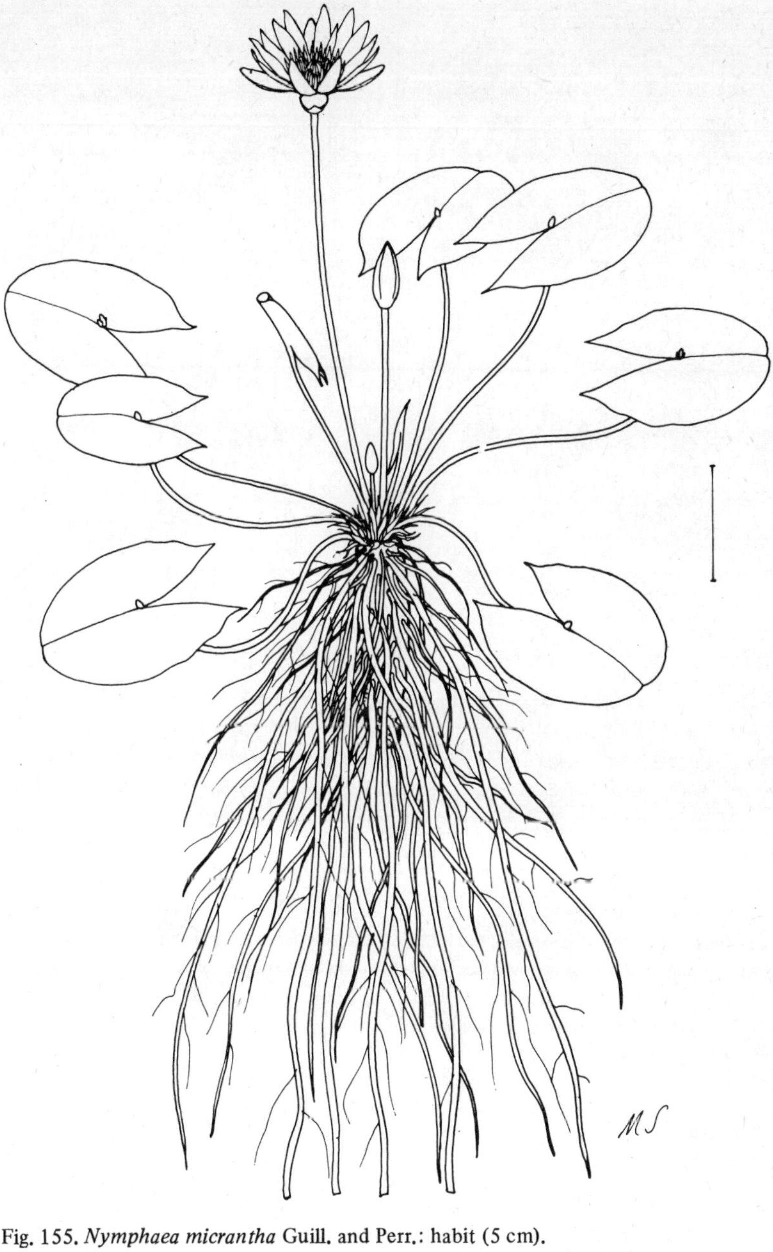

Fig. 155. *Nymphaea micrantha* Guill. and Perr.: habit (5 cm).

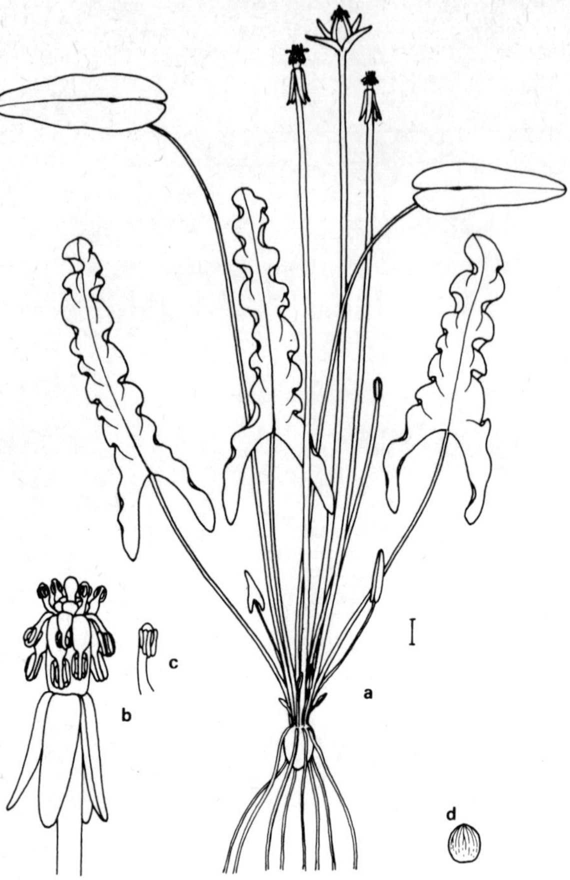

Fig. 156. *Ondinea purpurea* Hartog: a, habit (2 cm); b, flower; c, stamen; d, seed, after den Hartog.

Gonçalves, M. L. Revisao das Nymphaeaceae de Moçambique. Junta 25: 57–60 (1961)
Heslop-Harrison, Y. Biological Flora of the British Isles: Nymphaea L. em. Sm. (nom. cons.) Journ. Ecol. 43: 719–734 (1955)
Irvine, F. R. and Trickett, R. S. Water lilies as food. Kew Bull. 1953 (3): 363–370 (1953)
Radics, F. A. Revision of the Nymphaea material in the Hungarian Natural History Museum. Ann. Mus. Hung. 59: 135–145 (1964)
Trickett, R. S. A new tropical American waterlily Nymphaea belophylla. Kew. Bull. 26 (1): 29–31 (1971)

c. 40 species: through introductions now cosmopolitan. In still and flowing water up to 2 m or more deep in a wide variety of habitats. Many species have a dormant period and live through unfavourable conditions (cold or drought) as a dormant corm or rhizome. Many species and hybrids are cultivated as

Ondinea Hartog, Blumea 18 (2): 413 (1970)
Fig. 156.
Stem corm-like. Submerged leaf blades sagittate, translucent, with entire, wavy margin; floating leaf blades elongate-elliptic, leathery, with basal sinus. Sepals 4, linear or slightly spathulate, 9 to 17 mm long, 1.5 to 3 mm wide, pink to purple. Petals absent. Stamens 15, small and erect above, large and reflexed below; staminodes absent. Ovary 3 to 5 carpels, united, superior; stigma of 3 to 5, laterally united, lobes surrounding a central depression; seeds smooth, without aril.

Hartog, C. den, Ondinea, a new genus of Nymphaeaceae. Blumea 18 (2): 413–416 (1970)

1 species, *O. purpurea* Hartog; N. W. Australia. It occurs in small rivulets up to 60 cm deep which dry out for 5 to 7 months a year. During the dry period it aestivates as corms.

Victoria Lindley, Monog. 3 (1837)
Fig. 157.
Annual or rarely perennial, densely prickly herbs. Stem a short, thick rhizome. Mature leaves floating, peltate, orbicular, up to 2 m or more diam.; lower surface of leaf blade with spines, upper surface without spines; leaf margin turned upwards, forming a continuous rim around the leaf; juvenile leaves linear, sagittate, ovate to orbicular. Flowers borne above water surface, up to 40 cm diam. Sepals 4, with or without spines. Petals numerous, most longer than sepals, red, pink or white, passing into staminodes and stamens. Stamens numerous, up to 200. Ovary inferior, of numerous carpels, united; stigmatic disc obconical; seeds with pulpy aril.

Gessner, F. Zur Blattentwicklung von Victoria amazonica (Popp.) Sowerby. Ber. Deutsch. Bot. Ges. 82: 603–608 (1970)
Wagner, J. Die Königin der Seerosen, Brehm-Bücherei, Wittenberg Lutherstadt (1956)

2 species: S. America; *V. amazonica* (Pöpp.) Sowerby. Amazon Basin, Guianas and Mato Grosso; *V. cruziana* d'Orb., Parana, Uruguay, and N. Argentina. Occurs in slowly flowing rivers and backwaters. The seeds are very rich in starch and are locally called "Mays del Agua". Flour obtained from the seeds is used locally. Both species and their hybrid are widely cultivated as decorative plants.

decorative plants. In various parts of the world the seeds and rhizomes are eaten. The rhizomes are also used as a mordant in dyeing and tanning. Various parts of the plant are used locally in medicine but *Nymphaea* is not considered of primary importance. The seeds are a valuable part of the diet of many water birds.

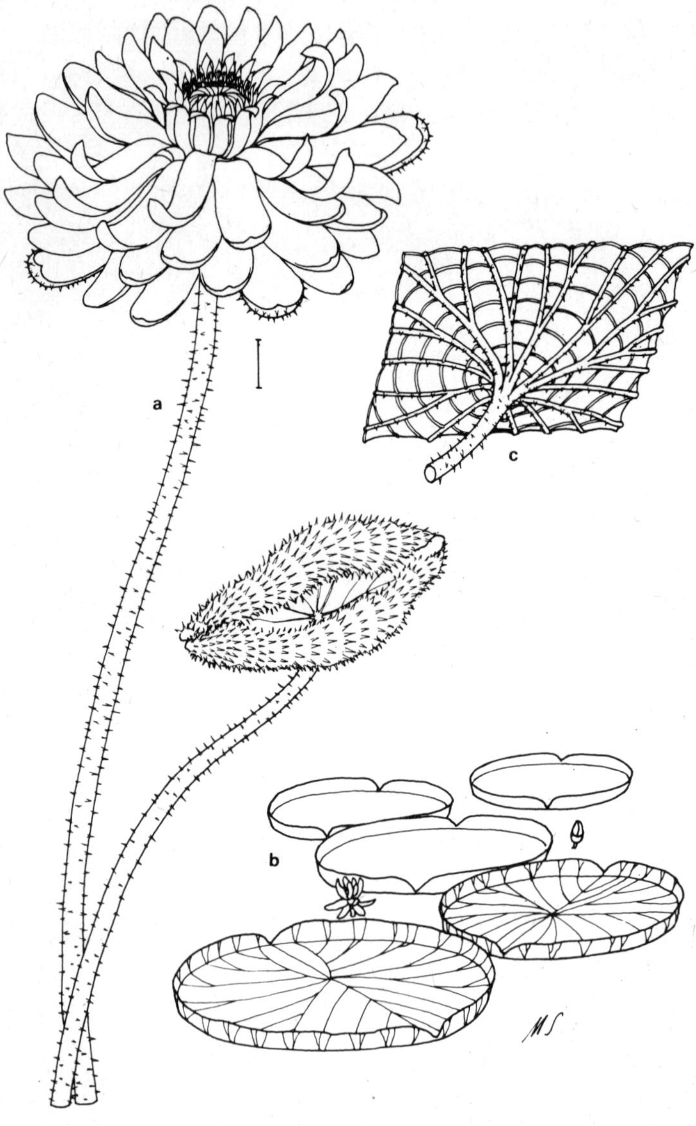

Fig. 157. *Victoria amazonica* (Pöppig) Sowerby: a, flower and young leaf (3 cm); b, habit; c, undersurface of leaf.

ONAGRACEAE

21 genera: *Boisduvalia* and *Ludwigia* contain aquatics.

1A Sepals not persisting in fruit; petals 2-lobed **Boisduvalia**

1B Sepals persisting in fruit; petals not 2-lobed **Ludwigia**

Boisduvalia Spach, Hist. Vég. Phan. 4: 383 (1835)
Fig. 158A.
Annuals. Stems erect or decumbent, branched from base. Leaves opposite below, alternate above, subsessile, elliptic to linear; margin minutely toothed or entire. Flowers bisexual, actinomorphic, borne singly in axils of leaf-like bracts. Sepals 4, erect, not persisting in fruit. Petals 4, purplish or pinkish to white, 2-lobed. Stamens 8; anthers versatile. Ovary inferior, 4-locular; placentation axile; stigma 4-lobed or capitate; fruit a capsule, dehiscing by longitudinal slits; seeds numerous.

Raven, P. H. and Moore, D. M. A revision of Boisduvalia (Onagraceae). Brittonia 17 (3): 238–254 (1965)

6 species: temperate regions of W. N. and S. America. All species germinate in wet places, *B. glabella* (Nutt.) Walp. and *B. cleistogama* Curran are usually submerged when juvenile. Not known to be of economic importance.

Ludwigia L., Sp. Pl. 118 (1753), [*Isnardia* L., *Jussiaea* L.]
Fig. 158B.
Herbs or shrubs. Stems erect, ascending or creeping and rooting at nodes, when submerged often swollen and spongy or bearing white spongy pneumatophores. Leaves cauline or in rosettes, alternate or opposite, mostly entire; stipules absent or reduced. Flowers bisexual, actinomorphic, solitary, clustered or arranged in an inflorescence. Sepals (3–) 4 (–7), adnate to ovary, persisting in fruit. Petals as many as sepals, or absent, yellow or white, entire, contorted in bud. Stamens usually as many or twice as many as sepals; anthers usually versatile, sometimes basifixed. Ovary inferior, number of locules equal to number of sepals; placentation axile; stigma slightly lobed, hemispherical or capitate; fruit a capsule, dehiscence irregular, by terminal pore or by flaps separating from valve-like top.

Munz, P. A. A revision of the New World species of Jussiaea. Darwiniana 4: 179–284 (1942), The American species of Ludwigia. Bull. Torrey Bot. Club 71: 152–165 (1944), Onagraceae – N. American Flora ser. II, 5: 1–278 (1965)

Fig. 158. A. *Boisduvalia glabella* (Nutt.) Walp.: a, habit (1 cm); b, fruit (1 mm); B. *Ludwigia stolonifera* (Guill. & Perr.) Raven: habit (1 cm).

Raven, P. H. The Old World species of Ludwigia (including Jussiaea), with a synopsis of the genus (Onagraceae). Reinwardtia 6 (4): 327–427 (1963)

c. 75 species: almost cosmopolitan, absent in the frigid zones. About 15 species are truly aquatic, the rest grow in wet or marshy places and may withstand flooding. *L. adscendens* (L.) Hara and *L. octovalvis* (Jacq.) Raven have been reported to be troublesome weeds in irrigated crops in Africa. Some species, such as *L. peploides* (Knuth) Raven, *L. stolonifera* (Guill. and Perr.) Raven, form floating mats of vegetation and contribute to the development of sudd.

* Since going to press we have found another genus with aquatic species: *Oxalis* L. (Oxalidaceae). For information see – Salter, T. M. The genus Oxalis. Journ. S. Afr. Bot. Suppl. 1, 1–355 (1944) and Survival Service Commission Red Data Book. Ang-38-6 (1971).

OXALIDACEAE

Oxalis L. c. 800 species; *O. natans* L. fil. and *O. disticha* Jacq. are aquatic.

Stems slender, floating, arising from an ovoid bulb. Leaves trifoliate, in a floating rosette; leaflets obcordate; petioles c. 1 cm long. Flowers stalked, solitary in leaf axils. Sepals 5, green. Petals 5, free, white or yellow. Stamens 10; anthers at 2 levels. Ovary 5-locular, superior; styles 5, free.

Found in shallow seasonal pools in S. Africa. *O. natans* is a rare and endangered species.

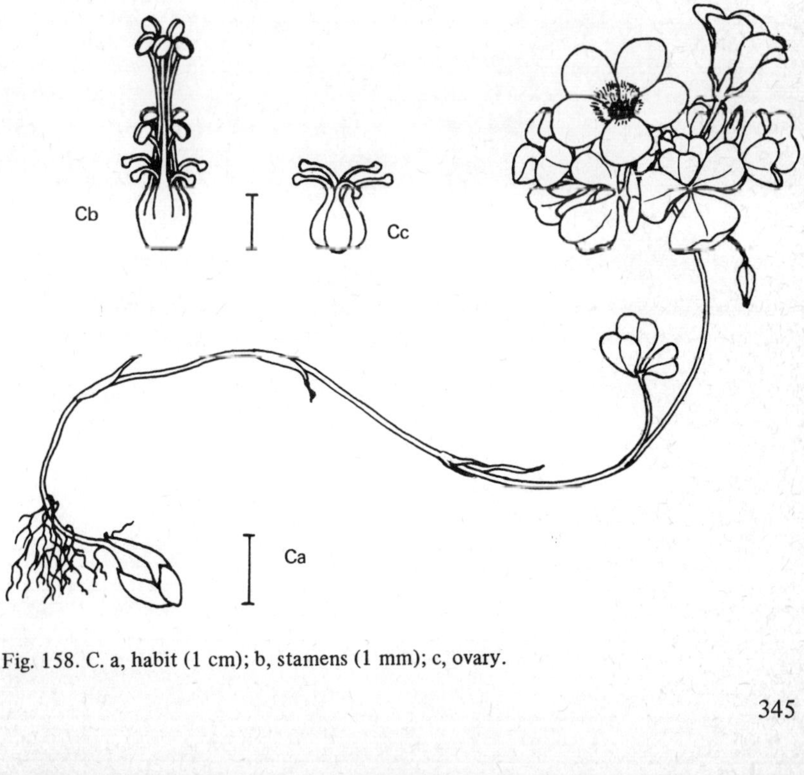

Fig. 158. C. a, habit (1 cm); b, stamens (1 mm); c, ovary.

PHILYDRACEAE

4 genera, of which 1 is aquatic.

Philydrum Banks and Sol. ex Gaertner, Fruct. Sem. Pl. 1: 62 (1788)
Fig. 159A.
Annuals or perennials. Stem short, corm-like. Leaves linear, in 2 ranks, crowded at stem base, equitant, monofacial, 40 to 80 cm long. Flowers zygomorphic, bisexual, sessile, solitary in the axils of spathaceous bracts; bracts reflexed during anthesis; flowering stems 1 m or more high. Perianth petal-like, yellow, 2-seriate, with 4 segments; outer 2 segments 12 to 15 mm long, *c.* 10 mm wide, acute at apex; inner 2 segments united below, *c.* 8 mm long, *c.* 2 mm wide. Stamen 1, inserted at base of outer, adaxial perianth segment; filament flattened; anther 2-locular, spirally twisted; pollen grains in tetrads. Ovary superior, 1-locular; style 1, simple, elongate; fruit a 3-valved capsule, triangular-oblong; ovules anatropous, parietal; seeds numerous.

Skottsberg, C. Bemerkungen über die Philydraceen. Bot. Jahrb. (Engler) 65 (2/3): 253–274 (1932)

1 species, *P. lanuginosum* Banks and Sol.: E and S E. Asia, N. E. Australia. Occurs in ponds, marshes and ricefields, but is also found in swamps and wet savannahs. The genus *Hemholtzia* from Australia and E. Malaysia has 2 species both of which occur in moist soil and may, perhaps be found as aquatics. The diagnostic features of *Hemholtzia* are: outer perianth segments united at base, anther straight, pollen grains single, ovary 3-locular, ovules axile, fruit indehiscent.

Fig. 159. A. *Philydrum lanuginosum* Banks & Solander: a, habit (3 cm); b, flower; B. *Littorella australis* Griseb.: habit (1 cm); see page 348.

PLANTAGINACEAE

3 genera: *Littorella* is aquatic.

Littorella Bergius, Vet. Acad. Handl. Stokh. 29: 341 (1768)
Fig. 159B.
A stoloniferous perennial, forming mats in water up to 4 m deep. Leaves linear, subulate or flattened, usually hemi-cylindrical, not septate, in basal rosettes. Inflorescence few-flowered, shorter than or equalling the leaves. Flowers actinomorphic, unisexual, *c.* 5 mm in diam. Petals 4, united at the base, dry and scarious. Male flower solitary, terminal; stamens 4, 1 to 2 cm long; anthers large. Female flowers in a few-flowered spike. Ovary 1-locular; style *c.* 1 cm; ovules 1 or 2; fruit indehiscent, hard.

Dietrich, H. Blütenmorphologie und palynologische Untersuchungen an Littorella. Feddes Repert. 82: 155–165 (1971)
Pilger, R. Littorella, in Engler, A. Pflanzenreich 102 (IV. 269): 433–437 (1937)

3 species: *L. uniflora* (L.) Aschers. in Europe, *L. americana* Fern. in N. America and *L. australis* Griseb. in temperate S. America. Flowering occurs only when the plant is exposed. The species are characteristic of the sandy or gravelly shores of acid lakes.

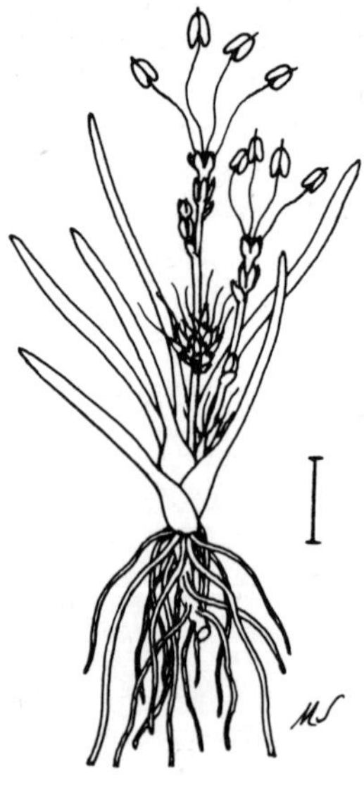

Fig. 159. B. *Littorella australis* Griseb: habit (1 cm).

POACEAE [GRAMINEAE]

Annual or perennial herbs, rarely shrubs or trees; stems erect, ascending or prostrate and creeping, usually branched at the base, in perennials forming sterile shoots and flowering stems (culms), in annuals only the latter present; culms cylindrical, rarely flattened, jointed, usually hollow in the internodes, often solid in tropical species, closed at the nodes. Leaves solitary at the nodes, sometimes crowded at the base of the stems, alternate and two-rowed, consisting of sheath, ligule and blade; sheaths encircling the culm, with the margins free and overlapping (open) or more or less united (closed), frequently swollen at the base; ligule placed at the junction of the sheath and the blade, membranous or reduced to a fringe of hairs, rarely absent; blades usually long and narrow, rarely broad, usually passing gradually into the sheath, rarely with a stalked base, flat, convolute, involte or terete, generally parallel-nerved. Inflorescence very variable (see "General Remarks"). The unit of the inflorescence is the spikelet. The axis of the spikelet (rachilla) usually bears at base two empty bracts (lower and upper glume). The axillary buds of the following bracts (lemmas) usually develop into a branchelet bearing a leaflet (palea) and the subsessile flower. The flowers are usually bisexual, sometimes unisexual, small and inconspicuous, usually consisting of 2 or 3 minute membranous or fleshy scales (lodicules) representing the perianth, stamens and pistil. Stamens hypogynous, 1 to 6, rarely more, usually 3, with delicate filaments and 2-locular anthers, the latter opening generally by a longitudinal slit. Ovary 1-locular, with one anatropous ovule often adnate to the adaxial side of the carpel; styles usually 2, rarely 1 or 3; stigmas generally plumose. The floral organs are referred to as being between lemma and palea. The whole, consisting of lemma, palea and flower, is called a floret. Sometimes the floret is reduced to a fully developed or scale-like, empty (sterile) lemma. Fruit mostly a caryopsis with a thin pericarp adnate to the seed, or rarely a nut or berry or a utricule with a free pericarp, with starchy endosperm and an embryo at the base on the abaxial side, ranging from $\frac{1}{4}$ to $\frac{3}{4}$ the length of the grain, or in a few species occupying the whole of one side of the grain (e.g. species of *Spartina*).

GENERAL REMARKS

INFLORESCENCE

The unit of the grass inflorescence is the spikelet. Although the variety of inflorescence-types found in the grasses is extraordinary, often most species of a genus have similar inflorescence-types. Sometimes this may even apply to systematic categories of higher level (subtribes, tribes). Thus the distinction between different inflorescence-types can be of great diagnostic importance.

RACEME: It is the simplest inflorescence, composed of the main axis (rachis) bearing pedicelled spikelets. Its occurrence in aquatic grasses is rather rare (*Pleuropogon*).

TRUE SPIKE: It is composed of the main axis (rachis) bearing sessile spikelets. Its occurrence in aquatic grasses is rather rare (*Orcuttia*).

SPIKE-LIKE RACEME: A raceme either with very shortly pedicelled (subsessile) spikelets (*Limnopoa*) or with pedicelled spikelets, which hardily spread and thus remain closely appressed to the axis. The term "spike-like raceme" is mainly used for describing the branches of some panicled inflorescences.

PANICLE: In this book we shall call "panicle" an inflorescence composed of the main axis with several racemes or spikes as 1st-order-branches and bearing the spikelets, or with 1st-order-branches, which are branched one or more times again before arriving to the nth-order-branch bearing the spikelets. The pedicels of the spikelets are not considered as branches. Particular attention should be payed to the following panicle-subtypes:

(a) Spike-like, cylindrical, densely flowered panicle (*Alopecurus, Sacciolepis*).
(b) Panicle composed of ascending, densely flowered, spike-like, panicled branches (*Hymenachne*).
(c) Panicle composed of one-sided, spike-like racemes or spikes (*Beckmannia, Brachiaria, Echinochloa, Paspalidium, Paspalum, Reimaria*).

In several genera (especially in the tribes Andropogoneae and Paniceae), usually with panicled inflorescences, the spikelets are borne in pairs. Usually each such pair consists of one sessile and one pedicelled spikelet (*Vossia, Saccharum*), rarely of unequally pedicelled spikelets (*Miscanthidium*). Each pair is considered as a unit, which means that neither the pedicelled nor the sessile spikelet is taken as an additional branch. Thus, for example, the inflorescence of *Limnopoa,* is called a raceme and not a panicle.

SPIKELETS: As a whole the spikelets may be terete, compressed laterally or compressed dorsally. At maturity the spikelet usually disarticulates. This can happen either below the glumes (*Panicum*) or above the glumes (*Agrostis*). In a few genera the upper part of the pedicel disarticulates with the spikelet (*Chikusichloa, Odontelytrum*). In a few genera with several to many-flowered spikelets the rachilla disarticulates between the florets (*Amphibromus*).

GLUMES: Generally there are 2 glumes; the one nearer the insertion of the spikelet on the branch is called the lower glume, the other one the upper glume. Usually the lower glume is inserted abaxially (*Panicum, Vossia*) but in some genera it is in an adaxial position (*Brachiaria, Oplismenopsis*). In some genera one or both glumes are prolonged into awns (*Oplismenopsis, Vossia*). In a few genera one or both glumes are rudimentary (*Oryza, Potamophila*) or even completely absent (*Paspalum*, lower one absent; *Hygroryza, Leersia, Reimaria,* both lacking).

LEMMAS: The number of lemmas in a spikelet is very variable within the family, but generally remains quite constant within the genera. The base of the floret, which sometimes is thickened, is called the callus. In several taxa either the apex of the lemma is prolongated into an awn (*Echinochloa*) or the lemma is awned from the back (*Alopecurus*). The number of lemmas indicates the number of florets a spikelet is composed of. A spikelet with 3 lemmas, for example, is called 3-flowered (has 3 florets) even if 2 of the lemmas are scale-like and sterile (*Oryza*). Sterile florets (lemmas) may be present below (*Panicum, Echinochloa*), or rarely above (*Phalaris*) the fertile ones. (*Phalaris* has 3 florets, the lower 2 are sterile and the single terminal one is bisexual.)

Many grasses are found in areas that are flooded for short times after severe rainfalls or they grow by streams, rivers or waterfalls where they are frequently flooded or bathed in spray. These grasses are not considered to be aquatic and are not included in this account. The following genera have species that are found in such habitats: *Eragrostis* Host., *Heteranthoecia* Stapf, *Hubbardia* Bor, *Jardinea* Steud., *Poa* L., *Robynsiochloa* Jaques-Félix, *Sorghastrum* Nash, and *Sorghum* Moench.

Arechavaleta, J. Las gramíneas uruguayas. Montevideo (1898)
Bews, J. W. The world's grasses. London (1929)
Bor, N. L. The grasses of Burma, Ceylon, India and Pakistan. Oxford, London, New York, Paris (1960)
Burbridge, N. T. Australian Grasses. Sydney, London, Melbourne (Vol. 2, 1968)
Burkart, A. Flora ilustrada de Entre Rios (Argentina). Vol. II. Gramíneas. Buenos Aires (1969)
Chipindall, Lucy K. A. A guide to the identification of grasses in South Africa, in "The grasses and pastures of South Africa", D. Meredith hon. ed. Cape Town (1955)
Hitchcock, A. S. Grasses of British Guiana. Contr. U.S. Nat. Herb. 22: 439–514 (1922)
Hitchcock, A. S. The grasses of Ecuador, Peru and Bolivia. Contr. U.S. Nat. Herb. 24: 291–556 (1927)
Hitchcock, A. S. The grasses of Central America. Contr. U.S. Nat. Herb. 24: 557–801 (1930)
Hitchcock, A. S. Grasses of Canton and vicinity. Lingnan Sci. Journ. 7: 177–266 (1929), 10: 339–340 (1931)
Hitchcock, A. S. and Chase, A. Manual of the grasses of the United States. 2. ed. Washington (1951)
Honda, M. Monographia Poacearum Japonicarum, Bambusoideis exclusis. Journ. Fac. Sci. Imp. Univ. Tokyo. Sect. III, Bot 3(1): 1–484 (1930)
Hsu, C. C. The Paniceae (Gramineae) of Formosa. Taiwania 9: 33–57 (1964)
Hubbard, C. E. Grasses. Penguin Book HA 295, 2. ed. London (1968)
Junge, P. Die Gramineen Schleswig-Holsteins. Mitt. Bot. Staatsinst. Hamburg 30: 101–330 (1913)

Klapp, E. Taschenbuch der Gräser. Berlin-Hamburg (1965)
Lamson, F. and Merrill, E. D. The Grasses of Alaska. Contr. U.S. Nat. Herb. 13: 47–92 (1909)
Lazarides, M. The Grasses of Central Australia. Canberra (1970)
Merrill, E. D. An Enumeration of Philippine Gramineae with keys to genera and species. Phillipp. Journ. Sci. 1 (Suppl.): 307–392 (1906)
Parodi, L. R. Gramíneas, in Enciclopedia Argentina de Agricultura y Jardinería, 1: 102–180. Buenos Aires (1959)
Pilger, R. Beiträge zur Flora von Südwest-Afrika. 1. Gramineae. Willdenowia 1: 198–274 (1954)
Pilger, R. Das System der Gramineae. Bot. Jb. 76: 281–384 (1954)
Pilger, R. Gramineae III, in Engler, A. and Prantl, K. "Die natürlichen Pflanzenfamilien". 2. ed. Vol. 14 e, Leipzig (1940)
Pilger, R. and Potztal, E. Gramineae II, in Engler, A. and Prantl, K., "Die natürlichen Pflanzenfamilien". 2. ed. Vol. 14 d, Berlin (1956)
Potztal, Eva. Graminales, in Engler's Syllabus der Pflanzenfamilien. 12. ed. Vol. 2: 561–579. Berlin (1964)
Raizada, M. B., Bharadwaja, R. C. and Jain, S. K. Grasses of the Upper Gangetic Plain. Part 1: Maydeae, Andropogoneae. Indian Forest Rec. Bot. 4: 171–277 (1961)
Rosengurt, B. Gramíneas uruguayas. Montevideo (1970)
Rotar, P. R. Grasses of Hawaii. Univ. Hawaii Press (1968)
Schmid, M. Flore agrostologique de l'Indochine. Agron. Trop. 13: 7–51, 143–237, 300–359, 459–522, 631–672, 687–703 (1958)

1A Leaf blades sharply serrulate (commonly 2 m long and 4 to 6 cm wide, forming great fan-shaped summits to the sterile shoots; plants very large, often 10 m high, dioecious; lemmas of female spikelets silky-hairy, lemmas of male spikelets glabrous; Trop. America)
Gynerium

1B Leaf blades not sharply serrulate

2A Leaves not differentiated into sheath and blade (loosely folded around the culm; upper part of spike-like panicle bearing lanceolate-linear, empty bracts, instead of spikelets; California)
Neostapfia

2B Leaves differentiated into sheath and blade

3A Leaf blades connected with leaf sheath by an up to 6 cm long stalk; upper surface of leaf blade with ribbon-like flanges or lamellae, up to 0.4 mm high (Zambia)
Hydrothauma

3B Leaf blades not stalked, lamellae or flanges on upper surface absent

4A Inflorescence a 1 to 5 cm long panicle, consisting of 3 to 6 nodes with verticillate, umbel-like, stalked and sessile clusters of spikelets (tiny annual; culms 5 to 6 cm long)
Coleanthus

4B Inflorescence not consisting of 3 to 6 nodes with verticillate, umbel-like, stalked and sessile clusters of spikelets

5A Leaf sheaths strongly inflated and usually functioning as floats (leaf blades 3 to 5 cm long and 0.7 to 2 cm wide, ovate-oblong or ovate-lanceolate, without prominent midrib; ligules absent; glumes absent; spikelets awned; usually floating; India and S. E. Asia)

Hygroryza

5B Leaf sheaths not strongly inflated, not functioning as floats

6A Inflorescence a true spike (2 to 10 cm long, spikelets several-flowered; California)

Orcuttia

6B Inflorescence not a true spike

7A Inflorescence partly enclosed in a nut-like ovoid or cylindric utricule (sheath) which has a pore at apex only and sometimes bears a rudimentary blade; utricule hardened at maturity, enclosing grain; exserted parts of inflorescence bearing male spikelets only

Coix

7B Inflorescence not enclosed in a nut-like utricule

8A Panicle entirely enclosed by sheath (spikelets 1-flowered, compressed laterally, keeled, less than 6 mm long, awnless, glumes absent; temperate to subtemperate regions of N. Hemisphere)

Leersia

8B Panicle not enclosed by sheath

9A Inflorescence of fascicled or solitary spikes; spikes terminating the culms and branches, but branches often short and in clusters, so that several spikes may arise from each node of culm; spikes at base enclosed by uppermost leaf-sheath of branch; spikelets in opposite pairs

Hemarthria

9B Not as above

10A Inflorescence consisting of terminal and axillary racemes with few spikelets; terminal racemes male, axillary ones female; spikelets without glumes and sterile lemmas; grass slender, usually floating, freely branching; leaf sheath generally closed (S. E. N. America)

Hydrochloa

10B Not as above

11A Spikelets solitary, supported by a shortly pedicelled partial involucre of united, false bracts; inflorescence a raceme, sheathed at base and with leaf blade exceeding tip of raceme (Abyssinia)
Odontelytrum

11B Not as above

12A Branches of inflorescence exceeding their uppermost spikelet and ending in a 4 to 12 mm long bristle

13A Glumes present; all branches of inflorescence ending in a 5 to 12 mm long bristle
Pseudoraphis

13B Glumes absent; some branches only of inflorescence ending in a 4 to 7 mm long bristle
Reimaria

12B Branches of inflorescence without mentioned character, or bristle not reaching 4 mm length

14A Racemes ending in a 1 to 2 mm long bristle (inflorescence of 1-sided, dense, spike-like racemes; spikelets in 2 rows, compressed dorsally, awnless)
Paspalidium

14B Racemes not ending in a bristle

15A Spikelets or clusters of spikelets subtended by 1 to many distinct stiff bristles or bracts

16A Spikelets solitary, subtended by 1 long bristle; inflorescence a raceme
Paratheria

16B Spikelets solitary or in clusters, subtended by 2 to many bristles or bracts; inflorescence a dense, interrupted or uninterrupted panicle

17A Inflorescence an uninterrupted panicle cylindrical, spike-like panicle; spikelets 2-flowered; bristles very slender or capillary
Pennisetum

17B Inflorescence at least in lower parts an interrupted spike-like panicle, composed of separated globose or cylindrical clusters of spikelets on a central axis; spikelets 4- to 6-flowered; bristles bract-like
Elytrophorus

15B Spikelets not subtended by bristles

18A Inflorescence a large (up to 80 cm long), silky-hairy panicle

19A Spikelets in pairs, 2-flowered; lower floret sterile

20A Both spikelets pedicelled (unequally so); fertile lemma awned
Miscanthidium

20B One spikelet sessile (the other pedicelled); spikelets awnless
Saccharum

19B Spikelets solitary (2–) 3- to 7-flowered; lowest floret bisexual male or sterile

21A Lemmas with long silky hairs, rachilla glabrous (or shortly hairy); lowest floret bisexual; rhizomes knotty
Arundo

21B Lemmas glabrous, rachilla with long, silky hairs; lowest floret sterile or male; rhizomes stout, not knotty
Phragmites

18B Inflorescence not a large, silky-hairy panicle

22A Inflorescence a simple raceme with pedicelled spikelets (culms 5 to 20 cm long; spikelets 5- to 10-flowered awned, compressed laterally; Arctic and Sub-arctic regions)
Pleuropogon

22B Inflorescence not a simple raceme with pedicelled spikelets

23A Inflorescence 1 to 2.3 cm long, a simple, more or less spike-like raceme; culms up to 20 cm long; spikelets in pairs; usually floating (S. India)
Limnopoa

23B Inflorescence more than 3 cm long, not a spike-like raceme

24A Inflorescence consisting of dense, globose or cylindrical clusters of spikelets on a central axis, each cluster being subtended by 2 to many bracts or sterile spikelets; in the lower and middle parts of the inflorescence the clusters being separated by distinct internodes, in the upper parts an uninterrupted spike-like panicle sometimes being formed
Elytrophorus

24B Inflorescence not consisting of globose or cylindrical clusters of spikelets

25A Inflorescence a terminal, dense, cylindrical, spike-like panicle

26A Spikelets strongly compressed laterally, 1-flowered; lemma awned from the back; palea absent
Alopecurus

26B Spikelets not strongly compressed laterally, 2-flowered; lemma not awned from the back; palea present

27A Leaf blades amplexicaul at base; upper lemma and palea gaping at apex
Hymenachne

27B Leaf blades not amplexicaul at base; upper lemma and palea closed at apex

28A Spikelets compressed dorsally; lower glume very short and nerveless; grain enclosed by upper lemma and palea, and by lower lemma and upper glume **Thyridachne**

28B Spikelets terete to slightly compressed laterally; lower glume about $^1/_3$ to $^1/_2$ as long as spikelet, both prominently nerved; grain only enclosed by upper lemma and palea **Sacciolepis**

25B Inflorescence not a terminal, dense, cylindrical, spike-like panicle

29A Spikelets in pairs (usually one spikelet sessile and the other pedicelled, or both unequally pedicelled; in *Pseudovossia* the pedicel of the pedicelled spikelet consisting of 2 or 3 internodes)

30A Pedicels of pedicelled spikelets very long, consisting of 2 or 3 internodes (pedicelled spikelets compressed dorsally, sessile spikelets compressed laterally; lower glumes terminating in a long awn; Cambodia, Thailand) **Pseudovossia**

30B Pedicels of pedicelled spikelets not long, not more than 1 internode

31A Lower glume awned; glume and awn flat; awn up to 2 mm wide and 30 mm long, with scabrous margins; inflorescence consisting of digitate or solitary, erect, glabrous, spike-like racemes; plants large **Vossia**

31B Lower glume absent or, if present, not awned

32A Inflorescence a large, loose panicle, composed of numerous verticillate spike-like racemes or 1st order branches, which in turn bear spike-like racemes

33A Pedicelled spikelets 2-flowered, lower floret sterile, upper floret male; sessile spikelet slightly compressed laterally; lower glume keeled, 7-nerved, with numerous upwardly curved spiny bristles **Vetiveria**

33B Pedicelled spikelets sterile, consisting of 2 glumes, which sometimes enclose a reduced lemma; sessile spikelet compressed dorsally; lower glume 2-keeled, 8- to 11-nerved, scabrid on the keels **Hemisorghum**

32B Inflorescence of 1 to numerous simple, very 1-sided spike-like racemes along a main axis, or of (1–) 2 to 7 spike-like racemes arranged in pairs or sub-digitately

34A Lower glume usually absent; rachis continuous and tough; upper lemma leathery, awnless **Paspalum**

34B Lower glume present; rachis disarticulating at nodes; upper lemma membranous, usually awned

Ischaemum

29B Spikelets not in pairs

35A Pedicels disarticulating with spikelets, forming a stipe at base

36A Glumes absent; spikelets 1-flowered; stipe 2 to 4 mm long (Japan)

Chikusichloa

36B Glumes present; spikelets 2-flowered; stipe 8 to 15 mm long (S. W. Africa)

Oryzidium

35B Pedicels not disarticulating with spikelets

37A Inflorescence consisting of several ascending, densely flowered, spike-like, panicled branches; branches 1-sided or not

38A Spikelets strongly compressed laterally; glumes as long as spikelet; branches not 1-sided; spikelets disarticulating above the glumes

39A Glumes narrowly-lanceolate, 1-nerved; spikelets 1-flowered; sterile lemmas absent

Agrostis

39B Glumes broadly-lanceolate, 3-nerved; spikelets usually 3-flowered; sterile lemmas present

Phalaris

38B Spikelets subterete or slightly to distinctly compressed dorsally; lower glume much shorter than spikelet; branches usually 1-sided; spikelets disarticulating below the glumes

40A Spikelets awned or cuspidate or mucronate, very convex on the back, flat or slightly concave ventrally; branches with flat or triquetrous rachis

Echinochloa

40B Spikelets awnless; subterete or slightly compressed dorsally; rachis of branches only slightly flattened

41A Leaf blades amplexicaul at base; lemma and palea more or less open at apex of ripe fruit; spikelets subterete

Hymenachne

41B Leaf blades not amplexicaul at base; lemma and palea not open at apex of ripe fruit; spikelets slightly compressed dorsally

Panicum

37B Inflorescence not consisting of several ascending, densely flowered, spike-like, panicled branches

42A Inflorescence composed of several distant, solitary, pinnate or ternate, spike-like racemes; spikelets inserted on 2 sides of the rachis, 2- to 10-flowered

Diplachne

42B Inflorescence composed of several spike-like, 1-sided, racemes or spikes along a main axis; (lowermost "racemes" sometimes branched near base); or inflorescence a more or less open, all-sided panicle

43A Inflorescence of several spike-like, 1-sided racemes or 1-sided spikes along a main axis (lowermost "racemes" sometimes branched near base

44A Spikelets strongly compressed laterally

45A Spikelets 1-flowered, very closely crowded; upper glume awned or acuminate at apex; lower glume 3 to 10 mm long, upper one 6 to 25 mm long, including awn
Spartina

45B Spikelets 3- to 9-flowered, not densely crowded; upper glume with acute or shortly mucronate apex; lower glume 1 to 1.5 mm long, upper one 2 to 2.5 mm long
Leptochloa

44B Spikelets not compressed laterally

46A Ligules absent
Echinochloa

46B Ligules membranous or consisting of a fringe of hairs

47A Glumes and lemmas with a thickened, laterally compressed apical crest
Acroceras

47B Glumes and lemmas without a thickened, laterally compressed apical crest

48A Glumes 2, equal, as long as spikelet, boat-shaped, mucronate; spikelets 1-flowered, rarely 2-flowered and with both florets bisexual (ligules membranous, 3 to 6 mm long)
Beckmannia

48B Glumes 2 (unequal), or 1 or both absent; upper glume occasionally as long as spikelet; lower glume always much shorter than upper one, if present; spikelets 2-flowered; lower floret male or sterile (reduced to sterile lemma); upper floret bisexual

49A Both glumes absent (spikelets strongly compressed dorsally)
Reimaria

49B At least upper glume present

50A Lower glume absent, spikelets plano-convex, awnless
Paspalum

50B Lower glume present

51A Lower glume adaxial

52A Upper glume awned; awn up to 8 mm long; racemes dense (S. America)

Oplismenopsis

52B Upper glume mucronate or not, awnless; racemes dense, spike-like (widespread)

Brachiaria

51B Lower glume abaxial, or lower glume neither distinctly abaxial nor distinctly adaxial

53A Lower glume clearly abaxial, upper glume about $^2/_3$ as long as spikelet or more; racemes ending in a short, 1 to 2 mm long bristle

Paspalidium

53B Lower glume not clearly abaxial; upper glume as long as spikelet; racemes not ending in a bristle

Echinochloa

43B Inflorescence a more or less open, all-sided panicle

54A All spikelets unisexual (male and female differing in shape, size and colour)

55A Terminal and axillary panicles present

Luziola

55B Inflorescence of terminal panicles only

56A Male and female spikelets on different branches of same terminal panicle; upper branches ascending, bearing female spikelets; style-base not persistent

Zizania

56B Male and female spikelets together on same branches of panicle

57A Male spikelets below; female spikelets awned; style-base persistent

Zizianopsis

57B Female spikelets below; spikelets awnless; style-base not persistent

Luziola

54B All or most spikelets bisexual (not differing in shape, size and colour)

58A Glumes and lemmas tipped by a thickened, laterally compressed crest

Acroceras

58B Glumes and lemmas not tipped by a thickened, laterally compressed crest

59A Ligules reduced to a fringe of hairs

60A Spikelets awned; lower glume adaxial

61A Lower part of panicle enclosed in uppermost leaf-sheath (Central Africa)

Louisiella

61B Lower part of panicle not enclosed by uppermost leaf-sheath (S. W. Africa, S. America)

62A Upper glume with a 10 to 18 mm long awn (S. W. Africa)
Oryzidium

62B Upper glume with an up to 8 mm long awn (La Plata Region, Uruguay, Argentina)
Oplismenopsis

60B Spikelets awnless; lower glume abaxial

63A Glumes equal to subequal, broad, convex; spikelets terete to globose, small

64A Glumes about as long as spikelet or only slightly shorter; lower or upper floret sometimes male
Isachne

64B Glumes much shorter than spikelet; lower floret bisexual, upper one female
Coelachne

63B Glumes unequal, lower one much shorter, upper one as long as spikelet; spikelets slightly compressed dorsally
Panicum

59B Ligules membranous

65A Glumes as long as spikelet, rough above on the keel (spikelets 1-flowered)
Agrostis

65B Glumes not as long as spikelet, not rough above on the keel; sometimes glumes absent

66A Spikelets with at least 3, usually more, fully developed bisexual florets

67A Spikelets awned (New Zealand and Australia)
Amphibromus

67B Spikelets awnless

68A Lemmas 3-nerved; glumes nerveless
Catabrosa

68B Lemmas 5- to 9-nerved; glumes nerved

69A Leaf-sheaths closed; both glumes 1-nerved
Glyceria

69B Leaf-sheaths open; glumes 3- to 5-nerved (at least upper one)

70A Upper glume 3-nerved; spikelets usually 5- to 15-flowered
Torreyochloa

70B Upper glume 5-nerved; spikelets 3- to 4-flowered
Scolochloa

66B Spikelets with only 2 or 1 fully developed florets

71A Mostly 2 fully developed bisexual florets present; spikelets disarticulating above the glumes (glumes unequal, shorter than lowest floret, nerveless; lemmas 3-nerved; apex of glumes and lemmas often erose)
Catabrosa

71B Only 1 bisexual floret developed; spikelets disarticulating below the glumes (glumes absent or scale-like or strongly unequal, nerved; lemmas 5- or more-nerved; apex of glumes and lemmas rarely erose)

72A Lower glume much shorter than upper one; upper glume equalling sterile lemma, both as long as spikelet, enclosing the fertile floret (spikelets slightly compressed dorsally, awnless)

Panicum

72B Both glumes either absent or reduced to a 2-lobed rim, or scale-like and neither one as long as spikelet; if sterile lemmas present, then usually very small

73A Sterile lemmas absent; glumes reduced to an entire or more rarely obscurely 2-lobed, somewhat thickened rim, or glumes completely absent

74A Pedicels jointed about the middle, the upper part scabrous; spikelets subterete to slightly compressed laterally; glumes completely absent (S. E. Asia)

Chikusichloa

74B Pedicels not jointed; spikelets compressed laterally; glumes reduced to an entire or 2-lobed rim (widespread)

Leersia

73B Sterile lemmas present, but usually very small (glumes very small, sometimes reduced to a 2-lobed, rarely entire rim)

75A Lemma rounded, membranous, glabrous or pubescent but not stiffly hairy, awnless; grain crowned with persistent style-base (New S. Wales)

Potamophila

75B Lemma keeled, usually leathery, margins and back often fringed with stiff hairs, usually awned; grain not crowned with persistent style-base

76A Lemma with thickened apex of spongy aerenchyma (S. America)

Rhynchoryza

76B Lemma not as above

Oryza

Acroceras Stapf in Prain, Fl. Trop. Africa 9: 621 (1920)
Fig. 160.

Annuals and perennials, often stoloniferous; culms ascending, 50 to 180 cm long. Leaf blades lanceolate, usually cordate at base; ligules reduced to rim fringed with short hairs. Inflorescence a panicle of 1 to many, 1-sided racemes; racemes inserted on central axis, usually only spreading slightly. Spikelets disarticulating below the glumes (*c.* 6 mm long), subterete, awnless, 2-flowered. Lower floret male or sterile; upper floret bisexual. Glumes 2, each with a thickened, laterally compressed crest at apex; lower glume inserted adaxially, obtuse to acuminate, 3- to 5-nerved, usually shorter than

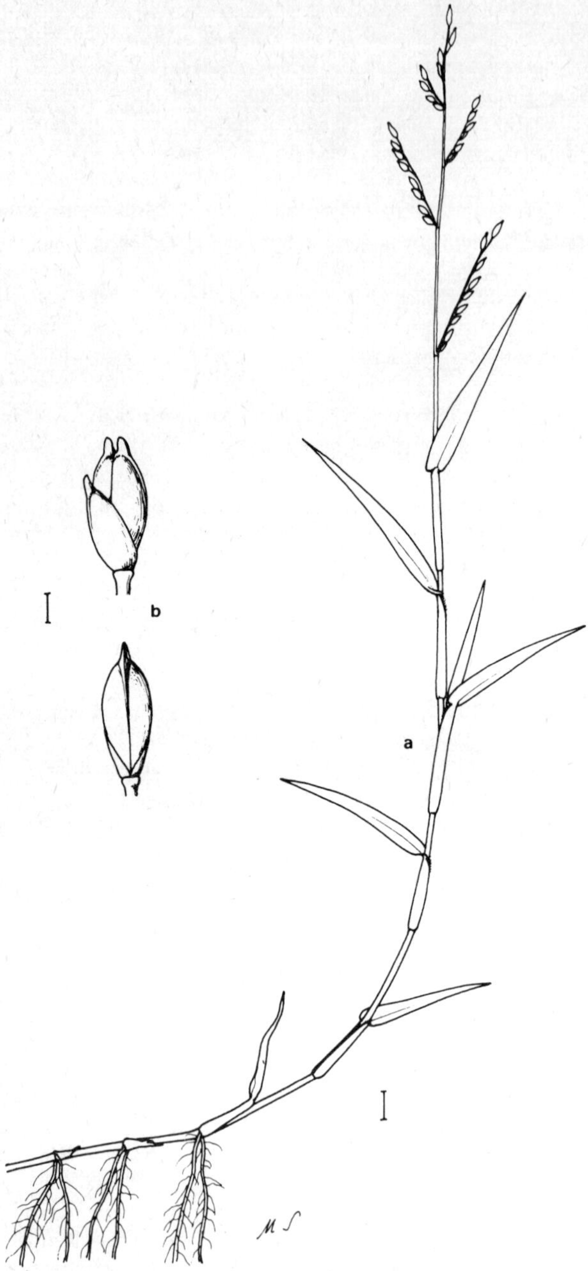

Fig. 160. *Acroceras zizanioides* (H.B.K.) Dandy: a, habit (1 cm); b, spikelet from front and back (1 mm).

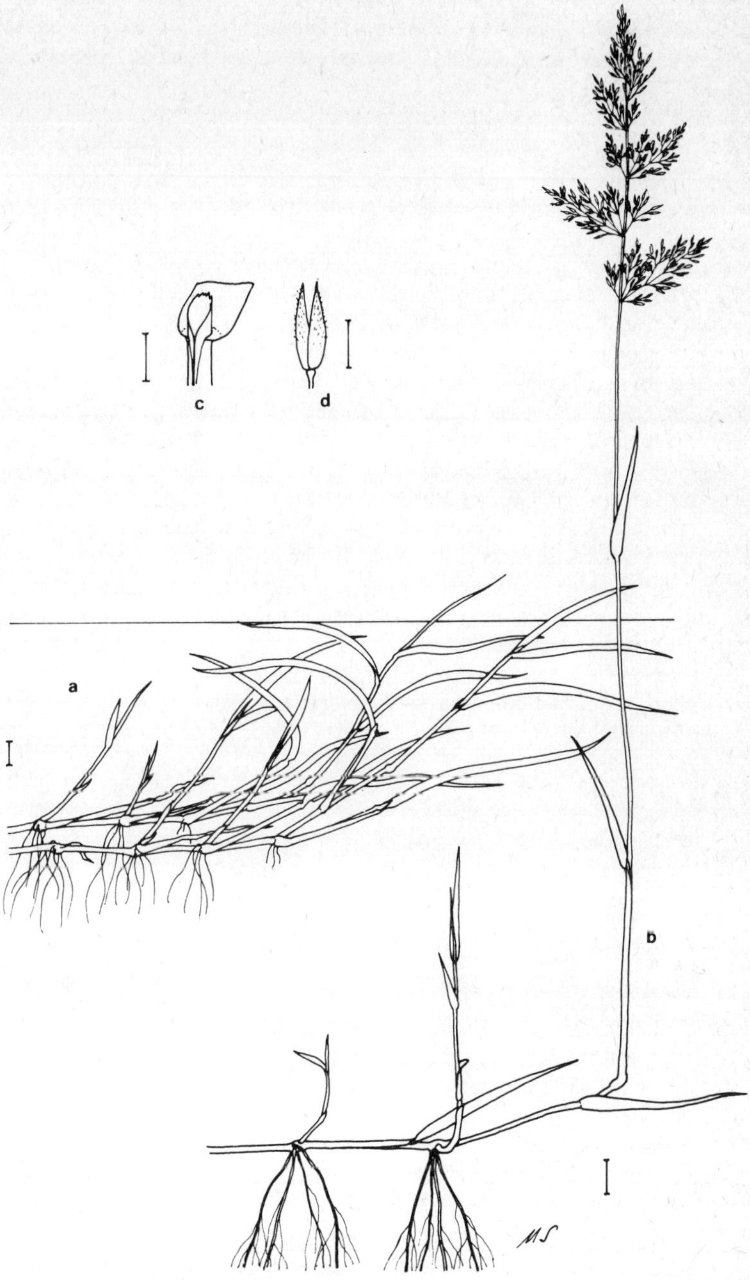

Fig. 161. *Agrostis stolonifera* L.: a, habit of sterile plant (1 cm); b, habit of fertile plant (1 cm); c, ligule (4 mm); d, spikelet (1 mm).

spikelet; upper glume 5- to 9-nerved. Lemmas similar to glumes. Paleas with more or less recurved callous tip. Stamens 3. Grain oblong, dorsally flattened, ventrally convex, tightly enclosed in hardened palea and lemma.

4 to 5 species: cosmopolitan in warmer regions. Generally semi-aquatic.

Agrostis L., Sp. Pl. 61 (1753)
Fig. 161.
Perennials or annuals, stoloniferous or rhizomatous, or caespitose; culms usually 20 to 90 cm long. Leaf blades 5.5 to 40 cm long, 0.3 to 1 cm wide, flat or folded, scabrous, rolled in the bud; sheaths open, rounded on the back; ligules membranous, up to 8 mm long. Inflorescence a panicle, 8 to 30 cm long, up to 10 cm wide, lanceolate to narrowly ovate in outline, open in flower, afterwards contracted; panicle sometimes consisting of ascending, densely flowered panicled branches (*A. natalensis* Stapf). Spikelets disarticulating above the glumes, compressed laterally, 2 to 3 (-3.5) mm long, 1-flowered, awnless or often awned. Glumes persistent, as long as spikelet, equal or slightly unequal, narrowly lanceolate, with acute to shortly cuspidate apex, 1-nerved, rough upwards on keels. Lemma up to $^3/_4$ the length of the glumes, thin, awnless or with a short awn from near the tip, 5-nerved, often hairy on the callus. Stamens 3. Grain enclosed between lemma and palea, oblong in outline, slightly compressed or subterete.

c. 200 species: cosmopolitan. Most species terrestrial, some races of *A. stolonifera* L. are aquatic; several semiaquatic species are recorded from S. Africa.

Alopecurus L., Sp. Pl. 60 (1753)
Fig. 162.
Perennials or rarely annuals, sometimes stoloniferous; culms often ascending, sometimes floating, up to 80 cm long. Leaf blades usually 2 to 15 cm long, 0.2 to 0.7 cm wide, flat; ligules membranous. Inflorescence a cylindrical, spike-like panicle, 1 to 7 cm long. Spikelets disarticulating below the glumes, 2 to 5 mm long, strongly compressed laterally, 1-flowered; awned (in *A. aequalis* Sobol. awn either included in glumes or very slightly protruding from them). Glumes equal or subequal, about as long as spikelet, 3-nerved, more or less united at base, acute to obtuse at apex, hairy on the keels and on the lateral nerves. Lemma from slightly shorter to barely longer than glumes, strongly flattened, united near the base, awned from the back, indistinctly 3- or 5-nerved; awn straight or twisted and geniculate, from shorter than glumes to more than twice their length. Palea absent. Lodicules absent. Stamens 3. Grain enclosed in lemma and glumes.

c. 25 species: 2 or 3 species aquatic, several semiaquatic. In temperate to subarctic regions of N. America and Eurasia, Australia and temperate and cold S. America (*A. antarcticus* Vahl).

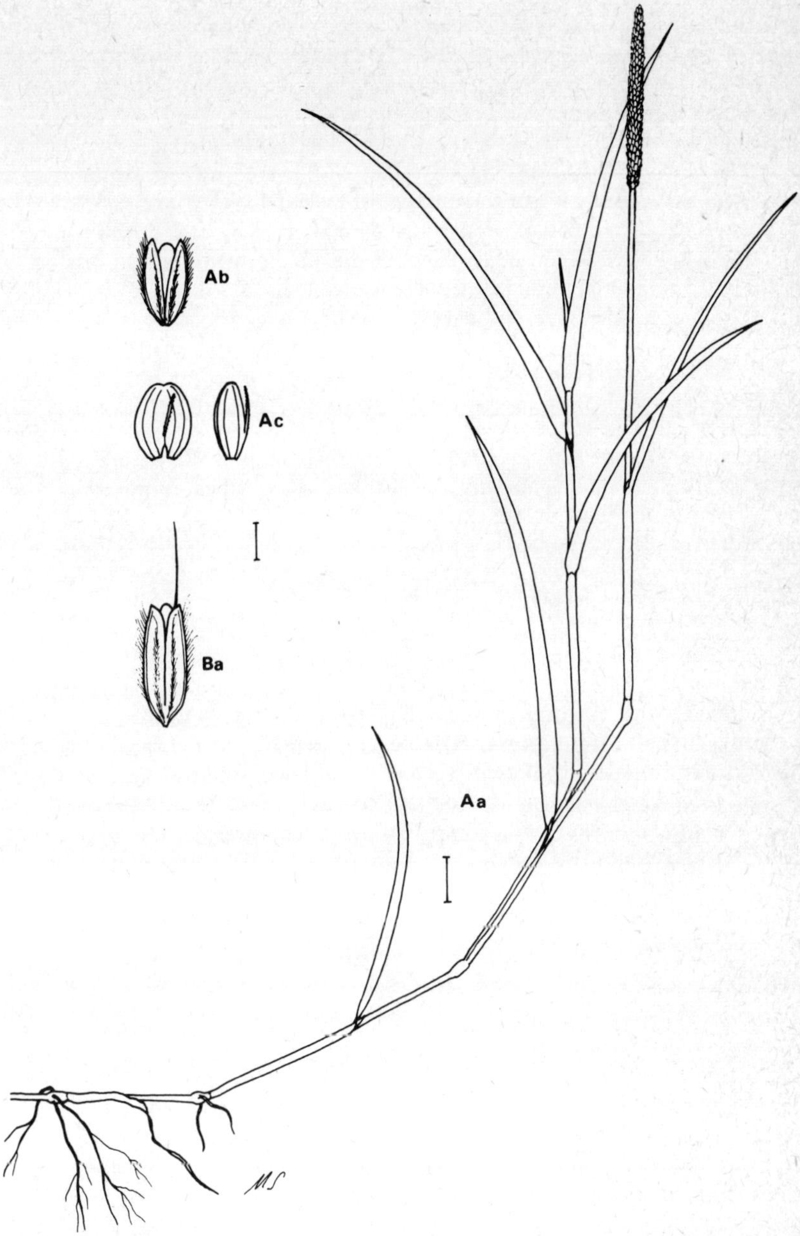

Fig. 162. A. *Alopecurus aequalis* Sobol.: a, habit (15 mm); b, spikelet (1 mm); c, lemma from back and side;
B. *Alopecurus geniculatus* L.: a, spikelet (1 mm).

Amphibromus Nees, Jour. Bot. (London) 2: 420 (1843)
Fig. 163.
Perennials, stoloniferous or tufted; culms 20 to 140 cm high, ascending or erect. Leaf blades 8 to 20 cm long, 0.1 to 0.5 cm wide, flat, lax or firm; ligules 4 to 15 mm long. Inflorescence a panicle, 10 to 40 cm long, the lower parts enclosed in the uppermost leaf-sheath, with rachilla disarticulating above the glumes and between the florets. Spikelets 3- to 7-flowered, awned, compressed laterally; all florets bisexual, or the uppermost male. Glumes 2, unequal; lower 3 to 5 mm long, 1- or 3-nerved; upper 4 to 6 mm long, 5-nerved. Lower lemma 5 to 6 mm long, 7-nerved awned from about the middle; awns straight or geniculate, 12 to 20 mm long. Paleas shorter than lemmas, 2-keeled. Stamens 3. Grain oblong in outline, glabrous, deeply grooved.

Swallen, J. R. The grass genus Amphibromus. Amer. Journ. Botany 18: 411–415 (1931)

c. 5 species: *A. fluitans* Kirk is aquatic in New Zealand; *A. neesii* Steud. is semiaquatic in Australia. *A. fluitans* is probably cleistogamous, fertilisation taking place before the panicle is extruded.

Arundo L., Sp. Pl. 81 (1753)
Fig. 164.
Robust perennial, with a thick creeping, knotty rhizome; culms 200 to 600 cm long, many-noded, hollow. Leaf blades up to 70 mm wide and up to 100 cm long, rounded at the base, tapering to a long, fine point, glabrous; ligule shortly membranous. Inflorescence a dense, erect panicle, 30 to 60 cm long, feathery, cream-coloured or brown. Spikelets pedicelled, solitary, disarticulating above the glumes and between the florets, 8 to 15 mm long, 2- to 7-flowered, awned, compressed laterally. All florets bisexual, successively smaller upwards (uppermost one reduced). Glumes 2, almost equal, 3- or 5-nerved, about as long as the spikelet, apex acute. Lemmas 5- to 9-nerved, 3 nerves percurrent, the middle one usually produced into a short awn (lower lemma about 10 mm long). The lemmas are silky-hairy but the rachilla is glabrous or shortly-hairy only. Stamens 3.

Perdue, R. E. Arundo donax – a source of musical reeds and industrial cellulose, Econ. Bot. 12: 368–404 (1958)

1 species, *A. donax* L., originally in warm regions of the Old World; has been introduced to America. It occurs along rivers, irrigation ditches, streams, on river banks, occasionally in marshes. The culms are used for lattices, mats and screens in the construction of adobe huts, and for making the reeds of clarinets and organ pipes.

Fig. 163. *Amphibromus fluitans* Kirk: a, habit (1.5 cm); b, spikelet (1 mm).

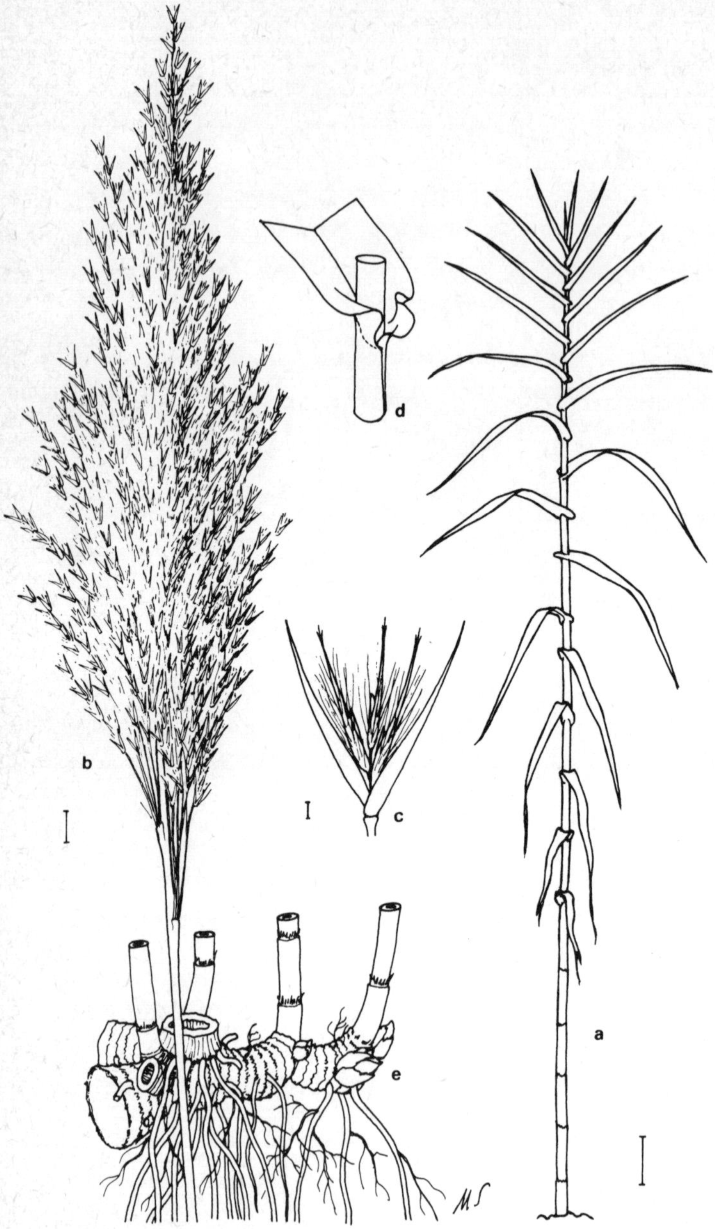

Fig. 164. *Arundo donax* L.: a, habit (10 cm); b, inflorescence (1 cm); c, spikelet (1 mm); d, leaf sheath; e, root stock.

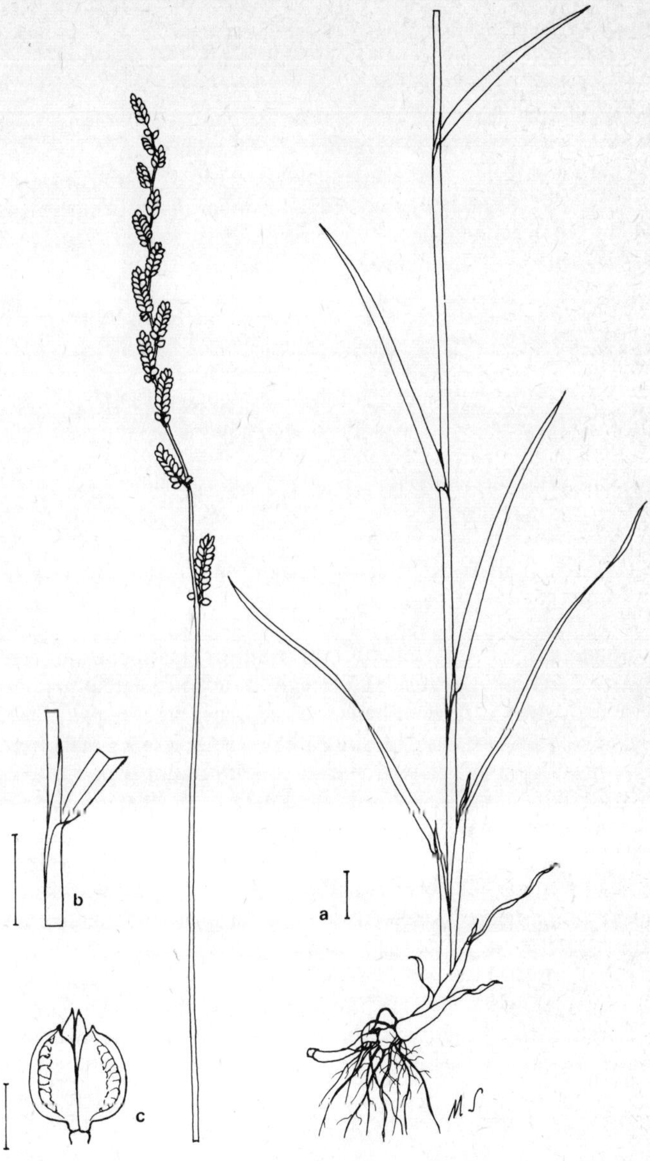

Fig. 165. *Beckmannia eruciformis* (L.) Host.: a, habit (1 cm); b, ligule (1 cm); c, spikelet (1 mm).

Beckmannia Host., Ic. Descr. Gram. Austr. 3: 5 (1805)
Fig. 165.

Annuals or perennials; culms erect, rather stout, up to 100 cm high. Leaf blades 7 to 20 cm long, 0.5 to 1 cm wide; ligules membranous, 3 to 6 mm long. Inflorescence a panicle, 10 to 35 cm long, composed of numerous, ascending, one-sided, 1 to 5 cm long spikes. Spikelets densely crowded, in 2 rows, disarticulating below the glumes, subterete, 3 to 4 mm long, 1- or 2-flowered; florets bisexual, awnless, acuminate or mucronate. Glumes equal, leathery, as long as spikelet, boat-shaped, obovate, inflated, transversely wrinkled, slightly united at base, 3-nerved; apex mucronate. Lemmas narrowly ovate, 5-nerved; apex acuminate or mucronate, slightly keeled. Stamens 3. Grain free, fusiform, obtusely angled.

Reeder, J. R. Affinities of the grass genus Beckmannia Host. Bull. Torrey Bot. Club 80: 187–196 (1953)

2 species: *B. eruciformis* (L.) Host. (2-flowered spikelets) and *B. syzigachne* (Steud.) Fern, (1-flowered spikelets). Both semi-aquatic; *B. syzigachne* has been reported in paddy fields in Japan. Both species in cooler and temperate regions of the N. Hemisphere.

Brachiaria Griseb. in Ledebour, Fl. Ross. 4: 469 (1853), [*Urochloa* sensu Nguyen pro parte]
Fig. 166.

Perennials or annuals, erect or sometimes creeping or floating; culms erect or ascending, 1 to 2 m long. Leaf blades linear to ovate-lanceolate, broad at base, with acute apex; ligules a fringe of hairs. Inflorescence a dense panicle, 15 to 25 cm long, composed of one-sided, spike-like racemes along a main axis; main axis terminating in a raceme. Spikelets solitary in upper parts of racemes, in lower parts usually in unequally pedicelled pairs or in clusters, disarticulating below the glumes, 2 to 4 mm long, awnless or mucronate, terete or dorsally compressed, 2-flowered; lower floret male or sterile; upper floret bisexual. Lower glume in solitary spikelets adaxial, from nearly as long as spikelet to much shorter, 0- to 7-nerved, with obtuse or acute apex; upper glume as long as spikelet, with obtuse or acute apex. Lower lemma as long as spikelet, shorter or absent; upper leathery, more or less hardened at maturity, slightly shorter than spikelet, with narrowly inrolled margins; apex mucronate or awnless. Stamens 3. Grain tightly enclosed by more or less hardened lemma and palea, dorsally compressed, broadly oblong to elliptic in outline.

c. 80 species. About 6 semiaquatic to aquatic (*B. mutica* (Forsk.) Stapf, *B. eruciformis* (J. F. Smith) Griseb.). In the warmer regions of both Hemispheres; most species in Africa.

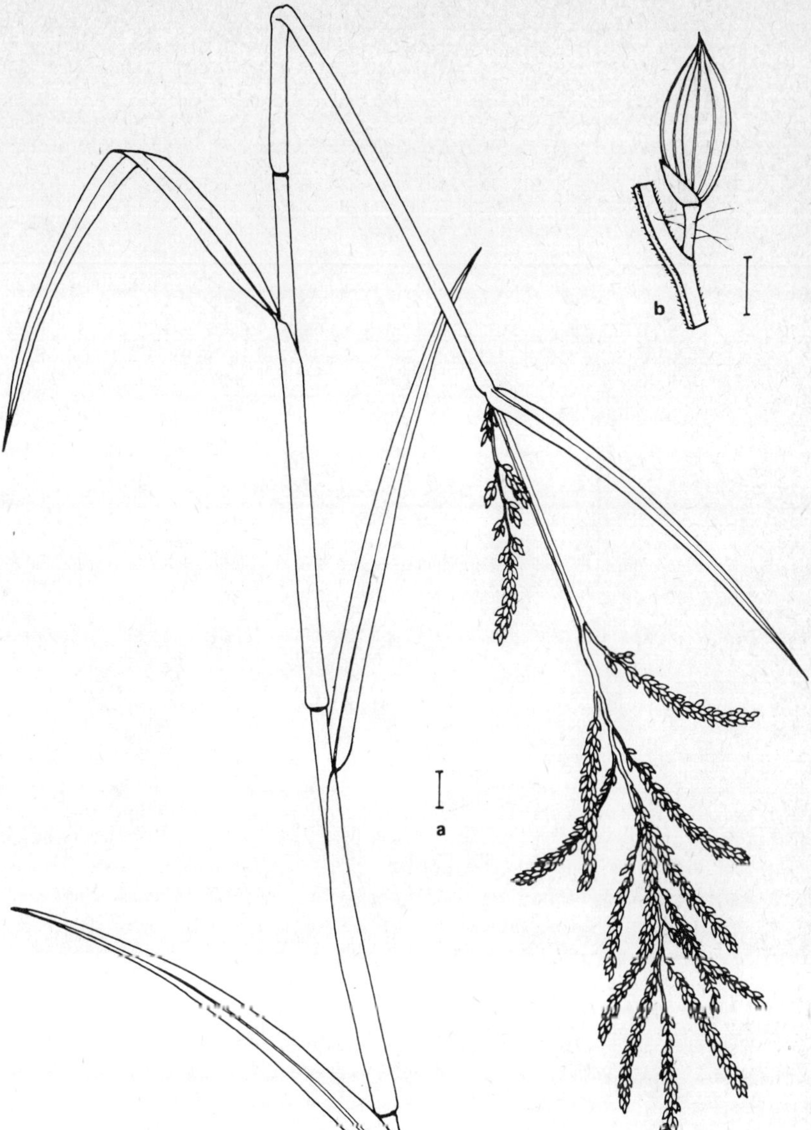

Fig. 166. *Brachiaria eruciformis* (J. F. Smith) Griseb.: a, inflorescence (1 cm); b, spikelet (1 mm).

Fig. 167. *Catabrosa aquatica* (L.) Beauv.: a, habit (1 cm); b, spikelet (1 mm).

Catabrosa Beauv., Agrost. 97 (1812)
Fig. 167.

Perennial or rarely annual, stoloniferous; culms, 10 to 70 cm high, ascending. Leaf blades flat, 0.2 to 1.3 cm wide and usually up to 17 cm long; ligules 2 to 8 mm long. Inflorescence a panicle, 5 to 20 cm long; primary branches fascicled in half-whorls. Spikelets disarticulating above the glumes, about 4 to 4.7 mm long, mostly 2-flowered rarely 1-, 3-, or 5-flowered; florets bisexual; awnless; more or less terete; upper floret caducous. Glumes 2, membranous, nerveless, with erose apex, unequal in length; lower 0.7 to 1.5 mm long; upper 1.5 to 2.2 mm long. Lemmas 2.5 to 3 mm long membranous, awnless, with 3 prominent, non-convergent nerves, and truncate and more or less erose apex. Stamens 3. Grain barely 2 mm long, fusiform.

Borrill, M. The affinities of Catabrosa aquatica. Yearb. Bot. Soc. Brit. Isles 1953: 59 (1953)

1 species: *C. aquatica* (L.) Beauv., aquatic to semiaquatic, along streams, ponds and lake-margins. In Eurasia and N. America and introduced into S. Africa.

Chikusichloa Koidz., Bot. Mag. Tokyo 39: 23 (1925)
Fig. 168.

Perennial, tufted, with short-branched rhizomes and ascending stems; culms 100 to 120 cm high. Leaf blades thin and firm, 0.8 to 1.2 cm wide and 30 to 70 cm long, scabrous; apex acuminate; ligules truncate, 2 to 3 mm long. Inflorescence an erect panicle, 40 to 50 cm long, rather open, the numerous very slender branches fasciculate, capillary, ascending to spreading; pedicels very slender, elongate, jointed about the middle, the upper part scabrous, disarticulating with the spikelet, forming a stipe 2 to 4 mm long. Spikelets 3 mm long, 1-flowered, awned or awnless, scabrous, somewhat laterally compressed or subterete. Glumes and sterile lemmas absent. Fertile lemma elliptic, thin, membranous, brownish, scabrous, 5- to 7-nerved, not keeled on back; sometimes with a terminal erect awn 3 to 6 mm long. Palea as long as lemma. Stamen 1. Grain fusiform, 2 to 2.5 mm long.

2 species: *C. aquatica* Koidzumi aquatic: in S. E. Asia, in shady sides of mountain streams. *C. mutica* Keng semiaquatic?: from S. W. China; differs from *C. aquatica* chiefly in having awnless spikelets.

Fig. 168. *Chikusichloa aquatica* Koidzumi: a, habit (2 cm); b, spikelet (1 mm).

Fig. 169. *Coix lacryma-jobi* L.: a, inflorescence (1 cm); b, spikelet (1 mm); c, upper glume of male spikelet.

Coix L., Sp. Pl. 972 (1753)
Fig. 169.
Annuals or perennials; culms solid, 1 to 3 m long, floating or erect and

branched. Leaf blades 8 to 100 cm long and 1.5 to 5 cm wide, linear to ovate-lanceolate, with midrib prominent, wide and pale towards the base; ligules short, ciliate; leaf sheaths glabrous or with long tubercled hairs at apex. Inflorescence partly enclosed in a nut-like hardened, ovoid or cylindric sheath (utricule) with united margins, with pore at apex only, sometimes bearing a rudimentary blade; the whole forming a false fruit, ultimately falling off with grain inside. Inflorescence probably consisting of 2 spike-like racemes; 1 inside the utricule, consists of 2 rudimentary, long-pedicelled spikelets and 1 sessile or shortly-pedicelled female spikelet; the other raceme, outside the utricule, consists of several male spikelets. Male spikelets in threes or twos, 2-flowered, or lower floret not fully developed, awnless, compressed dorsally. Female spikelets 2-flowered; lower lemma sterile; lower glume almost completely embracing spikelet. Stamens 3.

c. 1 to 4 species and several subspecies: in warmer regions of the world. *C. aquatica* Roxb. is an emergent aquatic in ditches and pools. *C. gigantea* Koenig ex Roxb. is reported to be a pest in ricefields.

Coelachne R. Br., Prodr. Fl. Nov. Holl. 187 (1810)
Fig. 170A.

Annual or perennial small and slender grasses; culms procumbent, ascending, 2.5 to 20 cm long, sometimes longer. Leaf blades narrow to ovate-lanceolate, 1 to 2 cm long; ligule a fringe of hairs. Inflorescence a panicle, not dense, loose, open or contracted, 5 to 10 cm long. Spikelets disarticulating above the glumes, about 2.5 mm long, 2-flowered, terete, awnless. Lower floret bisexual, upper floret female. Rachilla elongated between florets. Glumes membranous, subequal, much shorter than spikelet (0.8 to 1.5 mm long), ovate, obscurely 5-nerved, apex acute to obtuse. Lower lemma as long as spikelet, 3 to 5-nerved, somewhat hardened with palea at maturity. Upper lemma remaining membranous, narrower, shorter than lower one or subequal to it. Stamens of lower floret 2 or 3. Grain ovoid-oval, loosely enclosed by lemma and palea; embryo short.

c. 5 very similar species from Australia to Africa, India, S. China, Japan, Philippines. Generally semiaquatic, in wet soil, marshes, ditches.

Coleanthus Seidel, in R. and S. Syst. Veg. 2: 276 (1817)
Fig. 170B.

Tiny annual, spreading, more or less matted; culms about 5 to 6 cm long. Leaf blades 10 to 20 mm long, 0.5 to 1.5 mm wide; ligules 1 to 1.5 mm long; leaf sheaths inflated. Inflorescence a panicle, 1 to 5 cm long, consisting of 3 to 6 verticillate, peduncled and sessile, umbel-like clusters of spikelets. Spikelets 1-flowered. Glumes absent. Lemmas about 1 mm long, differentiated into an ovate basal portion and an awn-like tip, 1-nerved. Palea about 0.5 mm long, 2- or 4-toothed at the tip, 2-nerved, apex cuspidate. Stamens 2. Grain fusiform, exceeding lemma and palea.

1 species, *C. subtilis* (Tratt.) Seidel, originally from subtemperate Eurasia, introduced to N. America. In mud-flats and banks along rivers.

Fig. 170. A. *Coelachne simpliciuscula* Munro: a, habit (1 cm); b, spikelet (1 mm); B. *Coleanthus subtilis* (Tratt.) Seidel: a, habit (1 cm); b, inflorescence (1 mm).

Fig. 171. *Diplachne fusca* (L.) Beauv.: a, habit (1 cm); b, spikelet (1 mm).

Diplachne Beauv., Agrost. 80 (1812)
Fig. 171.
Perennials, tufted, somewhat coarse grasses; culms moderately long, up to about 130 cm. Leaf blades up to 30 cm or more long, narrow, flat or involute; ligules membranous, sometimes reduced to a fringe of hairs. Inflorescence a terminal, up to 30 cm long panicle, consisting of distant, solitary or pinnate or ternate spike-like racemes. Spikelets inserted on 2 sides of the rachis, turned up 2 sides, disarticulating above the glumes and between the florets, up to 15 mm long, 2- to 10-flowered, usually mucronate, terete or compressed dorsally. All florets bisexual or the uppermost reduced or sterile. Glumes shorter than or as long as spikelet, unequal or subequal, membranous, 1-nerved, keeled, apex acute; persistent. Lemmas membranous, with very narrow, incurved, hyaline margins, 3-nerved, with lateral nerves submarginal and percurrent or excurrent, midnerve prominent; apex usually distinctly notched and mucronate from the sinus; midnerve and margins in the lower part usually with silky-white, erect, appressed hairs. Paleas 2-keeled, subequal to lemmas. Stamens 3. Grain enclosed by slightly altered lemma and palea, oblong to obovoid-oblong, compressed dorsally, rarely terete.

c. 20 species: in warmer regions of the Old and the New World, but most species American. Usually semiaquatic to aquatic grasses. A widespread species (Africa, Asia, Australia) is *D. fusca* (L.) Beauv.

Echinochloa Beauv., Agrost. 53 (1812)
Fig. 172.
Perennials or annuals, frequently stoloniferous, often more or less succulent; culms pithy or hollow, ascending, up to 2.50 m or more long (but *E. pyramidalis* Hitchc. and Chase up to 4.50 m high). Leaf blades flat, 0.3 to 2.5 cm wide and 5 to 50 (–60) cm long; ligule a fringe of hairs, or absent. Inflorescence a panicle, usually up to 25 cm long, composed of crowded or loose, one-sided, spike-like racemes, along a main axis, not ending in a bristle: lower racemes solitary, fascicled or sometimes branched near base; rachis flattened or 3-angled. Spikelets solitary, subsessile or in irregular clusters, disarticulating from the pedicel, 3 to 5 mm long (rarely up to 8 mm long, in *E. polystachya* (Kunth) Hitchc.); 2-flowered; with lower floret male or sterile and upper one bisexual. Glumes 2, unequal; lower glume about half as long as spikelet, 3- or 5-nerved, acute cuspidate or short-awned, not clearly abaxial; upper glume as long as spikelet. Lower lemma as long as upper glume, flat or flattened on the back awned or cuspidate (awn up to 5 cm long), 5- to 7-nerved; upper lemma as long as spikelet, rounded on the back, not enclosing apex of palea, ovate-elliptic to lanceolate-oblong in outline, faintly 5-nerved; margins inrolled below. Stamens 3. Grain tightly enclosed by hardened lemma and palea, fusiform.

Hejný, S. Eine Studie über die Oekologie der Echinochloa-Arten. Biol. Pr. S. A. V. Bratislava 3 (5): 1–114 (1957)

Yabuno, T. Biosystematic studies of the genus Echinochloa. Jap. Journ. Bot. 19 (2): 277–323 (1966)

Yabuno, T. Biosystematic studies of Echinochloa stagnina and E. pyramidalis. Cytologia 33: 508–519 (1968)

Yabuno, T. Biosystematic studies of the genus Echinochloa. Proc. 12th Int. Genet. Cong. Tokyo 1: 184 (1970)

Fig. 172. *Echinochloa stagnina* (Retz.) Beauv.: a, habit (15 mm); b, spikelets on rachis (5 mm); c, spikelet (5 mm).

c. 30 species: cosmopolitan in warmer parts of world, probably introduced into America. About half the species are aquatic or semi-aquatic. Some species are cultivated for grain (Sawa, Sama, Sanwa or Millet) and forage, often in places unsuitable for rice. Some species of *Echinochloa* are often troublesome weeds in ricefields and irrigation channels; others are sudd-forming. For control in ricefields see Parker, C. and Byers, D. S. Proc. Afr. Weed Contr. Conf. 92–98 (1958). In Italy biological control using a smut fungus (*Sporosporium bullatum* Schr.) is being tried.

Elytrophorus Beauv., Agrost. 67 (1812)
Fig. 173.
Annuals; culms branched at base, 30 to 50 cm long. Leaf blades expanded, often overtopping the inflorescence; ligule membranous, sometimes minutely hairy. Inflorescence consisting of dense, globose or cylindrical clusters of spikelets on a central axis, the whole forming in the lower parts of the inflorescence a distinctly interrupted, in the upper parts a more or less uninterrupted spike-like panicle; the base of a cluster is formed by 2 to many bracts or sterile spikelets. Spikelets 4- to 6-flowered, strongly compressed laterally, 2 to 8 mm long, including awns; rachilla disarticulating between the florets. Glumes equal or slightly unequal, about as long as spikelet, awned or awnless, 1-nerved. Lemmas ovate-elliptic, 3- or 5-nerved, nerves and margins shortly hairy throughout or only on the upper part, the middle nerve produced

Fig. 173. *Elytrophorus globularis* Hack.: a, habit (1 cm).

Fig. 174. A. *Glyceria fluitans* (L.) R.Br.: a, habit (15 mm); b, spikelet (5 mm); B. *Glyceria maxima* (Hartm.) Holmb.: a, spikelet (5 mm).

into a short awn; uppermost lemma usually sterile. Paleas with keels conspicuously winged. Stamens 1 to 3. Grain small, ovate-elliptic to narrowly elliptic in outline; embryo almost as long as grain.

Schweickerdt, H. G. Ann. Natal Mus. 10(2): 191–214 (1942)

c. 4 species: India to S. China, Australia and Africa. Generally in pans, ditches and pools that dry out each year, often gregarious.

Glyceria R. Br., Prodr. Fl. Nov. Holl. 1: 179 (1810)
Fig. 174.
Perennials, rhizomatous or stoloniferous; culms frequently up to 1 m long (*G. maxima* (Hartm.) Holmb. up to 2.50 m long). Leaf blades 1.5 mm to 20 mm wide, flat or folded; ligules thin and usually prominent (up to 15 mm long); leaf sheaths closed. Inflorescence a panicle 4 to 45 cm long, narrow to open; branches solitary, in pairs or threes. Spikelets disarticulating above the glumes and between the florets, 5 to 35 mm long, up to 3.5 mm wide, (3-) 5- to 15-flowered, awnless, linear and subterete, or oblong to ovate in outline. Glumes 2, 1-nerved, shorter than the lowest lemma, membranous; apex generally acute to obtuse. Lemmas broad, rounded on the back, firm, with 1 to 9 rather prominent nerves; apex acute-rounded, truncate or obtuse, or 3-toothed. Stamens 3 or 2. Grain ovoid-oblong to obovoid, free.

Borrill, M. A biosystematic study of some Glyceria species in Britain. Watsonia 3: 291–306 (1956), op. cit. 4: 77–100 (1958)
Church, G. L. A cytotaxonomic study of Glyceria and Puccinella. Amer. Journ. Bot. 36: 155–165 (1949)
Lambert, J. Biological Flora of the British Isles: Glyceria maxima. Journ. Ecol. 34: 310–344 (1947)
Westlake, D. F. The biomass and productivity of Glyceria maxima. Journ. Ecol. 54: 745–753(1966)

c. 16 species, most of them semiaquatic to aquatic. In temperate and subarctic regions of the Northern Hemisphere, and in Australia, New Zealand, Tristan da Cunha and S. America.

Gynerium Willd. ex Beauv., Agrost. 138, 164 (1812)
Fig. 175.
Plants dioecious, very large, sometimes 10 m high. Culms clothed below with old sheaths. Leaf blades about 2 m long and 4 to 6 cm wide, forming a large fan-shaped summit to the sterile culms, with serrulate margins. Inflorescence a panicle, 1 m or more long, pale, plume-like; main axis erect; branches drooping. Spikelets usually 2-flowered, unisexual, with elongated internode between lower and upper floret. Male spikelets usually 3 to 4 mm long; glumes slightly unequal, upper one longer, about as long as spikelet or somewhat shorter, acuminate; lemmas glabrous, acuminate. Female spikelets about as long as male ones; glumes unequal, lower one as long as spikelet, thin; upper one produced in to about 12 mm long awn. Lemmas with long silky hairs.

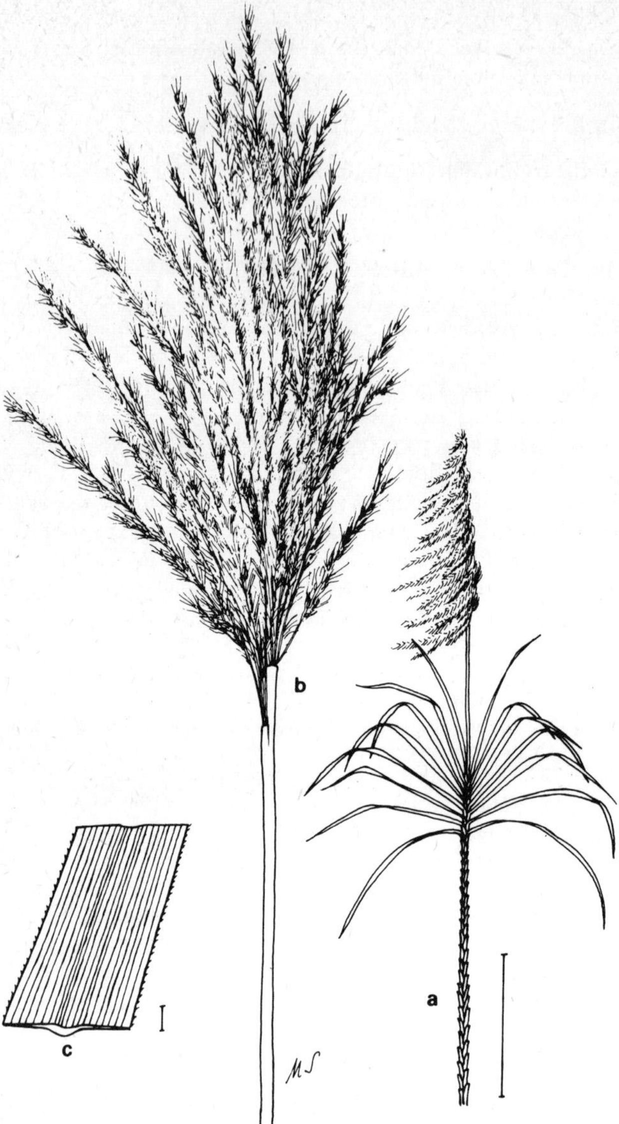

Fig. 175. *Gynerium sagittatum* (Aubl.) Beauv.: a, habit (1 m); b, inflorescence; c, segment of leaf (1 cm).

1 species, *G. sagittatum* (Aubl.) Beauv.: in swamps of Tropical America, often forming dense stands in or along stream banks. The stems are used as a substitute for bamboo.

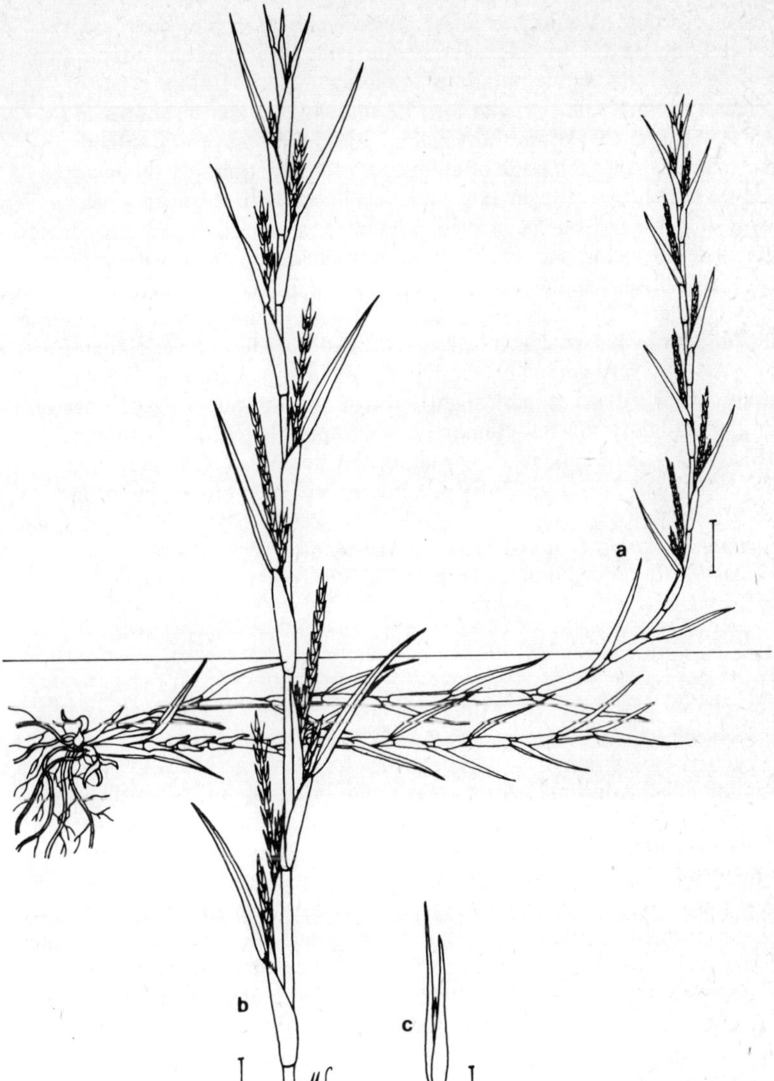

Fig. 176. *Hemarthria natans* Stapf: a, habit (3 cm); b, inflorescence (1 cm); c, spikelet (1 mm).

Hemarthria R. Br., Prodr. Fl. Nov. Holl. 1: 207 (1810)
Fig. 176.
Perennials, stoloniferous; culms either floating (3 to 8 m long), or ascending from long creeping base, or geniculate, up to 1 m high, compressed, many noded and branched. Leaf blades 3 to 8 mm wide, folded in bud, afterwards more or less flat; ligules membranous, very short, ciliate. Inflorescence of fascicled or solitary spikes; spikes 3 to 12 cm long, terminating culms and branches, but branches often short and in clusters so that several spikes may appear to arise from each node of culm; fascicles sometimes partly enclosed at base by leaf sheath of main axis. Spikelets in opposite pairs; each pair basically consisting of one pedicelled and one sessile spikelet; pedicels adnate to internodes of rachis, thus each opposite pair made up of a sessile spikelet and of the pedicelled spikelet of the next lower node; rachis tough, disarticulating tardily. Spikelets 5 to 7 mm long, 2-flowered, lower floret reduced to a sterile lemma, upper one bisexual; awnless; compressed dorsally. Lower glume leathery, usually very gradually acuminate, or acute, or obtuse, entirely or minutely 2-toothed, acute, with slightly rough margins, not winged, 7-nerved; upper glume slightly shorter, membranous except at hardened tip; glumes of primary pedicelled spikelets more elongated, acuminate. Sterile lemma 2-nerved. Stamens 3. Grain slightly compressed dorsally, oblong in outline.

c. 8 species, *H. natans* Stapf of Trop. E. Africa and Madagascar is aquatic. Other species are semiaquatic in Tropics and Subtropics of the Old and New World.

Hemisorghum C. E. Hubbard in Bor, Grasses of Burma, Ceylon, India and Pakistan 686 (1960)
Fig. 177.
Tufted perennial; culms tall, stout, erect, solid, terete, simple or sparingly branched. Leaf blades lanceolate linear, broad, flat; ligules very short, truncate, densely short-ciliate. Inflorescence a large, loose panicle, with numerous verticillate 1st order branches; the spikelets are borne in pairs of one pedicelled and one sessile spikelet in spike-like racemes (2nd order branches); at maturity the sessile spikelets disarticulate with the contiguous inferior internode of the rachis and the pedicelled spikelet; sessile spikelets 2-flowered, lower floret sterile, upper one bisexual, compressed dorsally, lanceolate to oblong-lanceolate, obtuse, awnless, contiguous to slightly imbricate; callus truncate, extremely short, bearded with a ring of minute hairs. Glumes dissimilar, equal or nearly so; lower one lanceolate, obtuse, flat or slightly convex on the back in the lower part, 2-keeled throughout, with narrow inflexed margins, closely serrately scabrid on the keels, coriaceous in the lower half, thinner above, 8- to 11-nerved; upper one lanceolate, acute, rounded on the back below the middle, slightly keeled towards the apex, coriaceous below, thinner upwards, 7-nerved. Lower lemma slightly shorter than glumes, lanceolate oblong, obtuse, faintly 2-nerved; palea absent; upper lemma nearly as long as the lower one and similar in texture, finely 1-nerved, with the nerve sometimes produced as a minute mucro

Fig. 177. *Hemisorghum mekongense* (A. Camus) C. E. Hubbard: a, inflorescence (2 cm); b, spikelets (1 mm).

between the 2 minute, apical lobes; palea equal to lemma, nerveless. Stamens 3. Grain obovate-oblong, compressed dorsally, tipped with the united style-bases; embryo about half the length of the grain. Pedicelled spikelet: sterile, smaller than the sessile one; lower glume linear to narrowly-lanceolate, 2-keeled, scabrid on the keels, 5 to 8 nerved; upper glume shorter than the lower, 5-nerved, keeled on the back towards the apex; lower lemma reduced or absent; upper lemma absent.

1 species, *H. mekongense* (A. Camus) C. E. Hubbard [*Sorghum halepense* (L.) Pers. var. *mekongense* A. Camus] in Laos, Thailand, Burma. On river banks.

Hydrochloa Beauv., Agrost. 135, 165 (1812)
Fig. 178.
Perennial (probably), very slender, usually floating and only inflorescence erect; culms up to 1 m or more long, freely branching, leafy. Leaf blades flat, 10 to 40 mm long, 1 to 7 mm wide; ligules very short; leaf sheaths usually closed. Inflorescence composed of terminal and axillary racemes with few spikelets; terminal racemes male; axillary racemes female. Spikelets disarticulating from the pedicel, 1-flowered, unisexual, awnless. Glumes and sterile lemmas absent. Fertile lemma thin, 7-nerved. Male spikelets with 2-nerved palea; stamens 6. Female spikelets with 5-nerved palea.

1 species: *H. caroliniensis* Beauv. In ponds and slow-flowing streams of S. E. N. America, usually floating, sometimes in sufficient abundance to become troublesome.

Hydrothauma C. E. Hubbard, in Hooker's Ic. Pl. 3458 (1947)
Fig. 179.
Annual; culms up to 20 cm high, geniculate, prostrate or ascending, branching at lower nodes. Leaf blades floating, up to 10 cm long and 0.3 cm wide, with ribbon-like flanges or lamellae on the upper surface, glabrous, connected to leaf sheath by an up to 9 cm long stalk; ligule about 1 mm long; leaf sheath of lower leaves with auricules. Inflorescence a panicle, about 6 cm long and 0.3 cm broad, densely spike-like. Spikelets disarticulating below the glumes, 2-flowered, awnless, compressed dorsally; lower floret male or sterile; upper bisexual. Glumes unequal; lower about half as long as spikelet, nerveless; upper equalling spikelet, 7- to 9-nerved. Lower lemma equalling upper glume, 7-nerved, dorsally flat; upper obscurely nerved, dorsally convex. Stamens 3. Grains small.

1 species: *H. manicatum* C. E. Hubbard, in shallow pans in Zambia.

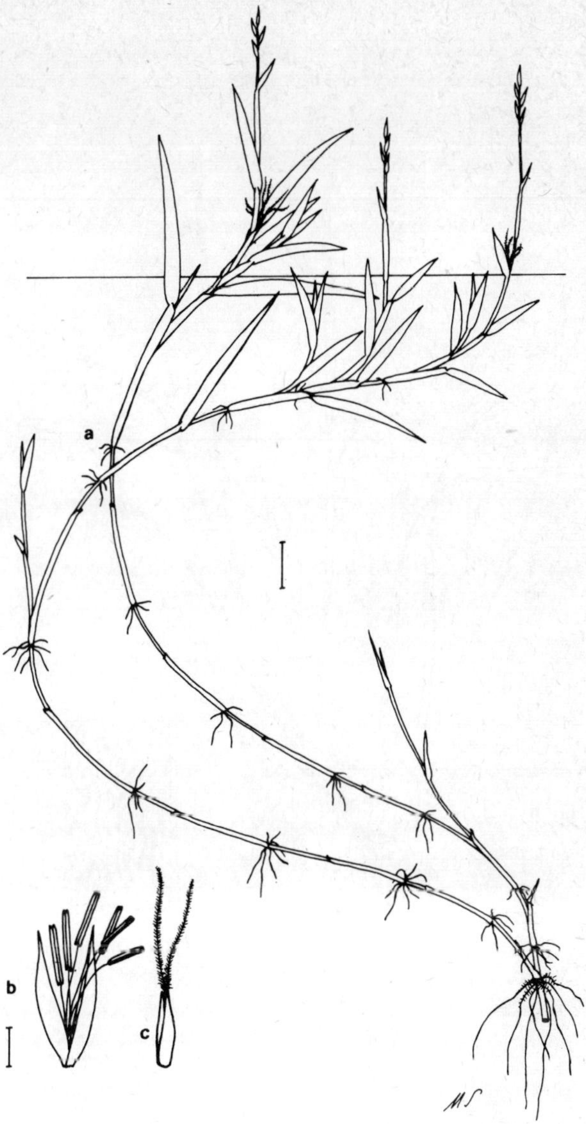

Fig. 178. *Hydrochloa caroliniensis* Beauv.: a, habit (1 cm); b, male spikelet; c, female spikelet (1 mm).

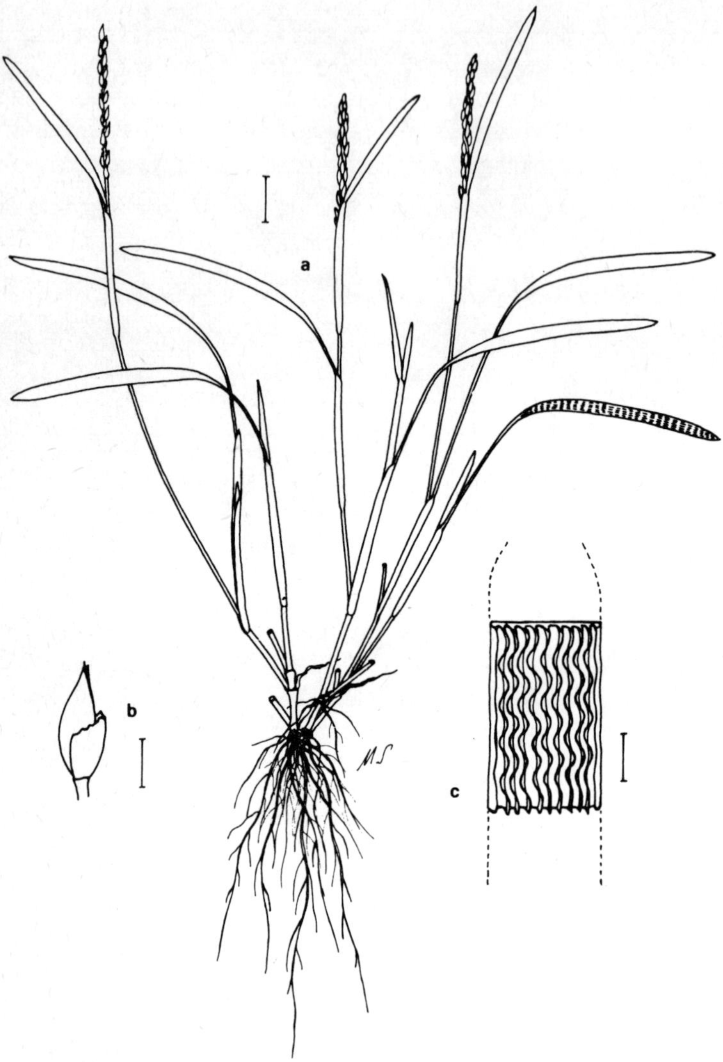

Fig. 179. *Hydrothauma manicatum* C. E. Hubbard: a, habit (1 cm); b, spikelet (1 mm); c, upper surface of leaf segment (1 mm).

Hygroryza Nees, in Wight et Arnott, Edinb. New Philos. Jour. 15: 380 (1833)
Fig. 180.
Perennial, stoloniferous; culms usually floating, spongy, branched, ascending, 30 to 50 cm long. Leaf blades 2 to 8 cm long and 0.5 to 1.7 cm wide, ovate to narrowly ovate-oblong; ligules absent; leaf sheaths open, usually strongly inflated. Inflorescence a panicle, 2 to 5 cm long, pyramidal, with few spikelets; pedicels long. Spikelets disarticulating from pedicel, 6 to 8 mm long, 1-flowered, awned, slightly compressed laterally. Glumes and sterile lemmas absent. Lemma thin, papery, 5-nerved, narrowed into an awn; awn 0.5 to 1.7 cm long. Palea as long as lemma, narrower, 3-nerved. Stamens 6. Grain narrowly oblong.

1 species: *H. aristata* (Retz.) Nees, floating in slow-moving streams, in ponds, tanks and lakes in India and S. E. Asia. It is readily eaten by cattle and the grains are said to be eaten by poor people.

Hymenachne Beauv., Agrost. 48, 165 (1812)
Fig. 181.
Perennials, stoloniferous; culms floating, creeping or ascending, 1 m or more long, sparingly branched, containing white pith. Leaf blades 7 to 35 cm long and 0.6 to 4 cm wide, linear or linear-lanceolate, amplexicaul at base, with acute apex; ligule membranous. Inflorescence a panicle, either very dense, spike-like, or long and narrow, with more or less appressed erect branches. Spikelets 1.5 to 5.5 mm long, subterete, awnless or with acuminate apex, 2-flowered, lower floret sterile, reduced to 1 lemma; upper floret bisexual awnless or acuminate, subterete. Glumes unequal; lower much shorter than spikelet, with acute apex; upper slightly shorter than spikelet, with acute to slightly acuminate apex. Lemmas acute to acuminate; sterile one as long as spikelet, 3- or 5-nerved; fertile one embracing palea except at apex. Stamens 3. Grain free; lemma and palea not hardened, more or less open at apex.

c. 8 species, aquatics and semiaquatics, in the Tropics of the Old and New World.

Isachne R. Br., Prodr. Fl. Nov. Holl. 196 (1810)
Fig. 182.
Annuals or perennials, often stoloniferous; culms usually geniculately ascending from a prostrate base, sometimes erect, up to 80 cm long. Leaves often with tubercle-based hairs, or hairless; leaf blades linear to lanceolate or ovate, usually closely and prominently nerved on the upper surface; ligule a fringe of hairs. Inflorescence a panicle, 2.5 to 12 cm long, open or somewhat contracted. Spikelets pedicelled, disarticulating above the glumes, 1.3 to 4 mm long, ellipsoid to obovoid, or globose, 2-flowered; both florets mostly bisexual, occasionally 1 floret unisexual, awnless. Glumes 2, tardily deciduous, equal or nearly so, as long as spikelet or nearly so, membranous, 5- to 9-nerved, or lower one 3-nerved; apex generally obtuse or truncate. Lemmas similar or dissimilar, narrowly elliptic to orbicular, obtuse or rounded at apex, as long as spikelet or exceeded by glumes, finely membranous to thinly

Fig. 180. *Hygroryza aristata* (Retz.) Nees: a, habit (1 cm); b, spikelet (5 mm).

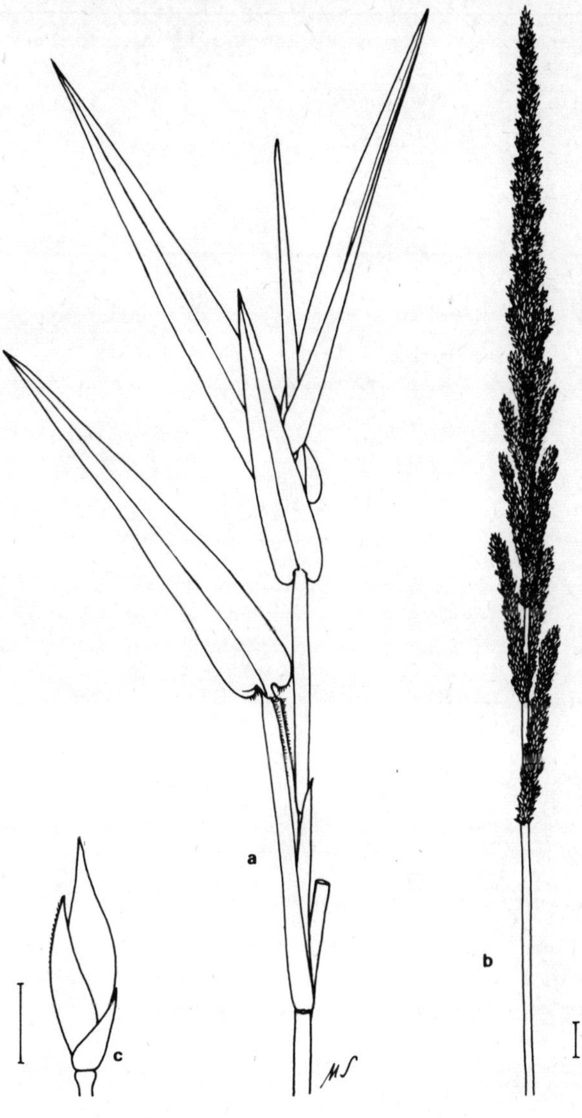

Fig. 181. *Hymenachne donacifolia* Chase: a, shoot apex; b, inflorescence (1 cm); c, spikelet (1 mm).

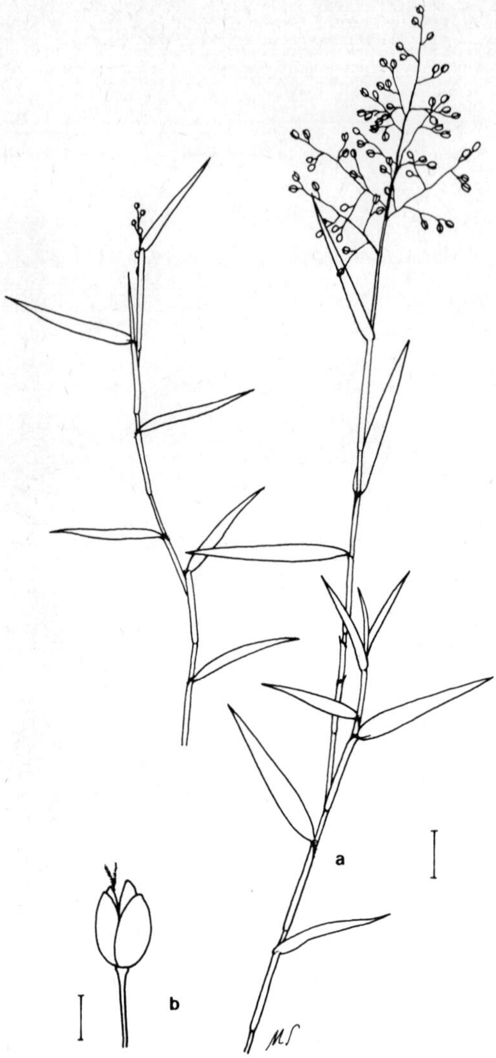

Fig. 182. *Isachne globosa* (Thunb.) O. Ktze.: a, habit (1 cm); b, spikelet (1 mm).

coriaceous, glabrous or minutely hairy, obscurely 5- to 7-nerved. Paleas similar in shape and texture to lemmas. Stamens 3. Grain enclosed between lemma and palea, plano-convex; obovate, elliptic or orbicular.

Pai-Chieh, K. Revision of the genus Isachne R. Br. (Gramineae) of China. Acta Phyto. Sinica 10: 6 (1965)

c. 60 species, some of them semiaquatic to aquatic, growing in swampy, marshy places, sometimes as weeds in ricefields; in tropical regions, most of them in tropical Asia.

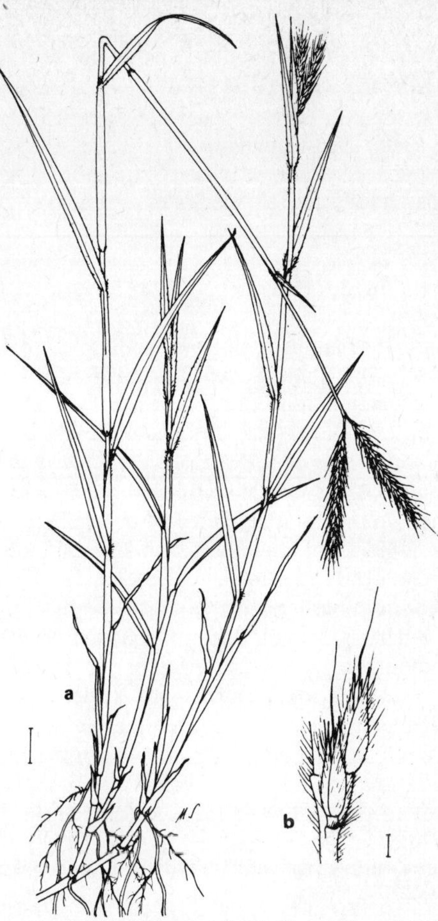

Fig. 183. *Ischaemum arcuatum* (Nees) Stapf: a, habit (1 cm); b, spikelets on rachis (1 mm).

Ischaemum L., Sp. Pl. 1049 (1753)
Fig. 183.
Perennials, with creeping, sometimes extensive, rhizomes; culms 30 to 120 cm long, simple or branched. Leaf blades flat, 4 to 16 mm wide, with apex tapering abruptly to a point; ligules well developed; leaf sheath with auricules. Inflorescence of (1) 2 to 7 spike-like racemes, arranged in pairs or groups. Spikelets in pairs, one always pedicelled, the other usually sessile; sessile or subsessile spikelet 5 to 6 (–8) mm long, awned or awnless 2-flowered. Lower floret male or sterile; upper bisexual. Pedicelled spikelet like sessile one, or showing all stages of reduction, sometimes represented by a small glume, especially in the upper part of the racemes. Glumes usually hard and rigid, lanceolate; lower glume awnless, usually 2-winged near margins; upper glume awnless or awned, 1-winged or keeled. Upper lemma 2-lobed and awned from between the lobes; awn bent and twisted. Paleas as long as lemmas. Stamens 3. Grain compressed dorsally.

Fujinoto, Y. Classification of Ischaemum L. based on vegetative characters, especially on the ligule. Hikobia 5: 104 (1967)

c. 40 species, some species semiaquatic to aquatic, e.g. *I. aristatum* L. in Asia, *I. santapaui* Bor in India, *I. arcuatum* (Nees) Stapf in S. E. Africa. Most species in the Tropics of the Old World, few species in America.

Leersia Swartz, Nov. Gen. Sp. Pl. Prod. 21 (1788)
Fig. 184.
Stoloniferous perennials or annuals; culms ascending, 20 to 150 cm long, occasionally longer when in floating mats. Leaf blades long, narrow, flat; ligules well developed, membranous. Inflorescence a panicle, usually exserted, terminal and more or less flaccid (in *L. oryzoides* (L.) Sw. also axillary panicles present; in some races of this species, particularly in northern regions, the panicles remain enclosed); spikelets often overlapping towards the end of the branches when young, looking then as if one spikelet were several-flowered. Spikelets compressed laterally, 1.5 to 8.5 mm long, usually awnless, 1-flowered. Glumes reduced to an entire or more rarely obscurely 2-lobed, somewhat thickened rim. Sterile lemmas absent. Fertile lemma ciliate-hispid to glabrous, 5-nerved, awnless (except in *L. nemostachya* Launert, *L. perrieri* (A. Camus) Launert, *L. stipiata* Bor and *L. tisserantii* (A. Chevalier) Launert). Palea awnless, occasionally awntipped, ciliate-hispid to glabrous, 3-nerved, subequal to lemma; margins clasping the inrolled margins of lemma. Stamens 1, 2, 3 or 6. Grain flat to only slightly compressed laterally.

Launert, E. A survey of the genus Leersia in Africa. Senck. Biol. 46 (2): 129–153 (1965)
Pyrah, G. Taxonomic and distributional studies in Leersia. Iowa State Journ. Sci. 44: 215–270 (1969)

c. 17 species, one of them, *L. hexandra* Sw., pantropical; 9 species in Africa, the rest in Asia, Europe, N. and S. America and Australia, in warm and temperate regions. *L. hexandra* is sometimes a weed in ricefields, irrigation channels etc. The delimitation of this genus from the genus *Oryza* is difficult, the most important character being the absence of sterile lemmas in *Leersia.* We follow here Launert who transferred the following 3 species of *Oryza* to *Leersia: Oryza angustifolia* C. E. Hubbard to *Leersia nemostachya* Launert, *O. perrieri* A. Camus to *L. perrieri* (A. Camus) Launert, *O. tisserantii* A. Chevalier to *L. tisserantii* (A. Chevalier) Launert.

Fig. 184. *Leersia hexandra* Swartz: a, habit (1 cm); b, spikelet (1 mm); c, stamens and styles.

Fig. 185. *Leptochloa aquatica* Scribn.: a, habit (1.5 cm); b, spikelet (1 mm).

Leptochloa Beauv., Agrost. 71 (1812)
Fig. 185.
Annuals or perennials; culms erect, usually up to 1.5 m long. Leaf blades flat. Inflorescence a panicle, composed of numerous, slender, 1-sided spikes (rachis at one side visible over the whole length). Spikelets disarticulating above the glumes and between the florets, up to 7 mm long, 3- to 9-flowered.

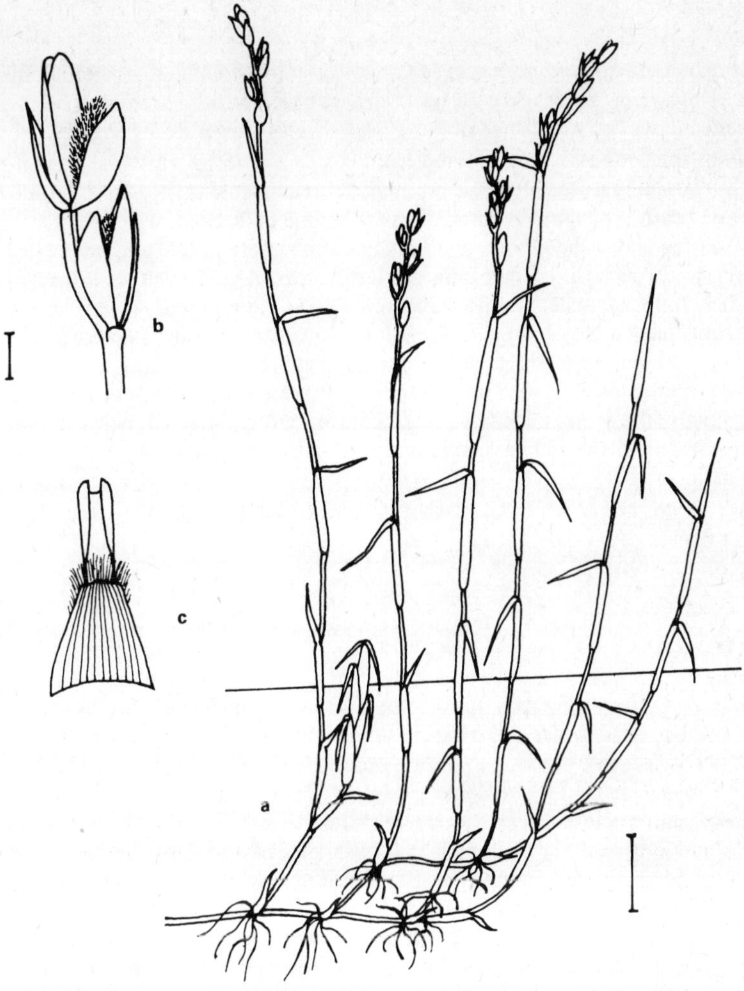

Fig. 186. *Limnopoa meeboldii* (C. E. C. Fischer) C. E. Hubbard: a, habit (1 cm); b, spikelets; c, ligule (1 mm), after Bor.

Florets awnless or mucronate, strongly compressed laterally. Glumes unequal to subequal, usually shorter than lowest lemma, 1-nerved, lanceolate to ovate, keeled, with acute or shortly mucronate apex; lower glume 1 to 1.5 mm long; upper glume 2 to 2.5 mm long. Lemmas obtuse or acute, sometimes 2-toothed and mucronate or short awned from between the teeth, 3-nerved, the nerves sometimes pubescent. Stamens 3. Grain ellipsoidal to obovate in outline.

c. 15 species: warmer regions of the world. *Leptochloa aquatica* Scribn. (Mexico) and *L. coerulescens* Steud. (Africa) are frequently found as aquatics in seasonal rivers and pools.

Limnopoa C. E. Hubbard, in Hook. Ic. 3432 (1943)
Fig. 186.
Annual or perennial, stoloniferous; culms delicate, prostrate and ascending, up to 20 cm high. Leaf blades 4 to 13 mm long, at base up to 1.5 mm wide, linear-lanceolate; ligules a fringe of hairs; leaf sheaths longer at the upper internodes than at basal ones, varying from 0.8 to 2 cm long. Inflorescence a solitary spike-like raceme, 1 to 2.5 cm long, emerging at maturity from uppermost leaf sheath. Spikelets in pairs, one sessile, one pedicelled, with rachilla disarticulating above the glumes and between the florets; sessile and pedicelled spikelets similar, 3.8 to 4.5 mm long, subterete, awnless, 2-flowered. Lower floret male; upper female. Glumes subequal, 2.7 to 3.5 mm long, glabrous; lower glume lanceolate-oblong; upper glume obovate to oblong. Lower floret with lemma as long as spikelet, membranous, ovate-elliptic or elliptic, glabrous, 5- to 7-nerved, palea almost as long as lemma, 2-keeled, stamens 3; upper floret with lemma densely and minutely pubescent, obscurely 5-nerved; palea similar to lemma. Grain enclosed between lemma and palea, plano-convex.

1 species, *L. meeboldii* (C. E. C. Fischer) C. E. Hubbard, floating in tanks in Kerala, India.

Louisiella C. E. Hubbard and Léonard, Bull. Jard. Bot. Bruxelles 22: 313–318 (1952)
Fig. 187.
Perennial, stoloniferous; culms up to several metres long, decumbent, becoming erect at tip. Leaf blades up to 20 cm long and 0.7 to 1 cm wide, linear or linear-lanceolate, apex tapering to a long point, basis abruptly contracted; ligules a fringe of hairs. Inflorescence a panicle, 8 to 10 cm long, 6 to 9 cm wide, lower part enclosed in uppermost leaf sheath, flaccid; primary branches up to 4.5 cm long, solitary or in pairs; pedicels 1 to 1.5 mm long. Spikelets disarticulating below the glumes, 7 to 8.5 mm long, including awn, almost terete, 2-flowered, lower floret sterile, reduced to 1 lemma; upper floret bisexual. Glumes very unequal; lower glume 1 mm long, adaxial, nerveless or 2-nerved; upper glume as long as spikelet, 7- to 9-nerved, acuminate, terminating in *c.* 4 mm long awn. Sterile lemma similar to upper glume, 7- to 9-nerved. Fertile lemma 4.5 mm long, piliferous, obscurely 7-nerved. Stamens 3. Grain oblong, plano-convex, compressed dorsally.

1 species, *L. fluitans* C. E. Hubbard and Léonard, in lakes and swamps of Central Africa; usually forming floating mats.

Fig. 187. *Louisiella fluitans* C. E. Hubbard and Léonard: a, habit (1 cm); b, spikelet (1 mm).

Fig. 188. *Luziola peruviana* Doell: a, habit (1.5 cm); b, male spikelet; c, female spikelet (1 mm).

Luziola Juss. Gen. 33 (1789)
Fig. 188.

Perennials, usually stoloniferous; culms usually ascending, 10 to 45 cm long. Leaf blades flat, 2 to 6 mm wide and as much as 40 cm long; ligules usually long. Inflorescence generally consisting of terminal and axillary panicles.

Spikelets pedicelled and disarticulating from pedicel, 1-flowered, unisexual, awnless; male spikelets mostly in the terminal panicles; female spikelets in the axillary panicles. (In *L. brasiliensis* (Trin.) Pilger male and female spikelets on the same branches of the terminal panicle, axillary panicles absent, female spikelets below.) Glumes and sterile lemmas absent. Male spikelets strongly compressed laterally, 4 to 8 mm long, early caducous, with lanceolate fertile lemma; stamens 6 to 16. Female spikelets ellipsoidal, 2 to 3 mm long, with usually oblong fertile lemma. Grain free, globose, finely grooved.

Swallen, J. H. The grass genus Luziola. Ann. Miss. Bot. Gard. 52: 472–475 (1965)

c. 11 species, most of them aquatic to semiaquatic, above all in marshes, swamps, lagoons. In America, from USA to S. Brazil, Uruguay, Argentina, Peru.

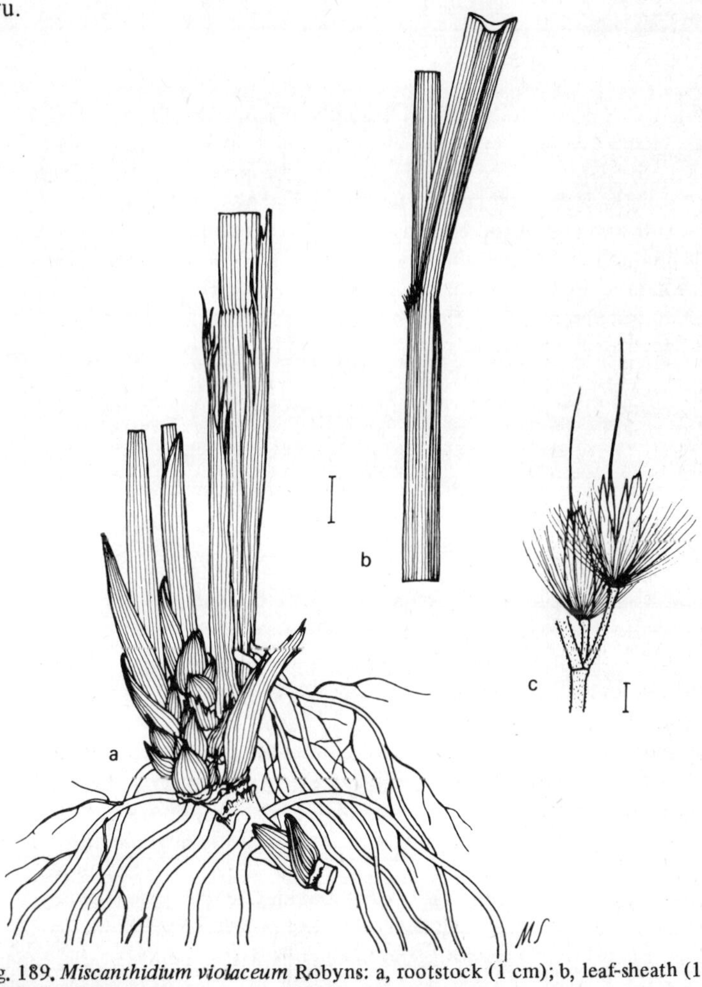

Fig. 189. *Miscanthidium violaceum* Robyns: a, rootstock (1 cm); b, leaf-sheath (1 cm); c, spikelet (1 mm).

Miscanthidium Stapf in Prain, Fl. Trop. Africa 9: 89 (1917)
Fig. 189.
Perennials, erect; culms up to 2.5 m long. Leaf blades up to 90 cm long or longer, narrow, midrib very stout or the whole blade terete; ligules membranous. Inflorescence a panicle, much branched, often large, silky. Spikelets borne in unequally pedicelled pairs, disarticulating from their pedicels, up to 5.5 mm long, awned, silky hairy and with involucre of hairs at base; 2-flowered. Lower floret reduced to an empty lemma; upper floret bisexual. Glumes 2, equal, leathery; lower glume dorsally flattened, more or less distinctly 2-keeled; upper glume keeled. Sterile lemma acute or mucronate; fertile lemma awned. Stamens 3. Grain oblong to linear-oblong.

c. 4 species: tropical and S. Africa. Large tufted grasses often found at the edges of pools, lakes and rivers. Frequently cultivated as a decorative plant outside Africa but apparently not becoming naturalised.

Neostapfia Davy, Erythea 7: 43 (1899)
Fig. 190.
Viscid annual; culms ascending from a decumbent base, 5 to 30 cm long. Leaves pale green, not differentiated into sheath and blade, loosely folded around culm, overlapping, 5 to 10 cm long, 1 to 1.2 cm wide at the middle, tapering to each end, minutely ciliate with raised glands on the margins and nerves. Inflorescence a spikelike panicle, 3 to 7 cm long, 0.8 to 1.2 cm wide, partially included in the uppermost leaves; upper part of axis bearing linear-lanceolate, empty bracts instead of spikelets. Spikelets subsessile 6 to 7 mm long, imbricate, usually 5-flowered. Glumes absent. Lemmas 5 mm long, fan-shaped, very broad, many-nerved, fringed with fine hairs.

Crampton, B. The grass genera Orcuttia and Neostapfia: a study in habitat and morphological specialisation. Madroño 15: 97–110 (1959)

1 species, *N. colusana* Davy: California, N. America. It is highly specialised and is apparently confined to vernal pools, it germinates as an aquatic but flowers as a terrestrial plant.

Odontelytrum Hackel, Oesterr. Bot. Zeitschr. 48: 86 (1898)
Fig. 191.
Perennial, rootstock submerged; culms partly submerged and partly floating, many-noded, branched. Leaf blades acute, up to 10 or 12 cm long and 3.5 mm wide, equally wide or only very slightly contracted at base, midrib very prominent above; ligule membranous 1 to 1.5 mm long; sheaths loose, with small auricles. Inflorescence a spike-like raceme, up to 7 cm long, sheathed at base, and with uppermost leaf blade exceeding tip of inflorescence. Spikelets solitary, supported by a shortly pedicelled partial involucre of united false bracts; involucre very irregularly 4- to 6-lobed, rough, purplish above, the entire portion up to $^2/_3$ the length of the spikelet, with 1 lobe free or nearly free to the narrow base, slightly keeled and produced above into a subulate and rough bristle exceeding the spikelet by 1 to 2 cm, the other lobes

Fig. 190. *Neostapfia colusana* Davy: a, habit (1 cm); b, detail of panicle; c, spikelet (1 mm).

triangular or linear and acute or slightly subulate, short and elongated and then only slightly exceeding the spikelet. Spikelets disarticulating with pedicel, up to 12 mm long, 2-flowered, awnless, slightly compressed dorsally; lower floret male, upper floret bisexual. Glumes very unequal; lower glume up to

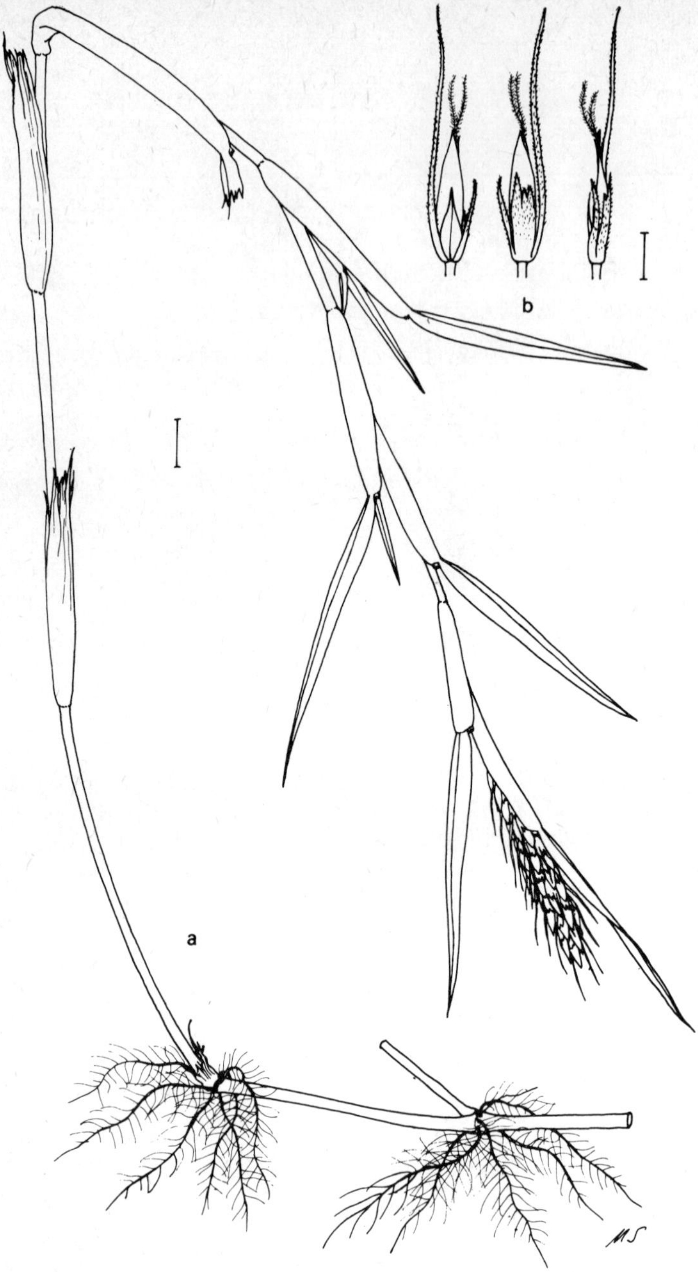

Fig. 191. *Odontelytrum abyssinicum* Hack.: a, habit (2 cm); b, spikelets (5 mm).

2 mm long, ovate, with obtuse to shortly acute apex; upper glume 5.2 to 8.7 mm long, ovate or ovate-lanceolate, with acute apex. Stamens 3.

1 species, *O. abyssinicum* Hackel, in stagnant water, in Abyssinia.

Fig. 192. *Oplismenopsis najadum* (Hackel and Arechav.) Parodi: a, habit (1.5 cm); b, spikelet (5 mm).

Fig. 193. *Orcuttia californica* Vasey: a, habit (1 cm); b, spikelet (1 mm).

differing in shape, either scale-like or linear, narrowly ovate to ovate, mostly shorter than fertile lemma. Fertile lemma rigid, keeled, 5-nerved: nerves prominent, building additional keels; margins and back of nerves usually fringed with stiff spreading hairs (absent in *O. meyeriana*), apex of lemma usually prolongated into a straight awn. Palea as long as fertile lemma, but narrower, keeled, 3- or 5-nerved. Stamens 6. Grain enclosed within lemma and palea, narrowly oblong.

Oplismenopsis Parodi, Not. Mus. La Plata Bot. 2: 2 (1937)
Fig. 192.
Perennial, rhizomes submerged or floating, up to 5 m long; culms 30 to 60 (–90) cm tall. Leaf blades 8 to 10 cm long, the middle ones 1 to 1.5 cm wide, lanceolate; ligules very shortly membranous at base, prolonged into a fringe of hairs; leaf sheaths striate, open. Inflorescence a panicle, consisting of several, more or less one-sided, ascending, alternate racemes; at the lower nodes often 2 or 3 racemes fascicled. Spikelets usually solitary, shortly pedicelled, 6 to 8 mm long, awned, more or less terete, 2-flowered; lower floret male, upper floret bisexual. Glumes 2, very unequal; lower glume distinctly adaxial, half as long as spikelet, 1-nerved; upper glume as long as spikelet, lanceolate, 7-nerved, produced in an up to 8 mm long awn. Lemmas similar but shorter than upper glume, 5- or 7-nerved, only awn-pointed, or shortly awned. Stamens 3.

1 species, *O. najadum* (Hackel and Arachav.) Parodi, in swamps, in the La Plata region (Argentina and Uruguay).

Orcuttia Vasey, Bull. Torrey Bot. Club 13: 219 (1886)
Fig. 193.
Annuals, with fibrous roots; culms tufted, fragile at the nodes, 5 to 20 (–30) cm long, viscid at maturity. Leaf blades rolling as they get older, ligules a fringe of hairs. Inflorescence a true spike, 2 to 10 cm long, rachilla not disarticulating. Spikelets up to 1.5 cm long, up to 20-flowered, compressed laterally. Glumes 2 to 5 (–10) mm long, cleft into 2 to 5 acute teeth at apex or sometimes apex acute and entire. Lemmas 11- to 15-nerved, with 5 to 9 acute or awn-like teeth at apex. Palea 2-keeled. Stamens 3. Grain small.

Hoover, R. F. The genus Orcuttia. Bull. Torrey Bot. Club 68: 149–156 (1941)
Crampton, B. The grass genera Orcuttia and Neostapfia: A study in habitat and morphological specialisation. Madroño 15: 97–110 (1959)

5 species: California, N. America; they are highly specialised and apparently confined to vernal pools.

Oryza L., Sp. Pl. 333 (1753)
Fig. 194A.
Annuals or rhizomatous or stoloniferous perennials; culms 1 to 2 m long. Leaf blades long, narrow, flat; ligules well developed, membranous. Inflorescence a panicle, 15 to 40 cm long, open or contracted, usually drooping. Spikelets pedicelled, disarticulating above the glumes (cultivated races of *O. sativa* L. do not disarticulate), 1.5 to 10 mm long and 3 to 4 mm wide, generally strongly compressed laterally, usually awned (except *O. glaberrima* Steud. and *O. meyeriana* (Zoll. and Mor. ex Steud.) Baillon), 3-flowered, lower 2 florets reduced to sterile lemmas, upper one bisexual. Glumes small, scale-like, rarely absent or reduced to a bilobed rim. Sterile lemmas 2, usually

Fig. 194. A. *Oryza sativa* L.: a, habit (1 cm); b, spikelet (2 mm);
B. *Rhynchoryza subulata* Baill.: a, part of inflorescence; b, spikelet.

Fig. 195. *Oryzidium bernardii* C. E. Hubbard & Schweickardt: a, inflorescence; b, lower part of stem (3 cm); c, spikelet (1 mm).

Chandler, R. (ed.) Rice genetics and cytogenetics. Elsevier Amsterdam: 1–274 (1964)
Mian, A. L. The wild and cultivated species of rice – their distribution and interrelationships. Pak. Journ. Sci. Ind. Res. 21: 96–105 (1969)
Mudra, A. "Reis", in Lehrbuch der Züchtung landwirtschaftlicher Kulturpflanzen. Ed. Hoffman, W., Mudra, A. und Plarre, W. Vol. 2: 158–170. P. Parey: Berlin and Hamburg (1970)
Sharma, S. D. and Shastry, S. V. S. Phylogenetic studies in the genus Oryza. L. Primitive and advanced characters. Il Riso 20: 127–136 (1971)
Sokolova, J. To the taxonomy of the genus Oryza. Bull. Appl. Bot. Gen. Plant Breeding. Leningrad 41: 117–147 (1970), non vidi.
Tateoka, T. Taxonomic studies of Oryza I., II., III. Bot. Mag. Tokyo 75: 418 (1962), 75: 455 (1962), 76: 165 (1963)
Tateoka, T. Notes on some grasses XVI. Embryo structure of the genus Oryza in relation to the systematics. Amer. Journ. Bot. 51 (5): 539–543 (1964)

19 species: warmer regions of the world. Most species are aquatic or semi-aquatic. *Oryza sativa* L. (rice) is the most important crop plant in the world and is usually cultivated as an aquatic annual. The literature on rice is enormous, only a few recent papers on the systematics have been chosen. The genera *Oryza* and *Leersia* are difficult to separate, the treatment of Launert (see under *Leersia*) has been adopted. The wild species of rice are particularly troublesome weeds in ricefields.

Oryzidium C. E. Hubbard and Schweickerdt, Kew Bull. 1936: 326 (1936)
Fig. 195.
Perennial, stoloniferous and partly floating in water; culms 90 to 120 cm long, smooth, soft and easily squashed. Leaf blades up to 30 cm long or longer and 0.6 to 0.8 mm wide, tapering to a long, fine point; ligule a fringe of hairs; leaf sheaths broad, papery. Inflorescence a panicle, 10 to 25 cm long, 2 to 4 cm wide, more or less contracted but not dense; branches whorled or solitary; pedicels solitary, 5 to 15 mm long. Spikelets disarticulating with pedicels, 8 to 10 mm long, awned, compressed dorsally 2-flowered, lower floret male, upper floret bisexual. Glumes 2, very unequal; lower glume adaxial, usually scale-like, white, 1 to 2 mm long, nerveless or 3-nerved; upper glume as long as spikelet, awned, prominently 7-nerved, scabrid on nerves, with 10 to 18 mm long awn. Lower floret with lemma 8 to 9 mm long, 5-nerved, smooth acute or produced into *c.* 1 mm long awn. Upper floret with lemma slightly leathery, 7-nerved; palea as long as lemma, 2-nerved. Stamens 3. Grain enclosed by lemma and palea, oblong in outline, strongly compressed dorsally.

1 species, *O. barnardii* C. E. Hubbard and Schweickerdt, in vleis and swamps, in S. W. Africa.

Panicum L., Sp. Pl. 55 (1753)
Fig. 196.
Annuals or perennials; tufted, stoloniferous or rhizomatous. Leaf blades mostly

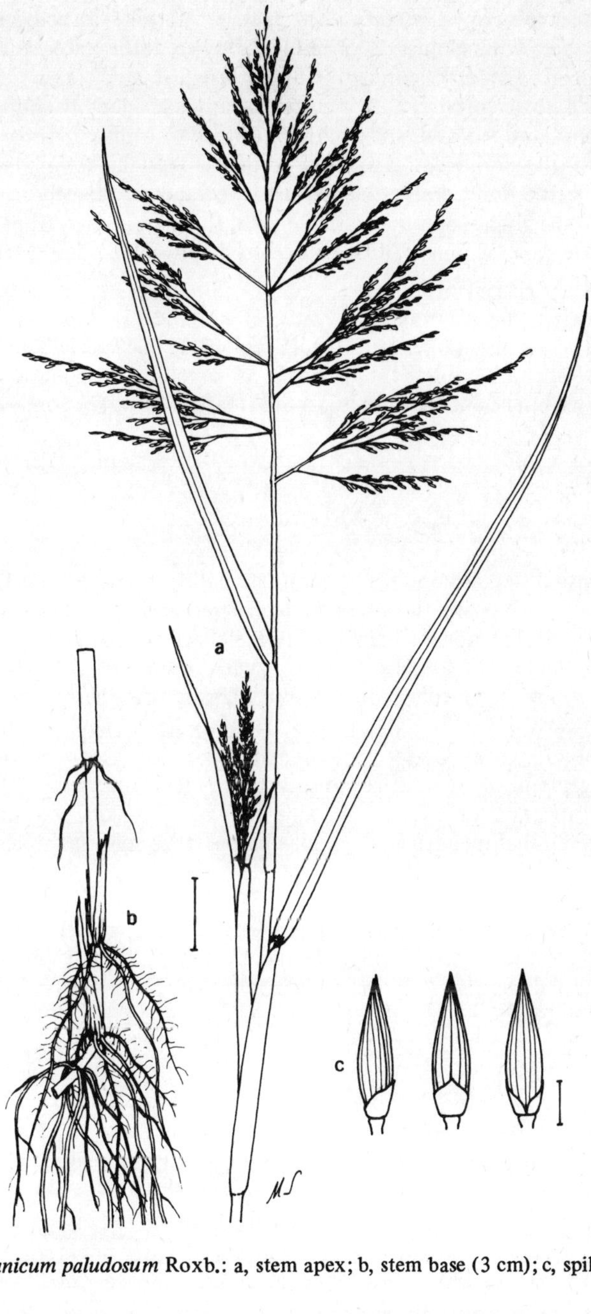

Fig. 196. *Panicum paludosum* Roxb.: a, stem apex; b, stem base (3 cm); c, spikelets (1 mm).

flat, rolled in bud, auricules absent; ligules ciliate or membranous; leaf sheaths open. Inflorescence an open or contracted panicle. All spikelets pedicellate, disarticulating below the glumes, 2- or rarely 1-flowered; the lower spikelet generally reduced to a sterile lemma; the upper bisexual, awnless, more or less compressed dorsiventrally. Lower glume usually considerably shorter than the upper, sometimes reduced; apex acute or obtuse to truncate; upper glume mostly equal to the sterile lemma and like it, the two usually completely enclosing the fertile floret. Fertile lemma hardened and very faintly nerved in comparison to the glumes and the sterile lemma; the margins overlapping the equally firm palea. Stamens 3. Grain enclosed between hardened lemma and palea, globose to ellipsoidal.

c. 500 species: cosmopolitan except in cold regions. A few species, such as *P. stagninum* Retz and *P. interruptum* Willd., are aquatic. The aquatic species are exclusively tropical and are usually found floating on water; some species are found as weeds in ricefields.

Paratheria Grisebach, Cat. Pl. Cub. 236 (1866)
Fig. 197.
Perennials, more or less stoloniferous; culms ascending, branched. Leaf blades linear, up to 15 cm long and 0.4 cm wide; ligules reduced to a hairy rim. Inflorescence a raceme, up to 15 cm long. Spikelets pedicelled and subtended by a 15 to 30 mm long bristle, disarticulating with the about 1.5 mm long pedicel, 6 to 9 mm long, awnless, compressed dorsally, with hairy base, 2-flowered; lower floret sterile, reduced to an empty lemma; upper floret bisexual. Glumes 2, very short, scale-like, membranous, nerveless. Sterile lemma as long as spikelet, lanceolate to subulate, 7- to 11-nerved; fertile lemma similar to sterile one, 7-nerved. Palea 2-nerved. Stamens 3. Grain enclosed between slightly hardened lemma and palea, compressed dorsally, narrowly elliptic-oblong in outline.

2 species, *P. prostrata* Griseb. and *P. glaberrima* C. E. Hubbard; the first aquatic to semiaquatic; the second apparently semiaquatic. *P. prostrata* is known from tropical Africa and tropical America, *P. glaberrima* only from Sierra Leone. The former has densely beared nodes while the latter is glabrous.

Paspalidium Stapf in Prain, Fl. Trop. Africa 9: 582 (1920)
Fig. 198.
Perennials, tufted or with many-noded, simple or branched, often floating stolons; or annuals; culms ascending, up to 1 m long. Leaf blades flattened or rolled; ligule a fringe of short hairs. Inflorescence a panicle, composed of several sessile, spikelike, one-sided racemes, arranged singly on alternate sides of a long central axis to which they are more or less appressed; main axis ending in a terminal raceme, usually longer than racemes immediately below it; lateral racemes with flattened rachis, abaxial side with zig-zag-shaped keel;

Fig. 197. *Paratheria prostrata* Griseb.: a, habit (1 cm); b, spikelet (1 mm).

Fig. 198. *Paspalidium geminatum* (Forsk.) Stapf: a, inflorescence (1 cm); b, spikelet (1 mm); c, rachis, abaxial view; d, rachis, adaxial view (1 mm).

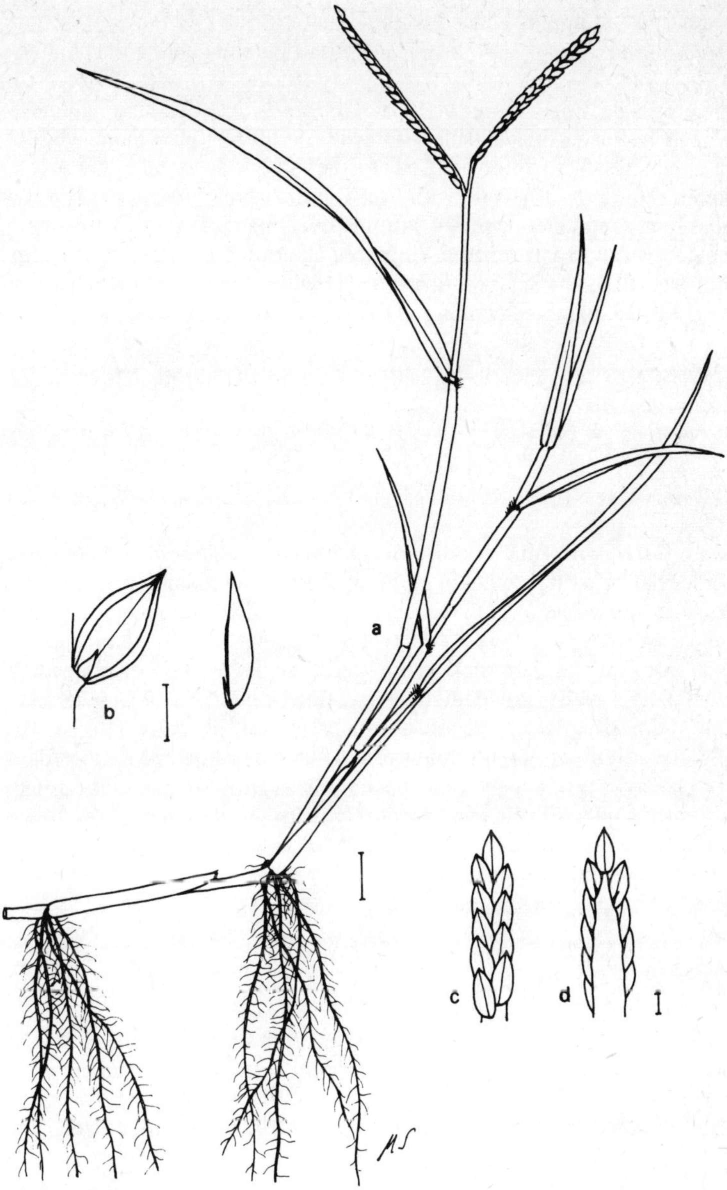

Fig. 199. *Paspalum distichum* L.: a, habit (1 cm); b, spikelets (1 mm); c, apex of rachis, abaxial view; d, rachis, adaxial view (1 mm).

all racemes ending in a short, 1 to 2 mm long bristle. Spikelets subsessile, solitary, on alternate sides of the rachis, close together, forming two dense rows, each spikelet appressed to a shallow cavity formed by the keel of the rachis and by the disc- or cup-like pedicel, disarticulating below the glumes, up to 3 mm long, awnless, nearly flat dorsally, convex ventrally, 2-flowered. Lower floret male; upper floret bisexual. Glumes 2, unequal; lower glume abaxial, much shorter than spikelet, truncate to obtuse; upper glume about $^2/_3$ as long as lemmas, 3- to 7-nerved. Lower floret absent or with lemma as long as spikelet, thin, 5- to 7-nerved, with broad, inflexed margins; upper floret with lemma as long as spikelet, leathery with incurved margins; apex acute, not mucronate; palea of same texture, embraced all around by margins of lemma; apex not free. Stamens 3. Grain tightly enclosed by the more or less hardened lemma and palea, plano-convex.

c. 12 species: generally aquatics to semiaquatics, in warmer regions of Eurasia, Africa and Australia.

Paspalum L., Syst. Nat. ed. 10, 855, 1359 (1759)
Fig. 199.
Perennials, tufted with fibrous roots, or rhizomatous; culms variable in height. Leaf blades flat or with inrolled margins; ligules usually membranous. Inflorescence composed of (1) 2 to many, 1-sided, spike-like racemes, racemosely arranged on all sides of the main axis; rachis of racemes flattened and keeled at abaxial side. Spikelets alternate along keel of rachis, solitary or in unequally pedicelled pairs, usually crowded, disarticulating below the glume, awnless, compressed dorsally, flat or concave abaxially, convex adaxially. Glumes 1, rarely 2; lower glume if present, minute, scale-like; upper glume and sterile lemma equal, as long as spikelet, or upper glume slightly shorter, with usually acute or obtuse apex. Fertile lemma leathery, with inrolled margins embracing palea all around, obscurely 3- to 7-nerved. Palea as long as lemma, 2-nerved, flat, membranous, inflexed margins. Stamens 3. Grain tightly enclosed between hardened lemma and palea, compressed dorsally.

Arber, A. The North-American species of Paspalum. Contrib. U.S. Nat. Herb. 28: 1–310 (1929)

c. 200 species: warmer regions of the world, with most species in America. A few species, such as *P. repens* Berg., *P. flaccidum* Nees and *P. scrobiculatum* L., are aquatic or semiaquatic. They are found in shallow water or floating on the surface. Several species have been reported to be weeds in ricefields.

Pennisetum Rich. in Pers., Syn. I 72 (1805)
Fig. 200.
Perennials or annuals, often stoloniferous or rhizomatous; culms simple or profusely branched. Leaf blades linear to lanceolate, flat, folded or rolled;

Fig. 200. *Pennisetum natalense* Stapf: a, habit (3 cm); b, spikelet (1 mm).

ligules a fringe of hairs, or membranous. Inflorescence a spike-like panicle, dense, cylindrical and bristly; branches numerous, usually very short and leaving stumps or scars on the rachis when falling. Spikelets 2-flowered, solitary or in clusters of 2 to 5, each spikelet or cluster surrounded by few to many bristles; bristles very slender or capillary, free or united at the base into a minute disc, very unequal, one bristle usually longer and stouter than the others. At maturity spikelet or spikelet-cluster disarticulates with the surrounding involucre of bristles. Lower floret male or sterile, rarely bisexual; upper floret bisexual; when spikelets are in clusters, then outer spikelets of cluster often male. Glumes 2, usually unequal; lower glume up to half as long as spikelet, 1- to 3-nerved; upper glume usually up to half, rarely almost as long as spikelet, 1, 3 or rarely 5-nerved. Lemmas somewhat similar to upper glume, 1- to 13-nerved. Stamens 3. Grain enclosed betweeen slightly hardened lemma and palea, mostly oblong or compressed dorsally. In cultivated forms the grain is free.

c. 130 species: *P. natalense* Stapf of S. Africa is an aquatic, some other species are semiaquatic. The genus is widely distributed in the warmer regions of both Hemispheres, the greatest number of species growing in tropical and subtropical Africa. Several non-aquatic species are widely cultivated for fodder and grain.

Phalaris L., Sp. Pl. 54 (1753)

Fig. 201.

Rhizomatous perennial or annual; culms 60 to 200 cm long, stout, erect, or bent at base. Leaf blades 10 to 35 cm long, 0.6 to 1.8 cm wide, often with short auricules; ligules membranous. Inflorescence a congested panicle, 5 to 25 cm long, 1 to 4 cm wide; panicle lobed into several dense branches, which spread slightly at anthesis. Spikelets very shortly pedicelled, disarticulating above the glumes, 4 to 6.5 mm long, (1-, 2-) 3-flowered. The lower 1 or 2 florets represented only by sterile and mostly very reduced lemmas; the uppermost floret bisexual. Glumes 2, more or less equal, as long as spikelet, broadly lanceolate to obovate, strongly compressed laterally and often strongly keeled, 3-nerved. Sterile lemmas 0.5 to 1.5 mm long, one each side of fertile lemma, short haired or absent. Fertile lemma 3 to 4 mm long usually hardened, mostly appressed hairy, with 5 indistinct nerves; apex obtuse to acute. Palea subequal to lemma, 2- or 3-nerved. Stamens 3. Grain tightly enclosed by hardened lemma and palea.

Anderson, D. E. Taxonomy and distribution of the genus Phalaris. Iowa State Coll. Journ. Sci. 36: 1–96 (1961)

c. 20 species, one of them, *P. arundinacea* L. is aquatic to semiaquatic, widespread, chiefly in temperate regions.

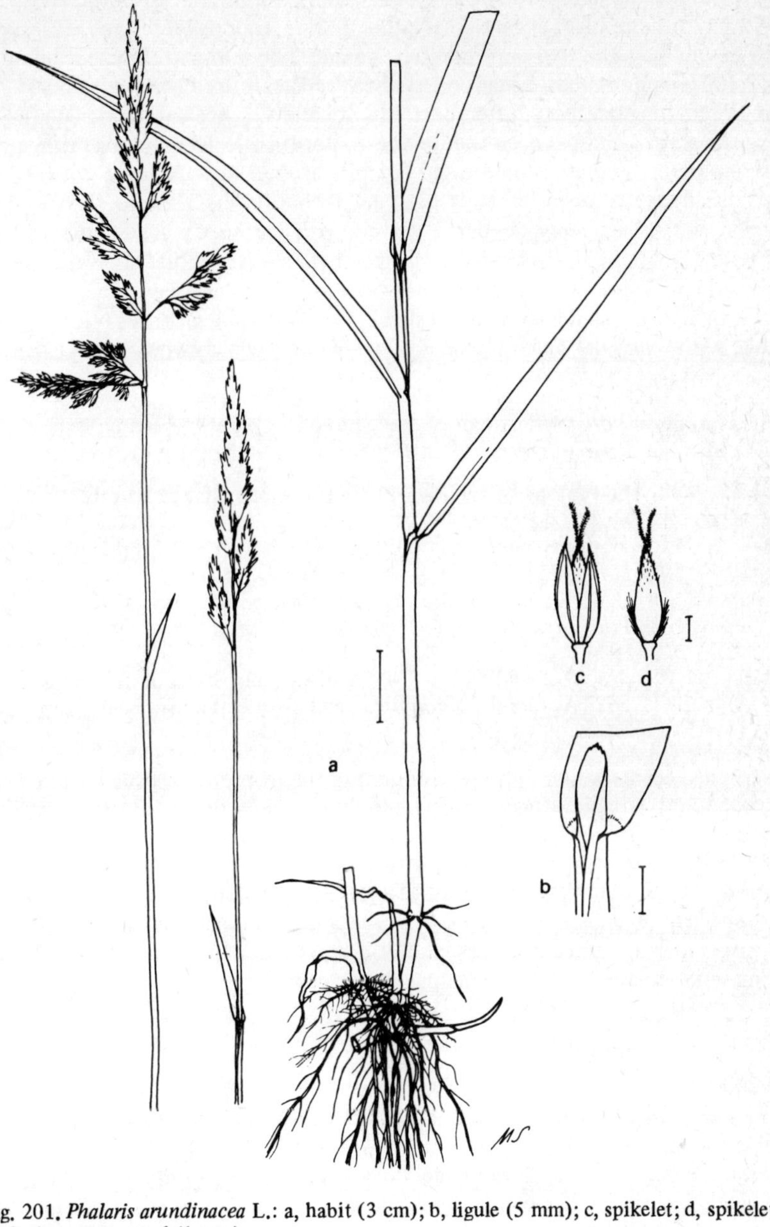

Fig. 201. *Phalaris arundinacea* L.: a, habit (3 cm); b, ligule (5 mm); c, spikelet; d, spikelet with glumes removed (1 mm).

Phragmites Adanson, Fam. 2: 34 (1763)
Fig. 202.
Perennials, with creeping stolons and/or rhizomes; culms hollow, simple, 60 to 400 cm or more high; extravaginal culms arising from stolons. Leaf blades 20 to 60 cm long, 0.8 to 4 cm wide, flat, broadly linear to narrowly lanceolate; ligule short, membranous, usually fringed with hairs. Inflorescence a panicle, 15 to 75 cm long, feathery, often purplish. Spikelets pedicelled, disarticulating above the glumes and between the florets, solitary, 10 to 15 mm long, awned, subterete before anthesis, compressed laterally after, 3 to 7 flowered. Lowest floret usually sterile or male, rarely bisexual; higher florets bisexual. Glumes 2, rounded on the back, very slender, 3-nerved, with acute apex; lower glume 3 to 6 mm long; upper one longer, but shorter than lowest floret. Lowermost lemmas 8 to 12 mm long, awnless, 3-nerved; upper lemmas progressively shorter but mostly slenderly awned-tipped, 1- to 3-nerved, glabrous but rachilla with long silky hairs. Stamens 2 or 3. Grain free, crowned by style-base.

The literature on *Phragmites* is enormous, the following have been chosen as a guide into the bibliography.

Fanshaw, D. B. The biology of the reed-Phragmites mauritianus Kunth. Kirkia 8 (2): 147–150 (1972)
Haslam, S. M. Community regulation in Phragmites communis Trin. Journ. Ecol. 59: 65–73, 75–78 (1971)
Nikolajevskij, V. G. Research into the biology of the common reed (Phragmites communis Trin) in the U.S.S.R. Folia Geobot. Phytotax. (Praha) 6: 221–230 (1971)
Rudescu, L., Niculescu, C. and Chivu, I. P. Monographia Stufului din Delta Dunari. 1–541. Academiei Republicii Socialiste Romania, Bucharesti (1965)

c. 3 species: cosmopolitan but less common in the tropics. It is usually found in dense stands fringing streams, rivers, ponds and lakes, in marshes and in estuaries. The commonest and most widespread species is *P. australis* (Cav.) Trin. ex Steud. [*P. communis* Trin.], it is usually called the common reed. *Phragmites* stands are very important as a refuge for wildlife. *P. australis* and other species are widely used by man for thatching and matting and in some countries, such as Rumania and Poland, it is harvested in large quantities as raw material for the paper and chemical industries.

Pleuropogon R. Br., Chlor. Melv. 31 (1823)
Fig. 203.
Perennials, with very slender rhizomes, about 1 mm in diameter; culms erect, 5 to 20 cm long. Leaf blades 2 to 9 cm or in aquatic, floating forms up to 35 cm long, mostly 1.5 to 2 mm wide, linear, gradually or abruptly tapering into an acute apex. Inflorescence a loose raceme; pedicels 1.5 to 3 mm long. Spikelets deflexed, 10 to 17 mm long, awned, compressed laterally, 5- to 10-flowered; rachilla disarticulating between the florets and above the glumes. Glumes 2, subequal, persistent, 1 to 2 mm long; lower glume 1-nerved; upper

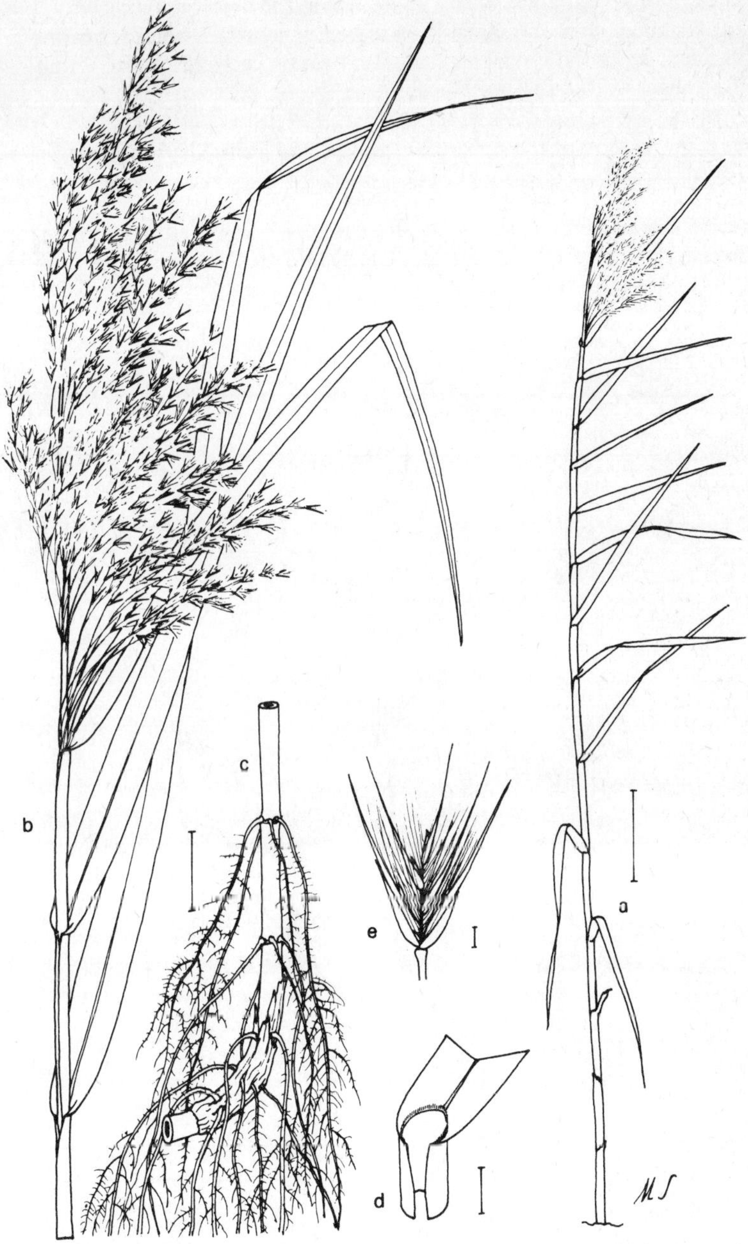

Fig. 202. *Phragmites australis* (Cav.) Trin. ex Steud.: a, habit (20 cm); b, stem apex; c, stem base (3 cm); d, ligule (1 cm); e, spikelet (1 mm).

glume faintly 3-nerved. Lemmas about 4 mm long, rounded on the back, terminating abruptly at apex, with a short awn at the apical notch; awn equalling terminal lobes of lemma. Palea equalling lemma, 2-keeled, bearing on each side 2 broad, lateral awns; the lower awn on each side 1.3 to 2 mm long, the upper awn 0.5 to 1 mm long.

Benson, L. A. revision of the semaphore grasses; the genus Pleuropogon. Amer. Journ. Bot. 28: 358–360 (1941)

c. 5 species, one of them, *P. sabinii* R. Br., palustrine or sometimes aquatic, the others are subpalustrine to terrestrial. *P. sabinii* grows in the Arctic regions of Eurasia and N. America.

Fig. 203. *Pleuropogon sabinii* R.Br.: a, habit (1 cm).

Fig. 204. *Potamophila parviflora* R. Br.: a, inflorescence (3 cm); b, spikelet (1 mm).

Potamophila R. Br., Prodr. Fl. Nov. Holl. 221 (1810)
Fig. 204.
Perennials, rhizomatous; culms erect, 1 to 1.5 m tall. Leaf blades up to 15 cm long, flat, with acuminate apex; ligules long, thin. Inflorescence a panicle, up to 40 cm long, much branched. Spikelets disarticulating above the glumes, 3.5 to 6 mm long, awnless, slightly compressed laterally, 3-flowered. Lower two florets reduced to small, sterile, empty lemmas; upper floret bisexual or unisexual, with organs of one sex rudimentary. Glumes 2, unequal, scale-like, about 1 mm long, nerveless, oblong-ellipsoidal; apex emarginate. Sterile lemmas reduced, awl-shaped, rounded or lobed, about 1 to 1.5 mm long. Fertile lemma thin, 5-nerved. Stamens 6. Grain obovoid, compressed or terete, crowned by thickened bases of the 2 styles, enclosed between lemma and palea, but free.

1 species, *P. parviflora* R. Br.: New South Wales, Australia. It is found in flowing water in streams and rivers.

Pseudoraphis W. Griffith, Not. Pl. Asiat. 3: 29 (1851)
Fig. 205.
Perennials; culms slender moderately low, glabrous. Leaf blades linear to lanceolate; ligule membranous, short or well developed. Inflorescence a panicle, (2.5–) 4 to 15 cm long, composed of slender, spike-like racemes, or consisting of one raceme only; racemes arranged on all sides of central axis, simple, all ending in a bristle (5 to 12 mm long), strongly compressed. Spikelets 4 to 10 mm long, acuminate or awned, terete or slightly compressed dorsally, 2-flowered. Lower floret sterile or male; upper floret bisexual or female. Glumes 2, very unequal; lower glume very small, membranous, nerveless, with obtuse or slightly praemorse apex; upper glume longer than lemmas, lanceolate, leathery, with acuminate apex or tapering into an awn, 5 to many nerved, usually scabrid with short bristle-like hairs. Lower lemma slightly shorter than upper glume; upper lemma much shorter than lower one, thinly membranous, hyaline. Stamens 3. Grain tightly enclosed between lemma and palea, linear oblong in outline, slightly compressed dorsally.

c. 5 species, aquatics or semiaquatics. In S. E. Asia, India, Malaysia, Australia. In ditches and shallow pools, usually gregarious.

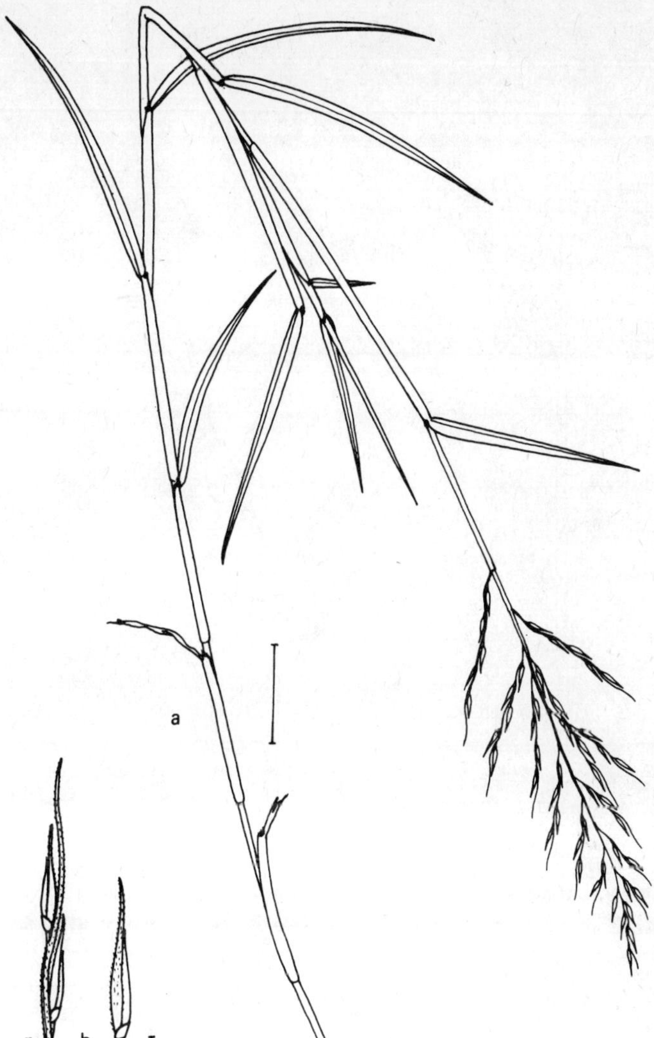

Fig. 205. *Pseudoraphis spinescens* (R.Br.) Vickery: a, habit (2 cm); b, spikelets (1 mm).

Fig. 206. *Pseudovossia cambogiensis* A. Camus: a, habit (1 cm); b, pedicelled spikelet (5 mm); c, section of pedicelled spikelet; d, sessile spikelet; e, section of sessile spikelet (5 mm).

Pseudovossia A. Camus, Bull. Mus. Hist. Nat. (Paris) 26: 665 (1921)
Fig. 206.
Rhizomatous perennial; culms up to 1 m tall. Leaf blades linear, 30 to 50 cm long. Inflorescence a lax panicle. Spikelets in pairs, consisting of one sessile and one pedicelled spikelet, with pedicels of the latter very long, jointed, individual spikelets not appressed to axis; rachis not only articulated below sessile spikelets but also between them. Sessile spikelet 2.2 to 2.5 cm long, compressed dorsally, 2-flowered; its lower floret female; upper floret bisexual; its lower glume elongate, acuminate, 2-keeled with smooth back; upper glume; shorter than lower, bent outwards, keeled and winged at tip; lemmas short; stamens 3. Pedicelled spikelet longer than sessile, compressed

laterally, 2-flowered; its lower floret female; upper bisexual, or florets sterile; its lower glume compressed laterally, keeled, long cuspidate, bent outwards; upper glume shorter, boat-shaped, keeled, winged at apex.

1 species, *P. cambogiensis* A. Camus: Cambodia and Thailand. In shallow water at the edges of rivers and lakes.

Reimaria Humboldt and Bonpland ex Flügge, Monogr. Pasp. 213 (1810)
Fig. 207.
Perennials, stoloniferous; culms compressed, often long and decumbent, then ascending, 20 to 40 cm long. Leaf blades flat, 2 to 4 mm wide; ligule a fringe of hairs. Inflorescence a panicle, consisting of 1 to 20 spike-like racemes, subpalmately or racemosely arranged along a short main axis, stiffly spreading at maturity; racemes in some species (e.g. *R. acuta* Flügge) often ending in a 4 to 7 mm long bristle. Spikelets alternate in 2 rows along one side of a narrow, flattened axis, not very close together, about 5 mm long, awnless, strongly compressed dorsally, lanceolate, 2-flowered. Lower floret sterile, reduced to lemma; upper floret bisexual. Glumes absent or sometimes a small rim present, or the upper glume sometimes present in the terminal spikelet. Sterile lemma about equalling the fruit, with acuminate or shortly cuspidate apex; fertile lemma scarcely hardened, acuminate, faintly nerved. Palea nearly half as long as fertile lemma. Stamens 2. Grain enclosed by palea, but free, oblong-linear in outline.

c. 4 species, *R. acuta* Flügge and *R. oligostachya* Munro semiaquatic to aquatic: in the New World, from Florida to Brazil.

Rhynchoryza Baillon, Bull. Soc. Linn. Paris 1063 (1892)
Fig. 194B.
Like *Oryza*, but 2 to 3 m tall; sterile lemma shortly 3-lobed and fertile lemma thickened with spongy tissue towards apex, tapering into a thick, flattened awn 2 to 2.5 cm long.

1 species: *R. subulata* Baill., in swamps, from S. Brazil to Paraguay and N. Argentina.

Saccharum L., Sp. Pl. 54 (1753)
Fig. 208.
Perennials, strongly tufted, robust, the aquatic species with long rhizomes; culms erect, 1.5 to 5.0 m tall (in woody vegetation up to 12 m tall). Leaf blades various, often narrow; ligules well developed. Inflorescence a large panicle, 15 to 80 cm long, composed of more or less branched, spikelike racemes. Spikelets usually surrounded by long silky hairs from the base, in pairs, one sessile, one pedicelled; the pedicelled spikelet disarticulating from the pedicel; the sessile spikelet disarticulating with the inferior internode

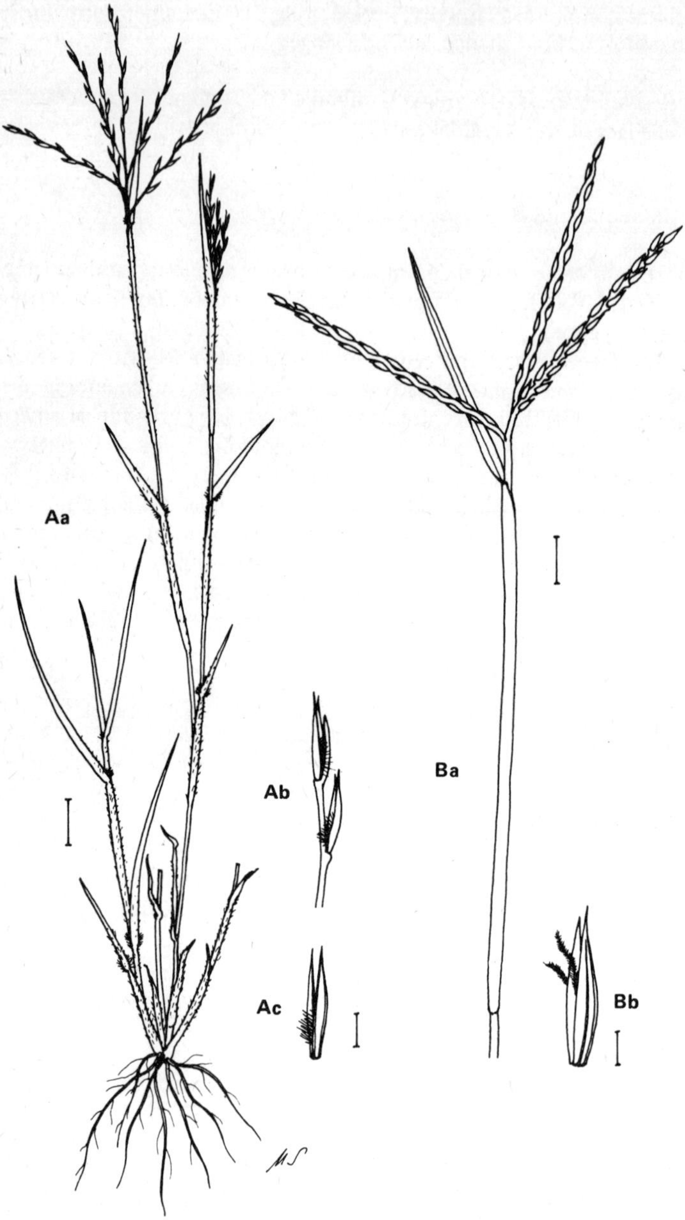

Fig. 207. A. *Reimaria acuta* Flügge: a, habit (1 cm); b, rachis; c, spikelet (1 mm); B. *Reimaria oligostachya* Munro: a, inflorescence (1 cm); b, spikelet (1 mm).

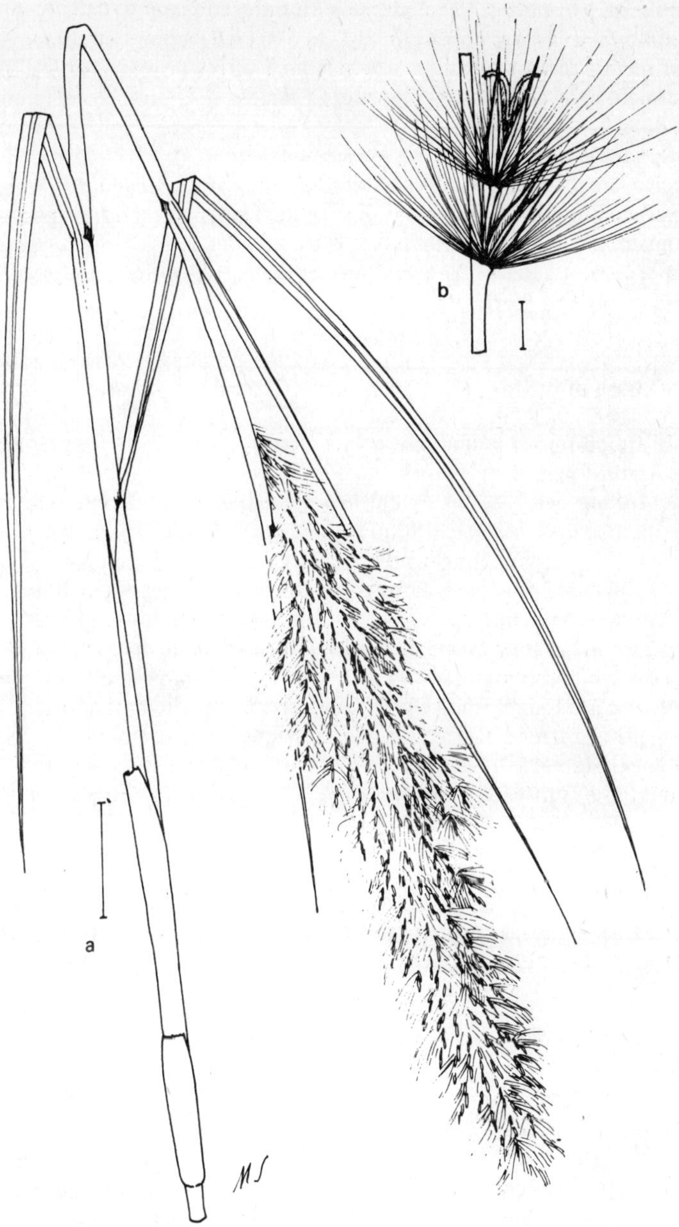

Fig. 208. *Saccharum spontaneum* L.: a, habit (3 cm); spikelets (2 mm).

and the pedicel of the pedicelled spikelet, awned or awnless, 3.5 to 5.5 mm long; 2-flowered. Lower floret sterile; upper floret bisexual. Glumes 2, as long as spikelet, acuminate; lower glume with inflexed margins, usually with an even number of nerves; upper glume 1- to 5-nerved, acuminate. Lemmas absent, or membranous or hyaline; upper lemma with a terminal, bristle-like, usually straight awn, or pointed or awnless. Stamens 3. Grain oblong in outline, or subglobose.

c. 10 species, one of them, *S. spontaneum* L., semiaquatic to aquatic, at least in some varieties and in some regions (e.g. in the Upper Nile); other species partly semiaquatic (e.g. *S. ravennae* L.). The cultivated species of *Saccharum* (sugar cane) are not aquatic. The wild species are native of the Tropics and Subtropics of the Old World.

Sacciolepis Nash in Britton, Man. 89 (1901)
Fig. 209A.
Annuals of stoloniferous perennials; culms spongy, 5 to 150 cm long, sometimes prostrate at base. Leaf blades 2 to 12 mm wide, flat or rolled; ligules very short. Inflorescence dense, cylindrical, spike-like, up to 30 cm long (1 to 2.5 cm long in *S. huillensis* Stapf). Spikelets pedicelled, disarticulating from pedicel, 2 to 3.5 mm long, awnless, terete or slightly compressed laterally or globose, 2-flowered. Lower floret male or sterile; upper floret bisexual. Glumes 2, unequal, prominently 5- to 11-nerved; lower glume about third to half as long as spikelet, with obtuse to acute apex; upper glume as long as spikelet, lanceolate, round on back. Lower lemma similar to upper glume, but almost straight on the back, lanceolate; upper lemma shorter, papery, slightly hardened at maturity, with opaque margins embracing palea, closed at apex. Stamens 3. Grain enclosed between lemma and palea, but free within. False fruit with knob-like stipe.

Simon, B. K. Three new species of Sacciolepis from the Flora Zambesiaca area. Kirkia 8: 85–90 (1971)

c. 30 species, generally semiaquatic to aquatic, in swampy places in the Tropics and Subtropics of both Hemispheres.

Scolochloa Link, Hort. Berol. 1: 136 (1827)
Fig. 210.
Perennial, strongly rhizomatous; culms erect, 2 m or more long, usually several sterile stems present. Leaf blades 0.5 to 1.0 cm wide, up to 55 cm long, firm, narrowed gradually to a slender tip; ligules 2 to 6 mm long, laciniate; leaf sheaths open. Inflorescence a panicle, often over 30 cm long, open; the branches ascending, scabrid. Spikelets pedicelled, disarticulating above the glumes, 8 to 11 mm long, awnless, compressed laterally closely 3- to 4-flowered. Florets bisexual. Glumes membranous-scarious, rounded on the back, often irregularly torn at apex; lower glume 4 to 6 mm long, shorter than lowest

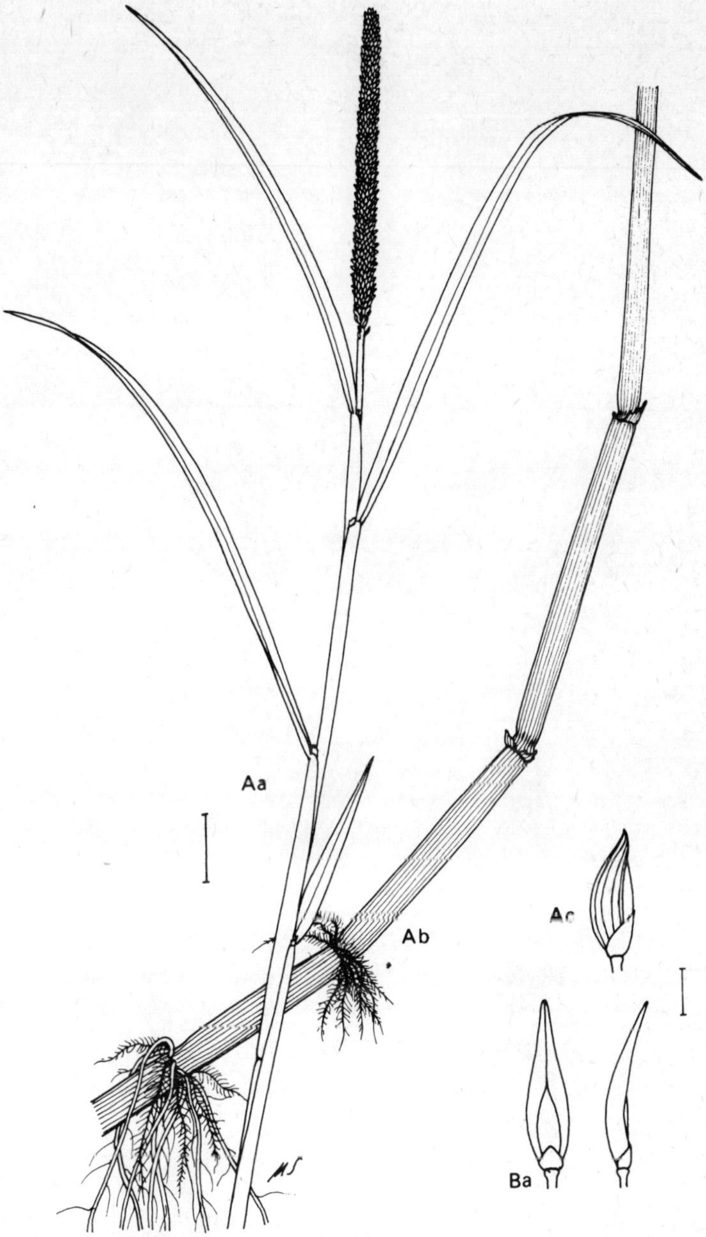

Fig. 209. A. *Sacciolepis africana* C. E. Hubbard & Snowden: a, stem (2 cm); b, stem base (1 cm); c, spikelet (1 mm);
B. *Thyridachne tisserantii* C. E. Hubbard: a, spikelets (1 mm).

Fig. 210. *Scolochloa festucacea* Willd.: a, habit (3 cm); b, spikelet; c, lower glume (2 mm).

floret, 3-nerved; upper glume 6 to 8 mm long, about equalling lowest lemma, 5-nerved. Lemmas about 6 mm long, rounded on the back, firm, irregularly toothed at apex, with 7 converging nerves, strongly bearded-villous on the callus. Palea subequal to lemma. Stamens 3.

2 species, *S. festucacea* Willd. is an aquatic, it occurs in marshes and at the edges of lakes, usually standing in water; it is native to Eurasia, mainly in the temperate to subarctic regions, but introduced to N. America. *S. festucacea* resembles *Glyceria maxima* (Hartm.) Holmb. but *Scolochloa* has open leaf sheaths and longer ligules.

Spartina Schreber, Gen. 43 (1789)
Fig. 211.
Perennials, rhizomes either absent or firm with closely imbricate scales; culms up to 2.5 m high, terete, hardened, glabrous. Leaf blades flat or involute, margins scabrous, with acuminate apex; ligules a fringe of hairs. Inflorescence a panicle, 8 to 50 cm long, consisting of 3 to 50 racemose arranged spikes; spikes 1.5 to 15 cm long, 1-sided, with a 3-angled rachis, usually spreading. Spikelets close together, in two rows, disarticulating below the glumes, 6 to 25 mm long, awned, or apex acuminate (*S. bakeri* Merr.), strongly compressed laterally, 1-flowered, bisexual. Glumes scabrid to hispid; lower glume very narrow, about 3 to 10 mm long including awn, nearly equalling lemma; upper glume awned or with acuminate apex (6 to 9 mm long in *S. bakeri* Merr. and 10 to 25 mm long in *S. pectinata* Link). Lemma keeled; apex acute to obtuse or bilobed. Palea thin and papery, slightly exceeding the lemma. Stamens 3.

Marchant, C. J. Evolution in Spartina. Journ. Linn. Soc. London Bot. 60: 1, (1967); 381, 411 (1968)
Mobberley, D. G. Taxonomy and distribution of the genus Spartina. Iowa State Coll. Journ. Sci. 30: 471–574 (1956)

c. 16 species: mostly temperate America, some species on the coasts of Europe, Africa and Tristan da Cunha. Most species are found on alkaline soil or in brackish water. *S. bakeri* Merr. and *S. pectinata* Link are frequently found in fresh water in N. America.

Thyridachne C. E. Hubbard, Kew Bull. 1949: 363 (1949)
Fig. 209B.
Like *Sacciolepis*, but spikelets compressed dorsally; lower glume very short and nerveless; grain enclosed between upper lemma and palea and between hardened upper glume and lower lemma.

1 species, *T. tisserantii* C. E. Hubbard; Central Africa: aquatic to semiaquatic.

Fig. 211. *Spartina pectinata* Link: a, habit (3 cm); b, spikelet (2 mm).

Fig. 212. *Torreyochloa fernaldii* (St John) Church: a, habit (1 cm); b, spikelet (1 mm).

Torreyochloa Church, Amer. Journ. Bot. 36: 163 (1949)
Fig. 212.
Like *Glyceria*, but leaf-sheaths free and overlapping; upper glume always 3-nerved; lemmas 5-nerved; apex of lemma broadly obtuse or blunt, membranous and irregularly toothed; grain oblong in outline, hilum short.

c. 6 species, 5 in N. America, 1 in Eurasia. Aquatic or semiaquatic grasses.

Vetiveria Bory, Lem. Bull. Soc. Philom. 43 (1822)
Fig. 213.
Perennials with strong, thick rhizomes; culms up to 3 m high, erect or ascending, more or less compressed at base. Leaf blades 0.4 to 1.5 cm wide, 30 to 130 cm long, flat towards apex, slightly folded and not narrowed at base, scabrid on margins; ligules very short, upper part consisting of hairlike strips; leaf sheaths compressed, lower ones strongly keeled. Inflorescence a panicle, 15 to 40 cm long, composed of numerous whorls of simple, rarely branched, slender racemes on a central axis. Spikelets in pairs, one sessile, one pedicelled; the sessile spikelet disarticulating at least with the accompanying pedicel (of the pedicelled spikelet), often with the entire accompanying pedicelled spikelet and the inferior internode. Sessile spikelet 4 to 7 mm long, awnless or shortly awned, slightly compressed laterally 2-flowered, its lower floret sterile, upper floret bisexual; its glumes equal, as long as spikelet, narrowly lanceolate; lower glume convex, with incurved (not inflexed) margins, with numerous upwardly directed or curved firm, spiny bristles, 7-nerved; upper glume boat-shaped, with bristles, awnless, pointed or awned, 3-nerved; its lemmas awnless, mucronate or awned; sterile lemma as long as spikelet, thinly membranous, 2-nerved; fertile lemma shorter than sterile one, 1-nerved. Pedicelled spikelet usually slightly shorter than sessile one, 2-flowered; lower floret sterile, upper floret male; compressed dorsally; glumes thinner, not so spiny as in sessile spikelet; lemmas usually awnless. Stamens 3. Grain oblong in outline.

c. 6 species: Tropics of the Old World. *V. nigritana* (Benth.) Stapf from Africa is aquatic. *V. zizanioides* (L.) Nash is semiaquatic and frequently cultivated for its aromatic oil.

Vossia Wall. et Griff., Journ. As. Soc. Bengal 5: 572 (1836), nom cons.
Fig. 214.
Perennial, with stout creeping, submerged or floating stolons; culms very long, up to 4.5 m, spongy, thick. Leaf blades up to 90 cm long and 2.8 cm wide, linear, with apex tapering to a fine point; ligules very short, densely hairy. Inflorescence consisting of solitary or palmately arranged, erect, spike-like racemes, 15 to 30 cm long, at the tip of the culms. Spikelets in pairs, one sessile, one pedicelled; sessile spikelets in cuplike cavities, appressed to central part of axis; racemes almost terete; sessile spikelets disarticulating very tardily,

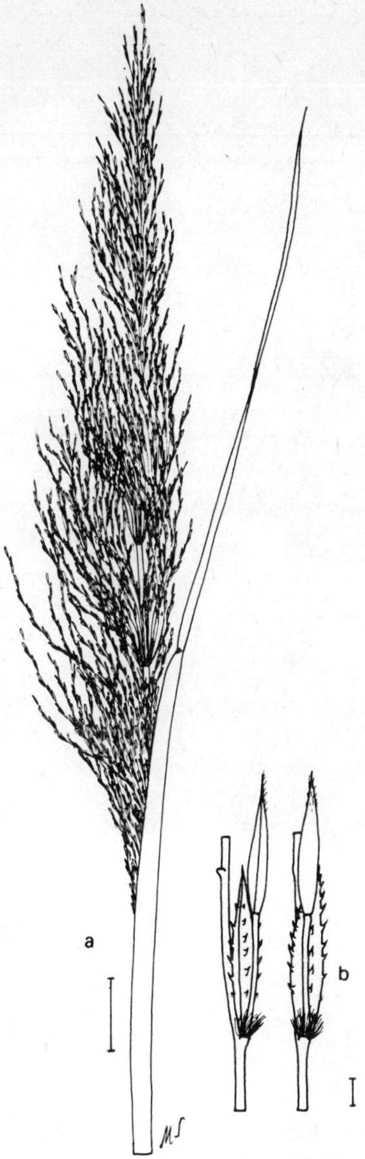

Fig. 213. *Vetiveria nigritana* (Benth.) Stapf: a, inflorescence (3 cm); b, pairs of spikelets (1 mm).

Fig. 214. *Vossia cuspidata* (Roxb.) Griff.: a, stem base (2 cm); b, inflorescence; c, rachis with spikelets, adaxial view; d, rachis with spikelets, abaxial view (1 cm); e, spikelet (1 mm).

together with inferior internode of rachis and pedicel of pedicelled spikelet; sessile spikelets awned, back almost flat, 2-flowered; lower floret male, upper floret bisexual. Pedicelled spikelets similar to sessile, usually slightly smaller and often both florets male. Glumes 2, unequal; lower glume broad, ovate, flattened, leathery, narrowed into a rigid, green awn, up to 30 mm long and 2 mm wide at base, with scabrous margins; upper glume up to 8 mm long, ovate oblong in outline, mucronate, 5- to 9-nerved. Lemmas membranous, of lower floret 2-nerved, of upper faintly 3-nerved. Stamens 3.

1 species, *V. cuspidata* (Roxb.) Griff., aquatic, in standing or slowly flowing water, in tropical Africa and India. It is often regarded as a pest in Africa because large populations occasionally block waterways.

Zizania L., Sp. Pl. 991 (1753)
Fig. 215.
Annuals or perennials with strongly developed rhizomes and stolons; culms usually 2 to 3 m long, erect. Leaf blades large, flat, up to 2.5 cm wide; ligules prominent, membranous. Inflorescence a panicle, 30 to 50 cm long; lower branches spreading or ascending, bearing pendulous male spikelets; upper branches ascending and at maturity erect, bearing appressed female spikelets. Male spikelets caducous. Spikelets disarticulating from the pedicel, 1-flowered, unisexual; glumes and sterile lemmas reduced to a small collar-like ridge at tip of pedicel. Male spikelet about 7 to 12 mm long, not hardened, somewhat compressed laterally; lemma 5-nerved; acuminate to awn-tipped; palea about as long as lemma, 3-nerved; stamens 6. Female spikelet hardened, subterete but strongly angled, usually 3 to 4 cm long, or longer (including awn); lemma papery, terminating in a long, slender awn (usually about 2 cm long), 3-nerved; palea 2-nerved, closely clasped by lemma. Grain cylindric, 1 to 4 cm long; style-base not persistent.

Dore, W. G. Wild Rice. Ottawa (1969)

4 species: *Z. aquatica* L., *Z. palustris* L. and *Z. texana* Hitchc. occur in marshes, borders of streams, lakes and ponds, usually in shallow water, in E. N. America. *Z. latifolia* (Griseb.) Stapf occurs in similar habitats in China, Japan, S. Russia, N. E. India and Burma. According to Dore (1969, p.6) the wild-rice, as *Zizania* is vernacularly called, is the native cereal of Canada and the northern parts of the USA. Its grains are used for human food. *Zizania* is also important as a food and shelter for water-fowl. *Z. latifolia* is partly cultivated in sawahs by Chinese. The plants are attacked by a fungus which hinders stem-elongation and flowering and causes a thickening of the stem. The infected plants are cooked and eaten like asparagus (after C. A. Backer and R. C. Bakhuizen van den Brink jr. in Flora of Java).

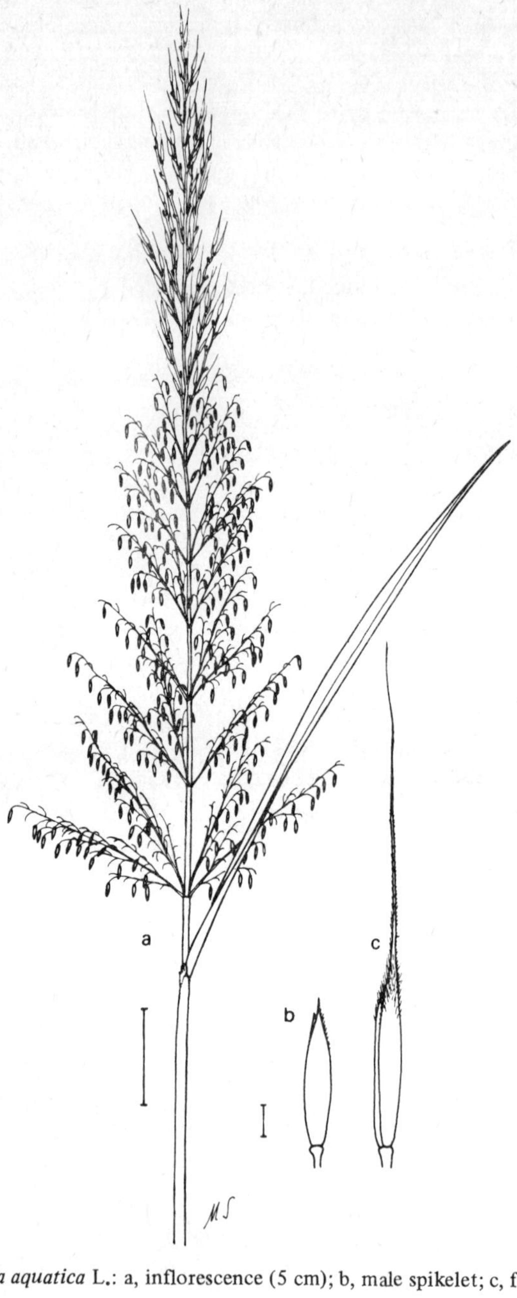

Fig. 215. *Zizania aquatica* L.: a, inflorescence (5 cm); b, male spikelet; c, female spikelet (2 mm).

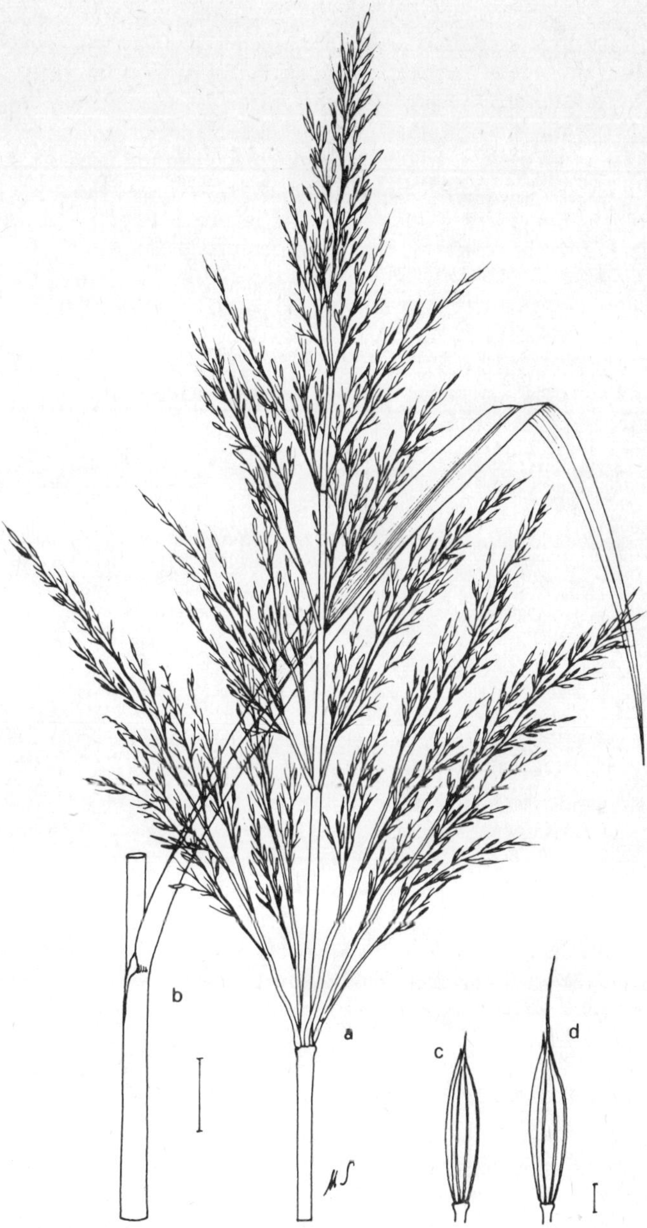

Fig. 216. *Zizaniopsis miliacea* (Michx.) Döll & Aschers.: a, inflorescence; b, leaf sheath (3 cm); c, male spikelet; d, female spikelet (1 mm).

Zizaniopsis Döll and Ascherson ex Döll in Martius, Fl. Bras. 2 (2): 12 (1871) Fig. 216.
Perennials with stout creeping rhizomes, culms usually 1 to 3 m high. Leaf blades flat, 1 to 2 cm wide, with acuminate apex; ligules papery elongated, up to 37 mm long. Inflorescence large, panicles 30 to 120 cm long, open but somewhat narrow, with numerous branches. Spikelets all pedicelled and disarticulating from the pedicel, 1-flowered, unisexual. Male and female spikelets together on the same branches of the panicle, the male below. Glumes and sterile lemmas absent; fertile lemma 7-nerved. Male spikelets 3.5 to 10 mm long, awnless; lemma acute to slightly acuminate; stamens 6. Female spikelets 10 to 15 mm long, with lemma tapering into an awn (2 to 8 mm long). Grain obovoid, free from lemma and palea, beaked with the inferior part of the persistent style.

c. 3 species; semiaquatics to aquatics, mostly in marshes, creeks and river banks. In N. and S. America.

PODOSTEMACEAE

Roots capillary, ribbon-shaped, thalloid or foliose, attached to rocks in swiftly flowing water, usually photosynthetic. Stems absent or present, small or large, simple or branched. Leaves absent or present or reduced to 1 stipel, extremely variable in shape and size; stipules absent or present. Flowers in spike-like inflorescences, cymes, irregular clusters or solitary, completely enveloped in membranous spathella or spathella absent. Perianth either present, 3- or 5-lobed, free or united at base or reduced to scale-like tepals; tepals 2 to many, usually alternating with stamens. Stamens 1 to many, in 1 or 2 complete whorls, in an incomplete whorl or confined to one side of flower, free or united, frequently 2 or rarely 3 or 4, borne on a distinct stalk (andropodium); pollen 1- to many-celled. Ovary superior, erect or sometimes obliquely inserted on pedicel, 1- to 3-locular; fruit a capsule, with equal or unequal valves; styles 1 to 3, very variable in shape and size; seeds axillary, 2 to numerous.

Engler, A. Podostemonaceae. Nat. Pflanzenfam. 18a: 1–68 (1930)
Gessner, F. and Hammer, L. Oekologisch-physiologische Untersuchungen an den Podostemonaceen des Caroni. Int. Rev. Ges. Hydrobiol. 47 (4): 497–541 (1962)
Royen, P. van. The Podostomaceae of the New World. Med. Bot. Mus. Utrecht 107: 19–150 (1951), 115: 1–21 (1953), 119: 215–263 (1954).
Willis, J. C. The evolution of the Tristichaceae and Podostemaceae. Ann. Bot. 40: 349–367 (1926).

c. 46 genera; *c.* 260 species: cosmopolitan in warmer regions. All species are aquatic herbs found growing attached to rocks, stones or wood in or near rapids and waterfalls, flowering when the water subsides. The sterile and fertile phases are often morphologically very dissimilar. Most of the species are confined to small geographical areas and some species are apparently confined to a single cataract or river. A notable exception, however, is *Tristicha trifaria* (Bory ex Wild.) Sprengel which is found in America from Mexico to Argentina and in tropical Africa, Madagascar and the Mascarene Islands. The genus *Podostemum* has species in America and India, otherwise all other genera are confined to a single continent (America 17 endemic genera, Africa including Madagascar 15, and Asia including India and Australasia 11). Some species are eaten by humans and their livestock but on the whole the family is of no great economic importance.

The genera *Blandowia* Willd. and *Carajaea* (Tul.) Wedd. are very poorly known (see Royen, op. cit. 119: 253–255, 1954) and of doubtful status so they are not included in this treatment.

1A Young flowers enveloped by a few leaves but not enclosed in a membranous spathella; perianth 3- or 5-lobed; ovary with 2 or 3 loculi
  2A Perianth segments 5, free; stamens 5 to 25; ovary 2-locular (S. America)

Weddellina

2B Perianth segments 3, free or united; stamens 1 or 3; ovary 3-locular
3A Stamens 1, rarely 2 (Africa and America)
Tristicha
3B Stamens 3
4A Roots thalloid, foliose, closely attached to rocks (India)
Dalzellia
4B Roots long, slender, much branched, partly creeping, partly floating (India and S. E. Asia)
Indotristicha

1B Young flowers totally enclosed in a membranous spathella; perianth consisting of scale-like or filamentous tepals; ovary with 1 or 2 loculi
5A Spathella containing 10 to 20 flowers (S. America)
Macarenia
5B Spathella containing a single flower
6A Flowers in a branched or unbranched, 2-sided, spike-like inflorescence or if not so then leaves very rough on upper surface
7A Stamens united in a ring (S. America)
Tulasneantha
7B Stamens free, not united in a ring
8A Styles flattened, crested; filaments widened (S. America)
Lonchostephus
8B Styles linear or spathulate, not crested; filaments linear, not widened (S. America)
Mourera
6B Flowers not in a 2-sided, spike-like inflorescence; leaves never rough on upper surface
9A Capsule strongly asymmetrical, borne perpendicular to stalk; upper capsule valve almost free from pedicel; flowers borne in cavities on surface of thallus (S. America)
Castelnavia
9B Capsule regular or slightly asymmetrical; not borne perpendicular to stalk; both valves distinctly attached to pedicel; flowers not borne in cavities on surface of thallus
10A Flowers inverted within unruptured spathella (Africa)
11A Style cock's comb-like, flattened with serrated edge (Africa)
Macropodiella
11B Style linear or conical, without serrated edge
12A Ovary flattened; midribs winged; wings longer than capsule (Africa)
Winklerella
12B Ovary not flattened, terete; wings absent

13A Flowers 3 to 20, in cymes, borne regularly along the stem, opposite the leaves; styles conical; capsule partly cylindrical (Africa)
Dicraeanthus

13B Flowers solitary or occasionally in fascicles, not borne regularly opposite leaves; styles linear; capsule ellipsoidal to globose

14A Tepals 3, one each side of andropodium and one borne between the 2 stamens; capsule ribs 12 to 18, those nearest the commissures shortest and not extending the length of the capsule (Africa)
Stonesia

14B Tepals 2, on each side of andropodium; capsule ribs 6 to 10, all extending the length of the capsule

15A Flowering stems bearing linear or subulate, simple leaves

16A Stamens 1; ovary 6-ribbed (Africa)
Monadriella

16B Stamens 2; ovary 10-ribbed (Africa)
Ledermanniella

15B Flowering stems either without or with leaves but when present at least some leaves divided or lobed into several segments

17A Capsule valves 2, unequal, the larger persistent after dehiscence; ribs on ovary wall 8, broad, flattened; stamen 1 (Africa)
Sphaerothylax

17B Capsule valves 2, equal, both persistent after dehiscence; ribs on ovary wall 6 to 10, narrow, rounded; stamens 2, rarely 1, 3 or 4 (Africa)
Inversodicraea

10B Flowers erect within unruptured spathella (Widespread)

18A Stamens 1 (see p. 449)

19A Tepals 3 to 7, in one complete or incomplete whorl, not all of them closely associated with the base of the stamen

20A Capsule with 6 ribs; flowers solitary (S. America)
Jenmanniella

20B Capsule with 12 ribs; flowers in groups (S. America)
Wettsteiniola

19B Tepals 2 or 3, not in whorl, all of them closely associated with the base of the stamen

21A Tepals 3, one each side of stamen base and one on back (abaxial side) of stamen

22A Shoots in opposite or sub-opposite pairs; at regular intervals along the root; anthers with extrorse dehiscence; capsule ribbed (C. and S. America)
Oserya

22B Shoots solitary or in groups of 2 to 4, irregular; anthers with introrse dehiscence; capsule smooth (S. America)
Devillea

21B Tepals 2, one each side of stamen base
23A Capsule smooth, without ribs
24A Capsule valves equal, both persistent in fruit (Africa)
Letestuella

24B Capsule valves unequal, the larger persistent in fruit or capsule indehiscent
25A Roots thalloid, foliose; flowers shortly stalked; both ovary loculi fertile, containing more than 4 seeds (E. and S.E. Asia)
Cladopus

25B Roots ribbon- or thread-like; flowers sessile; one ovary loculus sterile, the other containing 4 or less seeds (India)
Farmeria

23B Capsule with 6 to 14 ribs
26A Capsule with 6 ribs
27A Root filamentous, bearing opposite or sub-opposite stems at more or less regular intervals; pollen 1-celled (S. America)
Jenmanniella

27B Root thalloid, foliose, bearing scattered stems on upper surface; pollen 2-celled (Africa)
Saxicolella

26B Capsule with 8, 10, 12 or 14 ribs
28A Capsule with 14 ribs; pollen 1-celled, with ring of warts around the centre (S. America)
Monostylis

28B Capsule with 8, 10 or 12 ribs; pollen 2-celled, without ring of warts around the centre
29A Leaf bases swollen and enclosing the spathella; tepals as long as filament; styles subulate, erect (S.E. Asia)
Polypleurella

29B Leaf bases not swollen and not enclosing spathella; tepals very much shorter than filament; styles not subulate and not erect
30A Styles 2, filamentous, half or more as long as ovary; leaves repeatedly forked, not in rosettes (Africa)
Pohliella

30B Styles 2, flattened, elliptic to ovate or oblong-ovate in outline; leaves not forked

31A Thallus foliose
Zeylanidium

31B Thallus ribbon-like
32A Stems indistinct, very short, leaves in rosettes (Africa)
Butumia

32B Stems distinct, occasionally branched; leaves not in rosettes, distinctly in 2 rows, alternate (New Guinea, Australia)
Torrenticola

18B Stamens 2 or more
33A Stamens united below, or borne on an andropodium
34A Stamens 3 (rarely 4); style 1, erect, with semi-globose stigma (Africa)
Angolaea

34B Stamens 2, borne on andropodium; styles 2; stigmas not semi-globular
35A Tepals 3, one each side of andropodium base, and one at tip of andropodium between the two stamens
36A Ovary smooth, not ribbed; styles divided into several segments (S. America)
Mniopsis

36B Ovary 6- or 8-ribbed; styles simple (America, India)
Podostemum

35B Tepals 2, one each side of andropodium base
37A Midrib of each capsule valve running into the style; roots worm-like, up to 2 mm diam. (S. America)
Ceratolacis

37B Midrib of each capsule valve not running into the style; roots not worm-like
38A Leaves of fertile shoots in 4 distinct rows (India)
Willisia

38B Leaves of fertile shoots in 2 rows, variable or absent
39A Capsule smooth, without ribs
40A Capsule opening by 2, equal valves; valves persistent in fruit; spathella after rupturing more or less campanulate, with revolute teeth (Africa)
Letestuella

40B Capsule opening by 2, unequal valves, the larger valve persistent in fruit; spathella after rupturing not campanulate, without revolute teeth
41A Styles divided into several segments (S. America)
Mniopsis

41B Styles simple (Old World)
42A Filaments short, less than quarter as long as andropodium; tepals short, less than half as long as ovary (Africa)
Leiothylax
42B Filaments long, half as long as andropodium or longer; tepals long, more than half as long as ovary
43A Ovary sessile or very shortly stalked; fruit remaining enclosed by surrounding leaves; spathella boat-shaped, enclosed by surrounding leaves (India)
Hydrobryopsis
43B Ovary stalked; stalk more than 1.5 mm long; fruit lifted above surrounding leaves; spathella tubular or funnel-shaped, extending beyond surrounding leaves (India and S. E. Asia)
Griffithella
39B Capsule with ribs
44A Capsule valves 2, unequal in size and shape
45A Capsule 14-ribbed; ribs nearest the commissures shorter, not extending throughout length of capsule; styles equal in length (Madagascar)
Endocaulos
45B Capsule 8-ribbed; all ribs equally long, extending throughout length of capsule; styles unequal in length (India)
Zeylanidium
44B Capsule valves 2, equal in size and shape
46A Roots filamentous; stems regularly opposite or sub-opposite; capsule 6-ribbed (S. America)
Jenmanniella
46B Roots thalloid, flattened, ribbon-like or foliose; stems not regularly opposite or sub-opposite; capsule rarely 6-, usually 8- to 22-ribbed (Old World)
47A Roots ribbon-like, at least partly floating; spathella rupturing irregularly at apex; capsule 8- to 12-ribbed (India and Africa)
Polypleurum
47B Roots foliose, closely attached to rocks, not partly floating; spathella splitting more or less regularly down one side; capsule rarely 6-, usually 12- to 22-ribbed (India and Asia)
Hydrobryum
33B Stamens free
48A Style 1, bearing semi-globose stigma (Africa)
Angolaea

48B Styles 2, not bearing semi-globose stigmas (America)
49A Styles cock's comb-like, flattened, with serrate margin (S. America)
Lophogyne
49B Styles not cock's comb-like, not flattened, without serrate margin
50A Capsule smooth, without ribs (S. America)
Apinagia
50B Capsule with ribs
51A Capsule laterally flattened; midrib of each capsule valve winged, other ribs unwinged (S. America)
Rhyncholacis
51B Capsule terete; midrib of each capsule valve not winged or if winged then other ribs also winged
52A Roots filamentous; stems in opposite or sub-opposite pairs at regular intervals along the root; capsule 6-ribbed (S. America)
Jenmanniella
52B Roots thalloid, variously shaped; stems not in opposite or sub-opposite pairs at regular intervals along the root; capsule 6- to 14-ribbed
53A Pinnae with 1-sided stipels; capsule 12-ribbed (S. Brazil)
Wettsteiniola
53B Pinnae without stipels; capsule 6- to 14-ribbed
54A Capsule 10- to 14-ribbed (S. America)
Apinagia
54B Capsule 6- or 8-ribbed
55A Capsule 8-ribbed
56A Stems distinct, frequently branched (S. America)
Apinagia
56B Stems indistinct, not branched (C. and NW. S. America)
Marathrum
55B Capsule 6-ribbed
57A Capsule ribs winged; styles boat- or spoon-shaped, toothed (C. and NW. S. America)
Marathrum
57B Capsule ribs not winged; styles linear, not toothed (S. America)
Apinagia

Angolaea Weddell in DC., Prodr. 17: 300 (1873)
Fig. 218A.
Root unknown; stems branched, floating, up to 50 cm long. Leaves repeatedly forked, segments linear to filamentous. Flowers in cymes, erect in unruptured spathella; spathella ellipsoidal, stalked. Tepals 2, small, one each side of the stamens or andropodium. Stamens 3 or rarely 4, borne in a group together on one side of flower; andropodium very short or apparently absent; pollen

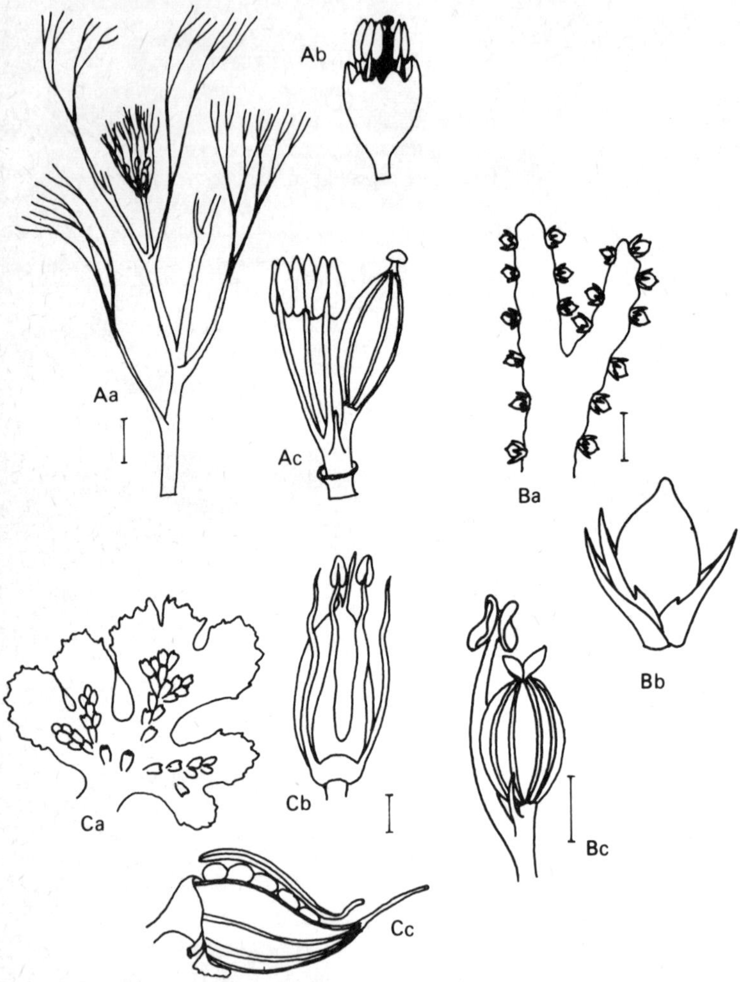

Fig. 218. A. *Angolaea fluitans* Wedd.: a, habit (1 cm); b, rupturing spathella; c, flower with spathella removed;
B. *Butumia marginalis* G. Taylor: a, habit (2 mm); b, spathella with leaves; c, flower (5 mm); after Taylor;
C. *Castelnavia princeps* Tul. and Wedd. emend. Warm.: a, habit; b, flower viewed from below (1 mm); c, capsule viewed from side.

2-celled. Ovary ellipsoidal, 8-ribbed (including commissural ribs); capsule opening by 2, equal valves; style 1, short, erect, bearing semi-globular stigma.

1 species, *A. fluitans* Wedd: River Quanza, Angola.

Apinagia Tul., Ann. Sci. Nat. Bot. sér. 3, II: 90, 97 (1849) em. v. Royen, Med. Bot. Mus. Utrecht 107: 25 (1951) [*Ligea* Tul., *Oenone* Tul.]
Fig. 217.
Very small to large colonies of opposite or sub-opposite individuals arising from branched roots; individuals with either single, unbranched base or thalloid base. Leaves in 2 rows, united at base in thalloid species, very variable in shape, often with tufts of filaments on adaxial surface; leaf stalk often distinctly sheathed at base. Flowers solitary or in groups, terminal or lateral. Tepals 2 to many, free or united, in complete or incomplete whorl or confined to 1 side of flower, sometimes 1 to 3 additional ones arising from staminal column. Stamens 1 to many, in 1 or 2 complete whorls or in an incomplete whorl or confined to 1 side of flower; filaments distinctly ribbed, sometimes keeled; anthers arrow-shaped, dehiscence introrse or extrorse; pollen 1-celled, 3-colpate. Ovary with 2 to 14 long ribs, with 2 long and 4 short ribs or without ribs, 2-locular; capsule ovoid, opening by 2 equal or slightly unequal valves; styles 2, cylindrical to linear.

c. 50 species: northern and central S. America.

Butumia G. Taylor, Bull. Brit. Mus. (Nat. Hist.) Bot. ser. 1: 55 (1953)
Fig. 218B.
Roots thalloid, ribbon-like, branched, closely attached to rocks; sterile stems absent; fertile stems simple, very short, borne on margin of thallus. Leaves in rosettes, sessile, subulate; inner leaves subtending flowers with stipule-like teeth. Flowers terminal, solitary, erect with unruptured spathella, ovoid, about 1 mm long, subsessile. Tepals 2, short, one each side of stamen. Stamen 1; filament ultimately about 1 to 2 mm long; anther about 0.5 mm long; pollen 2-celled. Ovary ovoid, about 1 mm long, 8-ribbed (including double commissural ribs); loculus 1; capsule opening by 2, equal valves; styles 2, flattened, ovate, to elliptic in outline, persisting in fruit.

1 species, *B. marginalis* G. Taylor: Tropical W. Africa, in the River Butum, S. Nigeria. This species is strikingly similar to *Inversodicraea minima* Engler in habit.

Castelnavia Tul. and Wedd., Ann. Sci. Nat. Bot. sér. 3, 9: 108 (1849)
Fig. 218C.
Roots thalloid, foliose, closely attached to rocks; stems short or absent. Leaves simple and linear or forked with linear segments or absent. Flowers numerous in cavities of thalloid base, distinctly zygomorphic, erect within

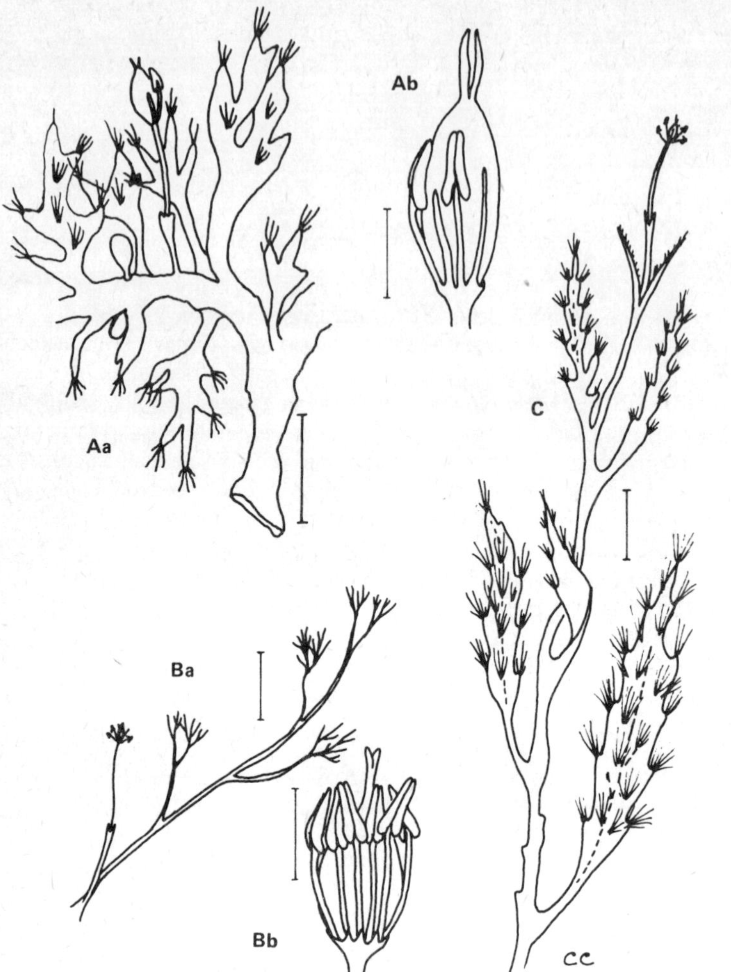

Fig. 217. A. *Apinagia pygmaea* (Bong.) Tul.: a, habit (1 cm); b, flower (1 mm); after v. Royen;
B. *Apinagia batrachifolia* (Mildbraed) v. Royen: a, habit (1 cm); b, flower (2 mm); after v. Royen;
C. *Apinagia surumuensis* (Engler) v. Royen: habit (2 cm); after v. Royen.

spathella; spathella tubular with many teeth at apex. Tepals 2 or 3, alternate with stamens. Stamens 1 to 3, free or united at base; filaments membranous cohering with base of ovary; anthers introrse; pollen 1-celled, 3-colpate. Ovary mostly perpendicular to stalk; loculus 1; carpel opening by 2, very unequal valves; the smaller valve almost free from pedicel, caducous, without ribs or 3- or 5-ribbed, with or without papillae; the larger valve saucer-like, persistent, 5- to 9-ribbed; styles 2, unequal, linear.

9 species: S.E. Brazil.

Ceratolacis (Tul.) Wedd. in DC., Prodr. 17: 66 (1873)
Fig. 219A.
Roots cylindrical or semicylindrical, up to 2 mm diam., worm-like, red, branched, closely attached to rocks; stems very short or absent. Leaves 2 to 3 mm long, either entire and linear or few times forked with linear segments.

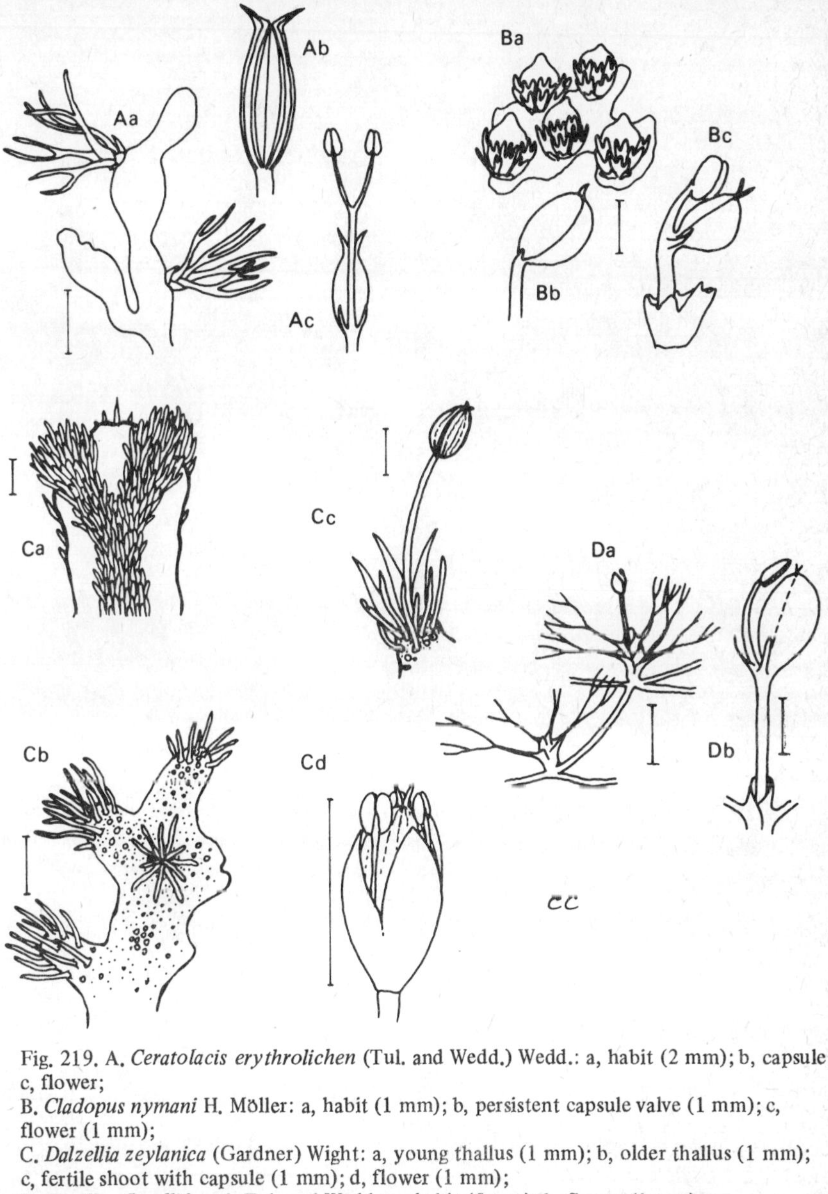

Fig. 219. A. *Ceratolacis erythrolichen* (Tul. and Wedd.) Wedd.: a, habit (2 mm); b, capsule; c, flower;
B. *Cladopus nymani* H. Möller: a, habit (1 mm); b, persistent capsule valve (1 mm); c, flower (1 mm);
C. *Dalzellia zeylanica* (Gardner) Wight: a, young thallus (1 mm); b, older thallus (1 mm); c, fertile shoot with capsule (1 mm); d, flower (1 mm);
D. *Devillea flagelliformis* Tul. and Wedd.: a, habit (5 mm); b, flower (1 mm).

Flowers 1 to few, erect in spathella; spathella obovate. Tepals 2 or 3, one each side of andropodium and one (when present) between the 2 filaments. Stamens 2, borne on andropodium; anther dehiscing introrsely. Ovary spindle-shaped, distinctly 6-ribbed; loculi 2, equal; capsule dehiscing by 2 equal, persistent, valves; midrib of each valve running into style; styles 2, linear.

1 species, *C. erythrolichen* (Tul. and Wedd.) Wedd.: Goyaz, Brazil.

Cladopus H. Möller, Ann. Jard. Bot. Buitenzorg, 16: 115 (1899), [*Hemidistichophyllum* Koidz., *Lawiella* Koidz.]
Fig. 219B.
Roots thalloid, foliose, pinnately branched, closely attached to rocks; stems very short, simple. Leaves on sterile shoots subulate or linear, up to 8 mm long; leaves on fertile shoots more or less scale-like, bifid or palmately lobed with linear lobes. Flowers shortly stalked, solitary, terminal; erect in unruptured spathella; spathella ovoid to globose, with apical papilla. Tepals 2, very small, one each side of stamen. Stamen 1; pollen 2-celled. Ovary smooth, ovoid, obliquely 2-locular; loculi unequal, containing more than 4 seeds; capsule sub-globose, opening by 2 valves, the larger of which is persistent; styles 2, linear.

c. 5 species: Java, Celebes, ? China, Japan.

Dalzellia Wight, Ic. Pl. India Or. 5: 34 (1852), [*Lawia* Griff. ex. Tul., *Mnianthus* Walp., *Terniola* Tul., *Tulasnea* Wight]
Fig. 219C.
Roots thalloid, frondose, flat, closely attached to rocks, up to 25 cm diam., usually irregularly lobed; stems simple, very short. Leaves numerous, simple, linear, occurring either on upper surface and edges of thalloid root or in closely packed rosettes on its older parts. Flowers terminal, solitary; spathella absent. Perianth 3, united below. Stamens 3, free. Ovary 9-ribbed, 3-locular; capsule opening by 3, equal valves; stigmas 3, linear.

Backhuizen, R. C. van den Brink. Lawia, Dalzellia, Mnianthus, Terniola and Indotristicha (Podostem.). Taxon 18 (5): 598–599 (1969)

1 species, *D. zeylanica* (Gardner) Wight: S. India and Ceylon.

Devillea Tul. and Wedd., Ann. Sci. Nat. Bot. sér. 3, 9: 107 (1849)
Fig. 219D.
Roots filamentous, branched, closely attached to rocks; stems short, in groups of 1 to 4, not opposite or sub-opposite. Leaves simple or repeatedly forked with filamentous segments, up to 2.5 cm long; base sheathing, usually

distinct from thallus. Flowers solitary, a few together, erect in spathella; spathella embedded in leaf bases. Tepals 3, one each side of stamen and one on back of stamen. Stamen 1; anther basifixed, introrse; pollen 1-celled, 3-colpate. Ovary smooth, ellipsoidal to globose; loculi 2, unequal; capsule up to 1.5 mm long, sub-globose, opening by 2, unequal valves, the larger persisting; styles 2, very short.

1 species, *D. flagelliformis* Tul. and Wedd.: Goyaz, Brazil.

Dicraeanthus Engler, Bot. Jahrb. 38: 94 (1905)
Fig. 220A.
Roots thalloid, creeping, usually star-shaped; stems elongate, branched, floating, up to 1 m long. Leaves linear or deeply divided into linear segments, borne regularly along one side of stem only. Flowers long-stalked, 3 to 20, in cymose inflorescences, borne opposite leaves, inverted within unruptured spathella; spathella elongate. Tepals 2, one each side of very short andropodium. Stamens 2, approximately half as long as ovary; filaments half as long as anther; pollen 2-celled. Ovary cylindrical, 10-ribbed, with all ribs running the length of capsule; loculi 2, equal; capsule when ripe opening by 2, persistent valves; styles 2, conical.

Wilde, J. J. F. E. de and Guillaumet, J.-L. Un nouveau Dicraeanthus africain (Podostemonaceae). Adansonia 4 (1): 91–98 (1964).

4 species: W. Africa. Used locally as salad.

Endocaulos C. Cusset, Adansonia ser. 2, 12 (4): 557–568 (1972)
Fig. 220B.
Roots thalloid, ribbon-like, infrequently branched, closely attached to rocks or partly floating; stems very short, simple, borne on margins of thallus. Leaves in 2 rows, simple, elongate, 2 to 3 cm long, inflated at base; base with 2, stipule-like lobes; at flowering time the elongate portions of the leaves become detached. Flowers terminal, solitary, erect with unruptured spathella; spathella opening irregularly; pedicel at anthesis 2 to 3 mm long. Tepals 2, one each side of andropodium. Stamens 2, borne on andropodium, somewhat longer than ovary; anthers introrse. Ovary 14-ribbed, with ribs nearest commissures shortest and not extending throughout length of capsule; loculi 2, unequal; styles 2, equal; capsule opening by 2, unequal valves, both persistent.

Cusset, C. Les Podostemaceae de Madagascar. Adansonia ser. 2, 12 (4); 557–568 (1972)

1 species, *E. mangorense* (Perr.) C. Cusset: Madagascar.

Farmeria Willis in Trimen, Handb. Fl. Ceylon 5: 386 (1900)
Fig. 220C.
Root, ribbon- or thread-like, closely attached to rocks; stems very short or absent. Leaves scattered in groups on upper surface and edge of root, linear,

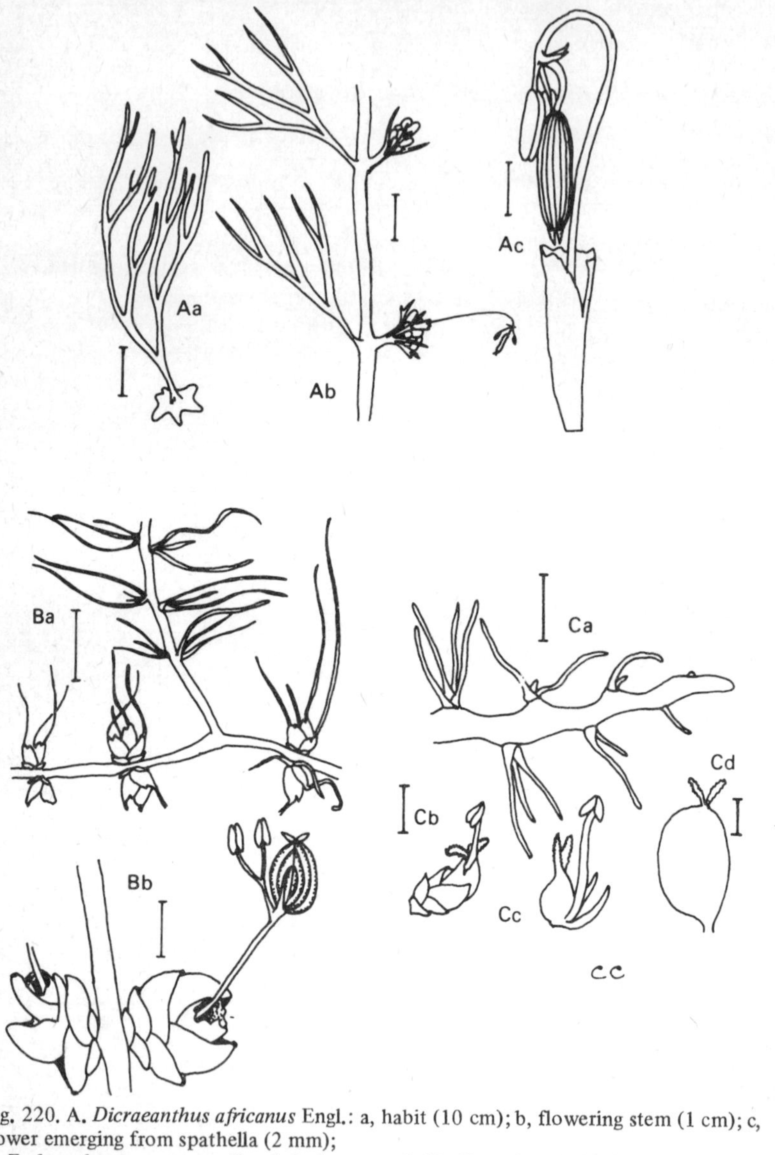

Fig. 220. A. *Dicraeanthus africanus* Engl.: a, habit (10 cm); b, flowering stem (1 cm); c, flower emerging from spathella (2 mm);
B. *Endocaulos mangorense* (Perr.) C. Cusset: a, habit (2 mm); b, fertile shoot with flower (1 mm); after Cusset;
C. *Farmeria indica* Willis: a, habit (5 mm); b, fertile shoot (1 mm); c, flower; d, capsule (1 mm).

up to 10 mm long, caducous. Fertile shoots short, with usually 6, scale-like leaves, with linear, caducous tips. Flowers terminal, solitary, sessile, erect in spathella, spathella splitting at tip. Tepals 2, linear, one each side of stamen. Stamen 1; filament slightly flattened laterally, exceeding ovary and styles. Ovary smooth or weakly 8- or 10-ribbed; loculi 2, very unequal, the lower aborting, the upper 2- or 4-seeded; capsule dehiscent or indehiscent; styles 2, linear.

2 species; *F. indica* Willis from S.W. India and *F. metzgerioides* (Trimen) Willis from Ceylon.

Griffithella (Tul.) Warming, Danske Vid. Selsk. Skrift. ser. 6, 11: 13, 65 (1901)
Fig. 221A.
Roots flat, closely attached to rocks, thalloid, foliose or ribbon-like, regularly or irregularly branched, up to 20 cm diam.; stems absent or present; sterile portion of root with simple, linear, up to 5 mm long leaves or leaf-like processes; fertile portion of root with short stems bearing 4 to 6, bract-like leaves. Bract-like leaves in 2 rows, flat or concave, with blunt or occasionally hair-like tip. Flowers solitary, erect in unruptured spathella; spathella tubular or funnel-shaped, splitting irregularly at tip, exceeding surrounding leaves. Tepals 2, linear, one each side of andropodium. Stamens 2, borne on andropodium. Ovary stalked, smooth, unribbed; loculi 2, unequal; capsule opening by 2, unequal valves, the larger persistent; styles 2, linear, rarely lobed.

1 species, *G. hookeriana* (Tul.) Warm.: S.W. India and ?Cochinchina.

Hydrobryopsis Engler in Engler and Prantl, Nat. Pflanzenfam. 18a: 60 (1930), [*Hydrobryum* Endl. pro parte]
Fig. 221B.
Roots flat, closely attached to rocks, thalloid, more or less regularly, pinnately branched, 3 to 10 mm wide; branches sometimes overlapping and giving root a foliose appearance; sterile stems and leaves not yet observed; fertile stems at edge of thallus, in branch "axils", with 5 to 7 bract-like leaves. Leaves in 2 rows, the lower small and flat, the upper larger and helmet-shaped. Flowers solitary, erect within spathella; spathella boat-shaped, somewhat rough, splitting at top, remaining embedded in surrounding leaves. Tepals 2, one each side of andropodium, linear, as long or longer than ovary. Stamens 2, borne on andropodium; andropodium as long or longer than ovary; filaments about as long as ovary. Ovary sessile or very shortly stalked, remaining embedded in surrounding leaves, smooth, unribbed; loculi 2, unequal; capsule opening by 2 valves, the larger persistent; styles 2, linear.

1 species, *H. sessilis* (Willis) Engler; S. India, known from 2 localities.

Hydrobryum Endl., Gen. Pl. Suppl. 1: 1375 (1841) [*Hydroanzia* Koidz.]
Fig. 221C.
Roots thalloid, foliose, hard, closely attached to rocks, up to 30 cm diam.; stems very short, simple or absent. Sterile leaves (or leaf-like processes) linear, up to 12 mm long, caducous, scattered in groups of 2 to 6 on upper surface of thallus; fertile leaves scale-like. Flowers terminal, solitary, erect within spathella; spathella usually splitting down one side. Tepals 2 (rarely 1, 3 or 4),

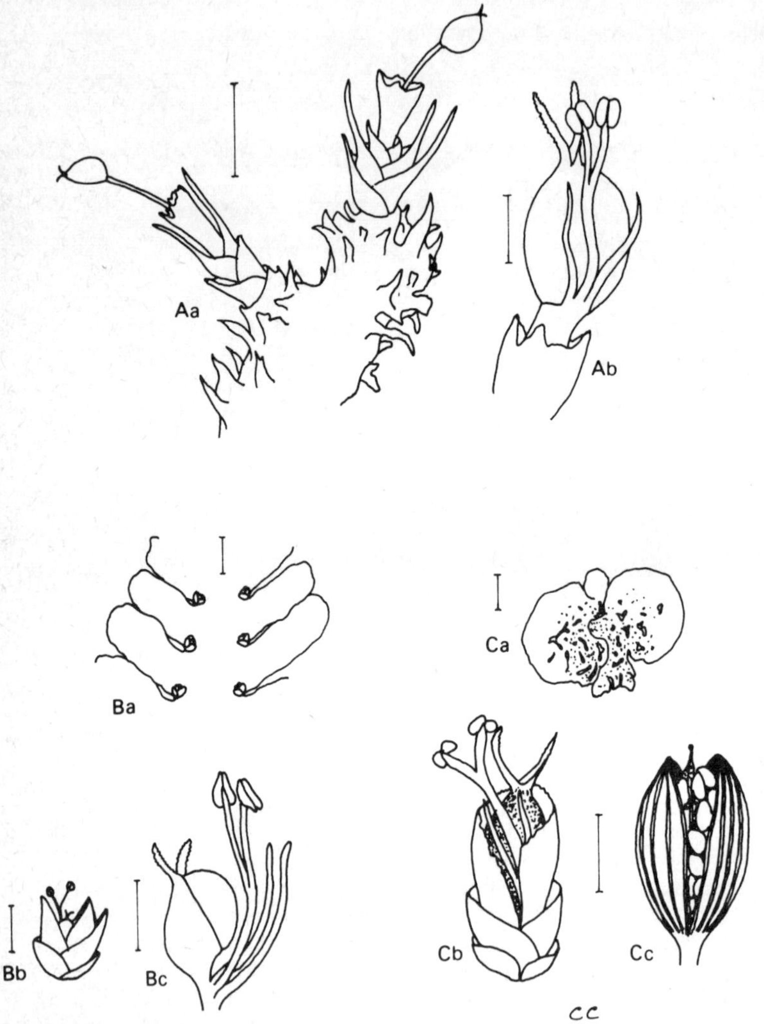

Fig. 221. A. *Griffithella hookeriana* (Tul.) Warm.: a, thallus with fruit (5 mm); b, flower (1 mm);
B. *Hydrobryopsis sessilis* (Willis) Engler: a, thallus (1 cm); b, fertile shoot (1 mm); c, flower (0.5 mm);
C. *Hydrobryum japonicum* Imamura: a, thallus (1 cm); b, flower (1 mm); c, capsule.

linear, one each side of andropodium. Stamens 2 (rarely 3), borne on andropodium. Ovary rarely 6-, usually 12- to 22-ribbed, loculi 2, equal; capsule opening by 2, equal, persistent valves; styles 2, linear, ovate or cuneate, entire or lobed.

c. 10 species (probably less): N.E. India, E. and S.E. Asia and Japan. Cusset, C. Adansonia sér. 2, 12 (2): 279–282 (1972) described a new, monotypic genus, *Diplobryum* from S. Viet-Nam. It is reputed to differ from *Hydrobryum* in having globose stigmas and ribbed, elongated seeds.

Indotristicha v. Royen, Acta Bot. Neerl. 8: 474–476 (1959), [*Dalzellia* Wight, pro parte]
Fig. 222A.
Roots long, slender, much branched, creeping or partly floating. Stems dimorphic, creeping or floating; floating stems simple or branched; branches of 2 kinds, long and leafless with bunches of photosynthetic filaments or short and leafy. Leaves in 3 rows, equal or unequal in length, usually overlapping. Flowers terminal, solitary; spathella absent. Perianth segments 3, united below. Stamens 3, free. Ovary 9-ribbed, 3-locular; capsule opening by 3, equal valves; styles 3, linear or flattened, with lobed margin.

Dransfield, J. and Whitmore, T. C. A Podostemacea new to Malaya. Blumea 18 (1): 153–155 (1970)

2 species: *I. malayana* Dransfield and Whitmore from Malaya and *I. ramosissima* (Wight) van Royen from W. India; perhaps congeneric with *Tristicha*

Inversodicraea Engler in Engler and Drude, Veg. d. Erde, Pflanzenwelt Africas 9 (3, 1): 271 (1915)
Fig. 222B.
Roots thalloid, foliose, closely attached to rocks; stems very short or elongate and floating. Leaves variable, simple, lobed or forked, linear with thread-like segments or scale-like with entire or toothed margin, at least some leaves divided. Flowers terminal, solitary, inverted with unruptured spathella; spathella opening irregularly at tip. Tepals 2, one each side of andropodium or stamen, small, rarely exceeding base of ovary. Stamens 2 (very rarely 3 or 4), borne on andropodium or 1, usually longer than ovary, bent and lying below ovary in unruptured spathella; pollen 1- or 2-celled. Ovary 6 to 10-ribbed, with all ribs running the length of the capsule, ovoid to ellipsoidal; loculi 2, equal; capsule when ripe opening by 2, persistent valves; styles 2, linear; seeds numerous.

c. 35 species: Tropical Africa and Madagascar. Cusset, C. Adansonia sér. 2. 12 (4): 557–568 (1972) has described a new genus *Thelethylax*. It is based on the presence of a small tepal, or staminode between the 2 stamens. In

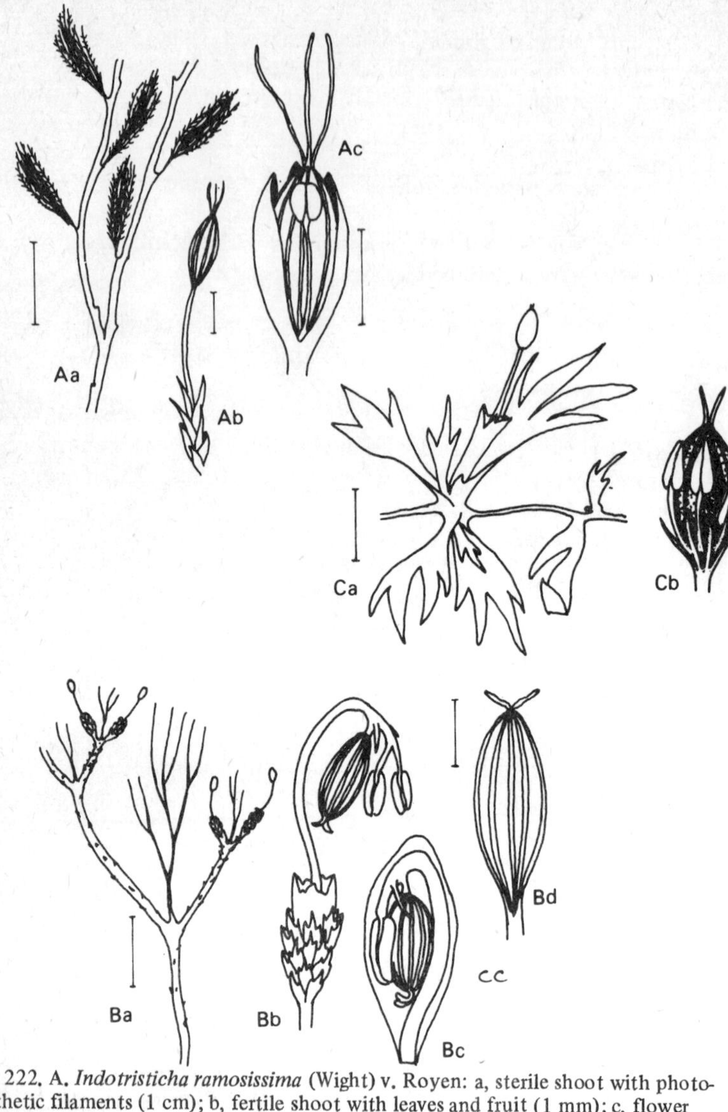

Fig. 222. A. *Indotristicha ramosissima* (Wight) v. Royen: a, sterile shoot with photosynthetic filaments (1 cm); b, fertile shoot with leaves and fruit (1 mm); c, flower (0.5 mm);
B. *Inversodicraea abbayesii* G. Taylor: a, upper portion of plant (1 cm); b, flower (1 mm); c, flower within spathella; d, fruit; after Taylor;
C. *Jenmaniella tridactylifolia* Engler: a, habit (5 mm); b, flower.

this respect it is similar to the W. African genus *Stonesia* but the capsule has 8 ribs and resembles that of *Inversodicraea. Thelethylax* has 2 species, *T. minutiflora* (Tul.) C. Cusset [*Inversodicraea minutiflora* (Tul.) Perr.] and *T. isalensis* (Perr.) C. Cusset [*I. isalensis* (Perr.) Perr.], both from Madagascar.

Jenmaniella Engler, Bot. Jahrb. 61, Beibl. Nr. 138: 7 (1927)
Fig. 222C.
Roots filamentous, branched, closely adhering to rocks; stems very short or absent. Leaves often united at base and almost indistinguishable from thallus, very variable, usually a few times forked or pinnate with forked segments; stipules occasionally present, particularly on leaves subtending flowers. Flowers solitary, erect within spathella; spathella club-shaped. Tepals 2 to 7, in complete or incomplete whorl. Stamens 1 to 7, free or sometimes 2 stamens borne by andropodium; number, shape and size of stamens varies from flower to flower on one individual, anthers introrse or extrorse; pollen 1-celled, 3-sulcate. Ovary shortly-stalked, ellipsoidal, 6-ribbed; loculi 2, equal; capsule opening by 2, equal valves; styles 2, elongate.

7 species: northeastern S. America.

Ledermanniella Engler, Bot. Jahrb. 43: 378 (1909)
Fig. 223A.
Roots thalloid, foliose, creeping; stems erect, simple or branched, 5 to 60 mm long. Leaves linear, up to 10 mm long, 0.2 to 0.4 mm wide, often one pair of teeth at base. Flowers long stalked, terminal, inverted within unruptured spathella; spathella ellipsoidal, occasionally constricted at centre. Tepals 2, one each side of andropodium. Stamens 2, rarely 3; filaments and andropodium elongate after anthesis. Ovary weakly 10-ribbed, with all ribs running the length of the capsule; loculi 2, unequal; capsule subglobose, opening by 2 valves, only one of which is persistent; styles 2, filamentous.

Hess, H. Podostemonaceae aus französisch Kamerun Ber. Geobot. Inst. Eidg. Techn. Hochschule Stift. Rübel 32: 189–190 (1961)

1 species, *L. linearifolia* Engler: W. Africa, Cameroons.

Leiothylax Warming, Danske Vidensk Selsk. Skrift. ser. 6, 9: 147, 154 (1899)
Fig. 223B.
Roots thalloid, closely attached to rocks or floating; stems branched, floating. Leaves up to 5 mm long, linear or forked with linear segments, often with stipule-like teeth at base. Flowers solitary or in groups, erect in unruptured spathella; usually cleistogamous; spathella ovoid, rupturing radially. Tepals 2, small, one each side of andropodium; andropodium *c.* 1 mm long. Stamens 2; filaments very short, less than ¼ as long as andropodium, pollen 1-celled. Ovary smooth, unribbed, sub-globose; loculi 2, unequal; capsule opening by 2 valves, the larger persistent; styles 2, linear.

Hess, H. Ueber die Familien der Podostemonaceae und Hydrostachyaceae in Angola. Ber. Schweiz. Bot. Gesell. 63: 369–370 (1953)

1 or 2 species: Equatorial W. Africa and perhaps E. Africa.

Letestuella G. Taylor, Bull. Brit. Mus. (Nat. Hist.) Bot. ser. 1. 57 (1953) Fig. 223C.

Roots probably thalloid, stems branching, up to 5 cm long. Leaves linear or forked with linear segments, up to 4 cm long, often with stipule-like teeth at base. Flowers terminal, solitary, erect within unruptured spathella; spathella ellipsoidal, subsessile or with up to 1 cm long stalk, becoming campanulate with revolute teeth after rupturing. Tepals 2, one each side of

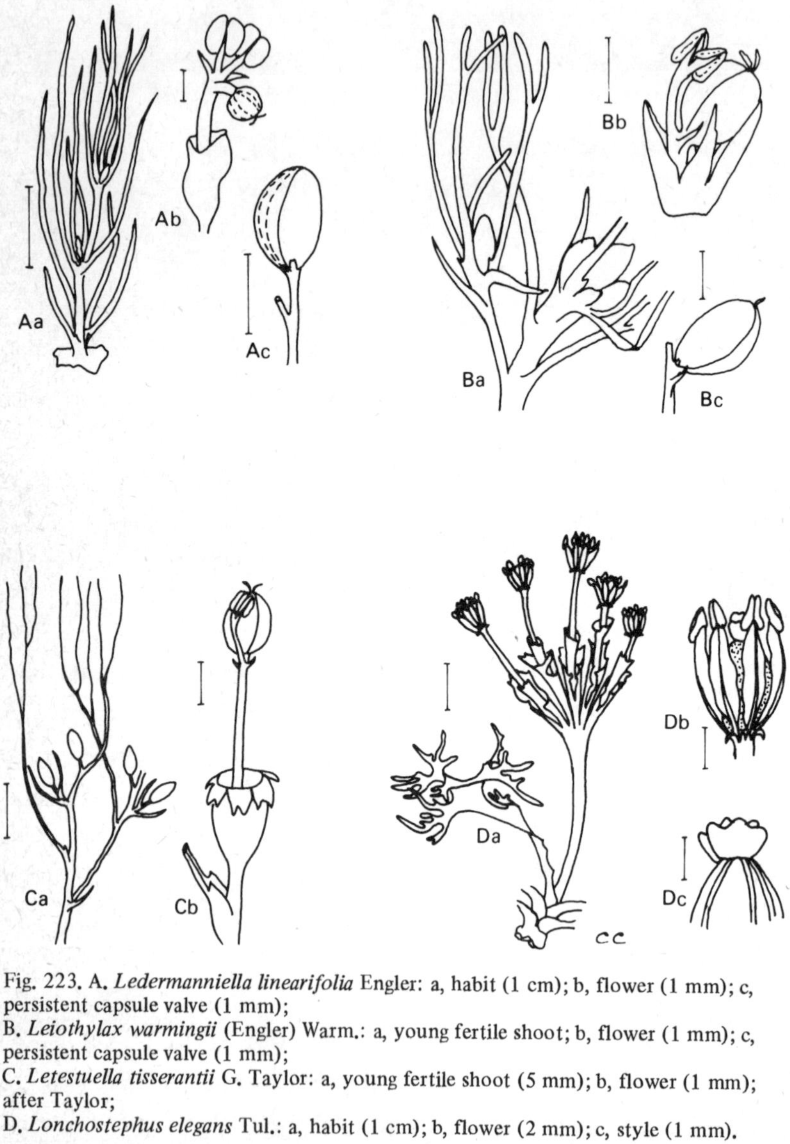

Fig. 223. A. *Ledermanniella linearifolia* Engler: a, habit (1 cm); b, flower (1 mm); c, persistent capsule valve (1 mm);
B. *Leiothylax warmingii* (Engler) Warm.: a, young fertile shoot; b, flower (1 mm); c, persistent capsule valve (1 mm);
C. *Letestuella tisserantii* G. Taylor: a, young fertile shoot (5 mm); b, flower (1 mm); after Taylor;
D. *Lonchostephus elegans* Tul.: a, habit (1 cm); b, flower (2 mm); c, style (1 mm).

andropodium. Stamens 2, rarely 1; pollen 1-celled. Ovary smooth, globose; loculi 2, equal; capsule opening by 2, equal, unribbed valves; styles 2, linear.

2 species, *L. tisserantii* G. Taylor and *L. chevalieri* G. Taylor: W. Africa.

Lonchostephus Tul., Arch. Mus. Nat. (Paris) 6: 198 (1852)
Fig. 223D.
Stemless herbs with *c.* 10 mm long and 2 to 5 mm wide base. Leaves 1.5 to 8.0 cm long; leaf stalk strap-like; leaf blade repeatedly forked; ultimate segments capillary, Flowers erect within spathella, alternating with boat- or leaf-like bracts, in 2-sided, unbranched, compressed, spike-like, 1.5 to 8.0 cm long inflorescences; spathella present. Tepals 5 to 8, free, lanceolate, *c.* 0.5 mm long. Stamens 5 to 8, in 1 whorl, free, 4.0 to 6.5 mm long; filaments widened, membranous, elliptic, 3 to 5 mm long, 1.0 to 1.5 mm wide. Ovary 8-ribbed, ovoid, 2-locular; capsule opening by 2, equal valves; styles 2, crested, flattened, persisting in fruit.

1 species, *L. elegans* Tul.: Upper Amazon, Brazil.

Lophogyne Tul., Ann. Sci. Nat. Bot. sér. 3, 11: 99 (1849)
Fig. 224A.
Roots thalloid, foliose or ribbon-like, closely attached to rocks or partly floating; stems very short, indistinct, sub-opposite. Leaves not distinctly separated from thallus at base, at top finely dissected or repeatedly forked; ultimate segments filamentous. Flowers solitary, erect within spathella; spathella embedded in thallus when young, ellipsoidal; mature spathella tubular. Tepals 2 to 5, in complete or incomplete whorl, lanceolate or linear acute. Stamens 2 to 4; anthers introrse, sometimes spirally coiled when dry; pollen 1-celled. Ovary ellipsoidal to ovoid, 6-ribbed; loculi 2, equal; capsule opening by 2, equal valves; styles 2, cock's comb-like, compressed, with serrate margin, persisting in fruit.

Tobler, F. Beiträge zur Oekologie und Biologie brasilianischer Podostemonaceen. Flora n.s. 28: 286–300 (1933)

2 species, *L. helicandra* Tul. and *L. arculifera* Tul. and Wedd.: eastern C. Brazil. Both species frequently grow intermingled.

Macarenia v. Royen, Med. Bot. Mus. Utrecht 107: 137 (1951)
Fig. 224B.
Stemless herbs with up to 7 mm wide, variable shaped base. Leaves up to 30 cm long; leaf stalk terete, up to 10 cm long, provided at base with up to 3 mm long, obtuse, membranous intrapetiolar stipule; ultimate leaf segments

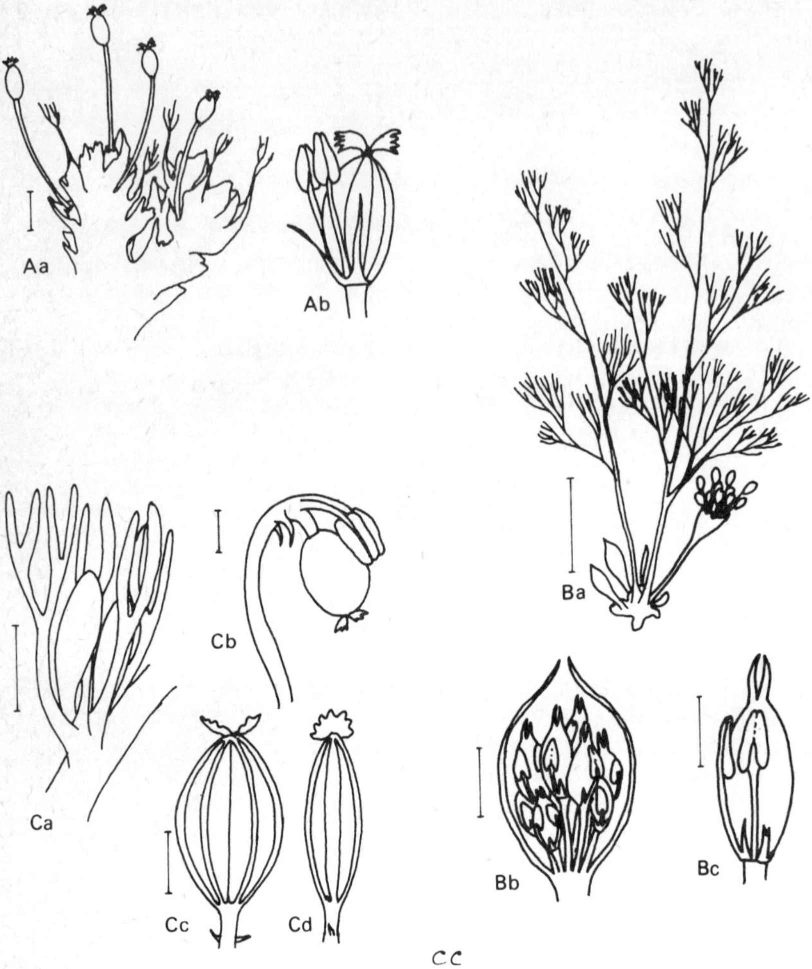

Fig. 224. A. *Lophogyne helicandra* Tul.: a, habit (1 mm); b, flower;
B. *Macarenia clavigera* v. Royen: a, habit (5 cm); b, young spathella opened (2 mm); c, flower (1 mm); after v. Royen;
C. *Macropodiella mildbraedii* Engler: a, leaves and unopened spathellas (5 mm); b, young flower (1 mm); c, capsule viewed from widest side (1 mm); d, capsule viewed from narrowest side (1 mm).

linear, up to 15 mm long, acute, with distinct nerve. Flowers 10 to 20, erect, in single, club-shaped spathella; spathella solitary or 2 to 3 together, enveloped at base by 2 membranous bracts; individual flowers without spathella. Tepals 3 to 5, lanceolate, *c.* 0.8 mm long. Stamens 2 to 4, up to 4 mm long. Ovary 6-ribbed; 2-locular; styles 2, simple.

1 species: *M. clavigera* v. Royen, Macarena Mountains, Colombia.

Macropodiella Engler, Bot. Jahrb. 60: 466 (1926)
Fig. 224C.
Roots filamentous, oval in section, creeping; stems in 2 rows, 10 to 15 mm long. Leaves divided, segments linear. Flowers inverted in unruptured spathella; spathella club-shaped. Tepals 2, very short, one each side of andropodium. Stamens 2; pollen 1-celled. Ovary ovoid, compressed, twice as wide as thick, weakly 6-ribbed; capsule opening by 2, equal valves; styles 2, cock's comb-like, flattened with serrated edges.

1 or 2 species: Equatorial W. Africa.

Marathrum Humb. and Bonpl., Pl. Aequinoct. 1: 39, t.II (1806), [*Lacis* Lindlay, pro parte]
Fig. 225A.
Small to medium sized colonies of individuals arising from branched roots; individuals with thalloid, irregular or branched bases; sterile individuals often larger and coarser than fertile ones; stems indistinct. Leaves in 2 rows or in 1 row along margin of base, either repeatedly forked or sub-entire with few lobes; intrapetiolar stipule sometimes present. Flowers erect within spathella, solitary or in groups. Tepals 3 to 25, in 1 complete or incomplete whorl, more or less adnate to base of stamens. Stamens 2 to 25, in 1 complete or incomplete whorl or confined to 1 side of flower; filaments 3-sided at base, sometimes branched; anthers dehiscing introrsely; pollen 1-celled, 3-colpate. Ovary 6- or 8-ribbed, with ribs occasionally winged; loculi 2, equal or slightly unequal; styles 2, linear, boat-shaped or spathulate, often notched at tip.

c. 25 species: C. America, W. Indian Islands and N. W. part of S. America.

Mniopsis Mart. and Zucc., Nov. Gen. Spec. Pl. Bras. 1: 3 (1822)
Fig. 225B.
Roots elongate, branching, closely attached to rocks; stems opposite or sub-opposite, distinct, simple or branched, usually with remains of leaves below and tuft of leaves at top. Leaves entire or few times forked, either all similar and provided with 1 or 2 stipules or of 2 kinds and then without stipules and in 4 rows. Flowers few to numerous, erect in spathella. Tepals 2 or 3, one each side of andropodium and one (when present) between or below the filaments. Stamens 2, borne on andropodium; anthers introrse; pollen 2-celled. Ovary globose to ellipsoidal, smooth, unribbed; loculi 2, unequal; capsule opening by 2, unequal valves, the larger persistent; styles 2, occasionally simple, usually branched.

c. 5 species: S. E. Brazil.

Monandriella Engler, Bot. Jahrb. 60: 457 (1926)
Fig. 225C.
Roots filamentous; stems erect up to 2 cm long. Leaves linear, up to 1 cm long, with tooth-like stipules at base. Flowers, terminal, solitary, inverted within unruptured spathella; spathella ellipsoidal. Tepals 2, short, pointed, one each side of andropodium. Stamen 1; filament as long or longer than anther; pollen 1-celled. Ovary 6-ribbed, ellipsoidal; 1 loculus; styles 2, linear.

1 species, *M. linearifolia* Engler: Equatorial W. Africa.

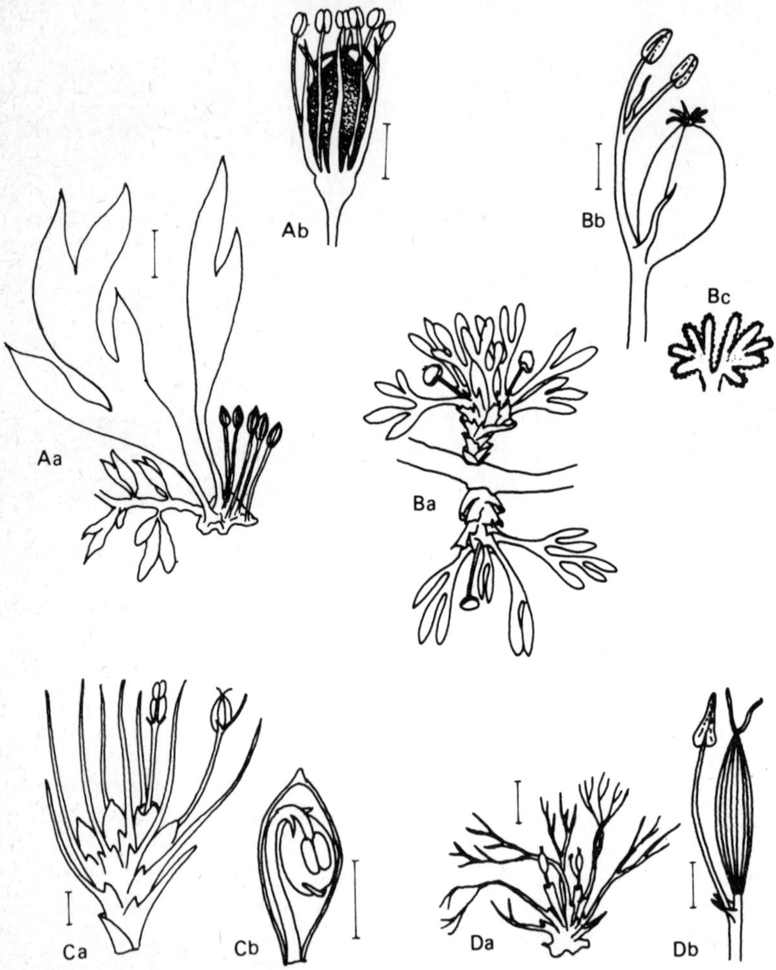

Fig. 225. A. *Marathrum utile* Tul.: a, habit (1 cm); b, flower (1 mm); B. *Miniopsis weddelliana* Tul.: a, habit; b, flower (1 mm); c, style; C. *Monandriella linearifolia* Engler: a. habit (1 mm); b, flower within spathella (1 mm); D. *Monostylis capillacea* Tul.: a, habit (1 cm); b, flower (1 mm).

Monostylis Tul., Arch. Mus. (Paris) 6: 201 (1852)
Fig. 225D.
Roots thalloid, 2 to 5 mm diam., closely attached to rocks; stems absent. Leaves united at base, up to 3 cm long, repeatedly forked; ultimate segments filamentous. Flowers solitary, erect in spathella; spathella tubular at maturity, up to 4 mm long. Tepals 2, one each side of stamen, about 1 mm long. Stamen 1.3 to 4 mm long; pollen 1-celled, constricted to one side with ring of warts around middle. Ovary ellipsoidal, 14-ribbed, borne on 1 mm long stalk; loculi 2, equal; capsule opening by 2 equal valves; styles 2, linear, persistent.

1 species, *M. capillacea* Tul.: Para, Brazil.

Mourera Aublet, Pl. Guiane Franç. 582 (1775)
Fig. 226A.
Stemless or with short stem formed by fusion of leaf bases. Leaves radical, very variable, 8 to 200 cm long, elliptical with fimbriate margin, cuneate, pinnately lobed or repeatedly forked with capillary segments, sometimes very coarse and provided with many rigid outgrowths on adaxial surface. Flowers alternating with bracts, in 2-sided, branched or unbranched, spike-like inflorescence or reduced to single flower, erect in spathella; bracts winged; spathella exceeding bracts. Tepals free, 5 to 20. Stamens 5 to 35, in 1 or 2 whorls; filaments lanceolate or linear. Ovary 6- to 14-ribbed, ovoid, 2-locular; styles 2, linear or spathulate.

Steude, H. Beiträge zur Morphologie und Anatomie von Mourera aspera. Beih. Bot. Centralbl. 53: 627–650 (1935)

c. 6 species: Northern S. America.

Oserya Tul. and Wedd., Ann. Sci. Nat. sér. 3, 9: 105 (1849)
Fig. 226B.
Roots filamentous or ribbon-like, branched, closely attached to rocks; stems short, usually opposite or sub-opposite. Leaves in 2 rows, simple or repeatedly forked, with filamentous segments, often with stipulate base, often indistinctly separated from thallus. Flowers solitary, erect within spathella; spathella tubular at maturity. Tepals 2 or 3, one each side of stamen and usually third one on back of filament. Stamen 1; anther basifixed, extrorse; pollen 1-celled, 3-colpate. Ovary ovate, 6- to 14-ribbed; loculi 2, unequal; capsule opening by 2 unequal valves, the larger persistent; styles 2, very short.

6 species: C. America, northern S. America and Central Brazil. The disjunction in distribution may be a reflection of poor collecting; *Oserya* is very small and easily overlooked.

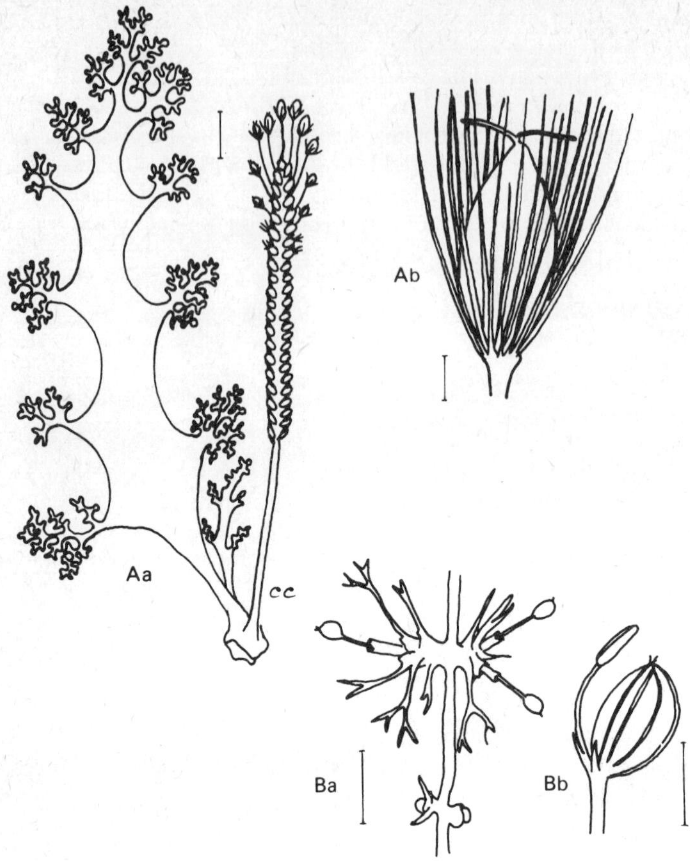

Fig. 226. A. *Mourera fluviatilis* Aubl.: a, habit, thallus with inflorescence (3 cm); b, young fruit (1 mm);
B. *Oserya minima* v. Royen: a, habit (5 mm); b, flower (1 mm); after v. Royen.

Podostemum Micheaux, Fl. Bor. Amer. 2: 164, t. 44 (1803)
Fig. 227A.
Roots elongate, branched, creeping; stems indistinct or distinct, arising laterally from root, branched or unbranched; fertile stems sometimes separate from sterile ones along the same root. Leaves in 2 rows, entire or repeatedly forked; intrapetiolar stipule usually present (often suppressed in lower leaves). Flowers axillary, solitary, erect within spathella; spathella splitting at top. Tepals 3, one each side of andropodium and one on back of andropodium or in the fork between the 2 filaments. Stamens 2 (rarely 1), borne on andropodium; anthers dehiscing introrsely; pollen 2-celled. Ovary

6- or 8-ribbed; loculi 2, unequal; capsule opening by 2 valves, the larger persistent; styles 2, simple, equal or unequal, linear, free.

c. 18 species: E. and S. N. America, W. Indies, C. America, S. and S. W. S. America, S. W. India and Ceylon.

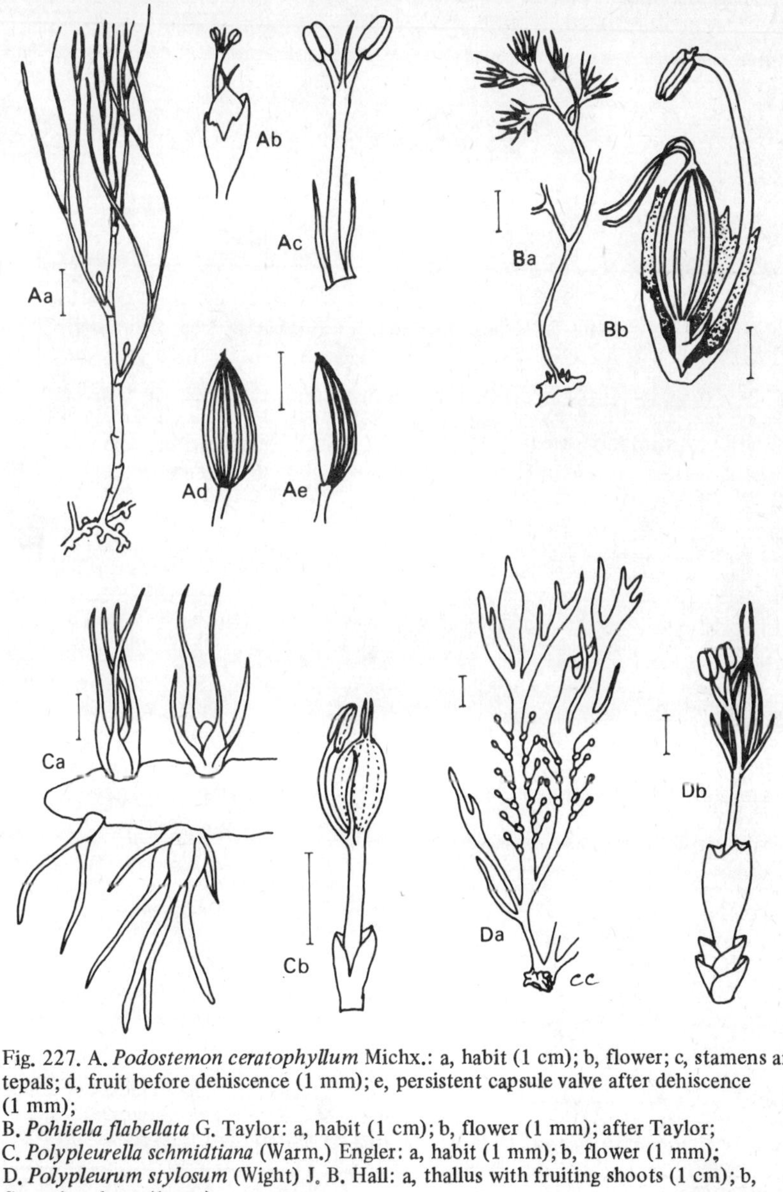

Fig. 227. A. *Podostemon ceratophyllum* Michx.: a, habit (1 cm); b, flower; c, stamens and tepals; d, fruit before dehiscence (1 mm); e, persistent capsule valve after dehiscence (1 mm);
B. *Pohliella flabellata* G. Taylor: a, habit (1 cm); b, flower (1 mm); after Taylor;
C. *Polypleurella schmidtiana* (Warm.) Engler: a, habit (1 mm); b, flower (1 mm);
D. *Polypleurum stylosum* (Wight) J. B. Hall: a, thallus with fruiting shoots (1 cm); b, flowering shoot (1 mm).

Pohliella Engler, Bot. Jahrb. 60: 457 (1926)
Fig. 227B.
Roots thalloid, closely attached to rocks; stems branched, slightly flattened, floating. Leaves repeatedly forked, with filamentous ultimate segments. Flowers borne on thallus and floating stems, erect within unruptured spathella; spathella ellipsoidal. Tepals 2, one each side of stamen. Stamen 1; pollen 2-celled. Ovary ellipsoidal, 10-ribbed (including commissural ribs); loculi 1 or 2; capsule opening by 2, equal valves; styles 2, filamentous, half or more as long as ovary.

2 species, *P. flabellata* G. Taylor and *P. laciniata* Engler: Tropical W. Africa.

Polypleurella Engler, Bot. Jahrb. 61, Beibl. Nr. 138: 9 (1927)
Fig. 227C.
Roots thalloid, ribbon-like, closely attached to rocks; sterile stems absent; fertile stems simple, very short, borne on margin of thallus. Leaves opposite, subulate, swollen at base. Flowers, terminal, solitary, erect within unruptured spathella; spathella ellipsoidal, subsessile, when young enclosed in swollen leaf bases. Tepals 2, linear, as long as filament. Stamen 1; pollen 2-celled. Ovary ovoid, about 1 mm long, weakly 10-ribbed; loculi 2, equal; capsule opening by 2, equal valves; styles subulate, erect, about half as long as ovary.

1 species, *P. schmidtiana* (Warm.) Engler: Siam.

Polypleurum (Tayl. ex Tul.) Warming. Kgl. Danske Vidensk. Selsk. Skr. ser. 6, Nat. Math. Afd. 11 (1): 4, 64 (1901), [*Dicraeia* Du Petit-Thouars]
Fig. 227D.
Roots thalloid, ribbon-like or thread-like, partly closely attached to rocks, partly floating; floating roots up to 50 cm long but usually shorter, much branched, often becoming somewhat woody at flowering time; stems short, produced at more or less regular intervals along margin of root; fertile stems with 2 to 8 or rarely more overlapping, scale-like leaves. Flowers terminal, solitary, erect within spathella; spathella opening irregularly at tip. Tepals 2, one each side of andropodium. Stamens 2, occasionally unequal. Ovary 8- to 12-ribbed; loculi 2, equal; capsule opening by 2, persistent valves; styles 2, linear.

Hall, J. B. New Podostemaceae from Ghana with notes on related species. Kew Bull. 26 (1): 125–136 (1971).

c. 5 species: India, Ceylon, Burma, Ghana (W. Africa) and Madagascar. Cusset, C. Adansonia ser. 2, 12 (4): 562 (1972), described a genus, *Paleodicraeia*, based on a single collection of Du Petit-Thouars, gathered before 1806. Time has not permitted an examination of this material but it is felt that it may well be a species of *Polypleurum*.

Rhyncholacis Tul., Ann. Sci. Nat. Bot. sér. 3, II: 95 (1894)
Fig. 228A.
Root thalloid, simple, branched or foliose; stems absent. Leaves repeatedly forked or pinnate, with filamentous ultimate segments; leaf base often with 2 sheaths. Fertile individuals frequently smaller and more slender than sterile ones. Flowers either solitary or in groups, erect in spathella; spathella tubular, rupturing at apex. Tepals 3 to 20, in complete or incomplete whorl, some-

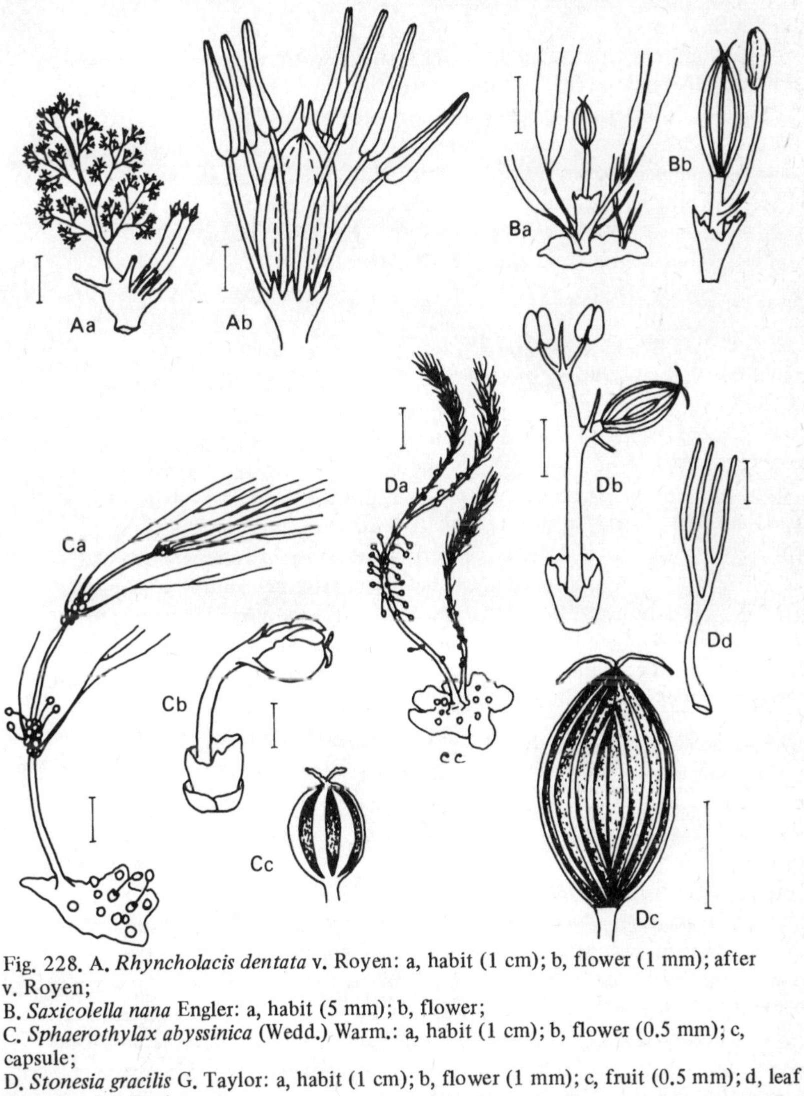

Fig. 228. A. *Rhyncholacis dentata* v. Royen: a, habit (1 cm); b, flower (1 mm); after v. Royen;
B. *Saxicolella nana* Engler: a, habit (5 mm); b, flower;
C. *Sphaerothylax abyssinica* (Wedd.) Warm.: a, habit (1 cm); b, flower (0.5 mm); c, capsule;
D. *Stonesia gracilis* G. Taylor: a, habit (1 cm); b, flower (1 mm); c, fruit (0.5 mm); d, leaf (1 mm); after Taylor.

times unequal in length. Stamens usually as many as tepals; anthers often with 2 or 3 teeth at top; pollen 1-celled. Ovary ellipsoidal to ovoid, compressed, 6- or rarely 8-ribbed; loculi 2, equal; capsule opening by 2, equal valves; midrib of each valve more or less winged; styles 2, subulate or clavate.

c. 25 species: northern part of S. America. Most species geographically restricted to a single cataract or river.

Saxicolella Engler, Bot. Jahrb. 60: 456 (1926)
Fig. 228B.
Root thalloid, closely attached to rocks; stems scattered on upper surface of root, branched or simple, floating. Leaves linear and simple or palmately divided in 2 or 3 linear segments. Flowers terminal, solitary, erect within unruptured spathella; spathella club-shaped. Tepals 2, small, one each side of stamen. Stamen 1; pollen 2-celled. Ovary ellipsoidal, 6-ribbed; loculus 1; capsule opening by 2 equal valves; styles 2, short, subulate.

1 species, *S. nana* Engler: Equatorial W. Africa.

Sphaerothylax Bischoff in Krauss, Flora 27: 426 (1844), [*Anastrophea* Wedd., *Leiothylax* Warming, pro parte, *Leiocarpodicraea* Engler, pro parte]
Fig. 228C.
Roots thalloid, tightly attached to rocks, either ribbon-like, about 1 mm wide, pinnately branched, or foliose, with rounded lobes; stems either very short and unbranched or long, branched and floating. Leaves scale-like, entire, rarely more than 2 on flowering stems or divided into linear segments on floating stems. Flowers terminal, solitary on individual stems but stems often produced in groups, inverted within unruptured spathella; spathella globose, rupturing irregularly. Tepals 2, linear, one each side of andropodium. Stamen 1, often 2-lobed; filament often flattened; pollen 2-celled. Ovary ovoid to sub-globose 8-ribbed; ribs wide; loculi 2, unequal; capsule opening by 2 valves, the larger persistent; styles 2, ovate to linear.

c. 10 species: Tropical and S. Africa.

Stonesia G. Taylor, Bull. Brit. Mus. (Nat. Hist.) Bot. ser. 1: 59 (1953)
Fig. 228D.
Roots thalloid, foliose, creeping; stems very short or elongate and floating. Leaves elongate, deeply divided with linear segments. Flowers terminal, solitary, inverted within unruptured spathella; spathella sessile on thallus. Tepals 3, one each side of andropodium, one at tip of andropodium between the stamens. Stamens 2, borne on andropodium. Ovary 12 to 18 ribbed, with ribs nearest commissures shortest and not extending throughout length of capsule, ellipsoidal to sub-globose; loculi equal; capsule when ripe opening by persistent valves; styles 2, linear; seeds numerous.

Cusset, C. Contribution à l'étude des Podostemaceae. III. Le genre Stonesia. Adansonia sér. 2, 13 (3): 307–312 (1973)

4 species: Tropical W. Africa, confined to a small region of Guinea.

Torrenticola Domin ex Steenis, Journ. Arnold Arb. 28: 241 (1947)
Fig. 229A.
Roots thalloid, ribbon-like, closely attached to rocks; stems erect, up to 5 cm long, simple or occasionally branched. Leaves distinctly in 2 rows, alternate, at base of stem scale-like, leathery, 2-lobed, towards apex of stem subulate to filamentous, with expanded, stipulate bases. Flowers solitary, shortly stalked, erect within unruptured spathella; spathella sub-globose, with apical papilla. Tepals 2, one each side of stamen. Stamen 1; pollen 2-celled. Ovary 10-ribbed, more or less globular; loculi 2, unequal; capsule valves unequal, the larger persistent; styles oblong-lanceolate.

1 species, *T. queenslandica* (Domin) Steenis: New Guinea and Queensland, N. Australia.

Tristicha Du Petit-Thouars, Gen. Nov. Madag. 3 (1806), nom. cons. *
Fig. 229B.
Roots flattened, branched, creeping; stems dimorphic, creeping or floating; floating stems up to 10 cm long. Leaves small, sessile, entire or divided, in 3 rows, 2 rows spreading, third row smaller and appressed. Flowers solitary, when young enveloped in leaves; spathella absent. Perianth segments 3, free or united at base. Stamen 1. Ovary 9-ribbed, 3-locular; capsule opening by 3, equal valves; styles 3, linear.

Jäger-Zürn, I. Morphologie der Podostemaceae I. Tristicha trifaria (Bory ex Willd.) Spreng. Beitr. Biol. Pflanzen 47: 11–52 (1970).

1 species, *T. trifaria* (Bory ex Willd.) Sprengel: Tropical America, Africa, Madagascar and Mascarene Islands. A highly polymorphic species with the widest distribution in the whole family. The plants form moss-like mats on rocks.

Tulasneantha v. Royen, Med. Bot. Mus. Utrecht 107: 9 (1951), [*Lacis* Lindley pro parte]
Fig. 229C.
Stemless herbs with 1 to 5 cm long, branched or unbranched base. Leaves 10 to 30 cm long; leaf stalk strap-like, smooth, 5 to 16 cm long; leaf blade fan-shaped, repeatedly forked. Flowers alternating with boat-shaped bracts, in

* See Aston, H. I. Aquatic Plants of Australia. Melbourne Univ. Press (1973) for *Tristicha* in Australia and note on status of *Indotristicha*.

2-sided, unbranched, compressed, spike-like, 8 to 30 cm long, inflorescences; spathella slightly exceeding bract. Tepals 6 to 10, lanceolate, *c.* 0.5 mm long. Stamens 6 to 10, united in a ring, 5.0 to 12.5 mm long. Ovary 8-ribbed, ovoid, 2-locular; capsule opening by 2, equal valves; styles 2, linear.

1 species, *T. monadelpha* (Bong.) v. Royen: W. Brazil.

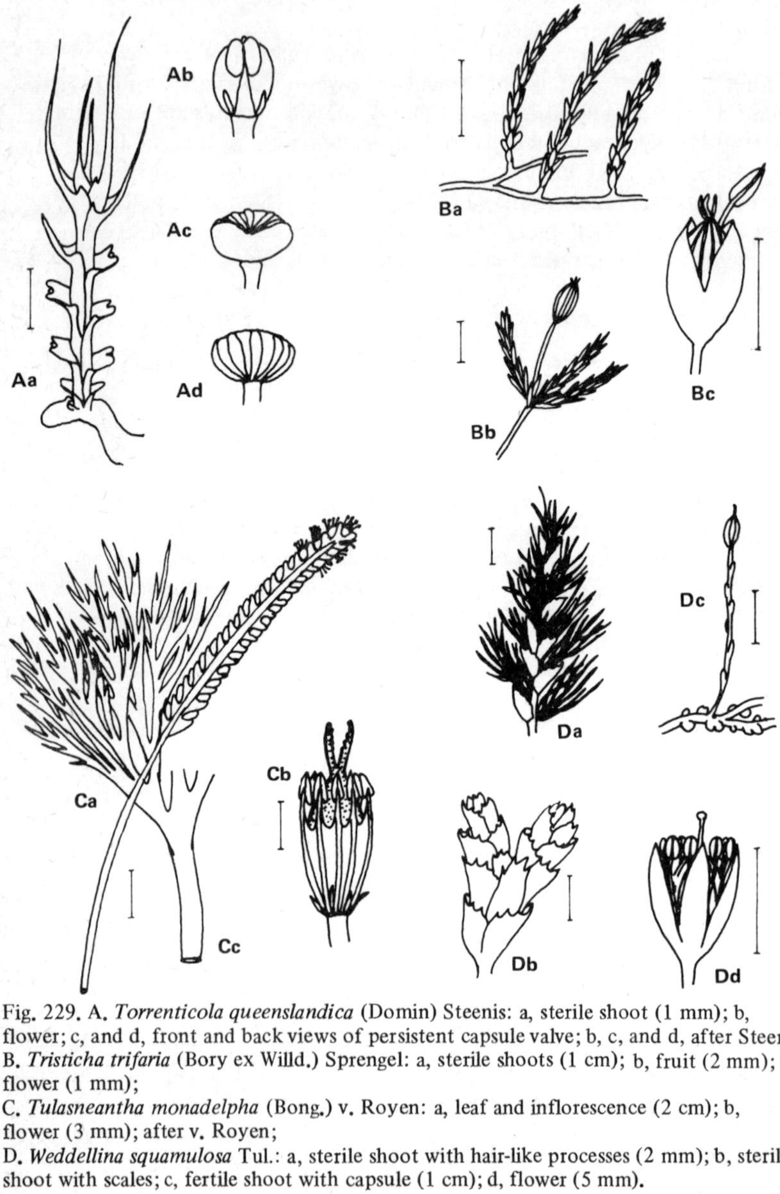

Fig. 229. A. *Torrenticola queenslandica* (Domin) Steenis: a, sterile shoot (1 mm); b, flower; c, and d, front and back views of persistent capsule valve; b, c, and d, after Steenis; B. *Tristicha trifaria* (Bory ex Willd.) Sprengel: a, sterile shoots (1 cm); b, fruit (2 mm); c, flower (1 mm);
C. *Tulasneantha monadelpha* (Bong.) v. Royen: a, leaf and inflorescence (2 cm); b, flower (3 mm); after v. Royen;
D. *Weddellina squamulosa* Tul.: a, sterile shoot with hair-like processes (2 mm); b, sterile shoot with scales; c, fertile shoot with capsule (1 cm); d, flower (5 mm).

Weddellina Tul., Ann. Sci. Nat. Bot. sér. 3, 11: 90, 113 (1849)
Fig. 229D.
Roots somewhat flattened, c. 1 mm wide, creeping, with numerous disc-like holdfasts; stems arising laterally from root; sterile stems floating, 2.5 to 80 cm long, irregularly pinnately branched; branches thickly covered with entire or divided scale-like leaves and numerous hair-like processes; fertile stems 2 to 12 cm long, unbranched, winged above, bearing scale-like leaves up to 2 mm long. Flowers solitary, when young enveloped in scale-like leaves; spathella absent. Perianth segments 5, free or slightly united at base, pink to lilac or white, 3 to 6 mm long. Stamens 5 to 25, in 1 whorl. Ovary 6-ribbed, 2-locular; capsule opening by 2, equal valves; style 1, linear; stigma discoid.

1 species, *W. squamulosa* Tul.: Colombia, Guiana, Surinam and N. Brazil.

Wettsteiniola Süssenguth, Feddes Repert. 39: 18 (1935)
Fig. 230A.
Roots thalloid, branched, closely attached to rocks; branches up to 1 cm wide, irregular; stems absent. Leaves either repeatedly pinnate or 2-pinnate with secondary pinnae repeatedly forked, up to 30 cm long; ultimate segments filiform; bases of pinnae with 1-sided stipel. Flowers in groups, erect in spathella. Tepals 3 to 6, in an incomplete whorl. Stamens 1 to 4 in incomplete whorl; anthers introrse. Ovary ellipsoidal to ovoid, 12-ribbed; capsule opening by 2 equal valves; styles linear.

2 species, *W. pinnata* Süssenguth and *W. accorsii* (Toledo) v. Royen: S. Brazil.

Willisia Warming, Danske Vid. Selsk. Skrift. ser. 6, 11: 13, 65 (1901)
Fig. 230B.
Roots flattened, thalloid, closely attached to rocks, irregularly lobed, 2 to 3 cm diam.; sterile stems few, often lacking, linear, simple or rarely branched, up to 50 cm long, cylindrical, up to 5 mm diam., almost leafless below, with numerous, up to 8 cm long, simple, filamentous leaves above; fertile stems, numerous, densely crowded, simple, rigid, erect, 2 to 10 cm long, closely covered with leaves. Leaves of fertile stems in 4, distinct rows, triangular, scale-like, rigid, with caducous, hair-like tips. Flowers solitary, terminal, sessile, erect within spathella; spathella forked at tip. Tepals 2, one each side of andropodium, linear. Stamens 2, borne on andropodium. Ovary smooth, unribbed or with 1 rib along centre of each valve; loculi 2, unequal; capsule opening by 2, unequal valves, the larger persistent and borne on a forked stalk; styles 2, linear.

1 species, *W. selaginoides* (Beddome) Warming: S. W. India, known only from the Anamallai Hills.

Winklerella Engler, Bot. Jahrb. 38: 97 (1905)
Fig. 230C.
Roots thalloid, foliose, closely attached to stones; stem simple or branched. Leaves divided, with linear segments, ultimate segments filamentous. Flowers

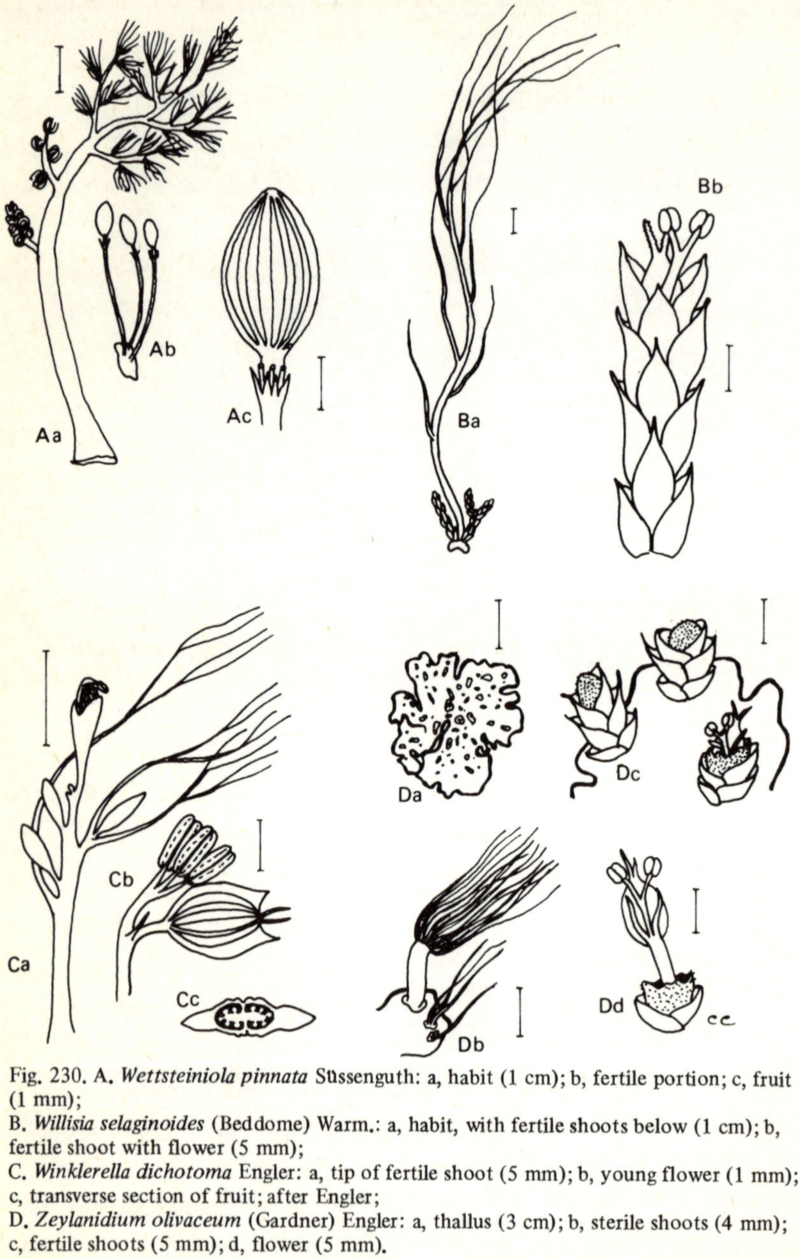

Fig. 230. A. *Wettsteiniola pinnata* Süssenguth: a, habit (1 cm); b, fertile portion; c, fruit (1 mm);
B. *Willisia selaginoides* (Beddome) Warm.: a, habit, with fertile shoots below (1 cm); b, fertile shoot with flower (5 mm);
C. *Winklerella dichotoma* Engler: a, tip of fertile shoot (5 mm); b, young flower (1 mm); c, transverse section of fruit; after Engler;
D. *Zeylanidium olivaceum* (Gardner) Engler: a, thallus (3 cm); b, sterile shoots (4 mm); c, fertile shoots (5 mm); d, flower (5 mm).

inverted within unruptured spathella; spathella ellipsoidal. Tepals 2, small, linear, one each side of andropodium. Stamens 3 or occasionally 2; pollen 1-celled. Ovary ovoid, flattened, with 2 lateral wings longer than the capsule and forming 2, tooth-like protuberances each side of styles; loculi 2; capsule splitting into equal 2-valves; styles 2, linear.

1 species, *W. dichotoma* Engler: W. Africa, Cameroons.

Zeylanidium (Tul.) Engler in Engler and Prantl, Nat. Pflanzenfam. ed. 2, 18a: 61 (1930) [*Hydrobryum* Endl. pro parte]

Fig. 230D.

Roots flat, closely attached to rocks, thalloid, ribbon-like or foliose, up to 25 cm diam.; stems short, simple. Leaves of sterile shoots in tufts, linear to lanceolate; leaves of fertile shoots bract-like, 4 to 8, with caducous, thread-like ends. Flowers terminal, solitary, erect within spathella; spathella splitting at top, distinctly rough in texture. Tepals 2, one each side of andropodium. Stamens 2 (rarely 1), borne on andropodium; andropodium as long or longer than ovary; anthers dehiscing introrsely. Ovary stalked, 8-ribbed; loculi 2, unequal; capsule opening by 2, unequal valves, the larger persistent; styles 2, unequal, linear of slightly flattened.

2 or 3 species: W. and S. India, Ceylon, Burma and Assam.

POLEMONIACEAE

15 genera of which 1, *Navarretia*, has aquatic representatives.

Navarretia Ruiz and Pavón, Fl. Peruv. Chil. Prodr. 20 (1794)
Fig. 231A.
Erect, spreading or prostrate annuals, 1 to 30 cm high, glandular, villous or almost glabrous. Leaves alternate, usually linear below, pinnate above, with spinous apices. Flowers sessile or subsessile, in dense spiny-bracteate heads. Calyx lobes 5, usually sharply pointed, united below by a scarious membrane. Petal-tube funnel-shaped or flat; lobes regular, usually 5, spreading, white, yellow, blue, violet or pink. Stamens 5, inserted on petal tube. Ovary 1- to 3-locular; stigma entire; capsule with 1 to many seeds in each loculus.

Crampton, B. Morphological and ecological considerations in the classification of Navarretia (Polemoniaceae). Madroño 12: 225–238 (1954)

30 species: California and western N. America and 1 species in Argentina and Chile. Several species are characteristic of vernal pools.

Fig. 231. A. *Navarretia bakeri* Mason: a, habit (1 cm); b, flower (1 mm); after Mason; B. *Polygonum foliosum* H. Lindb.: a, habit (3 cm); b, ochrea (1 cm).

POLYGONACEAE

c. 30 genera: *c.* 700 species. *Polygonum* contains aquatics.

Polygonum L., Sp. Pl. 359 (1753)

Fig. 231B.

Annual or perennial herbs. Stems erect, creeping or floating; nodes usually swollen, completely surrounded by a membranaceous or scarious sheath (the ochrea). Leaves alternate, variously shaped, always distinctly longer than wide, usually entire. Inflorescence of axillary fascicles of 1 to several flowers. Flowers actinomorphic, bisexual (often heterostylic) or unisexual. Perianth segments (2) 3 to 6, usually more or less equal, free or united at base, completely or partly enclosing the fruit. Stamens (3) 5 to 8 (9). Carpels (2) 3, united, superior; styles 2 or 3; ovule solitary, basal; fruit a lenticular or 3-angled nut.

Graham, S. A. and Wood, C. E. The genera of the Polygonaceae in the Southeastern United States. Journ. Arnold Arb. 46: 91–121 (1965)

Mitchell, R. S. Comparative leaf structure of aquatic Polygonum species. Amer. Journ. Bot. 58: 342–360 (1971)

Mitchell, R. S. A guide to aquatic smartweeds (Polygonum) of the United States. Virginia Polytechnic Institute, Water Resources Research Center, Bull. 41: 1–52 (1971)

c. 150 species: cosmopolitan. Many species of *Polygonum* grow in wet places and may withstand temporary flooding but relatively few (10 to 20) are truly aquatic. Most aquatic or nearly aquatic species are tropical. Several species are found as weeds in ricefields and irrigation channels (for example *P. acuminatum* H.B. and K., *P. barbatum* L., *P. celebium* Danser, *P. javanum* D.Br., *P. lanigerum* R. Br. and *P. limbatum* Meisn.).

Aston, H. I. Aquatic Plants of Australia. Melbourne Univ. Press (1973) describes another aquatic member of the Polygonaceae. It is *Rumex bidens* R. Br. and is found in southern Australia and northern Tasmania.

PONTEDERIACEAE

9 genera: all aquatic.

Annuals or perennials. Stems rhizomatous, stoloniferous or elongate, submerged, floating or creeping. Leaves sheathed and usually stipulate at base, alternate, linear or stalked; blades linear to orbicular, cuneate or cordate at base (in *Hydrothrix* hair-like in pseudowhorls). Flowers bisexual often heterostylic, solitary, paired, or in spike-like or panicle-like inflorescences, subtended by two spathes, one of which is sometimes leaf-like. Perianth tubular or lobed almost to base, white, blue or yellow, actinomorphic or zygomorphic; lobes usually 6 (4 or rarely 3 in *Scholleropsis*). Stamens 6, 3 or 1, similar or dissimilar, inserted on perianth. Ovary superior, 3-locular (occasionally 1 loculus only fertile); style 1, elongate; fruit a capsule or 1-seeded nutlet enclosed in perianth base; seeds numerous or solitary.

Castellanos, A. Las Pontederiaceae de Brazil. Arquivos Jard. Bot. Rio de Janeiro 16: 147–236 (1959)

Schultz, A. G. Las Pontederiaceas de la Argentina. Darwiniana 6: 45–82 (1942)

Schwarz, O. Pontederiaceae in Engler und Prantl, Natürl. Pfl. Fam. ed. 2, 15a: 181–188 (1930)

Singh, V. Vascular anatomy of the flower of some species of the Pontederiaceae. Proc. Indian Acad. Sci. ser. B, 56: 339–353 (1962)

1A Leaves hair-like, appearing whorled; stamen 1

Hydrothrix

1B Leaves not hair-like, not appearing whorled; stamens 3 or 6

- 2A Fruit a 1-seeded nutlet
  - 3A Outer wall of fruit with spine bearing ridges; perianth 2-lipped, with adaxial lip 1-lobed

    Reussia

  - 3B Outer wall of fruit with smooth, winged ridge; perianth 2-lipped, with adaxial lip 3-lobed

    Pontederia

- 2B Fruit a 3-locular, many seeded capsule
  - 4A Stamens 6
    - 5A Perianth distinctly tubular at base; anthers subequal in length, dorsifixed

      Eichhornia

    - 5B Perianth lobed almost to base: 1 anther much larger than the others, basifixed

      Monochoria

  - 4B Stamens 3
    - 6A Perianth of 4 (rarely 3) lobes

      Scholleropsis

    - 6B Perianth of 6 lobes

7A Filaments of the 2 short stamens obliquely globular-inflated; stipules absent, represented by a small, ligule-like structure; perianth zygomorphic **Eurystemon**

7B Filaments not markedly inflated; stipules free above the base; perianth actinomorphic or nearly so

8A Anthers all alike, coiled after anthesis; spathe sessile, enclosing perianth tube; seeds relatively few; perianth yellow **Zosterella**

8B Anthers dissimilar, not coiling; spathe stalked, not enclosing perianth tube; seeds very numerous; perianth white or blue **Heteranthera**

Eichhornia Kunth, Eichhornia, gen. nov. (1842), Enum Pl. 4: 129 (1843), nom. cons.
Fig. 232.
Annuals and perennials. Stems floating, creeping, often stoloniferous. Leaves submerged, floating or emergent, linear or stalked; stalk occasionally inflated; blade linear to orbicular. Inflorescence spike-like or paniculate; 2 spathes unlike, the lower leaf-like, the upper bract-like. Flowers di- or tristylic and perhaps monostylic. Perianth blue, tubular, 6-lobed; the adaxial lobe somewhat larger, bearing a yellow blotch. Stamens 6, at 2 levels; anthers subequal. Ovary 3-locular; style elongate long or short; fruit a capsule; seeds numerous.

c. 7 species: the majority in Tropical America but through introductions almost throughout the tropics and subtropics of the world. *E. crassipes* (Mart.) Solms-Laub., the water hyacinth, is a free-floating species. It has spread and become a serious pest in many tropical and subtropical regions. The literature on this species is enormous and there is even a journal entitled "Hyacinth Control Journal". For good accounts of this plant see: Sculthorpe, C. D. The biology of aquatic vascular plants. London (1967); Little, E. C. S. The control of water weeds. Weed Research 8: 79–105 (1968); Little, E. C. S. Handbook of utilization of aquatic plants. FAO Rome, PL: CP/20 (1968).

Fig. 232. *Eichhornia crassipes* (Mart.) Solms-Laub.: a, habit (3 cm); b, longitudinal section of flower (1 cm).

Eurystemon Alexander, N. Amer. Fl. 19 (1): 55 (1937)
Fig. 233.
Annual. Stems creeping or floating. Leaves linear, grass-like, 3 to 4 mm wide, without separated stipules, but with a ligular projection topping the vaginal portion. Inflorescence spike-like. Perianth blue and white, tubular, zygomorphic; adaxial lobes 5, erect, spreading 15 to 20 mm across; abaxial lobe 1, smaller, sharply deflexed. Stamens 3, unequal; the larger one with filament gradually expanded at the middle and anther *c.* 2 mm long; the 2 smaller ones with filaments nearly globosely inflated at the middle and anthers *c.* 1 mm long. Ovary 3-locular; style club-shaped; fruit a capsule 5 mm long, ovoid-ellipsoid; seeds numerous.

1 species, *E. mexicanum* (S. Watson) Alexander: America, southern USA and Mexico.

Fig. 233. *Eurystemon mexicanum* (S. Watson) Alexander: a, habit (1 cm); b, flower (2 mm).

Heteranthera, Ruiz and Pavón, Fl. Peruv. Chil. Prod. 9 (1794), nom. cons. Fig. 234A.
Annuals or perennials. Stems submerged, floating or creeping. Leaves linear or stalked; blades linear to ovate or reniform. Inflorescence 1-flowered or spike-like (lower flowers sometimes cleistogamous). Perianth blue or white, nearly actinomorphic, lobed almost to base; lobes 6. Stamens 3, one much larger than the others. Ovary 3-locular; style elongate persisting in fruit; fruit a capsule; seeds very numerous.

c. 10 species: tropical and subtropical America and Africa. Small herbs found in shallow water or on mud.

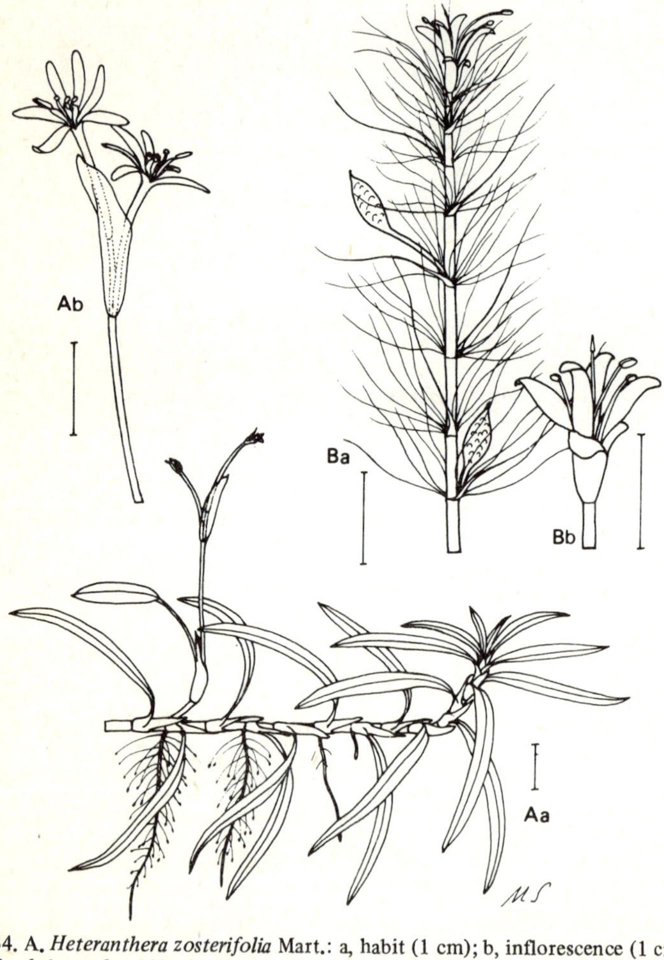

Fig. 234. A. *Heteranthera zosterifolia* Mart.: a, habit (1 cm); b, inflorescence (1 cm); B. *Hydrothrix gardneri* Hooker fil.: a, habit (1 cm); b, flower (1 cm).

Hydrothrix Hooker fil., Ann. Bot. 1 (2): 89 (1887)
Fig. 234B.
Submerged annual. Stems flaccid, elongate. Leaves simple, hair-like, appearing whorled (the long shoots have alternate capillary leaves within whose base is a large membranous sheath, inside the sheath is a short shoot with numerous similar unsheathed leaves). Flowers small, in pairs, in leaf axils, subtended by a 2-lobed membranous sheath; often cleistogamous. Perianth yellow, tubular, 6-lobed; lobes unequal. Stamen 1, inserted towards top of tube; staminodes 2. Ovary 3-locular but appearing 1-locular at maturity; style elongate, persistent; fruit an elongate capsule, opening by 3 valves; seeds numerous.

1 species, *H. gardneri* Hooker fil.: Ceara, Brazil.

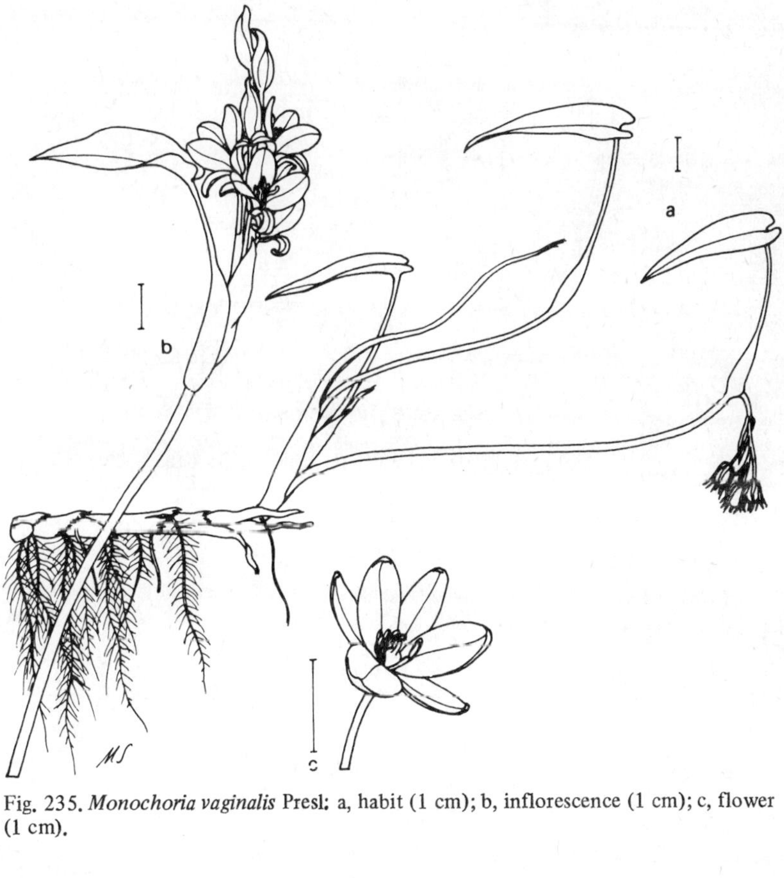

Fig. 235. *Monochoria vaginalis* Presl: a, habit (1 cm); b, inflorescence (1 cm); c, flower (1 cm).

Fig. 236. *Pontederia cordata* Lour.: a, habit (1 cm); b, inflorescence (1 cm); c, longitudinal section of flower (1 cm); d, fruit (1 mm).

Monochoria Presl., Rel. Haenk. 1: 127 (1830)
Fig. 235.
Annuals and perennials. Stems erect, ascending or occasionally creeping. Leaves emergent; petioles long or short; blades ovate to sagittate. Inflorescence elongate and racemose or umbel-like, arising from the leaf sheath. Perianth blue, 6-lobed; lobes free almost to base, subequal. Stamens 6, inserted on perianth tube; anthers unequal, 5 small and yellow, 1 large and blue. Ovary 3-locular; style elongate; fruit a 3-valved capsule; seeds numerous.

Verdcourt, B. The genus Monochoria in Africa. Kirkia 1: 80–83 (1961)

c. 5 species: tropical Africa, Asia and Australia. Usually found in shallow standing water in pools, ditches, canals and ricefields. The leaves are frequently eaten as a vegetable.

Pontederia L., Sp. Pl. 288 (1753)
Fig. 236.
Perennials. Stems submerged, floating, creeping or ascending. Leaves submerged, floating or emergent, linear or stalked; blades elliptic to ovate or sagittate. Inflorescence spike-like; the 2 spathe valves very different, the lower leaf-like, the upper bract-like; flowers tristylic. Perianth blue, tubular, externally hairy, 2-lipped; adaxial lobes 3, united for ½ their length, the middle lobe broader and with a yellow spot; abaxial lobes 3, equal, free. Stamens 6, at 2 levels. Ovary of 3 carpels, one of which is fertile; styles elongate, long or short; fruit a 1-seeded nutlet, enclosed in base of perianth tube, with smooth, winged, longitudinal ridges.

Lowden, R. M. Revision of the genus Pontederia L. Rhodora 75: 426–487 (1973)

Ornduff, R. The breeding system of Pontederia cordata L. Bull. Torrey Bot. Club 39: 407–416 (1966)

c. 5 species: America, from Canada to Argentina. Gregarious herbs found in shallow water or on mud. Lowden op. cit. has united the genera *Pontederia* and *Reussia*.

Reussia Endl. Gen. Pl. 1: 139 (1836), nom. cons.
Fig. 237.

Herbs. Stems submerged, floating, creeping or ascending. Leaves submerged, floating or emergent, linear or stalked; blades elliptic to orbicular. Inflorescence spike-like, with 3 or more flowers; flowers with the 2 spathe valves very different; the lower leaf-like, the upper bract-like; flowers tristylic. Perianth blue, tubular, 2-lipped; adaxial lip of 1 somewhat larger lobe, with a yellow spot; abaxial lip of 5, equal lobes. Stamens 6, at 2 levels. Ovary of 3 carpels, 1 of which is fertile; style elongate, short or long; fruit a 1-seeded nutlet enclosed in base of perianth tube, with spine bearing longitudinal ridges.

c. 4 species: Tropical C. and S. America. Gregarious herbs found in shallow water or on mud.

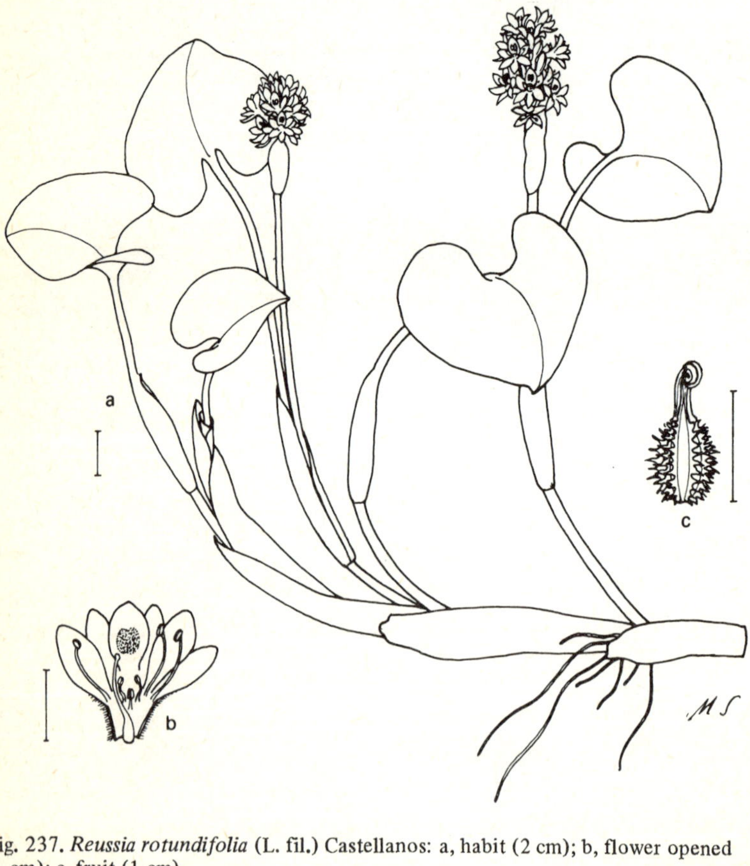

Fig. 237. *Reussia rotundifolia* (L. fil.) Castellanos: a, habit (2 cm); b, flower opened (1 cm); c, fruit (1 cm).

Fig. 238. *Scholleropsis lutea* H. Perr.: a, habit (1 cm); b, inflorescence (1 mm), after Letouzey.

Scholleropsis H. Perr., Not. Syst. (Paris) 5: 158 (1935)
Fig. 238.
Herb. Stems submerged or floating. Basal leaves submerged, linear, up to 7 cm long; cauline leaves linear or stalked, with blade elliptic to ovate, floating. Flowers solitary in leaf axils, sessile when submerged (probably cleistogamous), or stalked and borne above the water surface. Perianth enclosed below in *c.* 10 mm long sheath; perianth lobes 4 (or rarely 3), equal, yellow, Stamens 3, unequal, 1 large, 2 small. Ovary 3-locular; style elongate, persistent in fruit; fruit an irregularly dehiscent capsule; seeds very numerous.

Letouzey, R. Presence au Cameroun d'une Pontédériacée: Scholleropsis lutea H. Perr. endemique de Madagascar. Adansonia sér. 2, 7 (1): 33–37 (1967)

1 species, *S. lutea* H. Perr.: Madagascar and N. Cameroon by Lake Chad. A small aquatic herb that flowers as the flood waters recede.

Zosterella Small in Small and Carter, Fl. Lanc. Co. 68 (1913)
Fig. 239A.
Perennial or annual. Stems elongate, submerged. Leaves linear, grass-like. Flowers solitary. Perianth yellow, tubular, almost actinomorphic, partly enclosed within tubular spathe, 6-lobed; lobes subequal. Stamens 3, all alike, coiling downwards after anthesis; filaments expanded. Ovary of 3 carpels, appearing 1-locular; style elongate, club-shaped at apex; fruit an irregularly dehiscent capsule; seeds comparatively few.

c. 2 species: Temperate and subtropical N. America. Submerged herbs with emergent flowers, found in a variety of aquatic habitats. *Z. dubia* (Jacq.). Small is often cultivated as a decorative aquarium plant.

Fig. 239A. A. *Zosterelia dubia* (Jacq.) Small: a, habit (1 cm); b, style and stamens (3 mm); (see p. 493 for Fig. 239B).

PORTULACACEAE

19 genera: *c.* 500 species; cosmopolitan. Only the genus *Montia* contains aquatics.

Montia L., Sp. Pl. 87 (1753)
Fig. 239B.
Annual or perennial, somewhat fleshy, glabrous herbs. Stems creeping, floating or ascending. Leaves opposite (in aquatic species) or alternate, ovate, obovate, or spathulate, entire, lacking stipules. Flowers in terminal cymes, often inconspicuous, usually cleistogamous when submerged. Sepals 2, free or united below. Petals 3 to 5, often unequal in size, free or united below. Stamens 3 or 5. Ovary 1-locular; styles 3; fruit a globose capsule dehiscing with 3-valves; seeds 1 to 6, black.

Moore, D. M. The subspecies of Montia fontana L. Bot. Notiser 116 (1): 16–30 (1963)
Pedersen, A. De underarter af Montia fontana i Danmark. Bot. Tidsskr. 63: 368–371 (1968)

c. 50 species of which *M. fontana* L. sensu lato, is perhaps the only aquatic but other species such as *M. australasica* (Hooker fil.) Pax and Hoffm. from temperate regions of the southern hemisphere and *M. chamissoi* (Ledeb.) Dur. from N. W. America are found in wet places and are tolerant of flooding. *M. fontana* is cosmopolitan and is found in bogs, marshes or floating in pools; it is taxonomically complex and contains numerous micro- or sub-species; it is not known to be of any economic importance.

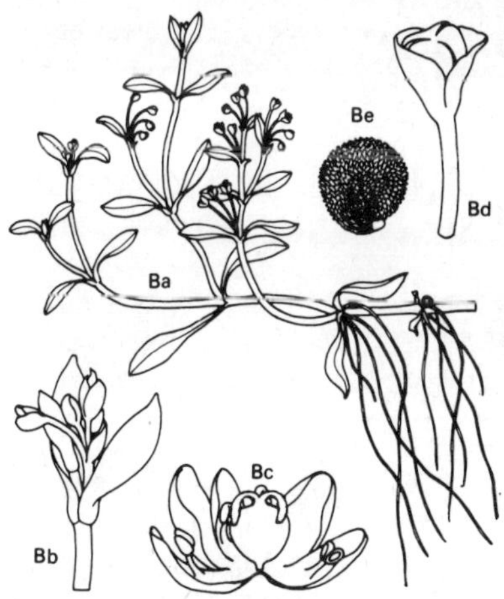

Fig. 239 B. *Montia fontana* L.: a, habit (1 cm); b, young inflorescence; c, flower opened; d, fruit; e, seed.

POTAMOGETONACEAE

2 genera: both aquatic.

Usually perennials, rarely annuals. Stems elongate, flexible, erect, floating or creeping; rhizomes slender, elongate, often producing specialised winter buds (turions). Leaves alternate or opposite, simple, entire, often heteromorphic; floating leaves broad; the submersed ones narrower, often linear or capillary; stipules present or absent. Inflorescence a stalked spike. Flowers bisexual, small, actinomorphic. Perianth of 4, free, bract-like, clawed scales, inserted opposite each stamen (the perianth is often considered to be a connective outgrowth). Stamens 4, adnate to perianth at base; anther sessile, 2-locular. Ovary superior, of 4 or rarely less, free or partly united carpels; style short or somewhat elongate; ovules solitary; fruit a nutlet or drupe.

Ascherson, P. and Graebner, P. in Engler, A. Pflanzenreich 31 (IV. II): 1–142 (1907)
Hagström, J. O. Critical researches on the Potamogetons. K. Svenska Vetensk. Akad. Handl. 55: 1–281 (1916)
Sattler, R. Perianth development of Potamogeton richardsonii. Amer. Journ. Bot. 52: 35–41 (1965)
Singh, V. Vegetative anatomy of some members of the Potamogetonaceae. Proc. Indian Acad. Sci. ser. B, 60: 214–231 (1964)
Singh, V. Vascular anatomy of the flower of Potamogetonaceae. Bot. Gaz. 126: 137–144 (1965)

1A Leaves mostly alternate, not folded longitudinally; fruit drupaceous (exocarp fleshy, endocarp bony)
Potamogeton
1B Leaves opposite, folded longitudinally; fruit a nutlet (pericarp thin)
Groenlandia

Groenlandia Gay, Compt. Rend. Acad. (Paris) 38: 703 (1854)
[*Potamogeton* L. pro parte]
Fig. 240A.
Submerged perennials. Stems slender, floating or creeping. Leaves submerged opposite, usually in pairs, rarely in whorls of 3, sessile, amplexicaul, very variable, 0.5 to 4 cm long, ovate-triangular to lanceolate, usually folded longitudinally and recurved, translucent; longitudinal veins 3 to 5; margin undulate and serrate, particularly near apex; stipules absent, except on leaves subtending the inflorescence then adnate to leaf base with 2 lateral auricules. Spikes emergent, usually 4-flowered, ovoid in fruit. Nutlets with thin pericarp (not fleshy or bony), *c.* 3 mm long, orbicular-obovate, strongly compressed, dorsal margin with keel.

Kohler, A. Zur Oekologie submerser Gefäss-Makrophyten in Fliessgewässern. Ber. Deutsch. Bot. Ges. 84 (11): 713–720 (1971)

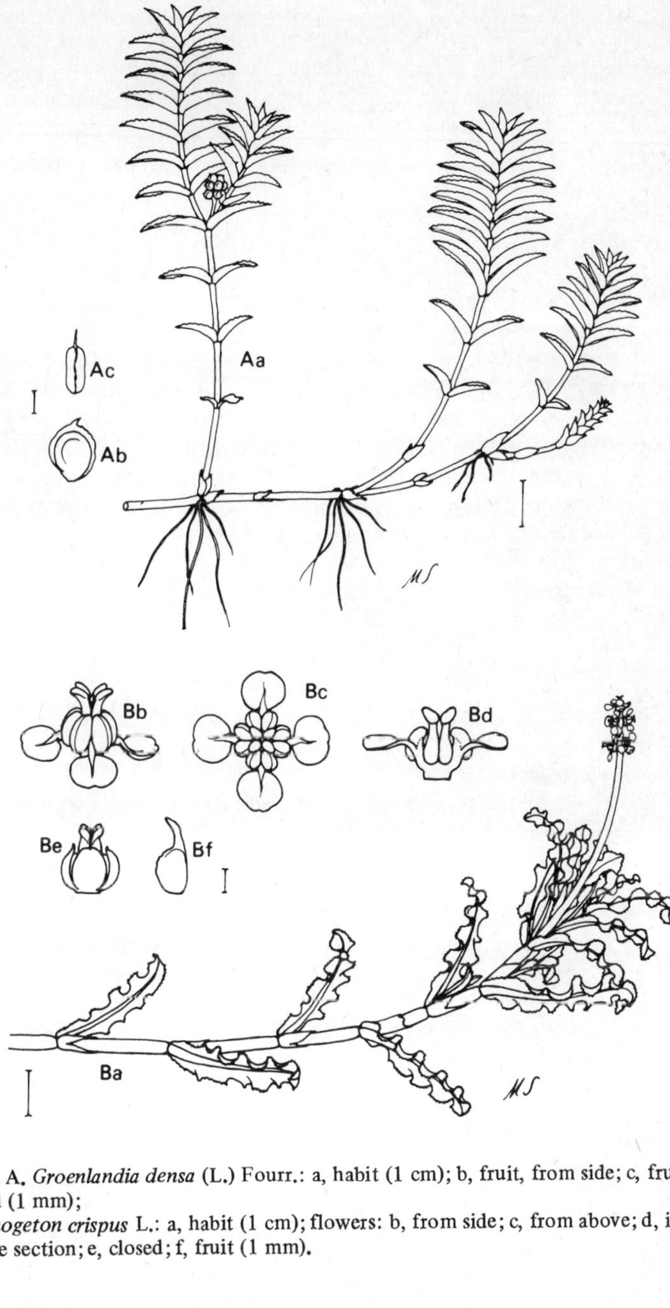

Fig. 240. A. *Groenlandia densa* (L.) Fourr.: a, habit (1 cm); b, fruit, from side; c, fruit, from end (1 mm);
B. *Potamogeton crispus* L.: a, habit (1 cm); flowers: b, from side; c, from above; d, in transverse section; e, closed; f, fruit (1 mm).

Posluszny, U. and Sattler, R. Floral development of Potamogeton densus. Canad. Journ. Bot. 51 (3): 647–656 (1973)

1 species, *G. densa* (L.) Fourr. [*Potamogeton densus* L., *P. serratus* L.]: W. Europe, S. W. Asia and N. Africa. It is often abundant and found in a wide variety of aquatic environments. *G. densa* shows a preference for eutrophic water but is sensitive to organic pollution; experiments are being carried out by Kohler (1971).

Potamogeton L., Sp. Pl. 126 (1753)
Fig. 240B.
Stems slender; rhizomes slender, perennating or developing specialised winter buds (turions). Leaves mostly alternate, rarely opposite; submerged leaves thin, translucent, linear or with broadened, sessile or stalked blade; floating leaves often lacking, leathery, opaque, narrowly to broadly elliptic-oblong; stipules usually present, free or adnate to leaf-base, sometimes sheathing the stem or united by their margins in front of leaf to form a ligule. Spikes emergent and wind pollinated or submerged and water pollinated, ovoid to cylindrical, dense, lax or interrupted. Fruit drupaceous, with fleshy exocarp and bony endocarp.

Dandy, J. E. The genus Potamogeton L. in tropical Africa. Journ. Linn. Soc. London (Bot.) 50: 507–540 (1937)
Dandy, J. E. Potamogeton and Ruppia in the Azores. Bol. Soc. Brot. ser. 2, 44: 1–7 (1970)
Fernald, M. L. The linear-leaved North American species of Potamogeton, section Axillares. Mem. Amer. Acad. Arts. Sci. 17: 1–183 (1932)
Haynes, R. R. Potamogeton in Louisiana. Proc. Louisiana Acad. Sci. 31: 82–90 (1968)
Ogden, E. C. The broad-leaved species of Potamogeton in North-America and Mexico. Rhodora 45: 57–105, 119–216 (1943)
Ogden, E. C. Key to the North American species of Potamogeton. Circ. New. York St. Mus. 31: 1–11 (1953)
For additional references see family description.

c. 100 species: cosmopolitan. Occurs in a large variety of aquatic habitats. Several species (i.e. *P. crispus* L., *P. nodosus* Poir., *P. octandrus* Poir., *P. pectinatus,* and *P. perfoliatus* L.) have been reported to be a nuisance in canals and ditches; many species, however, are important food plants for many animals.

PRIMULACEAE

c. 30 genera of which 4 contain aquatics; 1 *Hottonia* is exclusively aquatic.

Perennial or annual herbs. Flowers actinomorphic, 5-merous, often heterostylous. Petals united, forming rotate, campanulate or funnel-shaped corolla. Stamens inserted on the petal tube, opposite its lobes. Ovary superior, perigynous, or inferior, 1-locular, with free central placentation; style simple; ovules many; fruit a dehiscent capsule.

Channell, R. B. and Wood, C. E. The genera of the Primulales of the southeastern United States. Journ. Arnold Arb. 268–288 (1959)

1A All leaves submerged; inflorescence only emergent.
- 2A Leaves pinnate, the segments capillary, flowers in whorls. **Hottonia**
- 2B Leaves simple, flowers solitary **Anagallis**

1B At least some leaves on the aerial part of the stem
- 3A Leaves obovate, mostly radical; ovary inferior or half-inferior **Samolus**
- 3B Leaves ovate, all cauline; ovary superior
  - 4A Leaves alternate; petals small, white (plant delicate) **Anagallis**
  - 4B Leaves opposite; petals usually yellow (plant robust) **Lysimachia**

Anagallis L., Sp. Pl. 148 (1753)
Fig. 241A.
Small annuals or rarely perennials. Leaves alternate, ovate sometimes capillary, submerged. Inflorescence simple and emergent. Flowers axillary, solitary; pedicel without bracts. Sepals united only at base. Petals united only at base, up to 5 mm long, oblong or obovate, white or reddish. Stamens inserted at base of corolla tube; filaments pubescent. Capsule globose, dehiscing transversely.

Hess, H. Anagallis Kochii H. Hess, n. sp., eine neue Wasserpflanze aus Süd-Angola. Ber. Schweiz. Bot. Ges. 63: 213–215 (1953)
Hess, H. Anagallis Hürneri H. Hess, nov. spec., eine neue Wasserpflanze aus dem Belgischen Kongo. Ber. Schweiz. Bot. Ges. 67: 80–82 (1957)

c. 20 species of which 2 from West Africa. *A. kochii* Hess and *A. huerneri* Hess are submerged aquatics. *A. pumila* Schwarz is widespread in the tropics and occurs in swamps.

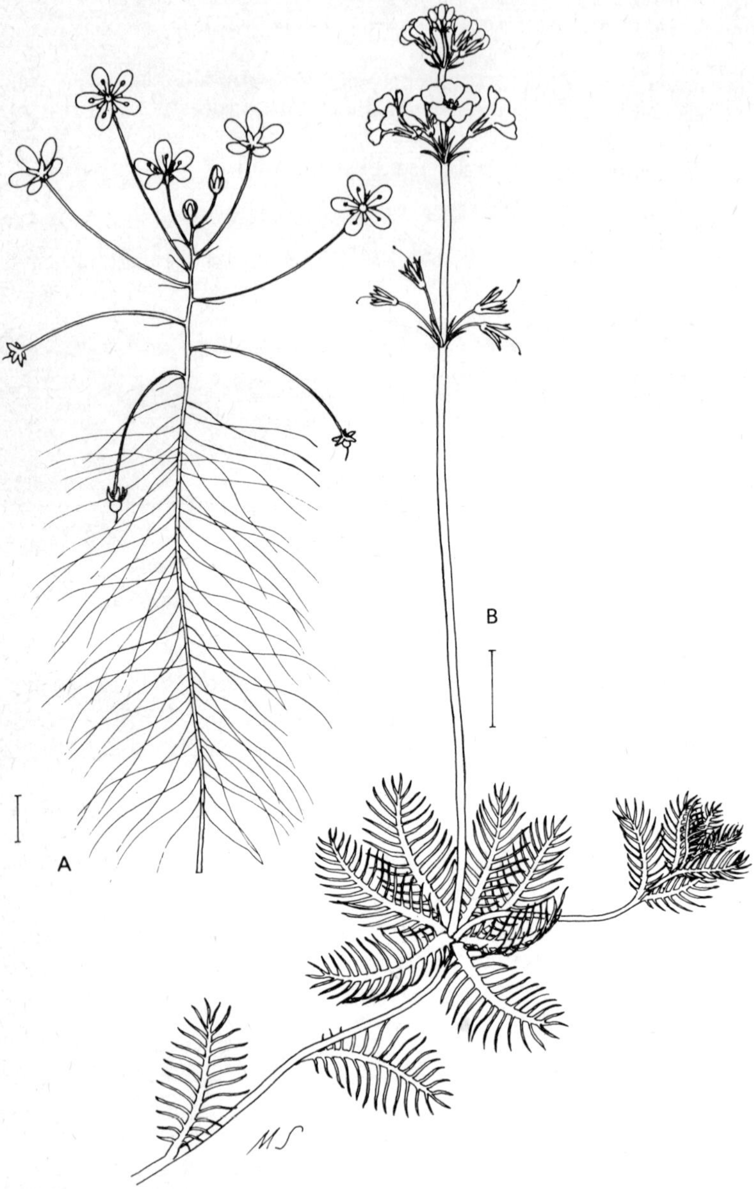

Fig. 241. A. *Anagallis kochii* H. Hess: habit (1 cm), (after Hess); B. *Hottonia palustris* L.: habit (3 cm).

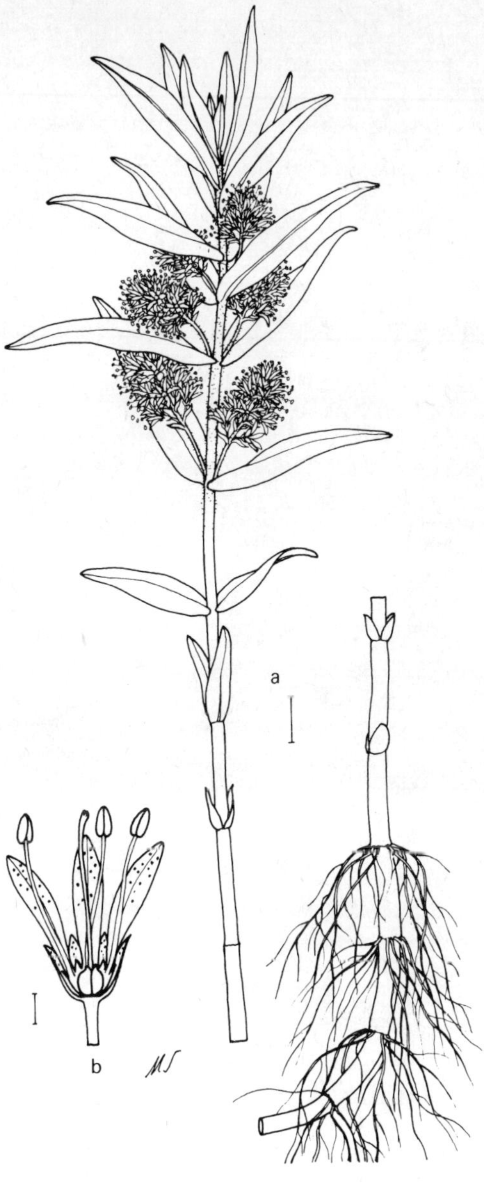

Fig. 242. *Lysimachia thyrsiflora* L.: a, habit (1 cm); b, flower (1 mm).

Fig. 243. *Samolus valerandi* L.: a, habit (1 cm); b, longitudinal section of flower (1 mm).

Hottonia L., Sp. Pl. 145 (1753)
Fig. 241B.
Submerged or floating aquatic annuals or perennials. Stems sometimes inflated. Leaves submerged, in rosettes or alternate, 1- or 2-pinnate, segments linear and occasionally forked. Inflorescence emergent, sometimes inflated. Flowers homostylous or heterostylous, pedicellate, in whorls or clusters; bracteoles subulate. Sepals linear, equalling or exceeding the petal-tube. Petals united into a short tube at the base, white or lilac with a yellow throat. Capsule dehiscing by 5 valves.

Prankerd, T. L. On the structure and biology of the genus *Hottonia.* Ann. Bot. 25: 253–267 (1911).

2 species: *H. palustris* L., in Europe and N. Asia, it has large showy flowers and the stem and inflorescence are not inflated; *H. inflata* Ell., in eastern N. America, it has small, self-pollinated flowers and very much inflated stem and inflorescence stalk. Both species are characteristically found in pools and ditches.

Lysimachia L., Sp. Pl. 146 (1753), [*Naumbergia* Moench.]
Fig. 242.
Erect or rarely creeping rhizomatous perennials, up to 2 m high. Leaves opposite or whorled, entire, often densely dotted with glands; the lower sometimes reduced and scale-like. Flowers borne singly in the leaf axils, or in a terminal, leafy panicle or in dense, many-flowered, spike-like axillary racemes around the middle of the stem. Sepals ovate to narrowly lanceolate. Petals usually yellow or orange, united at the base, forming a campanulate or rotate corolla. Capsule dehiscing by 5 valves; seeds smooth.

Ray, J. D. The genus Lysimachia in the New World. Illinois Bot. Monogr. 24 (3–4): 1–160 (1956)

c. 200 species: cosmopolitan. Mainly in swamps and shallow water.

Samolus L., Sp. Pl. 171 (1753)
Fig. 243.
Perennials, biennials or rarely annuals. Leaves mostly in basal rosettes, alternate, obovate to linear. Inflorescence racemose, simple or branched, leafy or leaves reduced to scales. Flowers homostylous, with or without bracts. Sepals 5, united below; lobes triangular. Petals white or pink, obtuse, united at base; tube shorter or longer than lobes. Stamens 5, inserted at base of petal tube, sometimes alternating with staminodes; filaments very short. Ovary inferior or half-inferior; capsule dehiscing by 5 valves.

Pax, F. and Knuth, R. in Engler, Pflanzenreich 22 (IV. 237): 336–344 (1905)

10 to 15 species: cosmopolitan, mainly S. Hemisphere. Occurs in wet places and shallow water, several species are found in brackish water.

RANUNCULACEAE

c. 48 genera: *c.* 1300 species: *Caltha* and *Ranunculus* contain aquatics.

1A Perianth segments alike; fruit a group of follicles
Caltha

1B Perianth segments differentiated into sepals and petals (honey-leaves); fruit a head of nutlets
Ranunculus

Caltha L., Sp. Pl. 558 (1753), [*Psychrophila* (DC.) Bercht. and Presl]
Fig. 244A.
Perennial herbs. Roots axillary, fibrous. Stem creeping, rhizomatous. Leaves alternate, simple or lobed, orbicular, cordate or sagittate (species of S. Hemisphere bear appendages on the base of the blade which may be free and upturned, incurved or adnate to the blade to appear as outgrowths from its upper surface); margin entire or toothed. Flowers in loose, cymose panicles or occasionally solitary, actinomorphic. Perianth segments alike, 5 or more, yellow or white. Stamens 8 to numerous. Carpels 2 to numerous, superior, free, nectar secreting; fruit a follicle bearing seeds along most of its ventral suture.

Hill, A. The genus Caltha in the southern hemisphere. Ann. Bot. (London) 32: 421–435 (1918)
Huth, E. Monographie der Gattung Caltha. Abh. Vortr. Ges. Naturw. (Berlin) 4: p. 33 (1891), reprinted in Helios 9: 55–78 (1892), 99–103 (1893)
Smit, P. G. A revision of Caltha (Ranunculaceae). Blumea 21: 119–150 (1973)

About 10 species: temperate and frigid regions of the world. Species delimitation in this genus is difficult. Most species are found in swamps or shallow water, some individuals occasionally remain submerged.

Ranunculus L., Sp. Pl. 548 (1753)
(*Ranunculus* contains *c.* 400 species of which *c.* 35 are aquatic. The following diagnosis refers only to aquatic species.)
Fig. 244B, C.
Annual or perennial herbs. Roots entirely axillary, fibrous. Stems erect, creeping or reduced to corm-like structures. Leaves whorled, opposite or alternate; very variable in shape, from linear to orbicular; entire, lobed or palmately divided; leaf segments often capillary; with or without stipules. Flowers solitary or in cymose panicles, actinomorphic. Sepals usually 5, green or tinged with blue, often caducous. Petals (honey-leaves) (0–) 5 (–12), yellow, white or white with yellow claw. Stamens 4 to numerous. Carpels 4 to numerous, superior, free; fruit a head of nutlets; nutlet with persistent glabrous style.

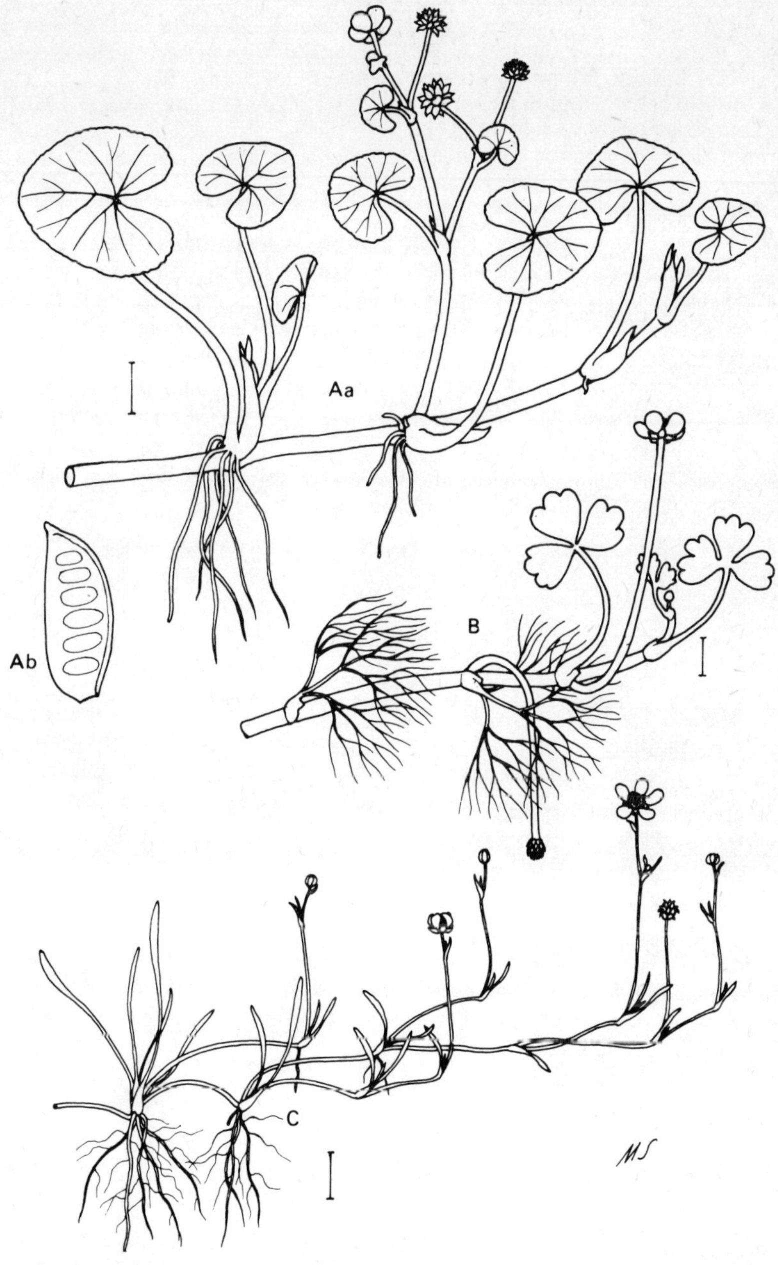

Fig. 244. A. *Caltha natans* Pallas: a, habit (1 cm); b, carpel (1 mm);
B. *Ranunculus baudotii* Godr.: habit (1 cm);
C. *Ranunculus reptans* L.: habit (1 cm).

Benson, L. A treatise on the North American Ranunculi. Amer. Midl. Nat. 40: 1–261 (194), Supplement, op. cit. 52: 238–369 (1954)
Cook, C. D. K. A monographic study of Ranunculus subgenus Batrachium (DC.) A. Gray. Mitt. Bot. Staatss. München 6: 47–237 (1966)
Glück, H. Biologische und morphologische Untersuchungen über Wasser- und Sumpfgewächse. 4: Jena (1924)
Lourteig, A. Ranunculaceae de Sudamerica templada. Darwiniana 9: 397–608 (1952)

c. 400 species, *c.* 35 are aquatic. The aquatic species are cosmopolitan but in the tropics they are confined to cold mountainous regions. They are found in a wide variety of aquatic habitats. They are of no great economic importance, but in certain regions, if uncontrolled, they may block water channels and cause local flooding, see Westlake, D.F. Proc 9th Brit. Weed Control Conf. 372–381 (1968). The majority of aquatic species are found in subgenus *Batrachium* (D.C.) A. Gray see Cook, C. D. K. loc. cit.; the other species are found in subgenus *Ranunculus* in the following sections: *Flammula* Webb, *Hecatonia* (Lour.) D.C., *Micranthus* (Ovcz.) A. Nyárády and *Xanthobatrachium* (Prantl) Ovcz.

RUBIACEAE

c. 500 genera, mainly tropical trees and shrubs. Some members of the herbaceous genera, e.g. *Galium, Oldenlandia* and *Diodia*, occur in marshes, but *Limnosipanea* appears to be the only one which contains true aquatics, *L. spruceana* growing in shallow water, the other 6 species in marshes.

Limnosipanea Hooker fil., in Hook. Ic. 1050 (1868)
Fig. 245.
L. spruceana Hooker fil.
Slender erect perennial. Leaves lanceolate to subulate, whorled, the submerged flaccid, the emersed ascending. Inflorescence cymose, repeatedly forked. Flowers bisexual, actinomorphic. Sepals 5, united below, adnate to ovary. Petals 5, pinkish, united in the lower half into a tube. Stamens 5, inserted in the mouth of the petal tube; anthers free. Ovary inferior; capsule 2-locular, many-seeded.

Limnosipanea spruceana occurs in tropical S. America by the Amazon in Brazil. It is remarkably *Hippuris*-like in vegetative characters.

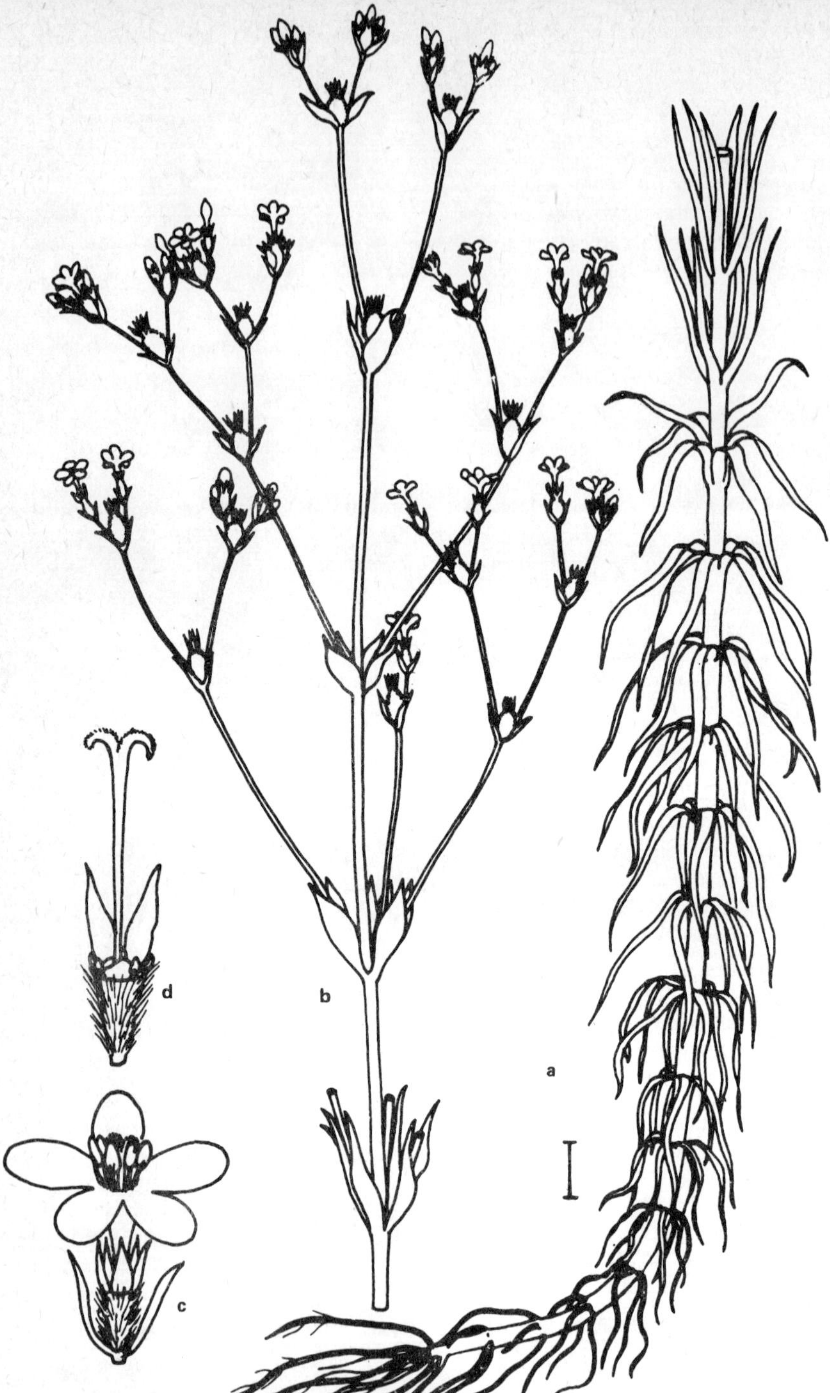

Fig. 245. *Limnosipanea spruceana* Hooker fil.: a, submerged stem (1 cm); b, inflorescence; c, flower; d, young fruit.

RUPPIACEAE

1 genus

Ruppia L., Sp. Pl. 127 (1753)
Fig. 246.
Submerged herbs. Stems slender; rhizomes slender, creeping. Leaves alternate or opposite, simple, capillary, somewhat toothed towards apex, sheathing at base. Inflorescence a short terminal raceme appearing umbel-like. Flowers bisexual, small, borne in pairs on slender, axillary stalks which at first are short and surrounded by a spathe-like base of enclosing leaf; at anthesis the stalks remain short or elongate, and after pollination the elongate stalks sometimes coil and submerge the floral structures. Perianth lacking or vestigial. Stamens 2, opposite each other; anthers sessile, 2-locular, longitudinally dehiscent. Ovary superior; carpels 4, or rarely more, free, at first sessile, in fruit stalked; style lacking; stigma peltate; ovules solitary; fruit a nutlet, occasionally drupaceous with a somewhat spongy exocarp.

Ascherson, P. and Graebner, P. in Engler, A. Pflanzenreich 31 (IV. 11): 142–145 (1907)
Gamerro, J. C. Observaciones sobre la biología floral y morfología de la Potamogetonáceae Ruppia cirrhosa (R. spiralis). Darwiniana 14 (4): 575–608 (1968)
Hartog, C. den. De Nederlandse Ruppia-soorten. Goteria 5: 148–153 (1971)
Mason, R. The species of Ruppia in New Zealand. New Zealand Journ. Bot. 5 (4): 519–531 (1967)
Ortu, A. M. Primary data on the germination of seeds of Ruppia maritima L. Giorn. Bot. Ital. 103 (6): 621 (1969)
Setchell, W. A. The genus Ruppia. Proc. Calif. Acad. Sci. ser. 4, 25 (18): 469–478 (1946)

c. 7 species, often regarded as 1 very polymorphic species: cosmopolitan. Usually grows in brackish water but occasionally found in fresh water near the sea. In New Zealand *R. polycarpa* R. Mason is frequently found in inland freshwaters even up to an altitude of *c.* 700 m. Pollination of all species takes place at the air–water interface, the pollen floats on the water surface and contacts floating stigmas.

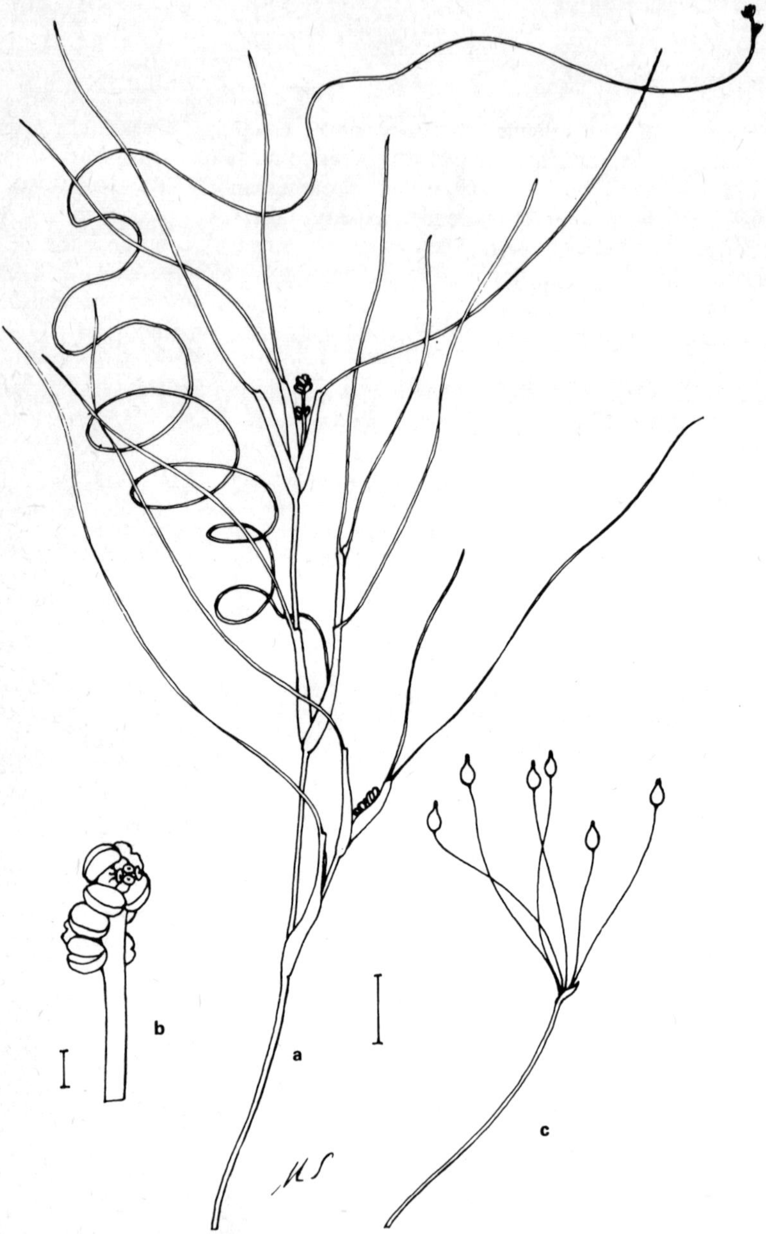

Fig. 246. *Ruppia cirrhosa* Grande: a, habit (1 cm); b, flowering inflorescence (1 mm); c, fruiting inflorescence (1 mm).

SAURURACEAE

5 genera: 3 in east Asia, 1 in California and 1 in Asia and eastern N. America. *Saururus* with 2 species and *Houttuynia,* monotypic, are primarily aquatic, the others grow in marshes.

Rhizomatous, perennial, usually aromatic herbs. Stems erect, jointed. Leaves alternate, entire with a lateral stipule on the petiole. Inflorescence spike-like. Flowers small, unisexual or bisexual, the lowest often subtended by a petal-like bract, the rest with a very small bract on the pedicel, or flowers sessile. Sepals and petals absent. Stamens 2 to 6 or more. Carpels 3 to 4, 1-seeded, free or united at the base; fruit a nutlet.

1A Inflorescence short, with four petaloid bracts at the base; stipules *c.* 1 cm long

Houttuynia

1B Inflorescence long; petaloid bracts absent; stipules very small, *c.* 3 mm long

Saururus

Houttuynia Thunb., Fl. Jap. 12 (1874)
Fig. 247A.
Stems up to 50 cm high. Leaves broadly ovate, reniform to cordate at the base, acuminate at the apex distinctly stipulate. Inflorescence 1 to 2 cm long. Petaloid bracts 4, oblong 1 to 1.5 cm long, white. Flowers sessile. Stamens 3; ovary of 3 partially united carpels; fruit subglobose, opening at the apex.

1 species: *H. cordata* Thunb. occurs in eastern Asia from India to Japan and Formosa, in wet places often as a weed of cultivation.

Saururus L., Sp. Pl. 341 (1753)
Fig. 247B.
Stems up to 1 m high. Leaves broadly ovate, reniform to cordate at the base, acuminate at the apex; stipules very small. Petaloid bracts absent. Inflorescence 10 to 30 cm long in fruit. Flowers stalked, each with a small (*c.* 2 mm long) bract on the pedicel. Stamens 6 or 4. Ovary of 3 or 4 carpels united at the base, each with 2 (or 4) ovules, only 1 developing. Fruit indehiscent, globose.

Hall, T. F. The biology of Saururus cernuus L. Amer. Midl. Nat. 24: 253–260 (1940)

Wood, C. E. The Saururaceae in the southeastern United States. Journ. Arnold. Arbor. 52 (3): 479–485 (1971)

2 species; *S. cernuus* L. occurs in swamps and shallow water in eastern N. America, *S. chinensis* (Lour.) Baill. in swamps and ricefields in southeast Asia. *S. cernuus* is considered a pest in S. E. USA as dense stands provide an ideal environment for the breeding of mosquitoes.

Fig. 247. A. *Houttuynia cordata* Thunb.: a, stem, fruiting (1 cm); b, flowering inflorescence; c, flower (1 mm);
B. *Saururus cernuus* L.: a, fruiting stem (1 cm); b, flowering inflorescence; c, flower (1 mm).

SCROPHULARIACEAE

c. 220 genera, of which 17 have aquatic species.

Herbs. Leaves opposite or whorled; stipules absent. Flowers actinomorphic or zygomorphic, when zygomorphic usually 2-lipped. Sepals tubular, 3- to 5-lobed. Stamens rarely 5, usually 4 or 2; 2 or rarely 3 staminodes occasionally present; anthers 2-locular, opening lengthwise or the loculi confluent at apex and opening by 1 continuous split. Ovary superior, sessile, usually 2-locular; septum horizontal; style terminal; fruit a capsule; ovules axile; seeds usually numerous with fleshy endosperm.

Many Scrophulariaceae are found in swamps, marshes and bogs. It has been found difficult to draw the line between the plants that may be found temporarily submerged following rains or floods and those that are regularly submerged for several months each year. The following genera contain species that withstand temporary floods but do not appear to flower in water and have thus been excluded from the accounts: *Artanema* D. Don, *Buchnera* L., *Calceolaria* L., *Conobea* Aubl. *Mercardonia* Ruiz and Pavon [*Pagesia* Rafin.], *Micrargeria* Bentham, *Scoparia* L., *Stemodia* L., *Striga* Lour.

Barroso, G. M. Scrophulariaceae indigénas e exoticas no Brasil. Rodriguésia 27: 9–108 (1952)

Dawson, G. Scrophulariaceas Bonaerensis. Rev. Mus. La Plata n.s. 8 (Bot.), 31: 1–62 (1956)

Pennell, F. W. Scrophulariaceae of Eastern Temperate N. America. Monog. Acad. Sci. Philadelphia 1: 1–650 (1935)

Pennell, F. W. Scrophulariaceae of Western Himalayas. Monog. Acad. Nat. Sci. Philadelphia 5: 1–163 (1943)

Pennell, F. W. A second summary of the Scrophulariaceae of New Guinea. Journ. Arnold Arb. 24: 243–274 (1943)

Thieret, J. W. The tribes and genera of Central American Scrophulariaceae. Ceiba 4: 164–184 (1954)

1A Petal tube very short; petal lobes flat, spreading; abaxial petal lobes external, overlapping in bud; stamens 2, spreading
Veronica

1B Petal tube short or long; petal lobes curved or hooded; adaxial petal lobes external, overlapping in bud; stamens 2 to 4, rarely 5, not spreading.

2A Abaxial filaments (of fertile or sterile stamens) appendaged at point of contact to the petals

3A Sepal tube 4-lobed (small creeping herbs)

4A Sepals united only at base; petal tube 2-lipped; adaxial petal lobes more or less united
Micranthemum

4B Sepals united at least ½ their length, except on abaxial side where parted to base; petal tube 1-lipped; adaxial petal lobes absent or much reduced
Hemianthus

3B Sepal tube 5-lobed
5A Leaves in basal rosette
Craterostigma
5B Leaves in opposite pairs along the stem
6A Fertile stamens 2
Lindernia
6B Fertile stamens 4
7A Appendages at base of abaxial filaments rounded and swollen
Craterostigma
7B Appendages at base of abaxial filaments spur-like or elongate
Lindernia
2B Abaxial filaments not appendaged at point of contact to petals
8A Sepal lobes 3 or 4
9A Sepal lobes subulate-lanceolate, much longer than sepal tube
Bythophyton
9B Sepal lobes broadly triangular, equal to or shorter than sepal tube
10A Stigma broadly spathulate, entire, curving over the stamens; stamens 2 or 4
Glossostigma
10B Stigma sub-capitate, almost entire or weakly 2-lobed, not longer than the stamens; stamens 4
Limosella
8B Sepal lobes 5
11A Petal lobes 4; capsule flattened laterally (S. E. USA)
Amphianthus
11B Petal lobes 5; capsule not flattened laterally
12A Sepal and petal lobes ciliate (minute herbs from Indomalaysia, Japan and N. Australia)
Microcarpaea
12B Sepal and petal lobes not ciliate
13A Stamen connective developed into a wide flap partly surrounding the anthers
Gratiola
13B Stamen connective not developed into a flap
14A Sepal tube 5-angled; abaxial petal lip with 2 hairy patches at base
Mimulus
14B Sepal tube not 5-angled, abaxial petal lip without 2 hairy patches at base
15A Sepals enlarged and globuse in fruit; stamens 2 (small creeping herbs, from N. Africa to India)
Peplidium
15B Sepals not enlarged and globuse in fruit, stamens usually more than 2

16A Petal tube not 2-lipped; petal lobes subequal
    17A Stigma subcapitate or weakly 2-lobed (usually tufted herbs, with or without stolons)
        Limosella
    17B Stigmas 2, free, occasionally joined at base (usually with elongated stems bearing leaves in pairs or whorls)
        Bacopa
16B Petal tube 2-lipped; petal lobes unequal
    18A Basal leaves in a rosette with stem leaves much reduced
        Dopatrium
    18B Basal leaves not in a rosette
        19A Petal tube yellow; stamens 4, 2 fertile and 2 sterile (Madagascar)
            Hydrotriche
        19B Petal tube white, blue or purple; stamens 2 to 5 all fertile
            20A Flowers in terminal or axillary spikes or racemes
                Limnophila
            20B Flowers solitary or in pairs in leaf axils
                21A Anther lobes free, usually divergent
                    Limnophila
                21B Anther lobes united, parallel
                    Bacopa

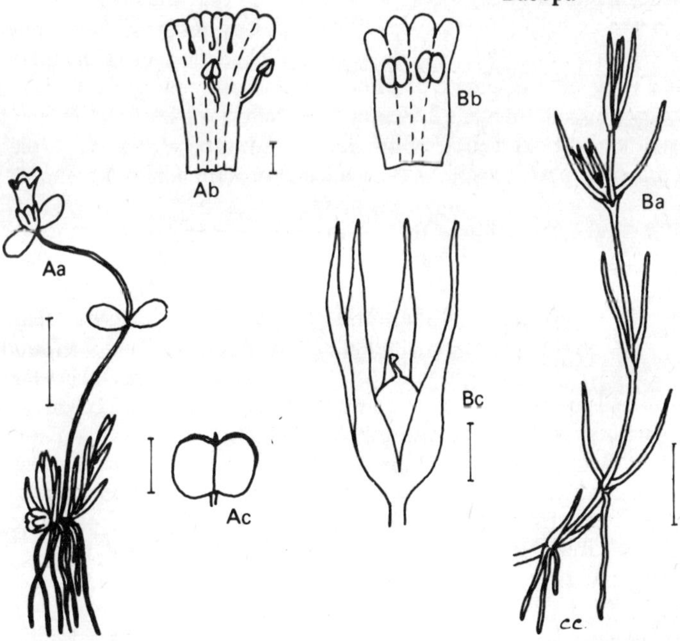

Fig. 248. A. *Amphianthus pusillus* Torrey: a, habit (1 cm); b, petal tube opened (1 mm); c, fruit (3 mm);
B. *Bythophyton indicum* Hooker fil.: a, habit (1 cm); b, flower, opened; c, sepals and fruit (1 mm).

Amphianthus Torrey, Ann. Lyc. New York 4: 82 (1837)
Fig. 248A.
Annual. Stems submerged or floating. Leaves opposite, dimorphic; submerged leaves lanceolate; floating leaves ovate. Flowers solitary, almost sessile, in leaf axils; submerged flowers cleistogamous; emergent flowers chasmogamous. Sepals tubular, 5-lobed; lobes subequal, longer than tube, with rounded apices. Petals white, *c.* 6 mm long, tubular, 4-lobed; lobes subequal, shorter than tube; adaxial lobe largest, with notched apex. Stamens 2. Ovary 2-locular; stigma elongate; style 2-lobed; capsule laterally flattened, loculicidally dehiscent.

1 species, *A. pusillus* Torrey: known only from a few pools on granite outcrops in Georgia and S. Carolina in the USA.

Bacopa Aublet, Pl. Guiane 128 (1775) nom. cons., sensu Pennell, Proc. Acad. Nat. Sci. Philadelphia 98: 83–98 (1946), [*Bramia* Lamark, *Herpestis* Gaertn., *Hydranthelium* H.B.K., *Hydrotrida* Small, *Macuillamia* Rafin., *Naiadothrix* Pennell]
Fig. 249A
Annuals and perennials. Stems erect, ascending, creeping or floating. Leaves opposite or apparently whorled, palmately or pinnately veined, simple or compound; segments variously lobed, divided or capillary. Flowers solitary or in pairs in leaf-axils; bracts just beneath the calyx or absent. Sepals tubular, 5-lobed; adaxial lobe usually larger (in the obligate aquatic species the lobes are sub-equal). Petals blue, purple or white, tubular, campanulate with 5, sub-equal lobes or 2-lipped. Stamens 2, 3, 4 or 5; anther loculi sessile, parallel. Ovary 2-locular, with or without a circle of bristles around its base; stigmas 2, free or somewhat joined at base; capsule with loculicidal or septicidal dehiscence.

Pennell, F. W. Reconsideration of the Bacopa-Herpestis problem of the Scrophulariaceae. Proc. Acad. Nat. Sci. Philadelphia 98: 83–98 (1946)

c. 100 species: warmer regions of the world, most species in America. Many species are found in wet and seasonally inundated regions and may be found submerged in water for several months each year. *B. monnieri* (L.) Wettstein is pantropical and frequently found as a weed in ricefields and irrigation ditches. Obligate aquatic species with capillary leaf segments are found in the section *Chaetodiscus* (characterised by having a ring of bristles surrounding the base of the ovary). The obligately aquatic species are: *B. myriophylloides* (Bentham) Wettstein from Brazil, *B. longipes* (Pennell) Standley from Cuba, *B. najas* Standley from Honduras and *B. reflexa* (Bentham) Edwall from Brazil. Several species are cultivated as decorative aquarium plants.

Bythophyton Hooker, Fl. Brit. India 4: 286 (1884)
Fig. 248B.
Probably annual. Stems up to 8 cm long, flaccid, loosely tufted, erect, usually wholly submerged. Leaves opposite, subulate, entire. Flowers axillary, shortly stalked. Sepals 4, equal, subulate-lanceolate, united at base. Petals tubular, 4-lobed; lobes subequal, shorter than tube. Stamens 2; filaments very short. Ovary 2-locular; style short, curved; stigma subcapitate; capsule much shorter than sepals, broadly oblong, compressed; seeds many, reticulate.

1 species, *B. indicum* Hooker fil.: Khasia in E. India and Luzon in the Philippines. Found in standing water in pools that usually dry out each year.

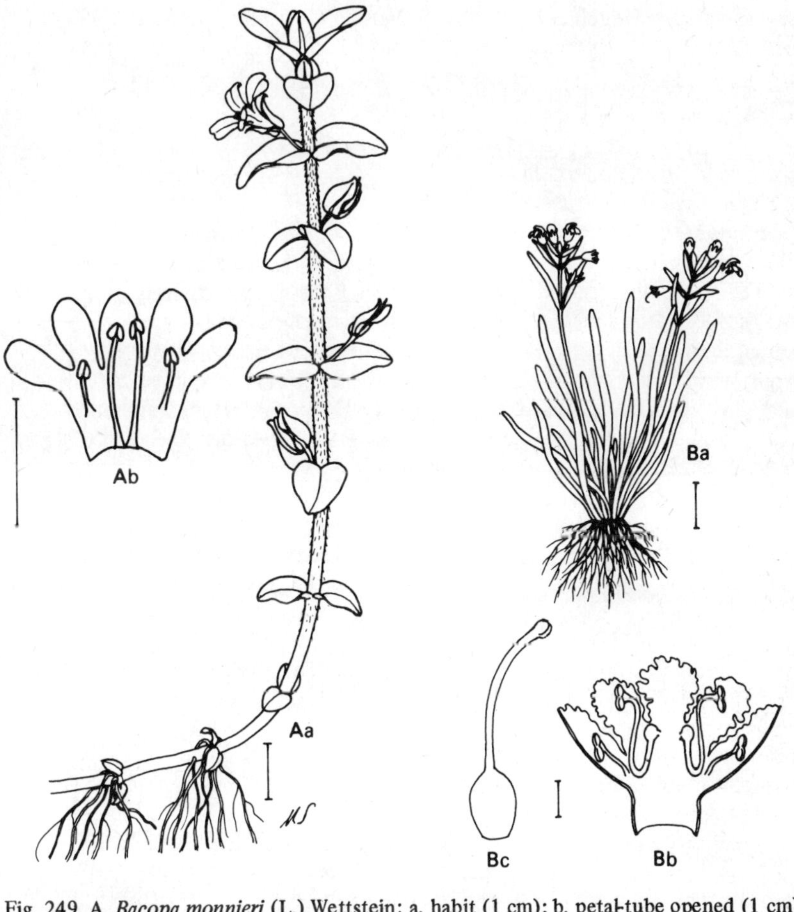

Fig. 249. A. *Bacopa monnieri* (L.) Wettstein: a, habit (1 cm); b, petal-tube opened (1 cm); B. *Craterostigma linearifolium* Engler: a, habit (1 cm); b, petal-tube opened (1 mm); c, ovary and style (1 mm).

Craterostigma Hochst. Flora 24: 668 (1841)
Fig. 249B.
Perennials. Stems reduced or elongate. Leaves in basal rosette or in pairs along the stem, entire. Flowers few in the axils or bract-like leaves or crowded in small terminal heads or rarely solitary. Calyx tubular, 5-lobed; lobes subequal, ribbed or winged, shorter than tube. Petals tubular, 2-lipped, adaxial lip concave, entire or notched; abaxial lip 3-lobed; lobes broad, subequal. Stamens 4; adaxial pair shorter, more or less included within tube with simple, elongate filaments; abaxial pair adnate to abaxial lip with a swollen appendage at the base of each filament; anther lobes divaricate. Ovary 2-locular; style elongate, often curved and persisting in fruit; stigma 2-lipped; capsule oblong, included within the sepals, 2-valved.

c. 20 species: Tropical and S. Africa and Madagascar. Mostly herbs of wet places, some species such as *C. monroi* S. Moore, *C. gracile* Pilger and *C. linearifolium* Engler are frequently found submerged. This genus is very similar to *Lindernia.*

Dopatrium F. Hamilton ex. Bentham, Bot. Reg. ad. 1770, (1835)
Fig. 250A.
Slender annuals. Stem erect, usually simple, usually somewhat succulent. Leaves opposite; basal leaves lanceolate to ovate, often forming a loose rosette; upper leaves usually reduced and scale-like. Flowers solitary in leaf axils, cleistogamous when submerged. Sepals tubular, 5-lobed; lobes equal or subequal, longer than tube. Petals tubular, strongly or weakly 2-lipped, 5-lobed, white, blue, purple or yellow, usually small but sometimes large and showy. Stamens 2, adaxial; anthers parallel, loosely joined; staminodes 2, abaxial. Ovary 2-locular; stigma more or less peltate or with 2 flattened lobes; capsule loculicidally dehiscent; seeds numerous.

c. 20 species: tropical Africa and Asia. Found in pools that usually dry out. *D. junceum* Buch.-Ham. is a common ricefield weed and has become naturalised in N. America.

Glossostigma Arn., Nov. Act. Nat. Cur. 18: 355 (1836), nom. cons.
Fig. 251A.
Small, usually gregarious annuals or short-lived perennials. Stems capillary, creeping, rooting at nodes. Leaves opposite or occasionally in whorls, sessile or stalked, linear to spathulate, rarely more than 1 cm long, entire. Flowers minute, solitary in axils of leaves. Sepals campanulate, persistent in fruit, 3- or 4-lobed; lobes broadly triangular. Petals shortly tubular, 5-lobed; lobes subequal. Stamens 2 or 4; anther cells diverging at base, confluent at apex. Ovary 2-locular; style elongate, flattened above into a broad spathulate stigma usually curving over the stamens; capsule globose, 2-valved.

Fig. 250. A. *Dopatrium junceum* Buch.-Ham.: a, habit (1 cm); b, flower; c, section of flower (1 mm); d, fruit;
B. *Lindernia hyssopioides* (L.) Benth.: a, habit (1 cm); b, petal-tube, opened; c, fruit (1 mm).

Duguid, F. C. Notes on Glossostigma elatinoides (Scrophulariaceae). Bull. Wellington Bot. Soc. 36: 15–17 (1969)

c. 5 species: Africa, India, Australia and New Zealand. Small prostrate plants often forming mats; found in shallow water, usually coming into flower as they become exposed. Cleistogamous flowers are frequent on submerged plants.

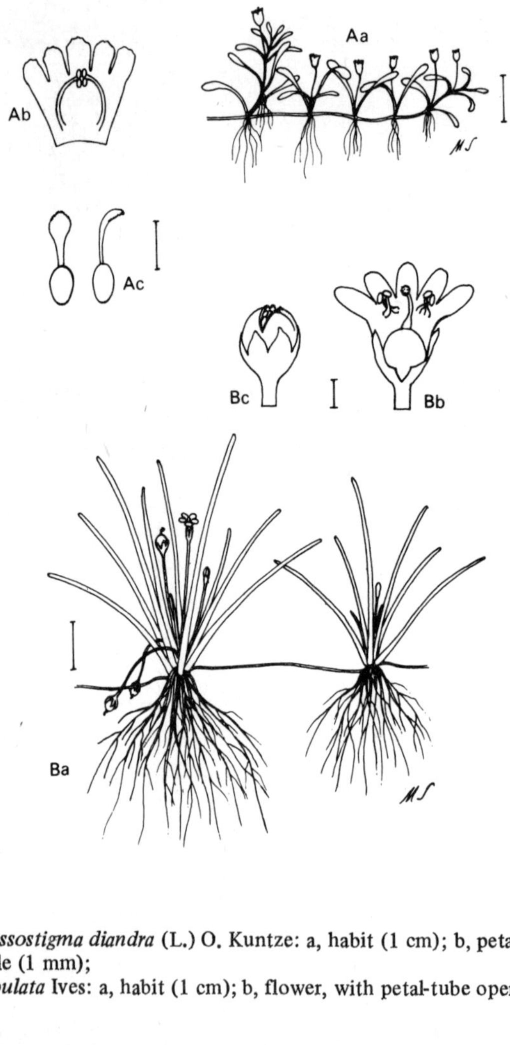

Fig. 251. A. *Glossostigma diandra* (L.) O. Kuntze: a, habit (1 cm); b, petal-tube, opened; c, ovary and style (1 mm);
B. *Limosella subulata* Ives: a, habit (1 cm); b, flower, with petal-tube opened; c, fruit (1 mm).

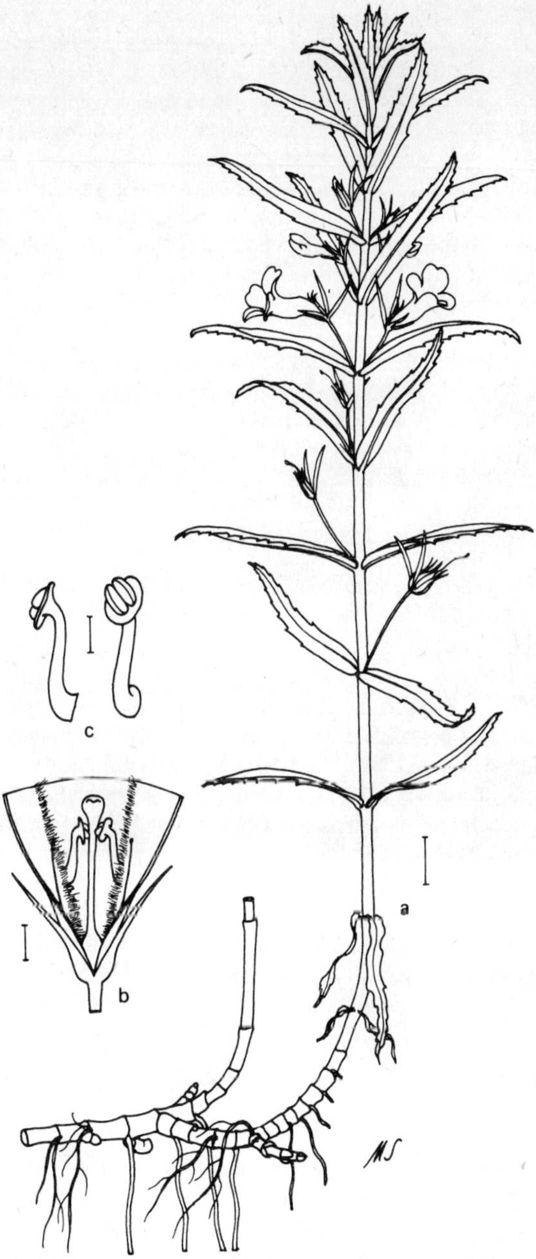

Fig. 252. *Gratiola officinalis* L.: a, habit (1 cm); b, base of flower opened (2 mm); c, stamens (1 mm).

Gratiola L., Sp. Pl. 17 (1753) sensu Pennell, Mong. Acad. Nat. Sci. Philad. 1: 69 (1935)
Fig. 252.
Annuals or perennials. Stems ascending or erect, rooting below, usually swollen when submerged. Leaves opposite and decussate, sessile, lanceolate to ovate; margin entire or toothed. Flowers solitary in leaf axils: Sepals tubular, 5-lobed; lobes much longer than tube, usually unequal in length and shape. Petals tubular, yellow or white, 2-lipped, 5-lobed; lips much shorter than tube; the inside of the adaxial lip and adaxial part of the tube hairy. Stamens 4 (either all fertile or 2 fertile and 2 sterile) or 2; connective developed into a wide flap partly surrounding the anthers. Ovary 2-locular; style elongate; stigmas 2, expanded; capsule 4-valved.

c. 20 species: widely distributed in N. and S. Temperate Zones and on mountains in the tropics. Most species are found in wet regions where they may be found in standing water for several months each year. *G. lutea* Rafin. from N. America and *G. ebracteata* Bentham and *G. heterosepala* Mason and Bacigalupi both from N. W. America are characteristically found in water.

Hemianthus Nutt., Journ. Acad. Nat. Sci. Philad. 1: 119 (1817)
Fig. 253A.
Small annuals or perennials. Stems creeping, rooting at nodes. Leaves opposite or whorled, usually sessile, lanceolate, spathulate to orbiculate, entire. Flowers minute, solitary in leaf axils, frequently cleistogamous. Sepals tubular, united for at least ½ their length, except on abaxial side where parted to base, 4-lobed. Petals white, 1-lipped (adaxial lobes absent or rudimentary); abaxial lip longer than sepals, distinctly 3-lobed. Stamens 2 or 4, appendaged near base. Ovary 1- or 2-locular, style linear; stigmas 2, linear ½ or less as long as style; capsule globose, 2-valved.

c. 10 species: Tropical America, mainly W. Indies and one species *H. micranthemoides* Nutt. extending northwards to Arctic America. Small creeping plants in shallow water,

Hydrotriche Zucc., Abh. Math.-Phys. Cl. Königl. Bayer.-Akad. Wiss. 1: 308 (1832)
Fig. 254.
Perennial. Stems elongate, often unbranched, up to 70 cm long. Leaves dimorphic; submerged leaves appearing as whorls of *c.* 16, simple, capillary segments, each 2 to 5 cm long (morphologically the leaves are sessile, in opposite pairs and each leaf is divided to the base into *c.* 8 capillary lobes); emergent leaves bract-like, entire or 3-toothed. Inflorescence erect, simply racemose, with 2 to 5, *c.* 1 to 5 cm long stalked flowers. Sepals tubular, *c.* 5 cm long, with 5, elongate-triangular lobes. Petals tubular, weakly 2-lipped;

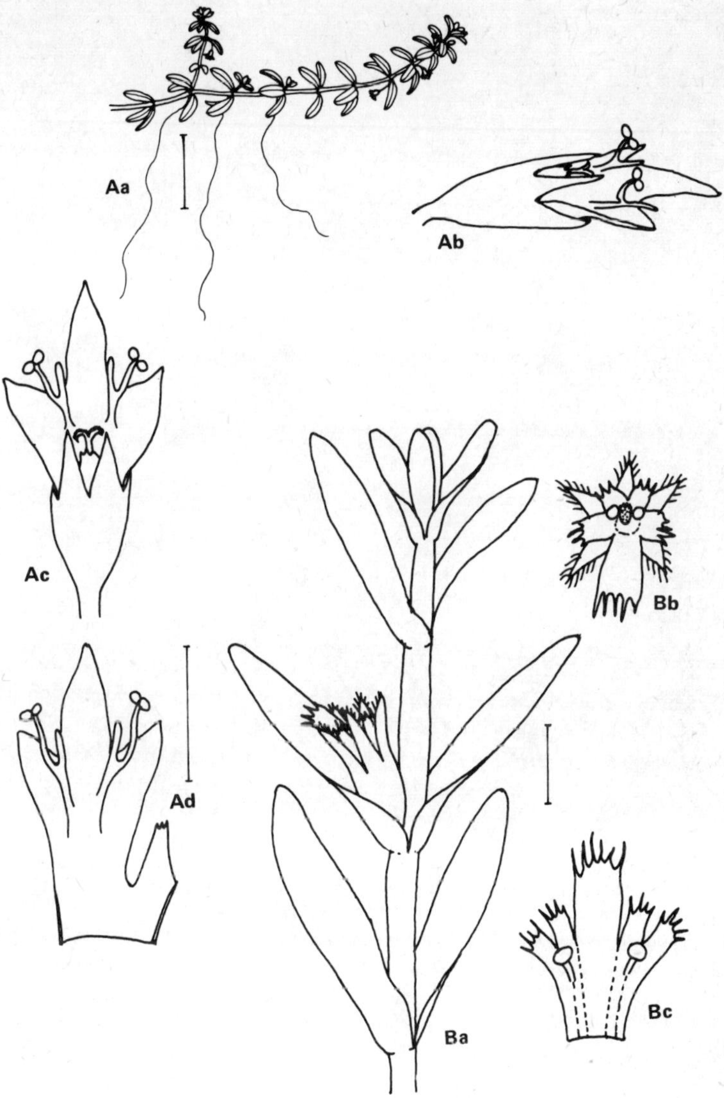

Fig. 253. A. *Hemianthus micranthemoides* Nutt.: a, shoot (1 cm); b, flower from the side; c, flower from above; d, petal-tube opened (1 mm);
B. *Microcarpaea minima* (Koen.) Merr.: a, habit (1 mm); b, flower viewed from front; c, petal-tube split adaxially.

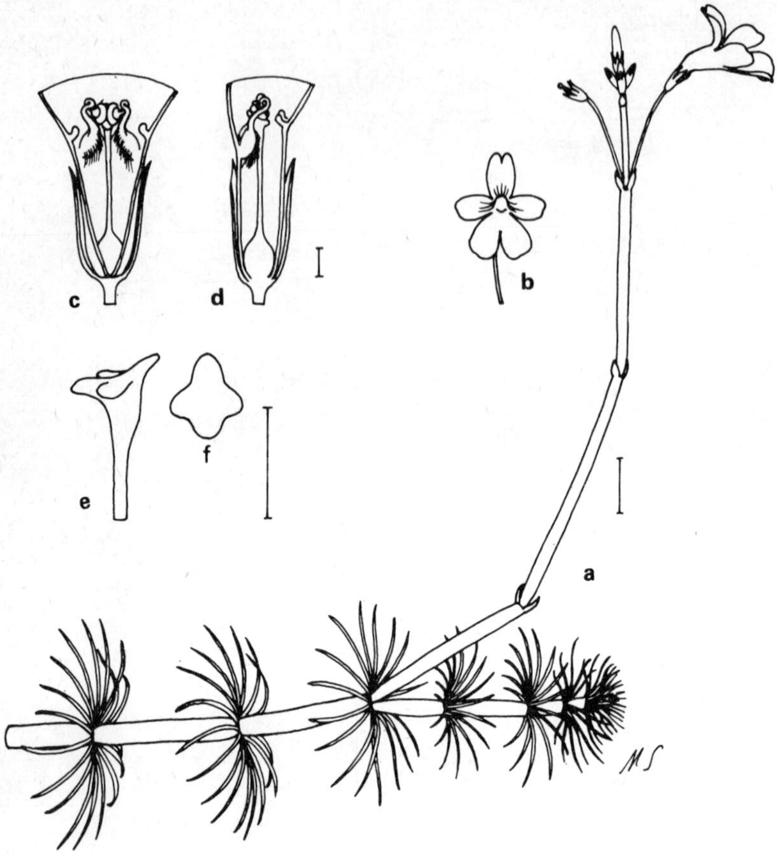

Fig. 254. *Hydrotriche hottoniiflora* Zucc.: a, habit (1 cm); b, flower viewed from front; c, base of flower, opened and viewed abaxially; d, base of flower, opened and viewed laterally (1 mm); e, stigma, viewed laterally; f, stigma viewed from front (1 mm).

tube *c.* 5 mm long, yellow; lips white to pale blue; adaxial lip 2-lobed; abaxial lip 3-lobed; lobes subequal, *c.* 12 mm long, the adaxial slightly smaller. Stamens 2, *c.* 2 to 5 mm long, exerted; filaments widened at base; staminodes 2, small. Ovary 2-locular; style *c.* 5 mm long; stigma subcapitate; capsule globose, opening by 2, loculicidal valves.

Heine, H. and Michel, P. Hydrotriche hottoniiflora Zuccarini (1832). Piscic. Franç. 17: 11–15 (1969)

1 species: *H. hottoniiflora* Zucc. known only from Madagascar. Occurs in water up to 70 cm or more deep. Frequently cultivated as a decorative aquarium plant.

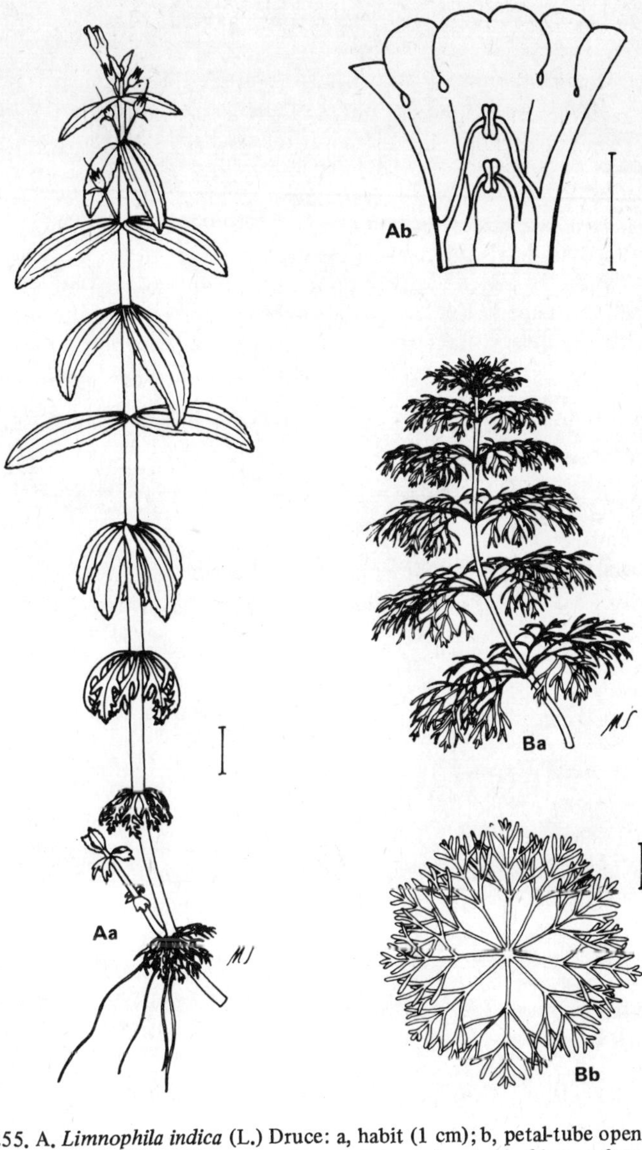

Fig. 255. A. *Limnophila indica* (L.) Druce: a, habit (1 cm); b, petal-tube opened (5 mm); B. *Limnophila sessiliflora* Blume: a, submerged shoot; b, whorl of leaves, from above (1 mm).

Limnophila R. Br., Prodr. 442 (1810), nom. cons., [*Ambulia* Lam.]
Fig. 255.
Annuals or perennials, frequently aromatic when bruised. Stems erect, ascending or creeping. Leaves polymorphic; submerged leaves in whorls, laciniate, pinnatifid to compoundly pinnate with capillary lobes; emergent leaves whorled or opposite, sessile or stalked, entire to serrate, laciniate or pinnately divided. Flowers solitary in leaf axils or in loose or compact, terminal or axillary spikes or racemes. Sepals tubular, 5-lobed; lobes sub-equal or with adaxial lobe enlarged. Petals tubular or funnel-shaped, blue or white, 5-lobed, 2-lipped; adaxial lip outside in bud, entire or 2-lobed; abaxial lip 3-lobed, erect or spreading. Stamens 4, included; adaxial pair shorter or rarely absent; anthers free, usually divergent. Ovary glabrous; style capillary, deflexed at tip; stigma 2-lipped; capsule ellipsoidal to globose, septicidally 4-valved; valves bifid; seeds small numerous.

Philcox, D. A taxonomic revision of the genus Limnophila R.Br. (Scrophulariaceae). Kew Bull. 24: 101–170 (1970)

c. 35 species: warmer parts of the Old World. About 13 species are truly aquatic and bear finely divided submerged leaves; the rest are essentially marsh plants but many of them grow in seasonally inundated regions. Many species have been recorded as weeds in ricefields. Some species are cultivated as decorative aquarium plants and have been recorded as escaping from cultivation in southern USA.

Limosella L., Sp. Pl. 641 (1753)
Fig. 251B.
Annuals or perennials. Stems reduced or stoloniferous. Leaves usually in rosettes rarely in pairs along the stem, simple, entire; blades (when distinguishable from the petiole) linear, elliptic to ovate; petioles linear or subulate. Flowers solitary, distant or crowded, long-stalked or almost sessile, chasmogamous or cleistogamous. Sepals tubular, campanulate, (rarely 4-) 5-lobed; lobes broadly triangular, much shorter than tube, usually equal, rarely 2-lipped. Petals tubular, rarely 4- or usually 5-lobed, usually exceeding the sepals; lobes shorter than tube, subequal, white, pink, blue or violet. Stamens 4, all fertile, subequal, not exceeding the petal-lobes. Ovary rarely 1- usually 2-locular; style linear, shorter or longer than ovary; stigma subcapitate, almost entire or weakly 2-lobed; capsule globose or globose-ellipsoidal, opening by 2 valves; seeds few or numerous.

Glück, H. Beiträge zur Systematik, Morphologie und Biologie der Gattung Limosella. Bot. Jahrb. (Engler) 66 (5): 488–566 (1934)

c. 11 species: cosmopolitan. Essentially small amphibious plants occurring in shallow, still or flowing water. Flowering usually takes place as the water level falls.

Lindernia All., Mél. Phil. Math. Soc. Turin (Misc. Taur.) 3 (1): 178 (1762?) [*Bonnaya* Link and Otto, *Chamaegigas* Dinter, *Ilysanthes* Rafin., *Vandellia* L.]

Fig. 250B.

Annuals or short-lived perennials. Stems erect, ascending or creeping, usually quadrangular in transverse section. Leaves opposite, simple, sessile or shortly stalked, entire, crenate or toothed. Flowers solitary or in terminal or axillary racemes or in umbel-like clusters. Sepals tubular, 5-lobed; lobes shallow or lobed almost to base; sepal tube 5-nerved, each nerve with an obscure or distinct rib, or minute wing. Petals tubular, 2-lipped; adaxial lip either entire or 2-lobed, suberect; adaxial lip 3-lobed, spreading, longer than adaxial lip. Stamens either 4, with all fertile or with abaxial pair sterile, or 2, both fertile the abaxial pair reduced to staminodes; filaments of adaxial pair frequently with a spur-like appendage at or near base; anthers free or joined. Ovary 2-locular; style capillary, erect; stigma 2-lipped; capsule globose, ovoid, obovoid, ellipsoid, or oblong to narrowly cylindric, with septicidal dehiscence, usually leaving the entire septum as a median plate, seeds numerous.

Hickel, B. Zur Kenntnis einer xerophilen Wasserpflanze. Chamaegigas intrepidus Dtr. aus Südwestafrika. Int. Rev. Ges. Hydrobiol. 52 (3): 361–400 (1967)

Mukerjee, S. K. Revision of the Indo-Burmese species of Lindernia Allioni. Journ. Indian Bot. Soc. 24: 127–134 (1945)

Philcox, D. Revision of the Malesian species of Lindernia All. Kew Bull. 22: 1–72 (1968)

Smook, L. Some observations on Lindernia intrepidus (Dinter) Oberm. Dinteria 2: 13–21 (1969)

Yamazaki, T. Notes on Lindernia, Vandellia, Torenia and their allied genera in Eastern Asia, pt. 1. Journ. Jap. Bot. 29: 299–305 (1954); pt. 2. Bot. Mag. Tokyo 69: 14–24 (1955); pt. 3 and 4. Journ. Jap. Bot. 30: 171–180, 359–364 (1955)

c. 120 species: warmer regions of the world. *Lindernia* is a large and taxonomically difficult genus, most species are found in swamps and may be found temporarily submerged. Numerous species occur as weeds in ricefields.

Micranthemum Michx., Fl. Bor. Amer. 1: 10 (1803) nom. cons.

Fig. 257A.

Annuals or perennials. Stems creeping, rooting at nodes. Leaves opposite or whorled, sessile, broadly ovate to orbicular, entire. Flowers solitary in leaf-axils, sessile or shortly stalked. Sepals 4, united at base. Petals tubular, weakly 2-lipped, white; 3 abaxial lobes about as long as tube, exceeding the sepals; 2 adaxial lobes united, appearing as 1, entire or bifid lobe, shorter than the sepals, about ½ as long as abaxial lobes. Stamens 2 (adaxial), appendaged near base. Ovary apparently 1-locular; style linear about equal to ovary in length; stigmas 2, about ¼ as long as style; capsule globose, rupturing irregularly.

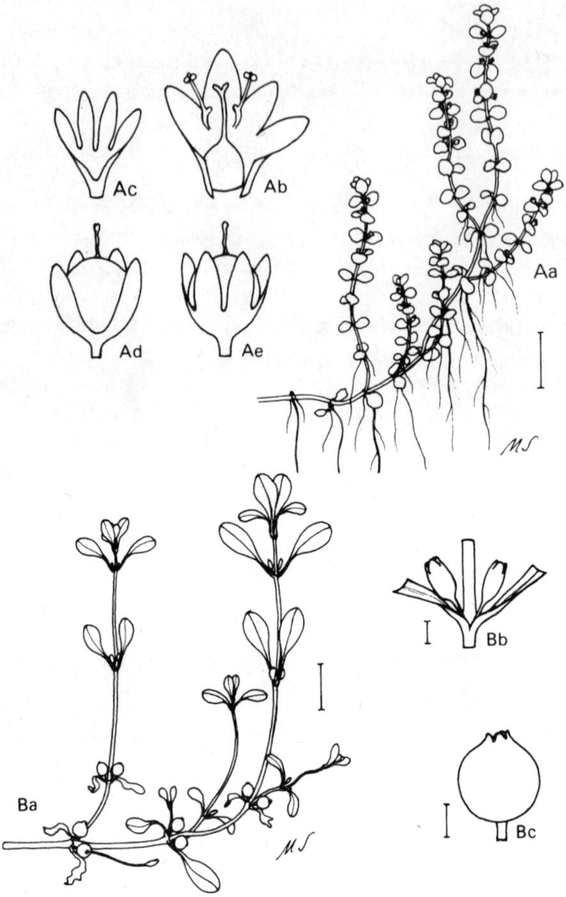

Fig. 257. A. *Micranthemum umbrosum* (Walt.) Blake: a, habit (1 cm); b, flower opened adaxially; c, sepal tube viewed adaxially; d, fruit viewed adaxially; e, fruit viewed abaxially; B. *Peplidium maritimum* Wettst.: a, habit (1 cm); b, flowers (1 mm); c, fruit (1 mm).

Williams, L. O. Tropical American Plants, XII. Micranthemum. Fieldiana (Bot.). 34 (8): 122–124 (1972)

4 species, *M. umbrosum* (Walt.) Blake: eastern N. America and S. America, *M. procerorum* L. Wms. in Mexico, *M. standleyi* L. Wms. in Guatemala and *M. rotundatum* Wright in Cuba. Small creeping plants forming mats of vegetation in seasonally inundated areas.

Fig. 256. *Mimulus ringens* L.: a, base of stem; b, apex of stem (1 cm).

Microcarpaea R. Br., Prodr. 435 (1810)
Fig. 253B.
Annual, usually gregarious and forming tight mats of vegetation. Stems ascending, up to 10 cm long but usually less. Leaves opposite, subsessile, linear to oblong-linear, 2 to 5 mm long, entire. Flowers solitary in 1 axil of each leaf pair, inconspicuous, cleistogamous when submerged. Sepals tubular, 1.5 to 3 mm long, 5-lobed; lobes slightly shorter than tube, acute, recurved, with ciliate margins. Petals light purple, tubular, 2-lipped, 5-lobed; lobes ciliate at apex; abaxial lobe longest, just exceeding the sepals. Stamens 2; anthers appearing 1-locular at maturity. Ovary 2-locular; stigma appearing as a spathulate tip of the style; capsule ellipsoidal to globose, 2-valved, loculicidally dehiscent.

1 species, *M. minima* (Koen.) Merr. [*M. muscosa* R. Br.]: Indomalaysia, Japan, N. Australia. Occurs in shallow water that normally dries out each year, frequently found as a weed in ricefields.

Mimulus L., Sp. Pl. 634 (1753)
Fig. 256.
Annuals or perennials. Stems erect or ascending, occasionally stoloniferous. Leaves opposite, sessile or petiolate, simple, entire, toothed or occasionally lobed. Flowers solitary, axillary or terminal, often in open racemes. Sepals tubular or campanulate, persistent and often enlarging with age, 5-angled, 5-toothed; teeth equal or unequal. Petals tubular, 2-lipped or equally lobed, blue, red, purple, yellow or rarely white; adaxial lip exterior in bud 2-lobed; abaxial lip mostly longer, 3-lobed, usually with 2 hairy patches at base. Stamens 4, filaments adnate to lower part of petal tube; anthers 2-locular, joined at apex, more or less spreading at base. Ovary 2-locular; style usually longer than stamens; stigma 2-lobed, with often sensitive, plate-like lobes; capsule oblong or linear, 2-valved, loculicidally dehiscent; seeds small, numerous.

Grant, A. L. A monograph of the genus Mimulus. Ann. Mo. Bot. Gard. 11 99–388 (1924)
Vickery, R. K. Barriers to gene exchange between members of the Mimulus guttatus complex. Evolution 18: 52–69 (1964)

c. 120 species: mainly New World but scattered in the Old World. Practically all species of section *Mimulus* (*c.* 5 species) and section *Simiolus* (*c.* 16 species) grow in shallow standing water, in swamps or on the banks of streams. Several species are cultivated as decorative plants.

Peplidium Delile, Fl. Egypte 50, 148 (1812)
Fig. 257B.
Small creeping herbs. Stems much branched, rooting at nodes. Leaves opposite obovate or spathulate, fleshy, entire. Flowers small, 1 to 3 in leaf axils, sub-sessile or shortly stalked. Sepals tubular, bluntly 5-toothed, enlarged and globose in fruit. Petals tubular, 5-lobed; lobes longer than tube, sub-equal. Stamens 2. Ovary 2-locular; stigma large, flat, tongue-shaped; curved; capsule globose to ovoid, dehiscing irregularly, or regularly with 2 or 4 valves.

2 species: *P. maritimum* Wettst. [*P. humifusum* Delile]; Egypt, Kurdistan, India and Ceylon: *P. muelleri* Bentham; N. E. Australia. In shallow water at the edges of pools and tanks; flowering as the water level drops.

Veronica L., Sp. Pl. 9 (1753)
c. 250 species: cosmopolitan in temperate zones. The aquatic species are confined to the relatively small section *Beccabunga* (Hill) Dum.
Fig. 258.
Perennial herbs. Stems creeping, ascending or erect. Leaves opposite, lanceolate to oblong, slightly to distinctly toothed. Flowers in opposite, axillary racemes. Sepals of 4 essentially distinct lobes. Petals very shortly, tubular, blue, pinkish or white, 4-lobed; adaxial lobe the largest; abaxial lobe the smallest. Stamens 2; filaments diverging, raising anthers above the petal lobes. Ovary 2-locular; style elongate; stigma capitate or 2-lobed; capsule laterally flattened, loculicidally dehiscent.

Pennell, F. W. "Veronica" in North and South America, Rhodora 23: 1–41 (1921)

Sect. *Beccabunga* (Hill) Dum. *c.* 12 species; widespread in temperate zones but most species in N. Temperate regions. Mostly in wet places some species such as *V. anagallis-aquatica* L., *V. catenata* Pennell, *V. glandifera* Pennell and *V. connata* Rafin. are characteristically found at least partly submerged.

Fig. 258. *Veronica anagallis-aquatica* L.: a, habit (3 cm); b, inflorescence (1 cm); c, petal-tube; d, capsule (1 mm).

SPARGANIACEAE

1 genus.

Sparganium L., Sp. Pl. 971 (1753)
Fig. 259.
Perennial herbs. Stems corm-like; rhizomes elongate. Leaves radical, distichous, linear, erect or floating, sheathing at base. Flowers unisexual, crowded into separate, globose heads; female heads towards the base in each inflorescence. Perianth of 3 to 6 radiate scales. Male flowers of 1 to 8 stamens; filaments occasionally united. Female flowers of 1, occasionally 2, or rarely 3 united carpels; style 1, elongate, persisting in fruit; stigmas linear, as many as carpels; ovules solitary; fruit a drupe with dry, spongy exocarp, and hard endocarp.

Beal, E. O. Sparganium in the southeastern United States. Brittonia 12 (3): 176–181 (1960)
Cook, C. D. K. Sparganium in Britain. Watsonia 5: 1–10 (1961)
Cook, C. D. K. Biological Flora of the British Isles: Sparganium erectum L. Journ. Ecol. 50: 247–255 (1962)
Graebner, P. in Engler, A. Pflanzenreich 2 (IV. 10): 1–26 (1900)
Kaul, R. B. Adaptive leaf architecture in emergent and floating Sparganium. Amer. Journ. Bot. 59 (3): 270–278 (1972)
Müller-Doblies, U. Ueber die Blütenstände und Blüten sowie zur Embryologie von Sparganium. Bot. Jahrb. 89: 359–450 (1969)

19 species are generally accepted, but it is believed there are probably no more than 12 species: mostly in Temperate and Arctic N. Hemisphere, 1 or 2 species in Malaysia and S. Australasia. The erect species are usually gregarious and characteristically found in reedswamp communities. The floating species are found in still or flowing water in a variety of aquatic habitats. Wildfowl make nests, roost and feed in *Sparganium* stands and the fruits form an important part of their diet in late Autumn and early Winter.

Fig. 259. *Sparganium erectum* L.: a, base of plant; b, inflorescence (3 cm).

SPHENOCLEACEAE

1 genus: 2 species: pantropical.

Sphenoclea Gaertn., Fruct. 1: 113 (1788), nom. cons.
Fig 260.
Annuals. Stems erect or decumbent, usually swollen and pithy at base. Leaves spirally arranged, simple, entire; stipules absent. Inflorescences terminal, densely spikate. Flower bisexual, actinomorphic. Sepals adnate to ovary; lobes 5, imbricate. Petals campanulate, perigynous; lobes 5, imbricate. Stamens 5, adnate to petals, alternating with petal-lobes. Ovary semi-inferior, 2-locular; style 1, short; stigma capitate; fruit a capsule, opening along a horizontal line, the valve coming off like a lid; seeds minute, numerous, attached to large, spongy, axile placentas.

2 species, *S. dalzielli* N. E. Brown from W. Africa and *S. zeylanica* Gaertn. which is almost pantropical but thought to be introduced in America and S. E. Asia. Often gregarious in wet places and frequently found in shallow water. *S. zeylanica* is often regarded as a troublesome weed in ricefields; however, in Java young plants are eaten.

Fig. 260. *Sphenoclea zeylanica* Gaertn.: a, habit (3 cm); b, inflorescence (1 cm); c, petal-tube, opened; d, capsule (1 mm).

TETRACHONDRACEAE

1 genus.

Tetrachondra Petrie ex D. Oliver in Hooker, Ic. 2250 (1892)
Fig. 261A.
Minute creeping perennials forming dense turf-like patches. Stems slender, rooting at the nodes. Leaves opposite, up to 6 mm long, with flat, united petioles somewhat sheathing below; leaf blade broadly ovate to lanceolate, with ciliate margin. Flowers solitary, in leaf axils or terminal, 4-merous. Sepals 4, united at the base, persistent, *c.* 1.2 mm long. Petals 4, united to form a subrotate, short-tubed corolla. Stamens inserted between the petal lobes; anthers 2-locular. Ovary 4-lobed to the base; style long, arising from near the base of the ovary; fruit of 4 hairy, rounded nutlets.

Skottsberg, C., *Tetrachondra patagonica* n. sp. und die systematische Stellung der Gattung. Bot. Jahrb. (Beiblatt 107) 48: 17–26 (1913)

2 species, *T. hamiltonii* Petrie ex Oliver from New Zealand and *T. patagonica* Skottsb. from Patagonia, occur in shallow water and wet places.

Skottsberg (l.c.) discusses the reasons for considering the *Tetrachondraceae* separate from *Boraginaceae, Lamiaceae* and *Veronica* section *Pygmaea.* He considers the not completely gynobasic style, the solitary terminal flowers and the regular 4-merous corolla with no sign of reduction from 5 lobes as the most important differences from the *Lamiaceae.* Although *T. hamiltonii* may have axillary flowers and a gynobasic style, Allan in Flora of New Zealand Vol. I p. 963 (1961) retained the Tetrachondraceae as a distinct family. *Tetrachondra* is similar in many characters to *Callitriche* and no-one appears to have commented on it as a possible connection between the petaloid *Lamiaceae* and the apetaloid *Callitrichaceae.*

Fig. 261. A. *Tetrachondra hamiltonii* Petrie ex Oliver: a, habit (1 cm); b, flowering shoot; c, petal-tube, opened; d, fruit;
B. *Trapella sinensis* Oliver: a, habit (3 cm); b, flower (1 cm); c, petal-tube, opened; d, fruit (1 mm), after Glück.

TRAPACEAE [HYDROCARYACEAE]

1 genus; *Trapa.*

Trapa L., Sp. Pl. 120 (1753)
Fig. 262.
Annuals or short-lived perennials. Stems rooted in mud, floating. Leaves opposite below, alternate above, dimorphic; submerged ones sessile, linear, entire, caducous; (after falling green, pinnately-branched, adventitious roots develop from the leaf-scars); floating leaves in a rosette, stalked, rhombic, with toothed margin; base broadly cuneate or almost truncate; leaf stalk often with an ellipsoidal, spongy swelling. Flowers bisexual, actinomorphic, inconspicuous, solitary, axillary, short-stalked. Sepals 4, triangular, adnate to ovary, persistent, developing into 2, 3 or 4, hard horns in fruit. Petals 4, white or lilac, caducous. Ovary half-inferior, 2-locular; ovules pendent, 1 in each loculus; fruit a large, woody or bony, variously sculptured, spinose "nut".

Janković, M. Beitrag zur Kenntnis der individuellen Entwicklung der Wassernuss (Trapa L.). Arkh. Biol. Nauka (Beograd) 7: 17–23 (1955), 8: 9–19, 81–86 (1956)
Malik, H. C. It pays to grow Singhara and Bhen. Indian Farming 11 (8): 23–24 (1961)
Miki, S. Trapa of Japan with special reference to its remains. Journ. Inst. Polytech. Osaka Cy. Univ. ser. D, 3: 1–29 (1952)
Srivastava, G. D. and Tandon, R. K. A study of the autecology of Trapa bispinosa Roxb. Proc. Nat. Acad. Sci. India 21: 57–66 (1951)
Vassiljev, V. Systematics and biology of the genus Trapa (in Russian). Sovetsk. Bot. 15: 343–345 (1947)
Vassiljev, V. Species novae Africanicae generis Trapa L. (in Russian). Nov. Sist. Vysš. Rast. (Leningrad) 1965: 175–194 (1965)

1, 3 or up to *c.* 30 species (there is little agreement on species delimitation in *Trapa*); throughout the Old World, introduced in N. America and Australia. The fruits of *Trapa* contain much starch and fat and form a staple food in continental Asia, Malaysia and India. In addition to cultivating their own crop several countries in this area, such as Indonesia, also import the fruits directly from China. *Trapa* is also cultivated in S. and S. E. Europe and the fruits are exported to Chinese communities in N. Europe and the USA. *Trapa* grows very quickly and forms stable surface-floating mats which hinder navigation; it has been reported as troublesome in Rumania, Iran, Africa and E. N. America.

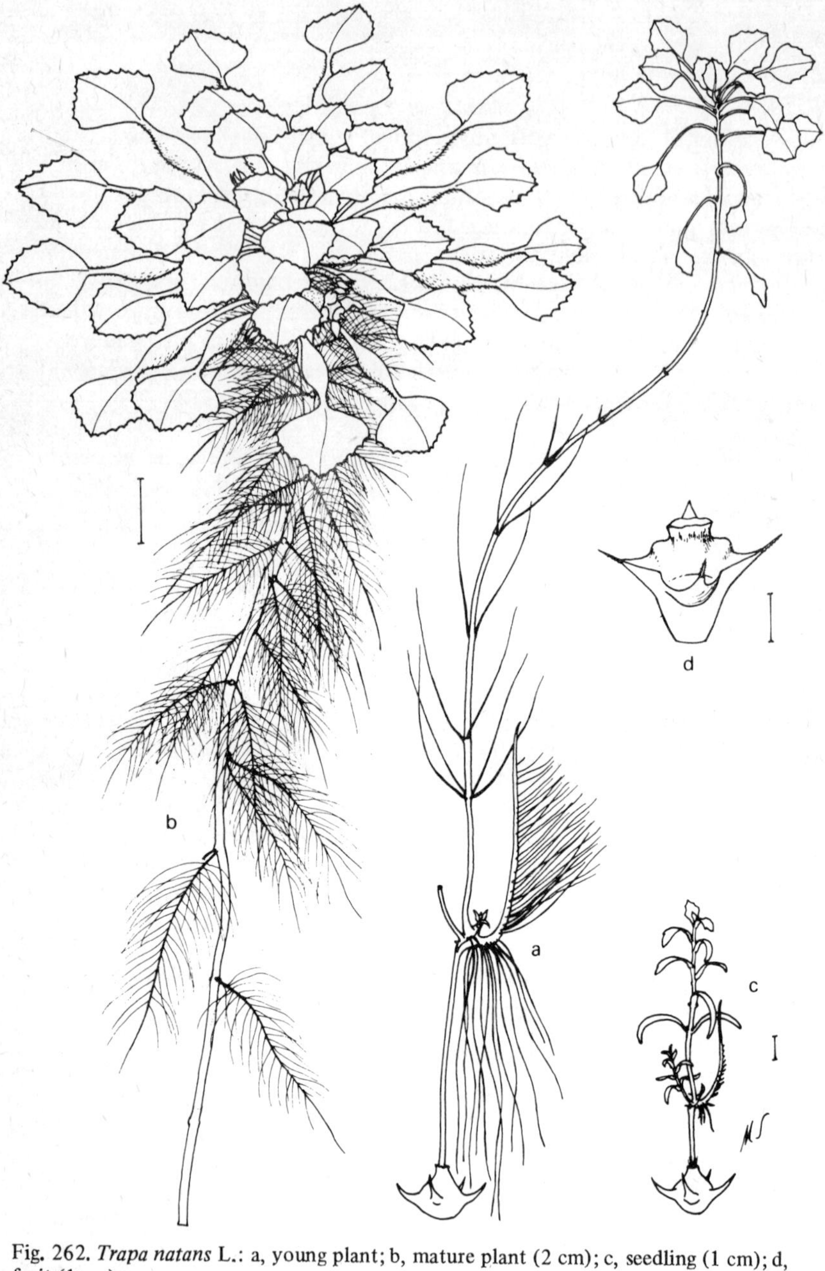

Fig. 262. *Trapa natans* L.: a, young plant; b, mature plant (2 cm); c, seedling (1 cm); d, fruit (1 cm).

TRAPELLACEAE

1 genus *Trapella,* earlier included in the Pedaliaceae.

Trapella F. W. Oliver in Hooker, Icon. Pl. 16: t.1595 (1887)
Fig. 261B.

Annuals. Stems bottom rooted, straggling under water. Leaves opposite, heteromorphic, without stipules; submerged leaves sessile, linear to lanceolate; floating leaves petiolate, with rounded, rhombic blade; margins toothed. Flowers solitary in leaf axils, often cleistogamous. Sepals tubular, partly adnate to ovary, enlarging in fruit, 5-lobed; lobes sharply triangular. Petals inserted at top of sepal tube, tubular, weakly 2-lipped, 5-lobed; abaxial lobes united through most of their length. Stamens 4, inserted on petal tube; adaxial pair fertile; abaxial pair sterile; anthers borne on a flattened, plate-like connective. Ovary about ¾ inferior, 2-locular; 1 loculus sterile the other fertile; style elongate; stigma of 2, unequal flaps; fruit a cylindrical, woody, 1-seeded, nutlet, bearing up to 5 processes at apex; the processes somewhat variable, either 5 long, spirally rolled at tips, or 3 long and spirally rolled and 2 thorn-like, or 2 or 3 wing-like; seed solitary, apical, with caruncle; endosperm reduced to single thin layer.

Glück, H. Die Gattung Trapella. Engl. Bot. Jahrb. 71: 267–336 (1940)
Ihlenfeldt, H.-D. Ueber die Abgrenzung und die natürliche Gliederung der Pedaliaceae R. Br. Mitt. Staatsinst. Allg. Bot. Hamburg 12: 43–128 (1967)
Li, H. L. Trapellaceae, a familial segregate from the Asiatic flora. Journ. Washington Acad. Sci. 44: 11–13 (1954)
Tralan, H. The genus Trapella Oliver in the Tertiary of Europe. Bot. Notiser 117 (2): 119–123 (1964), 118 (1): 21–24 (1965)

1 or 2 species: E. China, Japan and Manchuria. They are annuals with the habit of *Trapa* and occur in shallow, still or slowly flowing water. Fossil material has been found in Europe and C. Asia.

TYPHACEAE

1 genus.

Typha L., Sp. Pl. 971 (1753)
Fig. 263.
Robust perennials. Stems corm-like at base; rhizomes creeping. Leaves mostly radical, distichous, linear, erect, the lower part a long, open sheath closely enveloping the stem, upper sheaths tapering or often somewhat auriculate; upper part of leaf linear, elongate, obtuse to acute, semi-cylindrical to flat, or sometimes keeled below. Flowers unisexual, in cylindrical spikes, the male spikes above the female ones. Male flowers ephemeral, densely packed in the spike, and interspersed with simple or forked hairs or linear or linear-lanceolate bracts (perianth segments); filaments often united or branched. Female flowers on compound stalks interspersed with bracts and sterile flowers; ovary 1-locular, on a stalk bearing several tiers of long hairs (perianth); style linear; stigma linear-lanceolate; ovule solitary; fruit a dehiscent, "1-seeded capsule". Sterile flowers on an elongate hairy stalk terminating in a swollen, aborted ovary.

The literature on *Typha* is enormous. The following references and their bibliographies should serve as an insight on the work done on this genus.

Asplund, I. Embryological studies on the genus Typha. Svensk. Bot. Tidskr. 66: 1–17 (1972)

Bray, J. R. Estimates of energy budgets for a Typha (cattail) marsh. Science 136: 1119–1120 (1962)

Crespo, S. and Perez-Moreau, R. L. Revisión del género Typha en la Argentina. Darwiniana 14 (2–3): 413–429 (1967)

Hotchkiss, N. and Dozier, H. L. Taxonomy and distribution of North American cattails. Amer. Midl. Nat. 41: 237–254 (1949)

Lee, D. and Fairbrothers, D. A serological and disc electrophoretic study of North American Typha. Brittonia 21: 227–243 (1969)

McNaughton, S. J. Autotoxic feedback in relation to germination and seedling growth in Typha latifolia. Ecology 49 (2): 367–369 (1968)

McNaughton, S. J. Genetic and environmental control of glycolic acid oxidase activity in ecotypic populations of Typha latifolia. Amer. Journ. Bot. 56 (I): 37–41 (1969)

Müller-Doblies, D. Ueber die Verwandtschaft von Typha und Sparganium im Infloreszenz- und Blütenbau. Bot. Jahrb. 89 (4): 451–562 (1970)

Saha, S. The genus Typha in India. Bull. Bot. Soc. Bengal 22 (1): 11–18 (1968)

Smith, S. G. Experimental and natural hybrids in North American Typha. Amer. Midl. Nat. 78: 257–287 (1967)

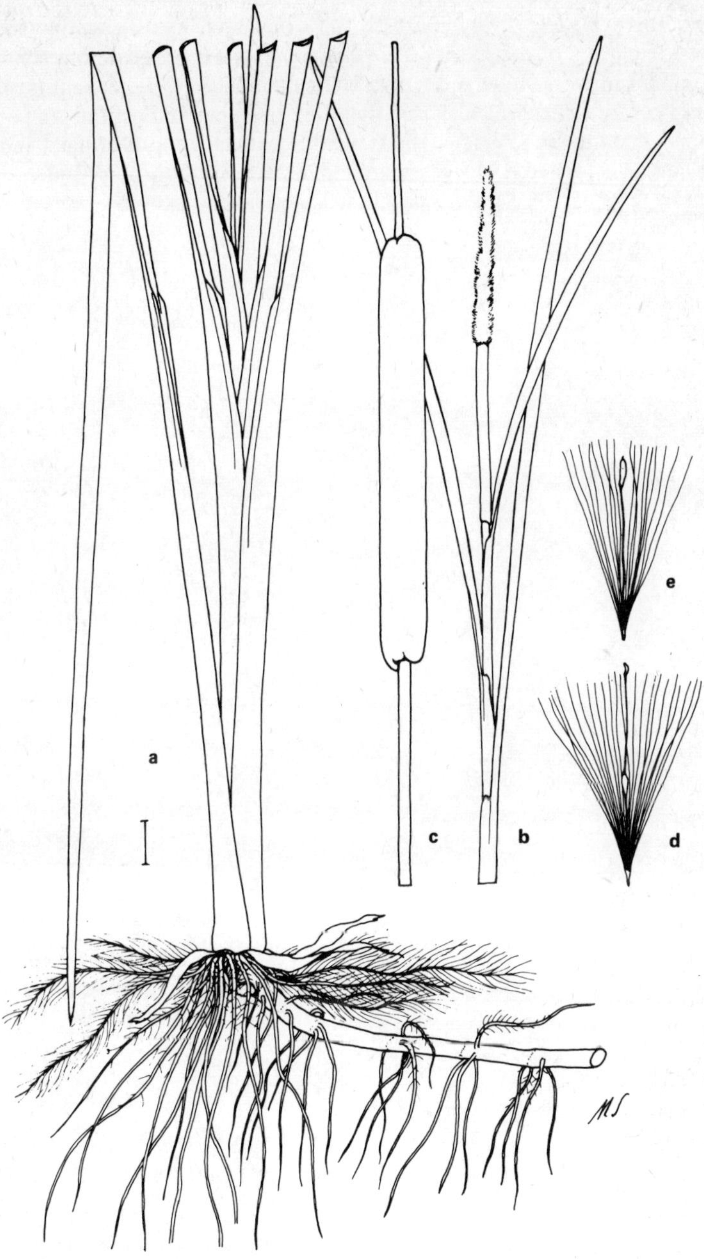

Fig. 263. *Typha latifolia* L.: a, plant base (2 cm); b, young inflorescence; c, fruiting inflorescence; d, fruit; e, sterile flower.

Probably about 10 species although many more names are in use: cosmopolitan. The taxonomy of *Typha* is very confused and a new revision on a world-wide scale is needed. Most of the species of *Typha* are gregarious reedswamp plants. A large number of water-fowl and semi-aquatic mammals are dependent on reedswamp communities which should therefore be conserved. In the temperate zones *Typha* is on the whole a beneficial plant but it can get out of hand and is considered a pest when it blocks canals and ditches. In the tropics and sub-tropics *Typha* can be a serious pest in irrigated land.

XYRIDACEAE [ABOLBODACEAE]

4 genera: *Abolboda* and *Xyris* contain aquatics.

Annuals or perennials. Stems reduced (in aquatics). Leaves mostly basal, simple, linear or capillary, sheathing at base. Inflorescence a simple, terminal, bracteate head. Flowers bisexual. Sepals 2 or 3, when 3 unequal (2 scale-like and 1 membranous forming a hood over the petals). Petals 3, almost free or united, yellow, blue or rarely white. Stamens 3, opposite petals; staminodes 3 or absent, alternating with stamens; anthers opening by longitudinal slits. Ovary superior, 1-locular, with 3 parietal or basal placentas; style simple or appendaged; ovules few to many; fruit a capsule enclosed in persistent petal tube.

Carlquist, S. Anatomy of Guayana Xyridaceae: Abolboda, Orectanthe and Achlyphila. Mem. New York Bot. Gard. 10 (2): 65–117 (1960)

Hamann, U. Merkmalsbestand und Verwandtschaftsbeziehungen der Farinosae. Willdenowia 2 (5): 639–768 (1961)

Idrobo, J. J. Xiridaceas de Colombia. Caldasia 6 (29): 185–260 (1954)

Maguire, B., Wurdack, J. J. and Collaborators. The botany of the Guayana Highland, pt. 3. Mem. New York Bot. Gard. 10 (1): 1–19 (1958); pt. 4, op. cit., 10 (2): 11–15 (1960)

Smith, L. B. and Downs, R. L. Xyridaceae from Brazil. Proc. Biol. Soc. Wash. 73: 245–260 (1960)

Smith, L. B. and Downs, R. L. Xyridaceae in Hoehne, F. C. Flora Brasilica 9 (12): São Paulo, pp. 217 (1968)

1A Petals united below, blue (rarely white); stylar appendages linear; inflorescence stalk with bracts; pollen grains with spines
Abolboda

1B Petals free to base; yellow (rarely white); stylar appendages absent; inflorescence stalk without bracts; pollen grains without spines
Xyris

Abolboda Kunth in Humb and Bonpl., Pl. Aequin. 2: 25 (1809)
Fig. 264A.
Leaves spiral or distichous and equitant. Inflorescence stalk bearing bracts; inflorescence bracts usually loosely packed. Sepals usually 2, when 3, 1 small. Petals distinctly united at base, blue, rarely white. Staminodia absent or simple, glabrous, subulate. Style with 2 or 3, linear, downwardly directed appendages; stigma disc-like or slightly 3-lobed.

For literature see after the family description.

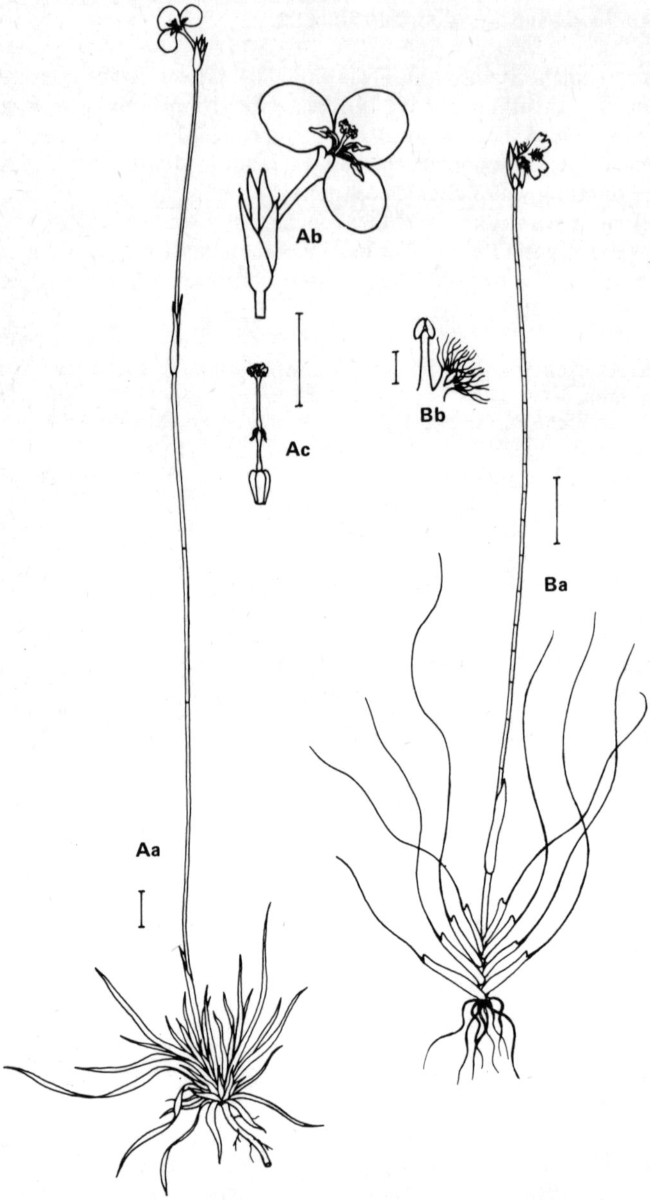

Fig. 264. A. *Abolboda pulchella* Humb. & Bonpl.: a, habit (1 cm); b, inflorescence; c, style (1 cm);
B. *Xyris exserta* Idrobo & Smith: a, habit (1 cm); b, stamen (1 mm).

c. 17 species; Tropical S. America. The majority of the species occur in marshy savanna which may be seasonally inundated.

Xyris L., Sp. Pl. 42 (1753)
Fig. 264B.
Leaves distichous and equitant. Inflorescence stalk naked; inflorescence bracts tightly packed. Sepals 3, 1 large and hooded. Petals almost free, yellow, rarely white. Staminodes 3, forked, ciliate. Style without appendages; stigmas 3, linear.

Duvigneaud, P. and Homes, P. Les Xyris du Bas-Congo et du Kwango. Bull. Soc. Bot. Belgique 87: 81–113 (1955)
Kral, R. The genus Xyris in Florida. Rhodora 62: 295–319 (1960)
Kral, R. Xyris of continental United States and Canada. Sida 2: 177–260 (1966)
Meikle, R. D. and Baldwin, J. R. Eriocaulaceae and Xyridaceae in Liberia. Amer. Journ. Bot. 39: 44–51 (1952)
Royan, P. van. Some new Australian species of Xyris. Blumea 7: 477–480 (1954)
For additional literature see after the family description.

c. 250 species: the largest concentration of species occurs in Tropical America, also found in N. America, Africa, Asia and Australia, but lacking in Europe. The majority of species are found in marshy or wet areas. *X. indica* L. and *X. pauciflora* Willd. are common ricefield weeds in Asia. *X. aquatica* Idrobo and Smith and *X. exserta* Idrobo and Smith from S. America are submerged aquatics.

ZANNICHELLIACEAE

3 genera: all aquatic. The Cymodaceaceae are recognised as a distinct family but being marine is excluded from this work.

Slender, submerged, dioecious or monoecious herbs. Stems slender, capillary, often creeping. Leaves opposite, pseudowhorled or alternate, linear, simple, entire, sometimes reduced to a sheath; at base sheathing with sheath free or adnate to leaves (or depending on morphological interpretation, stipules sheath-like, free or adnate to leaves). Flowers unisexual, axillary, solitary or clustered. Perianth a small cup (or sheath), or a few small scales (bracts) or absent. Stamens 1 to 3, when more than 1 then united; pollen globose. Ovary of 1 to 9, free carpels; style simple, usually persistent; stigma conspicuous, asymmetrically peltate to spathulate; ovule 1 pendulous; fruit a nutlet.

Graebner, P. in Engler, A. Pflanzenreich 31 (IV.II): 153–160 (1907)

1A Basal leaf-sheath (or stipules) not adnate to leaf; leaves usually opposite or in false whorls; plants monoecious (cosmopolitan)
Zannichellia

1B Basal leaf-sheath (or stipules) adnate to leaf for most of their length; leaves usually alternate; plants dioecious

2A Anthers 1-locular, without connectives (W. Old World)
Althenia

2B Anthers 2-locular, or appearing 6-locular, with distinct connectives (Australia and New Zealand)
Lepilaena

Althenia Petit in Saigey and Raspail, Ann. Sci. Obs. 1: 451 (1829)
Fig. 265A.
Stems slender, elongate; rhizomes slender, creeping. Leaves usually alternate, rarely opposite, capillary, up to 40 mm long and 0.5 mm wide; sheathing leaf-base auriculate (or stipules adnate to leaf for most of their length). Flowers solitary. Male flower of 1 stamen with a 1-locular anther. Female flower of usually 3 carpels; styles elongate, persisting in fruit, 2 to 3 mm long; stigma funnel-shaped to almost discoid; fruit ellipsoidal to oblong-ellipsoidal, almost symmetrical, *c.* 2 mm long (excluding style).

Onnis, A. Contributo alla conscenza dell'areale e della ecologia della Althenia filiformis Petit in Sardegna. Atti Soc. Tosc. Sci. Nat. Mem. B. 74: 1–20 (1967)

Onnis, A. Althenia filiformis Petit, nuova specie per la Toscana. Atti Soc. Tosc. Sci. Nat. Mem. B. 74: 71–75 (1967)

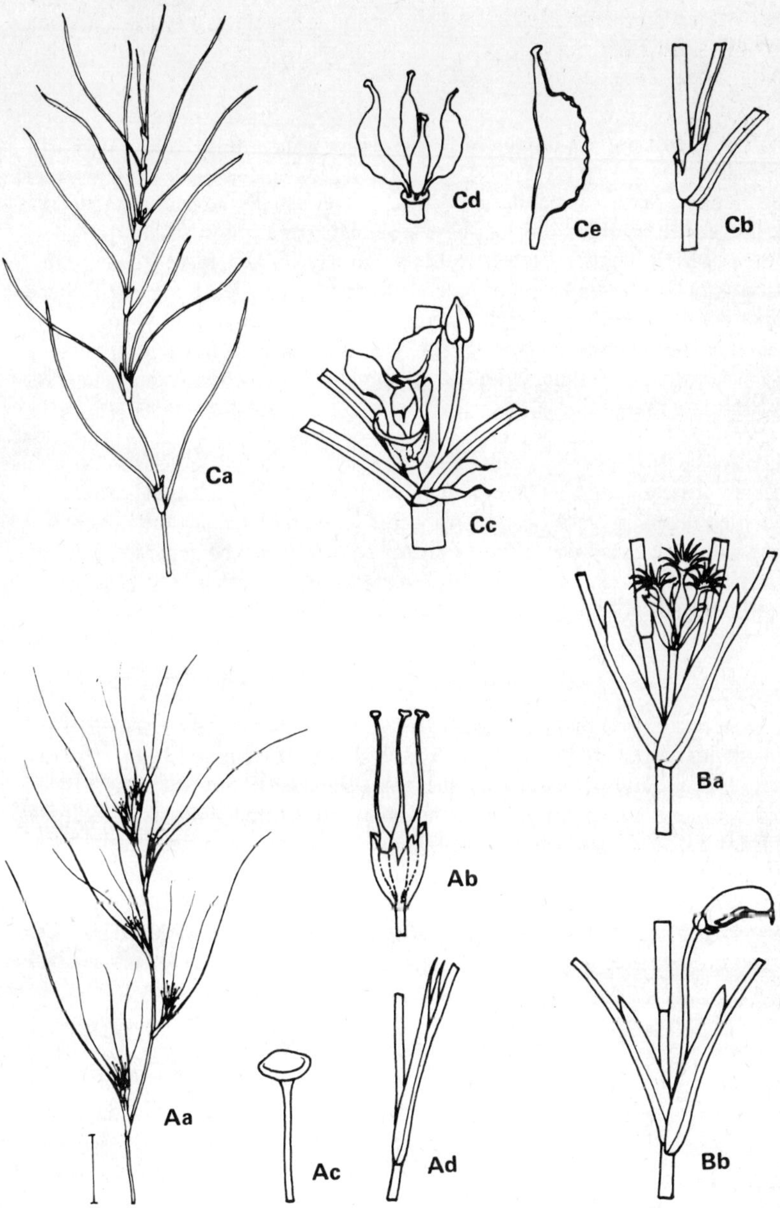

Fig. 265. A. *Althenia filiformis* Petit: a, habit (1 cm); b, fruits; c, style; d, stipules; B. *Lepilaena bilocularis* Kirk: a, female inflorescence; b, male inflorescence; C. *Zannichellia palustris* L.: a, habit (1 cm); b, leaf base; c, flowers; d, fruits (1 mm); e, fruit (1 mm).

1 species, *A. filiformis* Petit: C. and W. Mediterranean, Persia, S. Africa? A poorly known submerged aquatic usually found in brackish water but known to grow in freshwater.

Lepilaena J. Drumm. ex Harvey in Hooker, Kew Journ. Bot. 7: 57 (1855)
Fig. 265B.
Stems slender; rhizomes slender, creeping. Leaves usually alternate, rarely opposite, capillary; sheathing leaf-base auriculate (or stipules adnate to leaf for most of their length). Flowers solitary, shortly stalked. Male flower with 3, minute, perianth segments (or bracts); stamens 1 or 3; anthers sessile, united by their connectives, 2-locular, but when united appearing like 1, ovoid, 6-locular anther. Female flower of usually 3, distinct carpels, each tapering into a style; stigma oblique, cylindric or funnel-shaped, when funnel-shaped with fimbriate margin.

4 species: Australia and 1 (*L. bilocularis* Kirk) also in New Zealand. Slender submerged plants in fresh or brackish water. *L. bilocularis* has one 2-locular anther and is, in many ways, rather similar to *Althenia.* The genera *Lepilaena* and *Althenia* although geographically isolated should, perhaps, be totally or partially united.

Zannichellia L., Sp. Pl. 969 (1753)
Fig. 265C.
Stems slender, sparsely branched; rhizomes slender, creeping. Leaves mostly opposite, or in false whorls, rarely alternate, elongate-linear, flat, 1-nerved, 2 to 10 cm long, up to 2 mm wide; apex acute; sheathing leaf-base (or stipules) free from leaf. Flowers in bud surrounded by a sheath. Male flowers with 1 stamen; filament slender; anther 2- or 4-locular. Female flowers with 1 to 9 carpels; styles elongate, persistent; stigmas asymmetric-peltate; fruit an obliquely oblong, flattened, beaked nutlet, 3 to 6 mm long (including beak), toothed on the back.

Reinecke, P. A contribution to the morphology of Zannichellia aschersoniana. Graebn. Journ. S. African Bot. Soc. 30 (2): 93–101 (1964)

c. 5 or 1 polymorphic species, *Z. palustris* L.: cosmopolitan. It is a submerged plant found in a wide variety of aquatic habitats. It shows a preference for eutrophic waters and is frequently found in brackish or highly polluted water. Reinecke (1964) loc. cit., sheds some doubts on the above interpretation of the flower of *Zannichellia.*

ASTERACEAE (see page 155)

Cadiscus E. H. F. Meyer ex DC., Prodr. 7: 254 (1838)

Fig. 266

1 species, *C. aquaticus* E. H. F. Meyer: known only from South Africa where it grows in pans and vleis. It differs from *Hydropectis* in its ligulate outer florets, partially hairy achene, pappus of scales each divided at the apex into three or four bristles, and usually alternate leaves.

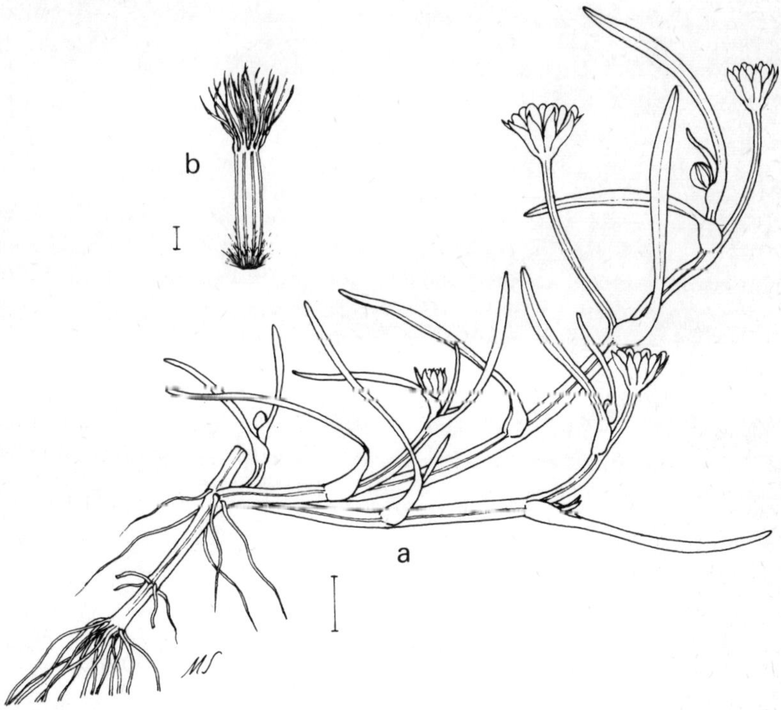

Fig. 266. *Cadiscus aquaticus* E. H. F. Meyer ex DC.: a, habit (1 cm); b, fruit (1 mm).

GLOSSARY

ABAXIAL With side or face originating away from the axis.
ACAULESCENT Stemless or apparently stemless.
ACHENE A small, dry, indehiscent, one-seeded fruit, originating from an inferior ovary.
ACTINOMORPHIC With radial symmetry.
ACUMINATE Tapering to a prolonged point.
ACUTE Distinctly and sharply pointed.
ADAXIAL With side or face originating next to the axis.
ADNATE With unlike parts congenitally grown together.
ALAR CELLS Cells at the basal angles of a moss leaf.
ALTERNATE Any arrangement of leaves or other parts not opposite or whorled; placed singly at different heights on the axis.
ANNUAL Completing its generative history in one year.
ANTHERIDIUM An organ or receptacle in which male sex cells are produced.
APEX Tip or extremity of an organ.
APICULATE Having a minute pointed tip.
APPRESSED Lying flat against an organ.
ARCHEGONIUM The structure containing the female sex cells.
ARCUATE Curved so as to form a ¼ of a circle or more.
ARIL An appendage or an outer covering of a seed.
ASCENDING Rising up; produced somewhat obliquely or indirectly upward.
ATTENUATE Long tapering.
AURICLE An ear; applied to ear-like lobes at base of leaf-blades.
AURICULATE With earlike appendages.
AWN A bristle-like appendage.
AXIL The upper angle formed between the axis and any organ that arises from it.
AXILE In the axis.
AXIS The main or central line of development of any plant or organ; the main stem.
BASIFIXED Attached or fixed by the base.
BERRY A simple fruit having a pulpy or fleshy ovary wall.
BILATERAL Arranged on opposite sides.
BISEXUAL Having both sexes.
BLADE Lamina; the expanded portion of a leaf.
BRACT A modified leaf subtending a flower or belonging to an inflorescence.
BRACT-CELL Special term in Characeae, see p. 40.
BRACTEATE With bracts.
BRACTEOLE A small bract inserted on the pedicel.
BRACTLET Special term in Characeae, see p. 40.
BRANCHLET Special term in Characeae, see p. 40.
BULBLET A little bulb produced in the leaf axils, inflorescence or other unusual places.
CADUCOUS Falling off early, or prematurely.
CALLUS An extension of the flowering-glume below its point of insertion, and adnate to the rachilla of the spikelet (in Poaceae).
CALYPTRA A hood or lid covering the capsule of a moss.
CAMPANULATE Bell-shaped.
CAPILLARY Hairlike; very slender.
CAPITATE Head-like.
CAPSULE A simple dry fruit, the product of a compound ovary splitting along 2 or more lines or the spore-containing organ of a moss.
CARPEL Female organ of a flower, consisting of ovary, style and stigma.
CHASMOGAMOUS With pollination taking place while the flower is open.
CILIATE With hairs projecting from the margin.
CLEISTOGAMOUS With pollination taking place within the unopened flower.
COLUMELLA Central sterile tissue of a moss capsule.
COMPOUND Similar parts aggregated into a common whole.
CONCAVE With surface rounded or bent inwards (cf. convex).
CONICAL Having the form of a cone.
CONNECTIVE A continuation of the filament bearing the anther.
CONTIGUOUS Touching each other at the ends or the edges.
CONTORTED Twisted or bent.

CONVEX With surface rounded or bent outwards, as if swollen.
CONVOLUTE Rolled together or coiled.
CORDATE Heart-shaped; said of leaves having the stalk at the broader and notched end.
CORM A solid bulb-like stem.
CORONULA Special term in Characeae, see p. 40.
CORTEX Special term in Characeae, see p. 40.
CORYMB An inflorescence in which the upper pedicels are shorter than the lower so that all the flowers are at approximately the same level.
CREEPING Running along on the ground and rooting.
CRENATE Regularly toothed, with the teeth very blunt or rounded.
CRUSTACEOUS Of hard and brittle texture.
CRYPTOGAM Non-flowering plants, in which spores rather than seeds are the main means of propagation, and the gametophyte phase is free-living.
CULM Flowering stem of grasses or sedges.
CUNEATE Wedge-shaped; triangular, with the narrow end at the point of attachment.
CYLINDRICAL Elongated with a circular transverse section.
CYME A broad determinate flower cluster, with the central flowers blooming first.
DECUMBENT Reclining or lying on the ground, but with the ends ascending.
DEHISCENCE The method of opening of a seed pod or an anther.
DIMORPHIC Occurring in two forms or states.
DIOECIOUS The male and female elements in different plants.
DISSECTED Deeply divided, or cut into many segments.
DIVIDED Characterised by a lobing or segmentation which extends to the base.
DORSAL Relating to the back; the surface away from the axis.
ELLIPSOID An elliptic solid.
ELLIPTIC A flat part or body that is oval and narrowed to rounded ends.
EMARGINATE With shallow notch at apex.
EMERSED Raised above and out of the water.
ENTIRE With even margin, without toothing or division.
EPIGYNOUS Borne on the ovary; said of floral parts in which the ovary is inferior.
EQUITANT Folded longitudinally and overlapping (of leaves).
EROSE Irregularly toothed as if eroded or gnawed.
FALCATE Sickle- or scythe-shaped.
FASCICLE A close cluster or bundle.
FILAMENT The part of a stamen that supports the anther.
FIMBRIATE With margin divided into a fringe.
FLACCID Limp and flabby.
FOLIOSE Leafy or leaf-like.
FREE CENTRAL A type of arrangement of ovules on a column or projection, arising from the middle of the base of the ovary, and not connected with the wall by a septum.
FRUIT A mature ovary or ovaries with or without closely related parts.
FURCATE Forked with terminal lobes like prongs.
GAMETANGIUM Special term in Characeae, see p. 40.
GAMETE Male or female sexual cell.
GAMETOPHYTE Gamete producing phase in the life cycle of a plant (usually Bryophyta).
GLABROUS Smooth, devoid of pubescence or hair of whatsoever form.
GLAUCOUS Bluish- or greyish-green, usually because of waxy coating.
GLOBOSE Spherical, globular.
GLUME Reduced, usually chaffy, bract of a grass or sedge flower.
GYNOBASIC A style which appears to arise from the base of the ovary.
HEAD A dense inflorescence of sessile or sub-sessile flowers.
HERB A plant naturally dying to the ground at the end of the growing season, lacking definite woody structure.
HETEROPHYLLOUS With leaves of more than one shape.
HETEROSTYLIC Having unlike or different length styles.
HILUM The scar or mark on a seed at the point of attachment to the placenta.
HISPID Coarsely and stiffly hairy.
HYALINE Thin and translucent or transparent.
HYPANTHIUM Cup-like receptacle

derived usually from the adnation of floral parts, and on which are seemingly borne sepals, petals and stamens.

HYPOGYNOUS Free from, but inserted beneath the ovary.

INDEHISCENT Not opening.

INFERIOR OVARY An ovary that is below the insertion of the other floral parts.

INFLORESCENCE Mode of flower bearing.

INTERNODE The part of the stem between two successive nodes.

INVOLUTE With the margins rolled upwards or inwards.

KEEL A central and longitudinal ridge.

LACINIATE Deeply and irregularly divided into narrow segments.

LANCEOLATE Lance-shaped, rather narrow, tapering to both ends with the broadest part below the middle.

LATERAL On or at the side.

LATEX A milky juice.

LEAFLET A single division of a compound leaf.

LEMMA The abaxial bract subtending a floret in the inflorescence of the grasses (Poaceae).

LIGULATE Strap-shaped or strap-like.

LIGULE A thin, scarious projection.

LINEAR Long and narrow with margins parallel, or nearly so.

LOCULE (LOCULUS) Compartment of ovary or anther.

LOCULICIDAL Said of capsules that are dehiscent along the loculus or back of the carpels.

LODICULE Small delicate scale at the base of the ovary in a grass flower (usually 2 in number).

LONGITUDINAL Running along the length of an organ (of ridge, groove, fold, etc.).

MEMBRANOUS Like a membrane, thin, usually transparent.

MERICARP A 1-seeded, nutlet-like fruit derived by the splitting of united carpels.

-MEROUS (2- or 3-merous) having the parts in twos or threes, or in multiples of two or three.

MONOECIOUS Having unisexual flowers with both sexes borne on the same plant.

MONOFACIAL Leaf in which adaxial face is reduced and then absent, and both sides of leaf are formed of abaxial surface [as in Iris].

MONOPODIAL Branching from one primary axis.

NERVE A leaf vein or slender rib.

NODE That point on a stem which normally bears a leaf or leaves.

NUT A dry indehiscent, 1-locular, 1-seeded fruit with a bony, leathery or papery wall.

NUTLET The diminutive of nut.

OBLANCEOLATE Inverted lanceolate.

OBLONG Longer than broad, with the margins nearly parallel.

OBOVATE Reversed ovate.

OBTUSE Blunt or rounded at the end.

OOGONIUM Special term in Characeae, see p. 40.

OPERCULUM Lid of the capsule (Bryophyta).

ORBICULAR Flat with a circular outline.

OVARY That part of the female organ which contains the ovules.

OVATE Shaped like a longitudinal section of a hen's egg, the broader end basal.

OVULE That which becomes a seed after fertilisation.

PALEA Inner (adaxial) bract of a grass (Poaceae) floret.

PALMATE Resembling a hand with the fingers spread; having lobes radiating from a common point.

PALMATISECT Palmately divided nearly to the midrib.

PANICLE A compound or branched inflorescence.

PAPILLA A small elongated projection.

PAPILLOSE Covered with papillae.

PAPPUS The bristle-like or hair-like structure at the junction of the achene and petal tube in the Asteraceae, often appearing terminal on the ripe fruit.

PARIETAL Borne on or belonging to a wall.

PEDICEL Stalk of a flower or spikelet.

PELTATE Shield-shaped, as a leaf attached by its lower surface to a stalk instead of by its margin.

PENDULOUS Hanging downwards.

PERENNIAL Continuing to live from year to year.

PERIANTH The floral envelope, of whatever form; the sepals and the petals.

PERICHAETIAL LEAVES Leaves immediately surrounding the base of the stalk of the capsule (Bryophyta).

PERIGYNOUS Borne around the ovary, neither above nor below.

PERISTOME Ring of teeth that surrounds the opening of a moss capsule.

PERSISTENT Remaining attached; not falling off.

PETIOLE The stalk of a leaf.

PINNA (plural PINNAE) Leaflet of a pinnate leaf.

PINNATE Feather-formed, as with leaflets of a compound leaf placed on either side of a stalk.

PINNATISECT Pinnately divided nearly to the midrib.

PITH Spongy tissue.

PLACENTA The place in the ovary where ovules are attached.

PLUMOSE Feather- or plume-like.

PROSTRATE Lying flat on the ground.

RACEME An indeterminate inflorescence consisting of a central stalk bearing a number of flowers, the lower blooming first.

RACHILLA The axis of the spikelet on which the florets are inserted.

RACHIS An axis bearing flowers (in the Poaceae).

RECEPTACLE That expanded portion of the axis which bears the floral organs.

RENIFORM Kidney-shaped.

RHIZOID Root-like organ in general appearance or function but not so in anatomy.

RHIZOME The rootstock or flattened stem having rootlike appearance, prostrate on or underground, sending off rootlets, the apex progressively sending up stems or leaves.

RHOMBOIDAL Approaching a rhombic outline, quadrangular, with the lateral angles obtuse.

RIGID Stiff, inflexible.

ROSETTE A cluster of spreading or radiating leaves.

ROTATE Wheel shaped: with short tube and lobes spreading at right angles to the axis.

SACCATE Bag-shaped.

SAGITTATE Enlarged at the base into two acute straight lobes, like the barbed head of an arrow.

SCABROUS Rough, with short hairs or teeth.

SCALE Any thin scarious body, usually a degenerative leaf.

SCAPE The flowering stem of a plant in which all leaves are radical.

SCARIOUS Thin, dry and membranous, not green.

SCHIZOCARP Ovary which splits into 1-seeded positions when mature.

SEGMENT One of the parts of an organ that is divided.

SEPTICIDAL Said of a capsule that dehisces through the lines of junction.

SEPTUM A wall which separates cavities.

SERRATE With sharp teeth.

SESSILE Without a stalk of any kind.

SHEATH Any long or more or less tubular structure surrounding an organ or part.

SIMPLE Of one piece; not compound.

SINUS The space between two lobes of a leaf.

SOLITARY Single, only one from the same place.

SORUS (plural SORI) A group of sporangia.

SPADIX Fleshy spike crowded with flowers, usually of Araceae.

SPATHE The bract or pair of bracts surrounding or subtending an inflorescence.

SPATHELLA Special term in the Podostemaceae, see p. 445.

SPATHULATE Of a leaf, widest near the apex, obtuse at apex, tapering into the stalk.

SPIKE An inflorescence consisting of a central stalk bearing a number of sessile or subsessile flowers.

SPIKELET The unit of the inflorescence in grasses.

SPINE Sharp outgrowth or apical point, usually woody or very stiff.

SPORANGIOPHORE A structure, not leaf-like, bearing sporangia.

SPORANGIUM Structure in which spores are produced.

SPORE Small asexual usually unicellular reproductive body.

SPOROPHYTE Spore-producing phase in the life history of a plant.

SPUR A tubular or saclike projection from a blossom, as of a petal or sepal.

STAMEN The pollen-bearing organ of the flower, the male organ in the flowering plants (consists of filament, connective and anthers).

STAMINODE A sterile stamen, or a

structure resembling such and borne in the staminal part of the flower.

STIGMA The part of the carpel or style which receives the pollen.

STIPE A short stalk

STIPEL Similar to a stipule but at the base of the leaflets of a compound leaf.

STIPULE An appendage borne at the base of the leaf in many plants.

STIPULODE Special term in Characeae, see p. 40.

STOLON A creeping stem of short duration produced by a plant which has a central rosette or erect stem.

STYLE The more or less elongated part of the female organ between the ovary and the stigma.

SUBULATE Awl-shaped; narrow and pointed.

SUFFRUITICOSE Woody at base.

SUPERIOR Growing or placed above.

SUPERIOR OVARY An ovary that is above the insertion of the other floral parts.

TEPAL A special term in the Podostemaceae, see p. 445.

TERETE Circular in transverse section.

THALLUS (plural THALLI) The entire cellular plant body without differentiation into stem and leaves.

TOOTHED Of the margin, with sharp teeth pointing outward.

TRICOLPATE With three grooves (of pollen).

TRUNCATE Ending abruptly, the base or apex nearly or quite straight across.

UMBEL An indeterminate inflorescence consisting of several stalked flowers having a common point of attachment.

UNDULATE Having a wavy margin.

UNISEXUAL Of one sex.

UNITED With like parts congenitally grown together.

VALVATE With the parts edge-to-edge (i.e. not overlapping).

VENTRAL Relating to front (or belly); the surface towards the axis.

VERSATILE Hung or attached near the middle and usually moving freely.

VISCID Sticky.

WHORL Cyclic arrangement of appendages at a node.

ZYGOMORPHIC Capable of division by only one plane of symmetry.

INDEX TO FAMILY AND GENERIC NAMES

The synonyms are printed in italics.